PRACTICAL APPLICATIONS OF FUZZY TECHNOLOGIES

THE HANDBOOKS OF FUZZY SETS SERIES

Series Editors
Didier Dubois and Henri Prade
IRIT, Université Paul Sabatier, Toulouse, France

FUNDAMENTALS OF FUZZY SETS, edited by Didier Dubois and Henri Prade
MATHEMATICS OF FUZZY SETS: ***Logic, Topology, and Measure Theory***, edited by Ulrich Höhle and Stephen Ernest Rodabaugh
FUZZY SETS IN APPROXIMATE REASONING AND INFORMATION SYSTEMS, edited by James C. Bezdek, Didier Dubois and Henri Prade
FUZZY MODELS AND ALGORITHMS FOR PATTERN RECOGNITION AND IMAGE PROCESSING, by James C. Bezdek, James Keller, Raghu Krisnapuram and Nikhil R. Pal
FUZZY SETS IN DECISION ANALYSIS, OPERATIONS RESEARCH AND STATISTICS, edited by Roman Slowinski
FUZZY SYSTEMS: ***Modeling and Control,*** edited by Hung T. Nguyen and Michio Sugeno
PRACTICAL APPLICATIONS OF FUZZY TECHNOLOGIES, edited by Hans-Jürgen Zimmermann

PRACTICAL APPLICATIONS OF FUZZY TECHNOLOGIES

edited by

Hans-Jürgen Zimmermann
Operations Research, RWTH, Aachen, Germany

KLUWER ACADEMIC PUBLISHERS
Boston/London/Dordrecht

Distributors for North, Central and South America:
Kluwer Academic Publishers
101 Philip Drive
Assinippi Park
Norwell, Massachusetts 02061 USA
Telephone (781) 871-6600
Fax (781) 871-6528
E-Mail <kluwer@wkap.com>

Distributors for all other countries:
Kluwer Academic Publishers Group
Distribution Centre
Post Office Box 322
3300 AH Dordrecht, THE NETHERLANDS
Telephone 31 78 6392 392
Fax 31 78 6546 474
E-Mail <orderdept@wkap.nl>

Electronic Services <http://www.wkap.nl>

Library of Congress Cataloging-in-Publication Data

Practical applications of fuzzy technologies / edited by Hans-Jürgen Zimmermann.
p. cm. -- (Handbooks of fuzzy sets series ; FSHS 6)
Includes bibliographical references.
ISBN 0-7923-8628-0
1. Automatic control. 2. Fuzzy systems. I. Zimmermann, H.-J. (Hans-Jürgen), 1934- II. Series
TJ213.P673 1999
629.8--dc21 99-40742
CIP

Printed on acid-free paper.

Printed in the United States of America

Contents

Series Foreword

Fuzzy sets were introduced in 1965 by Lotfi Zadeh with a view to reconcile mathematical modeling and human knowledge in the engineering sciences. Since then, a considerable body of literature has blossomed around the concept of fuzzy sets in an incredibly wide range of areas, from mathematics and logics to traditional and advanced engineering methodologies (from civil engineering to computational intelligence). Applications are found in many contexts, from medicine to finance, from human factors to consumer products, from vehicle control to computational linguistics, and so on. Fuzzy logic is now currently used in the industrial practice of advanced information technology.

As a consequence of this trend, the number of conferences and publications on fuzzy logic has grown exponentially, and it becomes very difficult for students, newcomers, and even scientists already familiar with some aspects of fuzzy sets, to find their way in the maze of fuzzy papers. Notwithstanding circumstantial edited volumes, numerous fuzzy books have appeared, but, if we except very few comprehensive balanced textbooks, they are either very specialized monographs, or remain at a rather superficial level. Some are even misleading, conveying more ideology and unsustained claims than actual scientific contents.

What is missing is an organized set of detailed guidebooks to the relevant literature, that help the students and the newcoming scientist, having some preliminary knowledge of fuzzy sets, get deeper in the field without wasting time, by being guided right away in the heart of the literature relevant for her or his purpose. The ambition of the HANDBOOKS OF FUZZY SETS is to address this need. It will offer, in the compass of several volumes, a full picture of the current state of the art, in terms of the basic concepts, the mathematical developments, and the engineering methodologies that exploit the concept of fuzzy sets.

This collection will propose a series of volumes that aim at becoming a useful source of reference for all those, from graduate students to senior researchers, from pure mathematicians to industrial information engineers as well as life, human and social sciences scholars, interested in or working with fuzzy sets. The original feature of these volumes is that each chapter is written by one or several experts in the concerned topic. It provides introduction to the topic, outlines its development, presents the major results, and supplies an extensive bibliography for further reading.

The core set of volumes are respectively devoted to fundamentals of fuzzy set, mathematics of fuzzy sets, approximate reasoning and information systems, fuzzy models for pattern recognition and image processing, fuzzy sets in decision analysis, operations research and statistics, fuzzy systems modeling and control, and a guide to practical applications of fuzzy technologies.

Didier DUBOIS Henri PRADE
Toulouse

Preface

Fuzzy set theory was conceived in 1965 as a formal theory which could be considered as a generalization of either classical set theory or of classical dual logic. In spite of the fact that Prof. Zadeh, when publishing his first contribution had already some applications in mind, fuzzy set theory for several reasons kept inside the academic sphere for more than 20 years. During these 20 years most of the basic concepts, which are nowadays used very successfully, have already been invented.

Starting at the beginning of the 80s Japan was the leader in using a part of fuzzy set theory - namely fuzzy control - for practical applications. Particularly improved consumer goods, such as video cameras with fuzzy stabilizers, washing machines including fuzzy control, rice-cookers etc., caught the interest of the media which led around 1989/1990 to the first "fuzzy boom" in Germany. Many attractive practical applications - not so much in the area of consumer goods but rather in automation and industrial control - led to the insight that the efficient and affordable use of this approach could only be achieved via CASE-tools. Hence, since the late 80s a large number of very user-friendly tools for fuzzy control, fuzzy expert systems, fuzzy data analysis etc. has emerged. This really changed the character of this area and started to my mind the area of **"fuzzy technology"**.

The next - and so far the last - large step in the development occurred in 1992 when almost independently in Europe, Japan and the USA the three areas of fuzzy technology, artificial neural nets and genetic algorithms joined forces under the title of **"computational intelligence"** or **"soft computing"**. The synergies which were possible between these three areas have been exploited since very successfully. Figure 1 shows these developments as a summary.

There is not only an invigorating influence of theory on applications, but the appearance of real applications in products visible and known to the broad public – such as video cameras, washing machines etc. - also triggered the "fuzzy boom" in Germany and other countries at the beginning of the 90s, which in turn increased the number of universities that were offering courses or graduate work in fuzzy sets in Germany from two to more than 20 within two years.

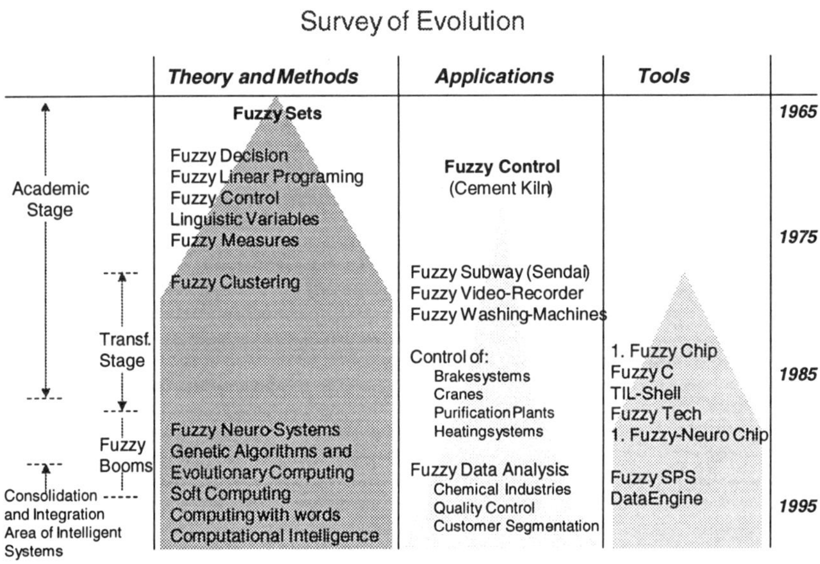

Fig. 1: From fuzzy set theory to computational intelligence

Figure 2 depicts the increase of publications in the last two decades of the lifetime of fuzzy set theory, exemplified by the publication output of "Fuzzy Sets and Systems":

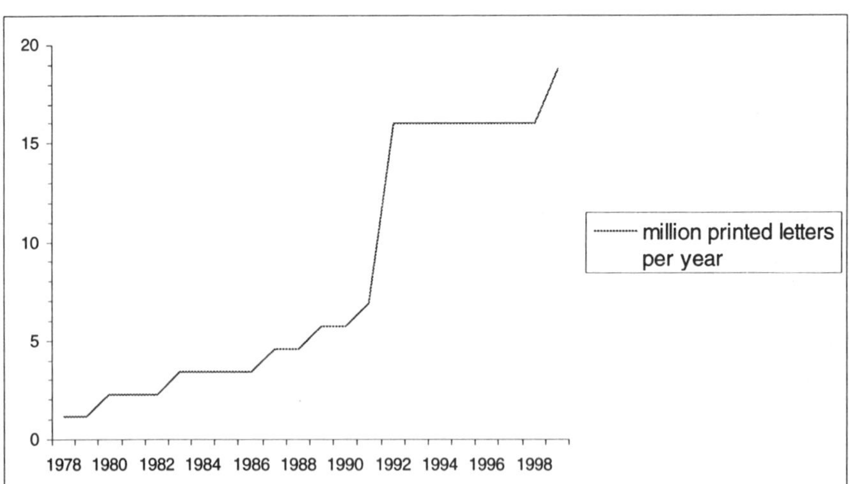

Fig. 2: FSS – in "digits" per year

So far I have used the term "applications" as if it was well defined. This is, however, certainly not true: applications might mean that one applies one theory to another. For instance, if one applies fuzzy set theory to topology or algebra or graph theory, one obtains fuzzy topology, fuzzy algebra or fuzzy graph theory. One might also apply it to dichotomous methods and one gets fuzzified methods (for instance, fuzzy linear programming, fuzzy clustering, fuzzy Petri nets etc.).

One can also apply methods (fuzzy or crisp) to models of problems (for instance fuzzy inventory models, fuzzy production control models etc.).

Eventually one can use methods or combinations of methods to solve real existing problems. In this volume we will not consider the first two kinds of "applications". They are the focus of other volumes of this series. This volume will focus on the last two types of applications: model and real applications. Even though these two types of applications are often not distinguishable in the literature, the reader should be aware of the distinguishing features of them:

A model application indicates very often how and where a certain technology could be used, it generally focuses on certain features of a problem and in most cases it is not verified. The author generally has the freedom to choose features and their character at his will.

A "real" application is different in a number of ways:

- The problem is given and **it** should be modeled and solved and not a modification of it that fits the available techniques.

- Very often not a single method is sufficient to solve the problem but a combination of several methods is needed.

- Generally the success of the solution involves very many details, a description of which would exceed the scope of a publication.

- Unsuccessful applications are not published for obvious reasons and successful applications are published to a very limited extent for competitive reasons.

- Often the "intelligent" part of the solution is a necessary but very small part and the rest is, for instance, good and professional software engineering.

It is the aim of this book to cover applications of fuzzy technology as comprehensively as possible. For the reasons mentioned above the coverage is, however, not complete and partly only exemplarily. The focus are applications of the two last kinds, i.e. model and real applications. Applications of fuzzy set theory to other theories and methods or the mathematics of fuzzy set theory itself are covered in other volumes of this book series and the interested reader is referred to them.

Due to the character of this volume the structure chosen is also not according to methods or theories but along the lines of large application areas. This has the disadvantage that upcoming new areas which cross the boundaries of application areas, such as nuclear engineering etc., are not yet included. Very important areas of fuzzy technology, such as data mining or fuzzy information processing, have not been included as extra chapters either because they have a kind of universal – and not area specific – applicability or they are considered to be rather methodological than application areas. The first short coming might be cured in the next edition of the book, when new areas can be included. The second weakness has been counteracted by extensive indexing and by including a chapter on software tools at the end of the book.

Apart from part V of the book, the volume is structured into four major parts, which significantly differ in character:

Part I: Engineering and Natural Sciences

Engineering applications in the form of fuzzy control were certainly the forerunners and gate openers for real applications of fuzzy set theory and fuzzy technology. Fuzzy control in Mamdani's fuzzy controller was primarily understood as the application of fuzzy expert system technology to problems of control engineering. Since then a number of things have changed:

First: Fuzzy control has developed from an expert system philosophy to an engineering design technology, i.e. fuzzy controllers and their components, such as number and shapes of membership functions, operators used, fuzzification and defuzzification methods employed have become calibration parameters, which the designer adjusts such that the controller - or the system it controls - behaves in the desired way.

Second: Applications have spread to many non-control problems of engineering, such as design, configuration, supervision, experimentation, modeling etc.

Third: Application areas have increased and cover nowadays almost all traditional kinds of engineering from mechanical over electrical to chemical and civil engineering.

Fourth: Some natural sciences have started to use fuzzy approaches. These areas are not engineering but they are closer to engineering than, for instance, to behavioral sciences. Therefore, they have been included in this part of the book exemplarily ecology and spatial analysis.

Fifth: The fuzzy methods employed in engineering and natural sciences have become much more diversified and contain beyond fuzzy control approaches fuzzy modeling tools, such as fuzzy Petri nets, fuzzy algorithmic approaches, such as fuzzy clustering, fuzzy optimization etc.

As mentioned above, the book tries to cover the most important application areas of fuzzy sets. The outstanding importance of engineering and natural science applications is acounted for by having 9 of 19 chapters assigned to this part. Still it is not claimed that all applications have been covered exhaustively.

Part II: Medicine

The oldest fuzzy applications in medicine are diagnostic systems, such as for instance CADIAG, or similar systems. Applications following this route are described in chapter 11. There exist only very few fuzzy applications in medical/consumer goods, such as fuzzy blood pressure measurement devices. But there exist applications which can be used by medical doctors to improve the medical support of patience. They are described in chapter 10. Eventually fuzzy methods are used increasingly in connection with advanced medical technologies, such as mammography, and have shown very good results. Exemplarily again this is shown in chapter 12 on image processing in medicine.

Part III: Management

Applications of fuzzy technology in management are as diverse as the different areas of management. In a recent survey of fuzzy applications in management in Germany it was shown that approximately 50 % of the applications were in the area of production and inventory control. Therefore, one of the largest chapters of this part of the book – namely chapter 15 – surveys this area.
Chapters 13 and 14 focus on very long-term and ill-structured planning and decision making problems on the strategic level, which are obviously very large and suitable potentials for the application of fuzzy technology.
Eventually chapter 16 exemplifies one of the youngest and fastest growing areas of applications of intelligent technologies, namely financial engineering. Chapter 16 focuses on actuarial applications but I am confident that the next edition of this volume will also contain other areas from financial engineering.

Part IV: Behavioral, Cognitive and Social Sciences

This part is dedicated to the applications which are probably closest to the original motivation of fuzzy set theory, namely scientific and engineering areas which are

concerned directly with the human beings themselves. Chapter 17 on applications in psychology is mainly concerned with human perception and communication.
Chapter 18 considers more but not exclusively the physical facets of humans. One should probably expect that many more applications can be found in this human-related areas but one should also recognize that implicitly the human aspects are also modeled with fuzzy sets in those applications described in many of the other chapters.
Nevertheless, an increased use of fuzzy set theory in psychology as well as the human factors area would certainly be desirable and possible.

Part V: Tools

As already mentioned above, chapter 19 surveys those software tools which transformed fuzzy set theory into fuzzy technology and which facilitate considerably the efficient building of fuzzy systems also by persons which are not specialists in fuzzy set theory. The author of chapter 19 did certainly an excellent job in selecting from the very numerous tools those that characterize the state of technology. The reader should realize, however, that this area is developing very fast and that new releases of the tools have appeared in the meantime and that it may be wise to inquire at the companies mentioned in this chapter, which release is the newest available at the time one considers using any of the tools.

This book will not only be useful for practitioners of fuzzy technology but also for scientists and students who are looking for applications of their models and methods, for topics of their theses, and even for venture capitalists that look for attractive possibilities for investments. I hope that all will find this book useful and I would also appreciate their comments for further editions of the volume.

Finally I want to thank very much all those who have made this book possible: the authors who patiently followed my always new requests for revisions, Katja Palczynski for never resigning but following the progress of the book for years and helping imagetively wherever possible, the referees of the different chapters and Kluwer Academic Publishers for their support. May all of them participate in a successful future of this volume.

Aachen Hans-Jürgen Zimmermann

Contributing Authors

Erik K. Antonsson
Engineering Design Research Laboratory
Division of Engineering and Applied Science
California Institute of Technology (Caltech)
1200 East California Blvd., Mail Code: 104-44
Pasadena, CA 91125, USA
e-mail: erik@design.caltech.edu

Kurt Becker
Jackstr. 2
52078 Aachen, Germany
e-mail: kbecker@debis.com

James C. Bezdek
Department of Computer Science
University of West Florida
Pensacola, FL 32514-5750, USA
e-mail: jbezdek@argo.cs.uwf.edu

Jean-Christophe Buisson
IRIT (Institut de Recherche en Informatique de Toulouse)
ENSEEIHT, 2 rue Carmichel
31071 Toulouse, France
and CTDIT, Ranguelil University Hospital
Toulouse, France
e-mail: jean-christophe.buisson@enseeiht.fr

Karen C. Chou
Civil & Environmental Engineering Department
223 Perkins Hall
The University of Tennessee
Knoxville, TN 37996-2010, USA
e-mail: kchou@utk.edu

Richard A. Derrig
Senior Vice President
Automobile Insurers Bureau of Massachusetts
Vice President, Research, Insurers Fraud Bureaus
of Massachusetts
101 Arch Street, Boston, MA 02110, USA

M. H. Fazel Zarandi
Dept. of Mechanical and Industrial Engineering
University of Toronto
5 King's College Road
Toronto, Ontario, M5S 3G8, Canada

Dominik Füssel
Institut für Automatisierungstechnik
FG Regelungstechnik und Prozeßautomatisierung
Technische Universität Darmstadt
Landgraf-Georg-Str. 4, 64283 Darmstadt, Germany

Jerzy Grobelny
Institute of Production Engineering and Management
Technical University of Wroclaw
Wybrzeze Wyspianskiego 27
50-370 Wroclaw, Poland

Rolf Isermann
Institut für Automatisierungstechnik
FG Regelungstechnik und Prozeßautomatisierung
Technische Universität Darmstadt
Landgraf-Georg-Str. 4, 64283 Darmstadt, Germany
e-mail: bhoppe@iat.tu-darmstadt.de

Jan Jantzen
Technical University of Denmark
Dept. of Automation
2800 Lyngby, Denmark
e-mail: jj@iau.dtu.dk

Waldemar Karwowski
Center for Industrial Ergonomics
Dept. of Industrial Engineering
University of Louisville
Louisville, Kentucky 40292, USA
e-mail: karwowski@louisville.edu

Kurt Konolige
Artificial Intelligence Center, SRI International
333 Ravenswood Ave
CA 94025 Menlo Park, USA
e-mail: konolige@ai.sri.com

Miroslawa Lasek
Faculty of Economic Sciences
Warsaw University
Dluga 44/55, 00-241 Warsaw, Poland
e-mail: mlasek@wne.uw.edu.pl

WookGee Lee
Center for Industrial Ergonomics
Dept. of Industrial Engineering
University of Louisville
Louisville, Kentucky 40292, USA

Yee Leung
Dept. of Geography
Center for Environmental Studies, and
Joint Laboratory for GeoInformation Science
The Chinese University of Hong Kong
Shatin, Hong Kong
e-mail: yeeleung@cuhk.edu.hk

Willi Meier
DECHEMA e.V.
Theodor-Heuss-Allee 25,
60486 Frankfurt, Germany
e-mail: meier@dechema.de

Gregg C. Oden
The University of Iowa
Dept. of Computer Science and Psychology
E11 Seashore Hall
Iowa City, Iowa 52242, USA
e-mail: oden@cs.uiowa.edu

Jens-Jørgen Østergaard
FLS Automation A/S
Høffdingsvej 77
2500 Valby, Copenhagen, Denmark
e-mail: jjoe@flsa.fls.dk

Krzysztof M. Ostaszewski
Actuarial Program Director
Dept. of Mathematics
University of Louisville
Louisville, KY 40292, USA
e-mail: kmosta01@homer.louisville.edu

Witold Pedrycz
Department of Electrical & Computer Engineering
University of Alberta
Edmonton, Alberta, Canada
e-mail: pedrycz@ee.ualberta.ca

Enrique H. Ruspini
Artificial Intelligence Center
SRI International
333 Ravenswood Ave
CA 94025 Menlo Park, USA
e-mail: ruspini@ai.sri.com

Alessandro Saffiotti
Applied Autonomous Sensor Systems (AASS)
Department of Technology and Science
University of Örebro
70182 Örebro, Sweden
e-mail: alessandro.saffiotti@ton.oru.se

Arkadiusz Salski
Ecology Center
Christian-Abrechts-University of Kiel
Schauenburgerstr. 112, 24118 Kiel, Germany
e-mail: arek@pz-oekosys.uni-kiel.de

Hans-Jürgen Sebastian
Dept. of Operations Research
Aachen Institute of Technology
Templergraben 64, 52062 Aachen, Germany
e-mail: sebasti@or.rwth-aachen.de

Michael Smithson
Division of Psychology
Australian National University
Canberra, A.C.T. 0200, Australia
e-mail: michael.smithson@anu.edu.au

Jens Strackeljan
Technische Universität Clausthal
Institut für Technische Mechanik
38678 Clausthal-Zellerfeld, Germany
e mail: tmjs@itm.tu-clausthal.de

Melanie A. Sutton
Department of Computer Science
University of West Florida
Pensacola, FL 32514-5750, USA
e-mail: msutton@uwf.edu

I. B. Türksen
Dept. of Mechanical and Industrial Engineering
University of Toronto
5 King's College Road
Toronto, Ontario, M5S 3G8, Canada
e-mail: türksen@mie.utoronto.ca

Henk B. Verbruggen
Dept. of Electrical Engineering
Delft University of Technology
P.O. Box 5031, 2600 GA Delft, The Netherlands
e-mail: h.verbruggen@its.tudelft.nl

Richard Weber
MIT Management Intelligenter Technologien GmbH
Promenade 9, 52076 Aachen, Germany
e-mail: rw@mitgmbh.de

Brigitte Werners
Department of Economics
Ruhr-University Bochum
44780 Bochum, Germany
e-mail: brigitte.werners@ruhr-uni-bochum.de

Felix S. Wong
Weidlinger Associates, Inc.
4410 El Camino Real, Suite 110
Los Altos, CA 94022-1049, USA
e-mail: fswong@worldnet.att.net

Yung-Nien Yang
Center for Industrial Ergonomics
Dept. of Industrial Engineering
University of Louisville
Louisville, Kentucky 40292, USA

James T. P. Yao
Dept. of Civil Engineering
Texas A&M University
College Station, TX 77843-3136, USA
e-mail: jtpyao@tamu.edu

Hans-Jürgen Zimmermann
Dept. of Operations Research
Aachen Institute of Technology
Templergraben 64, 52062 Aachen, Germany
e-mail: zi@or.rwth-aachen.de

I Engineering and Natural Sciences

1 FUZZY CONTROL IN THE PROCESS INDUSTRY: COMMON PRACTICE AND CHALLENGING PERSPECTIVES

Jan Jantzen
Henk Verbruggen
Jens-Jørgen Østergaard

Abstract: Control problems in the process industry are dominated by non-linear and time-varying behaviour, many inner loops, and much interaction between the control loops. Fuzzy controllers have in some cases nevertheless mimicked the control actions of a human operator. Simple fuzzy controllers can be designed starting from PID controllers, and in more complex cases these can be used in connection with model-based predictive control. For high level control and supervisory control several simple controllers can be combined in a priority hierarchy such as the one developed in the cement industry. For a design engineer, the chapter is a fairly detailed reference to various design configurations and design choices.

1.1 INTRODUCTION

Modern process control and production methods are confronted with a large number of requirements posed by increasing competition, environmental regulations, increase in energy and raw material costs and an increasing demand for high quality customer-tailored products. This leads to:

- An increasing demand for flexibility of the production process: operating the plant under varying throughput, different feed stock, varying product-mix and product grade. In other words: costumer-defined production in smaller amounts at the right due time, instead of producer-defined high-volume production.

- A strong demand for new production methods and new products. The production plants should be more compact to minimise energy consumption, the influence on the environment and the waste of materials. It is very important to introduce new products as fast as possible on the market (e.g. pharmaceutical industry), so the route from idea, first experiments, pilot plant and full-scale production should be as short as possible and less attractive alternatives should be excluded as soon as possible. This means the introduction of methods which are able to increase the complexity of the models of the process development in order to decide in an early stage how to proceed.
- The introduction of integrated information systems that are plant-wide and can handle various levels of automation in one concept: management, scheduling, planning, optimisation, supervision, fault detection and fault diagnosis, control. This requires the ability to handle qualitative and quantitative information with different levels of precision and complexity and the use of different interfaces to the people responsible for each automation level.

The result of these requirements for process control on different levels of process automation can be summarised as follows:

- Fast and extensive changes in operating conditions and operation modes leading to more emphasis on the highly non-linear and time-varying behaviour of the process.
- Many inner loops and utility feedbacks to decrease waste of material and energy consumption. This leads again to highly non-linear systems of high dimensions with much interaction between the control loops.
- The need for new models of the system which describe the non-linear time-varying behaviour of the new production methods and production units in a more unified way and not only based on first principle descriptions or on a black box approach, but based on a combination of knowledge of system experts, measurements and operational experience.
- The need for dynamic and reactive responses on all levels of automation.

The control engineering community developed a complete framework for linear systems which resulted in a number of universal analysis and design methods for closed and open loop systems. Although many attempts were made during the last decades to develop analysis and design methods for non-linear systems, there is no universal method available at the moment for these systems.

This means that modern control practice is asking for methods which have still to be developed, and in many cases one relies on process models linearized around a number of operating points.

Can fuzzy modelling and control play a role in solving these problems to a certain extent? First of all fuzzy control was introduced to mimic the control actions of the human operator who was well experienced in controlling difficult to analyse and to describe control problems, for instance in the control of cement kilns (Holmblad and Østergaard, 1982).

In these applications a priori knowledge of the operators is translated into a knowledge based system and the final controller performs as well as the best operators. The main problems involved are the knowledge acquisition problem and the maintenance of the system when the plant is updated to produce other products, to handle different feedstocks, etc.

There is still a demand to apply these expert-based control methods for systems which are difficult to describe and to analyse and for systems for which extensive experience in operating the process is available from human operators and system designers. A recent example is the control of the weight and moisture of washing powders (Setnes, Fleer, Bruijn, Lemke and Emck, 1997). It is expected that this kind of applications will continue to be introduced also in the near future either to advise the operator in control or fault diagnosis applications or as a real-time control alternative for direct control and supervision.

However, as will be shown in the next sections, fuzzy control can also be described as a non-linear mapping. In that case the controller acts as a non-linear controller which is a natural extension of the still extensively used classical PID-controller which is in 90% of all low-level control applications still the reliable workhorse.

We have to keep in mind, however, that the PID-controller has become also more and more sophisticated, including all kinds of additional measures for proper functioning, such as anti-reset windup, proportional kick up, retarded integral action, etc. the modern PID-controller is already the result of the marriage between continuous linear control and discrete control which added non-linear and state-dependent control actions to the original concept, which nowadays contains only 20% of the code of a modern PID-controller.

So, the introduction of a fuzzy controller based on the concept of a PID-controller was an option which has been introduced already many years ago (Årzén 1993; Li, Bruijn and Verbruggen, 1994; Zhao, Tomizuka and Isaka, 1993) allowing the addition of all kinds of local non-linearities. The tedious tuning of the parameters of a fuzzy controller is also alleviated by starting from the settings of a conventional controller.

A control strategy which has obtained an increasing popularity in process control is model-based predictive control (MBPC), because this strategy is able to control multi-input multi-output (MIMO) systems subjected to constraints in control variables or states and outputs of the system. A number of commercial software packages is available and many applications of MBPC are reported in industry, see a recent survey by Richalet (1993).

This strategy uses a model of the process to predict future process outputs depending on proposed control actions in the near future and also a model of the process is used to optimise these control actions given a certain objective function and a desired reference trajectory.

Dominant ingredients in this approach are the model of the process which should be in many cases non-linear and an objective function which could also contain qualitative requirements and constraints. Especially for modelling highly non-linear and partly known systems fuzzy techniques have shown to be very useful, both for designing the controller and modelling the system. It is clear that MBPC needs a model of the process, so it would be expected that fuzzy techniques will become an important tool for future non-linear MBPC developments.

Fuzzy models can be obtained using a number of techniques, such as linguistic models based on Mamdani fuzzy rules (Sugeno, Murofushi, Nishino and Miwa, 1992), on a Takagi-Sugeno approach (Babuška and Verbruggen, 1997; Takagi and Sugeno, 1985) or on relational models (Pedrycz, 1993b). A recent publication (Babuška, 1998) gives a nice overview of recent fuzzy modelling techniques with many applications.

As has been shown (Babuška, 1998), some of the fuzzy modelling techniques can be used in combination with first principle modelling, measurements and expert knowledge, resulting in a more or less uniform framework for non-linear process modelling.

Finally it should be stated that the combination of fuzzy modelling for a part of the system combined with a mechanistic model which describes the known part of the system is very promising. This semi-mechanistic model can be used for prediction and control purposes. Besides, it allows a better transparency to the user of the process characteristics and sometimes allows for faster computations which is of importance in some real-time applications of the model such as in MBPC and feedback linearization techniques.

The remainder of this chapter contains 3 sections which can be read more or less independently.

In section 1.2 a short engineering introduction is given for designing simple fuzzy controllers. The objective is to identify and describe the choices to be made by the designer, based on an international standard which is underway.

In section 1.3 an overview is given of some new developments in fuzzy modelling and control which might be of major importance in applying fuzzy techniques in the process industry. Emphasis is given to model-based predictive control, which is gaining influence in many modern process control systems.

In section 1.4 a short introduction is given to the use of fuzzy logic in high level process control. In sections 1.2 and 1.3 the introduction of fuzzy control is based on educational and research experience, however, in section 1.4 one of the first developers and users of fuzzy control presents his view on how fuzzy control can be implemented in high level control.

1.2 DESIGN OF SIMPLE FUZZY CONTROLLERS

While it is relatively easy to design a PID controller, the inclusion of fuzzy rules creates many extra design problems, and although many introductory textbooks explain fuzzy control, there are few general guidelines for setting the parameters of a simple fuzzy controller. The approach here is based on a three step design procedure, that builds on PID control:

1. Start with a PID controller.
2. Insert an equivalent, linear fuzzy controller.
3. Make it gradually nonlinear.

Guidelines related to the different components of the fuzzy controller will be introduced shortly. In the next three sections three simple realisations of fuzzy controllers are described: a table-based controller, an input-output mapping and a Takagi-Sugeno type controller. A short section summarises the main design choices in a simple fuzzy controller by introducing a check list. The terminology is based on an international standard which is underway (IEC, 1996).

Fuzzy controllers are used to control consumer products, such as washing machines, video cameras, and rice cookers, as well as industrial processes, such as cement kilns, underground trains, and robots. Fuzzy control is a control method based on fuzzy logic. Just as fuzzy logic can be described simply as "computing with words rather than numbers", fuzzy control can be described simply as "control with sentences rather than equations". A fuzzy controller can include empirical rules, and that is especially useful in operator controlled plants. Take for instance a typical fuzzy controller

1. If error is Neg and change in error is Neg then output is NB
2. If error is Neg and change in error is Zero then output is NM (1)

...

The collection of rules is called a *rule base*. The rules are in the familiar if-then format, and formally the if-side is called the *condition* and the then-side is called the *conclusion* (more often, perhaps, the pair is called *antecedent - consequent* or *premise - conclusion*). The input value "Neg" is a *linguistic term* short for the word *Negative*, the output value "NB" stands for *Negative Big* and "NM" for *Negative Medium*. The computer is able to execute the rules and compute a control signal depending on the measured inputs *error* and *change in error*. The objective here is to identify and ex-

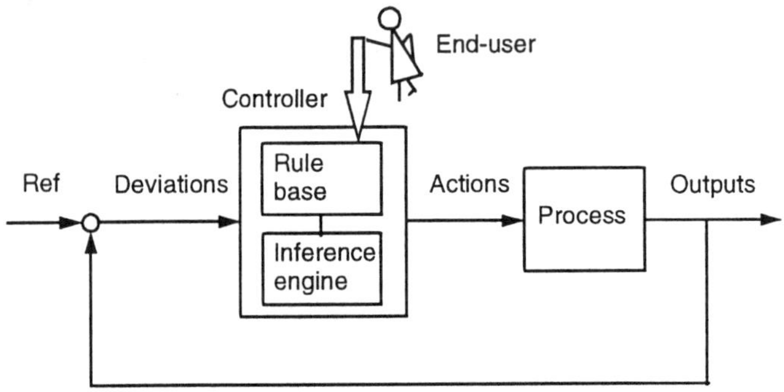

Figure 1 Direct control.

plain the various design choices for engineers.

In a rule based controller the control strategy is stored in a more or less natural language. The control strategy is isolated in a rule base opposed to an equation based description. A rule based controller is easy to understand and easy to maintain for a non-specialist end-user. An equivalent controller could be implemented using conventional techniques – in fact, any rule based controller could be emulated in, say, *Fortran* – it is just that it is convenient to isolate the control strategy in a rule base for operator controlled systems.

Fuzzy controllers are being used in various control schemes (IEC, 1996). The most obvious one is *direct control*, where the fuzzy controller is in the forward path in a feedback control system (Fig. 1). The process output is compared with a reference, and if there is a deviation, the controller takes action according to the control strategy. In the figure, the arrows may be understood as hyper-arrows containing several signals at a time for multi-loop control. The sub-components in the figure will be explained shortly. The controller is here a fuzzy controller, and it replaces a conventional controller, say, a *PID* (proportional-integral-derivative) controller.

In *feedforward control* (Fig. 2) a measurable disturbance is being compensated. It requires a good model, but if a mathematical model is difficult or expensive to obtain, a fuzzy model may be useful. Figure 2 shows a controller and the fuzzy compensator, the process and the feedback loop are omitted for clarity. The scheme, disregarding the disturbance input, can be viewed as a collaboration of linear and nonlinear control actions; the controller may be a linear PID controller, while the fuzzy compensator is a supplementary nonlinear controller

Fuzzy rules are also used to correct tuning parameters in *parameter adaptive control* schemes (Fig. 3). If a nonlinear plant changes operating point, it may be possible

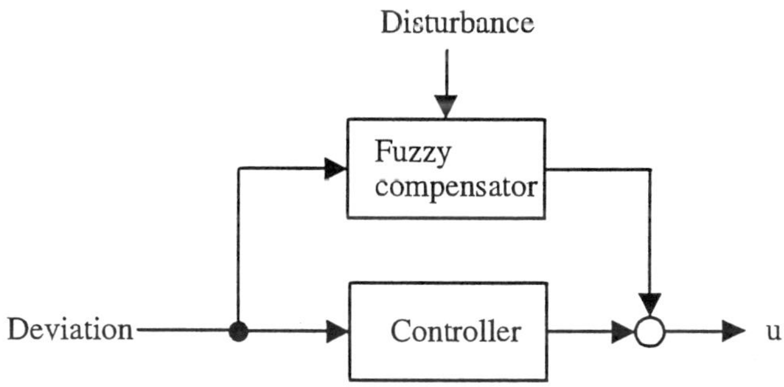

Figure 2 Feedforward control.

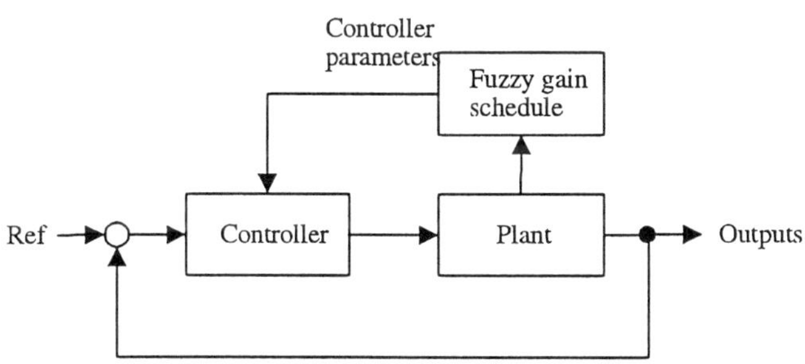

Figure 3 Fuzzy parameter adaptive control.

to change the parameters of the controller according to each operating point. This is called *gain scheduling* since it was originally used to change process gains. A gain scheduling controller contains a linear controller whose parameters are changed as a function of the operating point in a preprogrammed way. It requires thorough knowledge of the plant, but it is often a good way to compensate for nonlinearities and parameter variations. Sensor measurements are used as *scheduling variables* that govern the change of the controller parameters, often by means of a table look-up.

Whether a fuzzy control design will be stable is a somewhat open question. Stability concerns the system's ability to converge or stay close to an equilibrium. A *stable* linear system will converge to the equilibrium asymptotically no matter where the system state variables start from. It is relatively straight forward to check for stability in linear systems, for example by checking that all eigenvalues are in the left half of the complex plane. For nonlinear systems, and fuzzy systems are most often nonlinear, the stability concept is more complex. A nonlinear system is said to be *asymptotically stable* if, when it starts close to an equilibrium, it will converge to it. Even if it just stays close to the equilibrium, without converging to it, it is said to be *stable* (in the sense of Lyapunov). To check conditions for stability is much more difficult with nonlinear systems, partly because the system behaviour is also influenced by the signal *amplitudes* apart from the *frequencies*. The literature is somewhat theoretical and interested readers are referred to Driankov, Hellendoorn & Reinfrank (1993) or Passino & Yurkovich (1998). They report on four methods (Lyapunov functions, Popov, circle, and conicity), and they give several references to scientific papers. It is characteristic, however, that the methods give rather conservative results, which translate into unrealistically small magnitudes of the gain factors in order to guarantee stability.

Another possibility is to approximate the fuzzy controller with a linear controller, and then apply the conventional linear analysis and design procedures on the approximation. It seems likely that the stability margins of the nonlinear system would be close in some sense to the stability margins of the linear approximation depending on how close the approximation is. This paper shows how to build such a linear approximation, but the theoretical background is still unexplored. There are at least four main sources for finding control rules (Takagi & Sugeno in Lee, 1990).

Expert experience and control engineering knowledge. One classical example is the operator's handbook for a cement kiln (Holmblad & Ostergaard, 1982). The most common approach to establishing such a collection of rules of thumb, is to question experts or operators using a carefully organised questionnaire.

Based on the operator's control actions. Fuzzy *if-then* rules can be deduced from observations of an operator's control actions or a log book. The rules express input-output relationships.

Based on a fuzzy model of the process. A linguistic rule base may be viewed

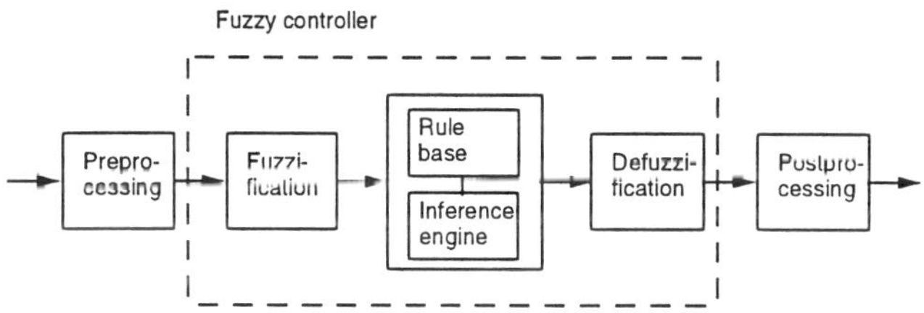

Figure 4 Blocks of a fuzzy controller.

as an inverse model of the controlled process. Thus the fuzzy control rules might be obtained by inverting a fuzzy model of the process. This method is restricted to relatively low order systems, but it provides an explicit solution assuming that fuzzy models of the open and closed loop systems are available (Braae & Rutherford in Lee, 1990). Another approach is *fuzzy identification* (Tong; Takagi & Sugeno; Sugeno – all in Lee, 1990; Pedrycz, 1993) or fuzzy model-based control (see later).

Based on learning. The self-organising controller is an example of a controller that finds the rules itself. Neural networks is another possibility.

There is no design procedure in fuzzy control such as root-locus design, frequency response design, pole placement design, or stability margins, because the rules are often nonlinear. Therefore we will settle for describing the basic components and functions of fuzzy controllers, in order to recognise and understand the various options in commercial software packages for fuzzy controller design.

There is much literature on fuzzy control and many commercial software tools (MIT, 1995), but there is no agreement on the terminology, which is confusing. There are efforts, however, to standardise the terminology, and the following makes use of a draft of a standard from the International Electrotechnical Committee (IEC, 1996). Throughout, letters denoting matrices are in bold upper case, for example **A**; vectors are in bold lower case, for example **x**; scalars are in italics, for example n; and operations are in bold, for example **min**.

1.2.1 Structure of a fuzzy controller

There are specific components characteristic of a fuzzy controller to support a design procedure. In the block diagram in Fig. 4, the controller is between a preprocessing block and a post-processing block. The following explains the diagram block by block.

Preprocessing The inputs are most often hard or *crisp* measurements from some measuring equipment, rather than linguistic. A preprocessor, the first block in Fig. 4, conditions the measurements before they enter the controller. Examples of preprocessing are:

- Quantisation in connection with sampling or rounding to integers;
- normalisation or scaling onto a particular, standard range;
- filtering in order to remove noise;
- averaging to obtain long term or short term tendencies;
- a combination of several measurements to obtain key indicators; and
- differentiation and integration or their discrete equivalences.

A *quantiser* is necessary to convert the incoming values in order to find the best level in a discrete *universe*. Assume, for instance, that the variable *error* has the value 4.5, but the universe is $\mathbf{u} = (-5, -4, \ldots, 0, \ldots, 4, 5)$. The quantiser rounds to 5 to fit it to the nearest level. Quantisation is a means to reduce data, but if the quantisation is too coarse the controller may oscillate around the reference or even become unstable.

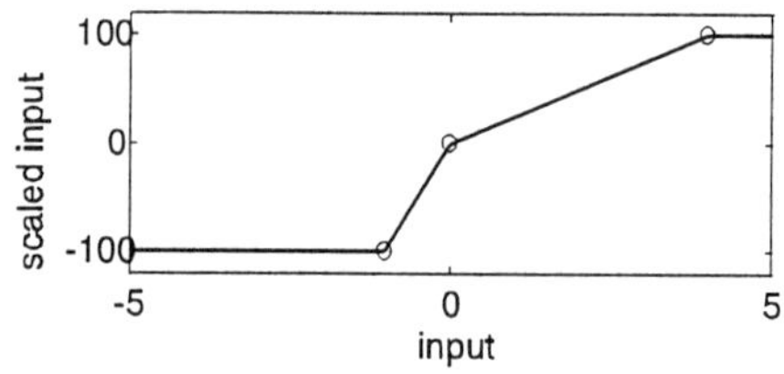

Figure 5 Example of nonlinear scaling of an input measurement.

Nonlinear *scaling* is an option (Fig. 5). In the *FL Smidth* controller the operator is asked to enter three typical numbers for a small, medium and large measurement respectively (Holmblad & Østergaard, 1982). They become break-points on a curve that scales the incoming measurements (circled in the figure). The overall effect can be interpreted as a distortion of the primary fuzzy sets. It can be confusing with both scaling and gain factors in a controller, and it makes tuning difficult.

When the input to the controller is *error*, the control strategy is a static mapping between input and control signal. A dynamic controller would have additional inputs, for example derivatives, integrals, or previous values of measurements backwards in time. These are created in the preprocessor thus making the controller multi-dimensional, which requires many rules and makes it more difficult to design.

The preprocessor then passes the data on to the controller.

Fuzzification The first block inside the controller is *fuzzification*, which converts

Table 1 Rules for a fuzzy PD controller.

No	Rule
1.	If error is Neg and change in error is Neg then output is NB
2.	If error is Neg and change in error is Zero then output is NM
3.	If error is Neg and change in error is Pos then output is Zero
4.	If error is Zero and change in error is Neg then output is NM
5.	If error is Zero and change in error is Zero then output is Zero
6.	If error is Zero and change in error is Pos then output is PM
7.	If error is Pos and change in error is Neg then output is Zero
8.	If error is Pos and change in error is Zero then output is PM
9.	If error is Pos and change in error is Pos then output is PB

each piece of input data to degrees of membership by a lookup in one or several membership functions. The fuzzification block thus matches the input data with the conditions of the rules to determine how well the condition of each rule matches that particular input instance. There is a degree of membership for each linguistic term that applies to that input variable.

Rule base The rules may use several variables both in the condition and the conclusion of the rules. The controllers can therefore be applied to both multi-input-multi-output (MIMO) problems and single-input-single-output (SISO) problems. The typical SISO problem is to regulate a control signal based on an error signal. The controller may actually need both the *error*, the *change in error*, and the *accumulated error* as inputs, but we will call it single-loop control, because in principle all three are formed from the error measurement. To simplify, this section assumes that the control objective is to regulate some process output around a prescribed setpoint or reference. The presentation is thus limited to single-loop control.

Rule formats Basically a linguistic controller contains rules in the *if-then* format, but they can be presented in different formats. In many systems, the rules are presented to the end-user in a format similar to the one in Table 1. The names *Zero*, *Pos*, *Neg* are labels of fuzzy sets as well as *NB*, *NM*, *PB* and *PM* (negative big, negative medium, positive big, and positive medium respectively). The same set of rules could be presented in a *relational* format, a more compact representation.

Error	Change in error	Output
Neg	Pos	Zero
Neg	Zero	NM
Neg	Neg	NB
Zero	Pos	PM
Zero	Zero	Zero
Zero	Neg	NM
Pos	Pos	PB
Pos	Zero	PM
Pos	Neg	Zero

(2)

The top row is the heading, with the names of the variables. It is understood that the two leftmost columns are inputs, the rightmost is the output, and each row represents a rule. This format is perhaps better suited for an experienced user who wants to get an overview of the rule base quickly. The relational format is certainly suited for storing in a relational database. It should be emphasised, though, that the relational format implicitly assumes that the connective between the inputs is always logical **and** – or logical **or** for that matter as long as it is the same operation for all rules – and not a mixture of connectives. Incidentally, a fuzzy rule with an **or** combination of terms can be converted into an equivalent **and** combination of terms using laws of logic (DeMorgan's laws among others). A third format is the tabular linguistic format.

		Change in error		
		Neg	Zero	Pos
Error	Neg	NB	NM	Zero
	Zero	NM	Zero	PM
	Pos	Zero	PM	PB

(3)

This is even more compact. The input variables are laid out along the axes, and the output variable is inside the table. In case the table has an empty cell, it is an indication of a missing rule, and this format is useful for checking completeness. When the input variables are *error* and *change in error*, as they are here, that format is also called a *linguistic phase plane*. In case there are $n > 2$ input variables involved, the table grows to an n-dimensional array; rather user-*un*friendly.

To accommodate several outputs, a nested arrangement is conceivable. A rule with several outputs could also be broken down into several rules with one output.

Lastly, a graphical format which shows the fuzzy membership curves is also possible (Fig. 7). This graphical user-interface can display the inference process better than the other formats, but takes more space on a monitor.

Connectives In mathematics, sentences are connected with the words *and*, *or*, *if-*

then (or *implies)*, and *if and only if*, or modifications with the word *not*. These five are called *connectives*. It also makes a difference how the connectives are implemented. The most prominent is probably multiplication for fuzzy **and** instead of minimum. So far most of the examples have only contained **and** operations, but a rule like "If error is very neg and not zero or change in error is zero then ..." is also possible.

The connectives **and** and **or** are always defined in pairs, for example,

$$\begin{array}{ll} \mathbf{a\ and\ b} = \mathbf{min}\,(\mathbf{a}, \mathbf{b}) & \text{minimum} \\ \mathbf{a\ or\ b} = \mathbf{max}\,(\mathbf{a}, \mathbf{b}) & \text{maximum} \\ \qquad\quad \text{or} & \\ \mathbf{a\ and\ b} = \mathbf{a} * \mathbf{b} & \text{algebraic product} \\ \mathbf{a\ or\ b} = \mathbf{a} + \mathbf{b} - \mathbf{a} * \mathbf{b} & \text{algebraic or probabilistic sum} \end{array} \tag{4}$$

There are other examples (e.g., Zimmermann, 1991, 31 – 32), but they are more complex.

Modifiers A linguistic *modifier*, is an operation that modifies the meaning of a term. For example, in the sentence "very close to 0", the word **very** modifies *Close to 0* which is a fuzzy set. A modifier is thus an operation on a fuzzy set. The modifier **very** can be defined as squaring the subsequent membership function, that is

$$\mathbf{very\ a} = \mathbf{a}^2 \tag{5}$$

Some examples of other modifiers are

$$\begin{array}{l} \mathbf{extremely\ a} = \mathbf{a}^3 \\ \mathbf{slightly\ a} = \mathbf{a}^{\frac{1}{3}} \\ \mathbf{somewhat\ a} = \mathbf{moreorless\ a\ and\ not\ slightly\ a} \end{array}$$

A whole family of modifiers is generated by $\mathbf{a}^p$ where p is any power between zero and infinity. With $p = \infty$ the modifier could be named **exactly**, because it would suppress all memberships lower than 1.0.

Universes Elements of a fuzzy set are taken from a *universe of discourse* or just *universe*. The universe contains all elements that can come into consideration. Before designing the membership functions it is necessary to consider the universes for the inputs and outputs. Take for example the rule

If error is Neg and change in error is Pos then output is 0

Naturally, the membership functions for *Neg* and *Pos* must be defined for all possible values of *error* and *change in error*, and a standard universe may be convenient.

Another consideration is whether the input membership functions should be continuous or discrete. A continuous membership function is defined on a continuous universe by means of parameters. A discrete membership function is defined in terms of a vector with a finite number of elements. In the latter case it is necessary to specify the range of the universe and the value at each point. The choice between fine and coarse resolution is a trade off between accuracy, speed and space demands. The quantiser takes time to execute, and if this time is too precious, continuous membership functions will make the quantiser obsolete.

Example 1 (standard universes) *Many authors and several commercial controllers use standard universes.*

- *The FL Smidth controller, for instance, uses the real number interval* $[-1, 1]$.
- *Authors of the earlier papers on fuzzy control used the integers in* $[-6, 6]$.
- *Another possibility is the interval* $[-100, 100]$ *corresponding to percentages of full scale.*
- *Yet another is the integer range* $[0, 4095]$ *corresponding to the output from a 12 bit analog to digital converter.*
- *A variant is* $[-2047, 2048]$, *where the interval is shifted in order to accommodate negative numbers.*

The choice of datatypes may govern the choice of universe. For example, the voltage range $[-5, 5]$ *could be represented as an integer range* $[-50, 50]$, *or as a floating point range* $[-5.0, 5.0]$*; a signed byte datatype has an allowable integer range* $[-128, 127]$.

□

A way to exploit the range of the universes better is scaling. If a controller input mostly uses just one term, the scaling factor can be turned up such that the whole range is used. An advantage is that this allows a standard universe and it eliminates the need for adding more terms.

Membership functions Every element in the universe of discourse is a member of a fuzzy set to some grade, maybe even zero. The grade of membership for all its members describes a fuzzy set, such as *Neg*. In fuzzy sets elements are assigned a *grade of membership*, such that the transition from membership to non-membership is gradual rather than abrupt. The set of elements that have a non-zero membership is called the *support* of the fuzzy set. The function that ties a number to each element x of the universe is called the *membership function* $\mu(x)$.

The designer is inevitably faced with the question of how to build the term sets. There are two specific questions to consider: (i) How does one determine the shape of the sets? and (ii) How many sets are necessary and sufficient? For example, the *error* in the position controller uses the family of terms *Neg*, *Zero*, and *Pos*. According to fuzzy set theory the choice of the shape and width is subjective, but a few rules of thumb apply.

- A term set should be sufficiently wide to allow for noise in the measurement.
- A certain amount of overlap is desirable; otherwise the controller may run into poorly defined states, where it does not return a well defined output.

A preliminary answer to questions (i) and (ii) is that the necessary and sufficient number of sets in a family depends on the width of the sets, and vice versa. A solution could be to ask the process operators to enter their personal preferences for the membership curves; but operators also find it difficult to settle on particular curves.

The manual for the TILShell product recommends the following (Hill, Horstkotte & Teichrow, 1990).

- *Start with triangular sets.* All membership functions for a particular input or output should be symmetrical triangles of the same width. The leftmost and the rightmost should be shouldered ramps.
- *The overlap should be at least 50 %.* The widths should initially be chosen so that each value of the universe is a member of at least two sets, except possibly for elements at the extreme ends. If, on the other hand, there is a gap between two sets no rules fire for values in the gap. Consequently the controller function is not defined.

Membership functions can be flat on the top, piece-wise linear and triangle shaped, rectangular, or ramps with horizontal shoulders. Fig. 6 shows some typical shapes of membership functions.

Strictly speaking, a fuzzy set A is a collection of ordered pairs

$$A = \{(x, \mu(x))\} \tag{6}$$

Item x belongs to the universe and $\mu(x)$ is its grade of membership in A. A single pair $(x, \mu(x))$ is a fuzzy *singleton*; *singleton output* means replacing the fuzzy sets in the conclusion by numbers (scalars). For example

1. If error is Pos then output is 10 volts
2. If error is Zero then output is 0 volts
3. If error is Neg then output is $-$ 10 volts

There are at least three advantages to this:

- The computations are simpler;
- it is possible to drive the control signal to its extreme values; and
- it may actually be a more intuitive way to write rules.

The scalar can be a fuzzy set with the singleton placed in a proper position. For example 10 *volts*, would be equivalent to the fuzzy set $(0, 0, 0, 0, 1)$ defined on the

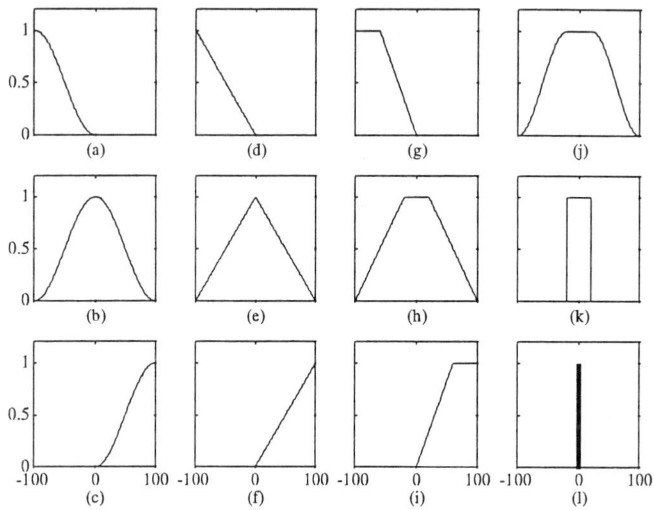

Figure 6 Examples of membership functions. Read from top to bottom, left to right: (a) s$-$function, (b) $\pi-$ function, (c) z$-$function, (d-f) triangular versions, (g-i) trapezoidal versions, (j) flat $\pi-$ function, (k) rectangle, (l) singleton.

universe $(-10, -5, 0, 5, 10)$ *volts*.

Example 2 (membership functions) *Fuzzy controllers use a variety of membership functions. A common example of a function that produces a bell curve is based on the exponential function,*

$$\mu(x) = \exp\left[\frac{-(x - x_0)^2}{2\sigma^2}\right] \tag{7}$$

This is a standard Gaussian curve with a maximum value of 1, x *is the independent variable on the universe,* x_0 *is the position of the peak relative to the universe, and* σ *is the standard deviation. Another definition which does not use the exponential is*

$$\mu(x) = \left[1 + \left(\frac{x - x_0}{\sigma}\right)^2\right]^{-1} \tag{8}$$

The FL Smidth *controller uses the equation*

$$\mu(x) = 1 - \exp\left[-\left(\frac{\sigma}{x_0 - x}\right)^a\right] \tag{9}$$

The extra parameter a *controls the gradient of the sloping sides. It is also possible to use other functions, for example the* sigmoid *known from neural networks.*

A cosine function can be used to generate a variety of membership functions. The s-curve *can be implemented as*

$$s(x_l, x_r, x) = \left\{ \begin{array}{ll} 0 & ; x < x_l \\ \frac{1}{2} + \frac{1}{2}\cos\left(\frac{x - x_r}{x_r - x_l}\pi\right) & ; x_l \leq x \leq x_r \\ 1 & ; x > x_r \end{array} \right\} \tag{10}$$

where x_l is the left breakpoint, and x_r is the right breakpoint. The z-curve *is just a reflection,*

$$z(x_l, x_r, x) = \left\{ \begin{array}{ll} 1 & ; x < x_l \\ \frac{1}{2} + \frac{1}{2}\cos\left(\frac{x - x_l}{x_r - x_l}\pi\right) & ; x_l \leq x \leq x_r \\ 0 & ; x > x_r \end{array} \right\} \tag{11}$$

Then the π-curve *can be implemented as a combination of the* s-curve *and the* z-curve, *such that the peak is flat over the interval* $[x_2, x_3]$

$$\pi(x_1, x_2, x_3, x_4, x) = \min(s(x_1, x_2, x), z(x_3, x_4, x)) \tag{12}$$

□

Inference engine Figures 7 and 8 are both a graphical construction of the algorithm in the core of the controller. In Fig. 7, each of the nine rows refers to one rule. For example, the first row says that if the *error* is negative (row 1, column 1) and the *change in error* is negative (row 1, column 2) then the output should be negative big (row 1, column 3). The picture corresponds to the rule base in Table 1. The rules reflect the strategy that the control signal should be a combination of the reference error and the change in error, a fuzzy proportional-derivative controller. We shall refer to that figure in the following. The instances of the *error* and the *change in error* $(0, 0)$ are indicated by the vertical lines on the first and second columns of the chart. For each rule, the inference engine looks up the membership values in the condition of the rule.

Aggregation The *aggregation* operation is used when calculating the *degree of fulfillment* or *firing strength* α_k of the condition of a rule k. A rule, say rule 1, will generate a fuzzy membership value μ_{e1} coming from the *error* and a membership value μ_{ce1} coming from the *change in error* measurement. The aggregation is their combination,

$$\mu_{e1} \textbf{ and } \mu_{ce1} \tag{13}$$

Similarly for the other rules. Aggregation is equivalent to fuzzification, when there is only one input to the controller. Aggregation is sometimes also called *fulfillment* of the rule or *firing strength*.

Activation The *activation* of a rule is the deduction of the conclusion, possibly

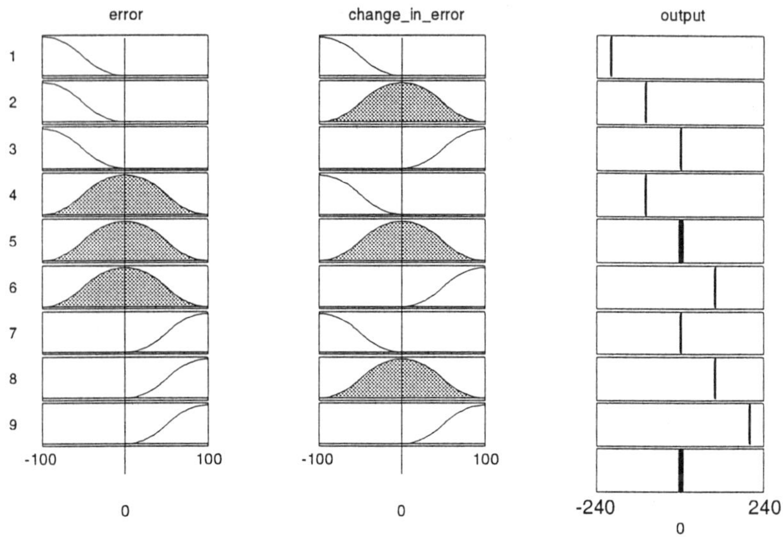

Figure 7 Graphical construction of the control signal in a fuzzy PD controller (generated by the Fuzzy Logic Toolbox for Matlab).

reduced by its firing strength. Thickened lines in the third column indicate the firing strength of each rule. Only the thickened part of the singletons are activated, and **min** or product (*) is used as the *activation operator*. It makes no difference in this case, since the output membership functions are singletons, but in the general case of $s-$, $\pi-$, and $z-$ functions in the third column, the multiplication scales the membership curves, thus preserving the initial shape, rather than clipping them as the **min** operation does. Both methods work well in general, although the multiplication results in a slightly smoother control signal. In Fig. 7, only rule five is active.

A rule k can be weighted a priori by a weighting factor $\omega_k \in [0, 1]$, which is its *degree of confidence*. In that case the firing strength is modified to

$$\alpha_k^* = \omega_k * \alpha_k. \tag{14}$$

The degree of confidence is determined by the designer, or a learning program trying to adapt the rules to some input-output relationship.

Accumulation All activated conclusions are *accumulated*, using the **max** operation, to the final graph on the bottom right (Fig. 7). Alternatively, **sum** accumulation counts overlapping areas more than once (Fig. 8). Singleton output (Fig. 7) and **sum**

accumulation results in the simple output

$$\alpha_1 * s_1 + \alpha_2 * s_2 + \ldots + \alpha_n * s_n \tag{15}$$

The alpha's are the firing strengths from the n rules and s_1 ... s_n are the output singletons. Since this can be computed as a vector product, this type of inference is relatively fast in a matrix oriented language.

There could actually have been several conclusion sets. An example of a one-input-two-outputs rule is "If e_a is **a** then o_1 is **x** and o_2 is **y**". The inference engine can treat two (or several) columns on the conclusion side in parallel by applying the firing strength to both conclusion sets. In practice, one would often implement this situation as two rules rather than one, that is, "If e_a is **a** then o_1 is **x**", "If e_a is **a** then o_2 is **y**".

Defuzzification The resulting fuzzy set (Fig. 7, bottom right; Fig. 8, extreme right) must be converted to a number that can be sent to the process as a control signal. This operation is called *defuzzification*, and in Fig. 8 the x-coordinate marked by a white, vertical dividing line becomes the control signal. The resulting fuzzy set is thus defuzzified into a crisp control signal. There are several defuzzification methods.

Centre of gravity (COG) The crisp output value u (white line in Fig. 8) is the abscissa under the centre of gravity of the fuzzy set,

$$u = \frac{\sum_i \mu(x_i)\, x_i}{\sum_i \mu(x_i)} \tag{16}$$

Here x_i is a running point in a discrete universe, and $\mu(x_i)$ is its membership value in the membership function. The expression can be interpreted as the weighted average of the elements in the support set. For the continuous case, replace the summations by integrals. It is a much used method although its computational complexity is relatively high. This method is also called *centroid of area.*

Centre of gravity method for singletons (COGS) If the membership functions of the conclusions are singletons (Fig. 7), the output value is

$$u = \frac{\sum_i \mu(s_i)\, s_i}{\sum_i \mu(s_i)} \tag{17}$$

Here s_i is the position of singleton i in the universe, and $\mu(s_i)$ is equal to the firing strength α_i of rule i. This method has a relatively good computational complexity, and u is differentiable with respect to the singletons s_i, which is useful in neurofuzzy systems.

Bisector of area (BOA) This method picks the abscissa of the vertical line that divides the area under the curve in two equal halves. In the continuous case,

$$u = \left\{ x \mid \int_{Min}^{x} \mu(x)\,dx = \int_{x}^{Max} \mu(x)\,dx \right\} \tag{18}$$

Here x is the running point in the universe, $\mu(x)$ is its membership, Min is the leftmost value of the universe, and Max is the rightmost value. Its computational complexity is relatively high, and it can be ambiguous. For example, if the fuzzy set consists of two singletons any point between the two would divide the area in two halves; consequently it is safer to say that in the discrete case, BOA is not defined.

Mean of maxima (MOM) An intuitive approach is to choose the point with the strongest possibility, i.e., maximal membership. It may happen, though, that several such points exist, and a common practice is to take the *mean of maxima* (MOM). This method disregards the shape of the fuzzy set, but the computational complexity is relatively good.

Leftmost maximum (LM), and rightmost maximum (RM) Another possibility is to choose the leftmost maximum (LM), or the rightmost maximum (RM). In the case of a robot, for instance, it must choose between left or right to avoid an obstacle in front of it. The defuzzifier must then choose one or the other, not something in between. These methods are indifferent to the shape of the fuzzy set, but the computational complexity is relatively small.

Postprocessing Output scaling is also relevant. In case the output is defined on a standard universe this must be scaled to *engineering units,* for instance, volts, meters, or tons per hour. An example is the scaling from the standard universe $[-1, 1]$ to the physical units $[-10, 10]$ volts.

The postprocessing block often contains an output gain that can be tuned, and sometimes also an integrator.

Example 3 (inference) *How is the inference in Fig.* 8 *implemented using discrete fuzzy sets?*

Behind the scene all universes were divided into 201 *points from* -100 *to* 100. *But for brevity, let us just use five points. Assume the universe* **u**, *common to all variables, is the vector*

$$\mathbf{u} = \begin{bmatrix} -100 & -50 & 0 & 50 & 100 \end{bmatrix}$$

A cosine function can be used to generate a variety of membership functions. The s-curve *can be implemented as*

$$s(x_l, x_r, x) = \left\{ \begin{array}{ll} 0 & , x < x_l \\ \frac{1}{2} + \frac{1}{2}\cos\left(\frac{x - x_r}{x_r - x_l}\pi\right) & , x_l \le x \le x_r \\ 1 & , x > x_r \end{array} \right\} \tag{19}$$

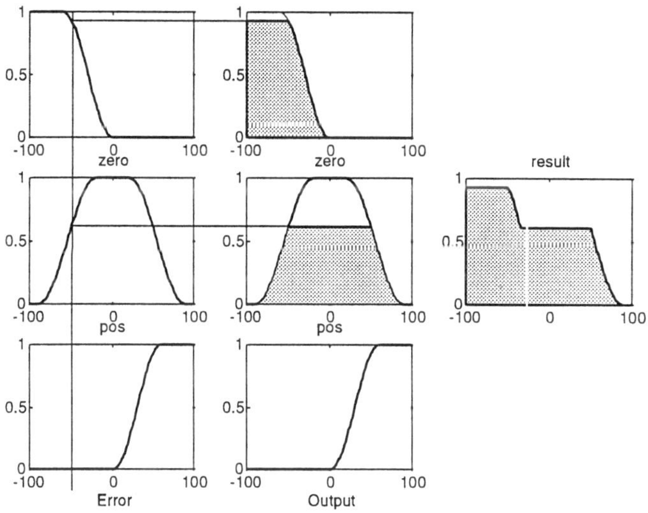

Figure 8 One input, one output rule base with non-singleton output sets.

where x_l *is the left breakpoint, and* x_r *is the right breakpoint. The* z-curve *is just a reflection,*

$$z(x_l, x_r, x) = \left\{ \begin{array}{ll} 1 & , x < x_l \\ \frac{1}{2} + \frac{1}{2}\cos\left(\frac{x - x_l}{x_r - x_l}\pi\right) & , x_l \leq x \leq x_r \\ 0 & , x > x_r \end{array} \right\} \tag{20}$$

Then the π-curve *(see for example Fig. 6(j)) can be implemented as a combination of the* s-curve *and the* z-curve, *such that the peak is flat over the interval* $[x_2, x_3]$

$$\pi(x_1, x_2, x_3, x_4, x) = \min(s(x_1, x_2, x), z(x_3, x_4, x)) \tag{21}$$

A family of terms is defined by means of the π*−function, such that*

$$\mathbf{neg} = \pi\,(-100, -100, -60, 10, \mathbf{u}) = \left[\begin{array}{ccccc} 1 & 0.95 & 0.05 & 0 & 0 \end{array}\right]$$

$$\mathbf{zero} = \pi\,(-90, -20, 20, 90, \mathbf{u}) = \left[\begin{array}{ccccc} 0 & 0.61 & 1 & 0.61 & 0 \end{array}\right]$$

$$\mathbf{pos} = \pi\,(-10, 60, 100, 100, \mathbf{u}) = \left[\begin{array}{ccccc} 0 & 0 & 0.05 & 0.95 & 1 \end{array}\right]$$

Above we inserted the whole vector ***u*** *in place of the running point* x; *the result is thus a vector. The figure assumes that* error $= -50$ *(the unit is percentages of full range). This corresponds to the second position in the universe, and the first rule contributes with a membership* $\mathbf{neg}(2) = 0.95$. *This firing strength is propagated to* *the conclusion side of the rule using* **min**, *such that the contribution from this rule is*

$$0.95\ \mathbf{min\ neg} = \begin{bmatrix} 0.95 & 0.95 & 0.05 & 0 & 0 \end{bmatrix}$$

The activation operation was **min** *here. Apply the same procedure to the two remaining rules, and stack all three contributions on top of each other,*

$$\begin{bmatrix} 0.95 & 0.95 & 0.05 & 0 & 0 \\ 0 & 0.61 & 0.61 & 0.61 & 0 \\ 0 & 0 & 0 & 0 & 0 \end{bmatrix}$$

To find the accumulated output set, perform a **max** *operation down each column. The result is the vector*

$$\begin{bmatrix} 0.95 & 0.95 & 0.61 & 0.61 & 0 \end{bmatrix}$$

The centre of gravity *method yields*

$$u = \frac{\sum_i \mu(x_i)\, x_i}{\sum_i \mu(x_i)} \tag{22}$$

$$= \frac{0.95 * (-100) + 0.95 * (-50) + 0.61 * 0 + 0.61 * 50 + 0 * 100}{0.95 + 0.95 + 0.61 + 0.61 + 0} \tag{23}$$

$$= -35.9 \tag{24}$$

which is the control signal (before postprocessing).

□

1.2.2 Table based controller

If the universes are discrete, it is always possible to calculate all thinkable combinations of inputs before putting the controller into operation. In a *table based controller* the relation between all input combinations and their corresponding outputs are arranged in a table. With two inputs and one output, the table is a two-dimensional look-up table. With three inputs the table becomes a three-dimensional array. The array implementation improves execution speed, as the run-time inference is reduced to a table look-up which is a lot faster, at least when the correct entry can be found without too much searching. Below is a small example of a look-up table corresponding to the rulebase in Table 1 with the membership functions in Fig. 7,

		change in error				
		-100	-50	0	50	100
	-100	-200	-160	-100	-40	0
	-50	-160	-121	-61	0	40
error	0	-100	-61	0	61	100
	50	-40	0	61	121	160
	100	0	40	100	160	200

(25)

A typical application area for the table based controller is where the inputs to the controller are the *error* and the *change in error*. The controller can be embedded in a larger system, a car for instance, where the table is downloaded to a table look-up mechanism.

Table regions Referring to the look-up table (25), a negative value of *error* implies that the process output y is above the reference Ref, because the error is computed as $error = Ref - y$. A positive value implies a process output below the reference. A negative value of *change in error* means that the process output increases while a positive value means it decreases.

Certain regions in the table are especially interesting. The centre of the table corresponds to the case where the *error* is zero, the process is on the reference. Furthermore, the *change in error* is zero here, so the process stays on the reference. This position is the stable point where the process has settled on the reference. The anti-diagonal (orthogonal to the main diagonal) of the table is zero; those are all the pleasant states, where the process is either stable on the reference or approaching the reference. Should the process move away a little from the zero diagonal, due to noise or a disturbance, the controller will make small corrections to get it back. In case the process is far from the reference and also moving away from it, we are in the upper left and lower right corners. Here the controller calls for drastic changes.

The numerical values on the two sides of the zero diagonal do not have to be antisymmetric; they can be any values, reflecting asymmetric control strategies. During a response with overshoot after a positive step in the reference, a plot of the point (*error*, *change in error*) will follow a trajectory in the table which spirals clockwise from the lower left corner of the table towards the centre. It is similar to a *phase plane* trajectory, where a variable is plotted against its derivative. A clever designer may adjust the numbers manually during a tuning session to obtain a particular response.

Bilinear interpolation If the resolution in the table is too coarse it will cause *limit cycling*, that is, oscillations around the reference. The table allows the *error* to drift away from zero until it jumps into a neighbouring cell with a nonzero control action. This can be avoided with *bilinear interpolation* between the cells instead of rounding to the nearest point. In the case of a two-dimensional table, an error E satisfies the relation $E_1 \leq E \leq E_2$, where E_1 and E_2 are the two neighbouring points. The change-in-error CE will likewise satisfy $CE_1 \leq CE \leq CE_2$. The resulting table value is then found by interpolating linearly in the E axis direction between the first pair $u_1 = (F(E_1, CE_1), F(E_2, CE_1))$ and the second pair $u_2 = (F(E_1, CE_2), F(E_2, CE_2))$, and then in the CE-axis direction between the pair (u_1, u_2).

n-Dimensional tables A three input controller has a three-dimensional look-up table. Assuming a resolution of, say, 13 points in each universe, the table holds 2197 elements. It would be a tremendous task to fill these in manually, but it is manageable with rules.

A three dimensional table can be represented as a two-dimensional table using a relational representation. Rearrange the table into three columns one for each of the three inputs (E_1, E_2, E_3) and one for the output (U), for example Table 2. Each input can take five values, and the table thus has $5 \times 5 \times 5 = 125$ rows. The table look-up is now a question of finding the right row, and picking the corresponding U value.

Table 2 Relation equivalent to a 3D look-up table.

E_1	E_2	E_3	U
−100	−100	−100	−100
−100	−100	−67	−89
−100	−100	0	−67
−100	−100	67	−44
−100	−100	100	−33
−100	−67	−100	−89
−100	−67	−67	−78
−100	−67	0	−56
...	...	...	...
100	100	100	100

1.2.3 Input-output mapping

Two inputs and one output results in a two dimensional table, which can be plotted as a surface for visual inspection. The relationship between one input and one output can be plotted as a graph. These plots are a design aid when selecting membership functions and constructing rules.

The shape of the surface can be controlled to a certain extent by manipulating the membership functions. In order to see this clearly, we will use the one-input-one-output case (without loss of generality). The fuzzy proportional rule base

1. If error is Neg then output is Neg
2. If eror is Zero then output is Zero (26)
3. If error is Pos then output is Pos

is the starting point for producing the six different mappings in Fig. 9. The rightmost column is the input-output mapping, and each row is a different controller. The controllers have the input families in the *if*-column and the output families in the *then* column. The results depend on the choice of design parameters, which in this case are

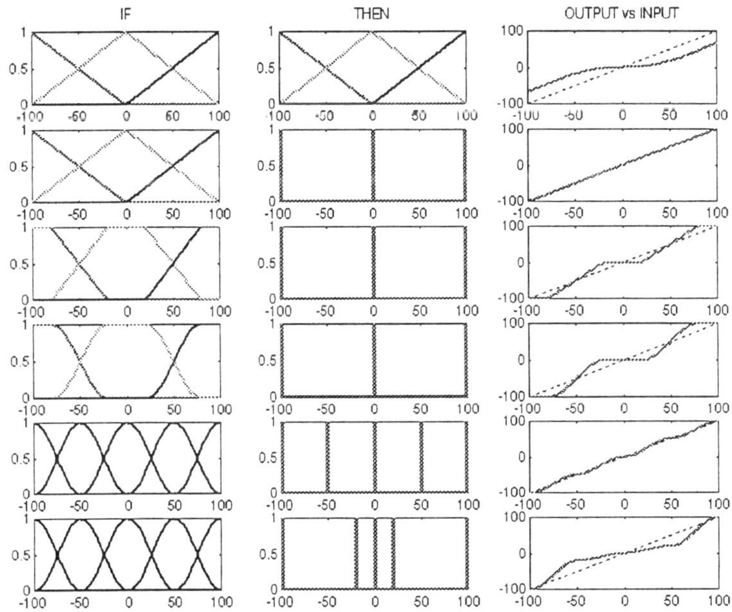

Figure 9 Input-output maps of proportional controllers. Each row is a controller.

the following: the * (*product*) operation for *activation* because it is continuous, the **max** operation for *accumulation* since it corresponds to set union, and *centre of gravity* for *defuzzification* since it is continuous, unambiguous, and it degenerates to *COGS* in the case of singleton output. If there had been two or more inputs, the * operation for **and** would be chosen since it is continuous. These choices are also necessary and sufficient for a linear mapping.

The following comments relate to the figure, row by row:

1. Triangular sets in both condition and conclusion result in a winding input-output mapping. Compared to a linear controller (dotted line) the gain of the fuzzy controller varies. A slight problem with this controller is that it does not use the full output range; it is impossible to drive the output to 100%. Another problem is that the local gain is always equal or lower than the linear controller.
2. Singleton outputs eliminate the problem with the output range. The set **pos** corresponds to 100, **zero** to 0, and **neg** to -100. The input terms are the same as before. Now the input-output mapping is linear.

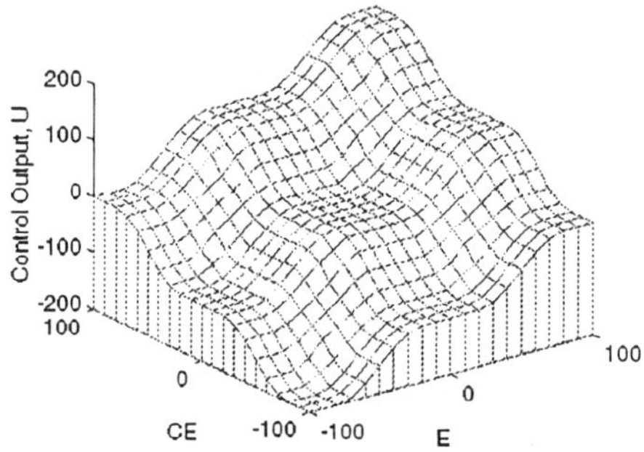

Figure 10 Example of a control surface.

3. Flat input sets produce flat plateaus and large gains far away from the reference. This is similar to a deadzone with saturation. Increasing the width of the middle term results in a wider plateau around the reference. Less overlap between neighbouring sets will result in steeper slopes.
4. If the sharp corners cause problems, they are removed by introducing nonlinear input sets. The input-output relationship is now smooth.
5. Adding more sets only makes the mapping more bumpy.
6. On the other hand with more sets it is easier to stretch the reference plateau by moving the singletons about.

The experiment shows that depending on what the design specifications are, it is possible to control, to a certain extent, the variation of the gain. Using singletons on the output side makes it easier. The results can easily be generalised to three dimensional surfaces. In all cases the activation operator was * (product), the accumulation operator was **max**, and the defuzzification method was *COG* or *COGS* – other operations may give slightly different results.

Control surface With two inputs and one output the input-output mapping is a surface. Figure 10 is a mesh plot of an example relationship between *error* E and *change in error* CE on the input side, and controller output u on the output side. The plot results from a rule base with nine rules, and the surface is more or less bumpy. The horizontal plateaus are due to flat peaks on the input sets. The plateau around the origin implies a low sensitivity towards changes in either *error* or *change in error* near the reference. This is an advantage if noise sensitivity must be low when the process is near the reference. On the other hand, if the process is unstable in open loop it is difficult to keep the process on the reference, and it will be necessary to have a larger gain around the origin.

There are three sources of nonlinearity in a fuzzy controller.

The rule base. The position, shape and number of fuzzy sets as well as nonlinear input scaling cause nonlinear transformations. The rules often express a nonlinear control strategy.
The inference engine. If the connectives **and** and **or** are implemented as for example **min** and **max** respectively, they are nonlinear.
The defuzzification. Several defuzzification methods are nonlinear.

It is possible to construct a rule base with a linear input-output mapping (Siler & Ying, 1989; Mizumoto, 1992; Qiao & Mizumoto; 1996). The following checklist summarises the general design choices for achieving a fuzzy rule base equivalent to a summation:

- ☐ Use triangular input sets that cross at $\mu = 0.5$;
- ☐ use the algebraic product (*) for the **and** connective;
- ☐ the rule base must be the complete **and** combination (cartesian product) of all input families;
- ☐ use output singletons, positions determined by the sum of the peak positions of the input sets;
- ☐ use *COG* defuzzification.

With these design choices the control surface degenerates to a diagonal plane (Fig. 11). A flexible fuzzy controller, that allows these choices, is two controllers in one so to speak. When linear, it has a transfer function and the usual methods regarding tuning and stability of the closed loop system apply.

1.2.4 Takagi-Sugeno type controller

We saw that the output sets can be singletons, but they can also be linear combinations of the inputs, or even a function of the inputs (Takagi & Sugeno, 1985). The general *Takagi-Sugeno* rule structure is

$$\text{If } f(e_1 \text{ is } \mathbf{A}_1, e_2 \text{ is } \mathbf{A}_2, \ldots, e_k \text{ is } \mathbf{A}_k) \text{ then } u = g(e_1, e_1 \ldots) \tag{27}$$

Here f is a logical function that connects the sentences forming the condition, u is the controller output, and g is a function of the inputs. A simple example is

$$\text{If error is Zero and change in error is Zero then } u = c$$

where c is a crisp constant. This is a *zero-order* model, and it is identical to singleton output rules. A slightly more complex rule is

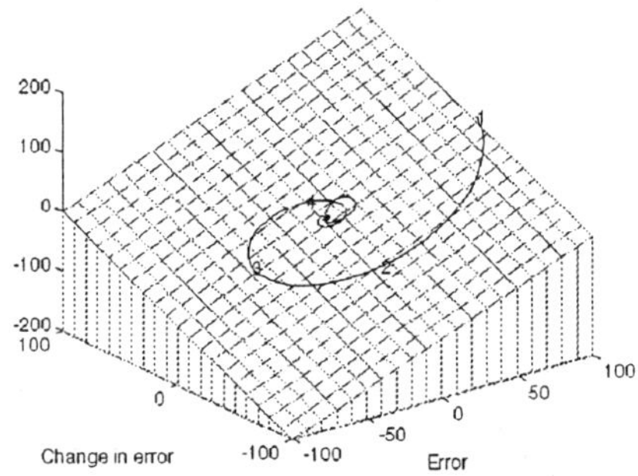

Figure 11 Linear surface with trajectory of a transient response.

If error is Zero and change in error is Zero then
$u = a * error + b * (change\ in\ error) + c$

where a, b and c are all constants. This is a *first-order* model. Inference with several rules proceeds as usual, with a firing strength associated with each rule, but each output is linearly dependent on the inputs. The output from each rule is a moving singleton, and the defuzzified output is the weighted average of the contributions from each rule. The controller interpolates between linear controllers; each controller is dominated by a rule, but there is a weighting depending on the overlap of the input membership functions. This is useful in a nonlinear control system, where each controller operates in a subspace of the operating envelope. One can say that the rules interpolate smoothly between the linear gains. Higher order models are also possible.

Example 4 (Sugeno) *Suppose we have two rules*

1. *If error is Large then output is Line1*
2. *If error is Small then output is Line2*

Line 1 is defined as $0.2 * error + 90$ *and line 2 is defined as* $0.6 * error + 20$. *The rules interpolate between the two lines in the region where the membership functions overlap (Fig.* 12*). Outside of that region the output is a linear function of the* error. *This type of model is used in neurofuzzy systems.*

□

In order to train a model to incorporate dynamics of a target system, the input is

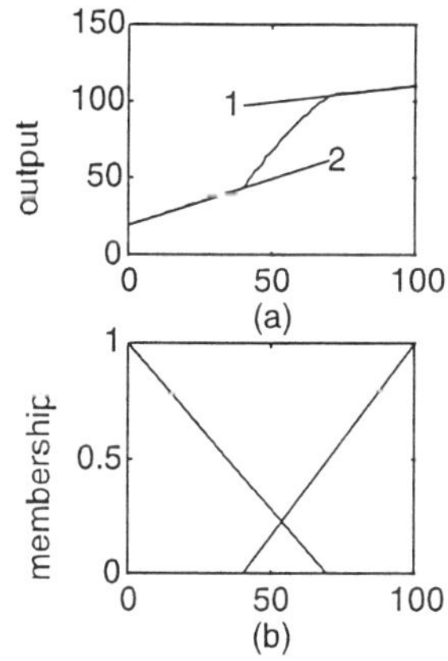

Figure 12 Interpolation between two lines (a), and overlap of rules (b).

augmented with signals corresponding to past inputs u and outputs y. In the time discrete domain the output of the model y^m, with superscript ^{m} referring to the model and ^{p} to the plant, is

$$y^m(t+1) = \widehat{\mathrm{f}}\left[y^p(t), \ldots, y^p(t-n+1); u(t), \ldots, u(t-m+1)\right] \qquad (28)$$

Here $\widehat{\mathrm{f}}$ represents the nonlinear input-output map of the model (i.e. the approximation of the target system f). Notice that the input to the model includes the past values of the plant output y^p and the plant input u.

1.2.5 Summary

In a fuzzy controller the data passes through a preprocessing block, a controller, and a postprocessing block. Preprocessing consists of a linear or non-linear scaling as well as a quantisation in case the membership functions are discretised (vectors); if not, the membership of the input can just be looked up in an appropriate function. When designing the rule base, the designer needs to consider the number of term sets, their shape, and their overlap. The rules themselves must be determined by the designer, unless more advanced means like self-organisation or neural networks are available. There is a choice between multiplication and minimum in the activation. There is also a choice regarding defuzzification; *centre of gravity* is probably most widely used. The postprocessing consists in a scaling of the output. In case the controller is incremental, postprocessing also includes an integration. The following is a checklist of design choices that have to be made:

- □ *Rule base related choices.* Number of inputs and outputs, rules, universes, continuous / discrete, the number of membership functions, their overlap and width, singleton output;
- □ *Inference engine related choices.* Connectives, modifiers, activation operation, aggregation operation, and accumulation operation.
- □ *Defuzzification method.* COG, COGS, BOA, MOM, LM, and RM.
- □ *Pre- and post-processing.* Scaling, gain factors, quantisation, and sampling time.

Some of these items must always be considered, others may not play a role in the particular design.

The input-output mappings provide an intuitive insight which may not be relevant from a theoretical viewpoint, but in practice they are well worth using. The analysis represented by plots is limited, though, to three dimensions. Various input-output mappings can be obtained by changing the fuzzy membership functions, and the chapter shows how to obtain a linear mapping with only a few adjustments.

The linear fuzzy controller may be used in a design procedure based on PID control:

1. Tune a PID controller.
2. Replace it with a linear fuzzy controller.
3. Transfer gains.
4. Make the fuzzy controller nonlinear.
5. Fine-tune it.

It seems sensible to start the controller design with a crisp PID controller, maybe even just a P controller, and get the system stabilised. From there it is easier to go to fuzzy control.

1.3 ADVANCED FUZZY CONTROL IN THE PROCESS INDUSTRY

In this section of the chapter an outline is given of some promising control methods based on fuzzy techniques which are expected to be potential solutions for problems arising in process control strategies in modern process industry. After an introduction, short descriptions are given of Fuzzy Inverse Control and Fuzzy Internal Model Control.

Next, a more extensive description of Fuzzy Model-based Predictive Control is given which is closely related to the approach accepted in the process industry for linear systems with constraints. Next, decision making in control and especially the application to model-based predictive control, is shortly introduced. Finally, problems arisirg when fuzzy logic is used for Multi-Input Multi-Output (MIMO) systems are shortly introduced. Multivariable control plays an important role in the process industry, because many control variables have to be controlled simultaneously and moreover, many interactions exist between manipulated variables and the different control variables. Much research has still to be done in this area, but some promising directions are already in progress.

1.3.1 Advanced control schemes based on simple controllers

In section 1.2 of the chapter a very simple scheme has been shown of a multi-local model controller. The control action consists of the blending of a number of local (in this case linear) controllers. The controllers can be synthesised based on linear model descriptions of the system in a number of operating points. By detecting the position of the system in relation to the different operating points, the control actions of the local controllers are weighted and constitute a varying influence of the various controllers on the operation of the system. The parameters of the controllers can be determined by local linear design methods of the same structure, e.g. PID-controllers.

It is, however, also possible to define different types of controllers depending on the operating conditions and objectives at a certain region of the process operation. This is sometimes called heterogeneous control. Examples are start up and shut down control actions as compared to stabilising control actions.

As already mentioned in the introduction a control strategy which is becoming quite popular in Process Control is the model-based approach. In the remainder of this section the explicit use of a model of the system as part of the control algorithm is treated.

1.3.2 Fuzzy inverse control

The most simple approach to design a controller is a completely open loop control strategy, in which the controller is the inverse of the process. In case we have a non-linear process model it is clear that a non-linear inverse controller will be found and that a perfect controlled system will be obtained. However, we have to keep in mind that the system can exhibit considerable delay times which should be taken into account and also other dynamic and static characteristics of the process should be well-known.

Besides, the remaining process model will never be an exact copy of the real process and there will be always disturbances acting on the process which will not be taken into account. However, many of these problems can be overcome using the control configuration called Internal Model Control.

When a fuzzy non-linear model is obtained from a (partly) unknown and highly complicated process, we have still the problem of inverting the fuzzy model.

Moreover, you have to keep in mind that these inversions should be computation-

(a) Global inversion *(b) Partial inversion*

Figure 13 Inversion of fuzzy models.

ally fast for its use in an in-line real-time control structure. This is simple for singleton fuzzy models. This type of models belongs to a general class of function approximations. Another type of model which can be inverted exactly is the Takagi-Sugeno type with affine inputs $u(k)$. However, constraining the model to an affine one reduces usually the accuracy of the model.

We should also keep in mind that inversion based approaches can only be applied to stable systems with a minimum phase character.

We normally speak about partial inversion and not about global inversion. This means in the case of a fuzzy model of a SISO system in which additional inputs are produced to get a dynamic model, that only one of the input variables of the model becomes the output of the inverted model and the output of the model becomes one of the inputs of the inverted model, see Fig. 13. The original model of the plant consists of n inputs of the fuzzy model $x_1, x_2, \ldots, x_n$ (previous inputs and outputs) and only one output y.

1.3.3 Fuzzy internal model control

To overcome some of the problems introduced by the open loop approach presented in the previous subsection (*Fuzzy Inverse Control*), a feedback approach was introduced by (Garcia and Morari, 1982). This approach called *Internal Model Control* (IMC) consists of three parts:

- a model to predict the effect of the control action on the system
- a controller based on the inverse of the process model
- a filter to increase robustness to model mismatch and disturbances.

When the model and the controller is a fuzzy system, this approach is called: *Fuzzy IMC*. A general scheme is depicted in Fig. 14.

It is clear that with a perfect model describing the process and no disturbances acting on the process the feedback signal becomes zero and open loop control is obtained. If a disturbance d is acting on the process, the feedback signal is equal to d and is not affected by the control action, but simply subtracted from the reference. So the effects

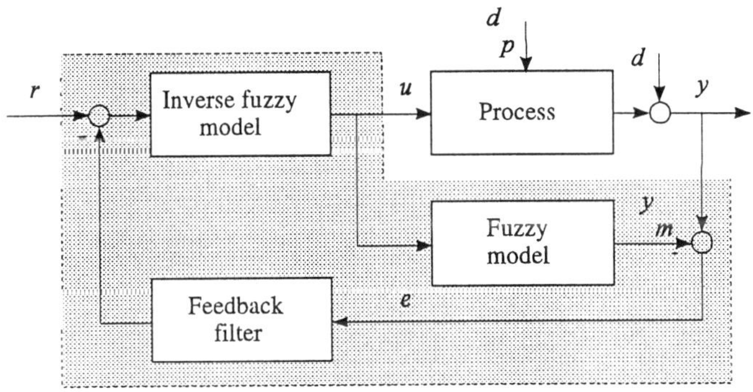

Figure 14 Fuzzy internal model scheme.

of output disturbances are completely cancelled.

It can be proved that under relatively weak circumstances (steady state of the control is inverse of the process) control without steady state errors can be obtained. A filter is introduced to increase the robustness of the system to model mismatch described by d_p and measurement noise.

It can reduce the loop gain to stabilise the system and smooth out noisy or rapidly changing signals, reducing the transient response of the IMC scheme. For non-linear systems the filter must be designed for the part of the system where the dynamics is fast.

However, depending on the model mismatch and the disturbances contaminating the system, it is not possible to design the filter in a predetermined way, when the process exhibits a non-linear behaviour, which is a major design problem.

1.3.4 Fuzzy model-based predictive control

The concept of Model-Based Predictive Control (MBPC) was introduced about two decades ago simultaneously by a number of authors (Richalet et al., 1978; Cutler and Ramaker, 1980). Although the concept is quite general a broad range of different methods are based on this concept depending on how the main ingredients of these methods are translated to specific solutions.

The basic concepts in all predictive control methods are the following:

- Use of a model to predict the process outputs at future discrete time instants over a certain prediction horizon (prediction model).
- Computation of a sequence of future control actions over a certain control horizon by minimising a certain objective function, which requires that the predicted

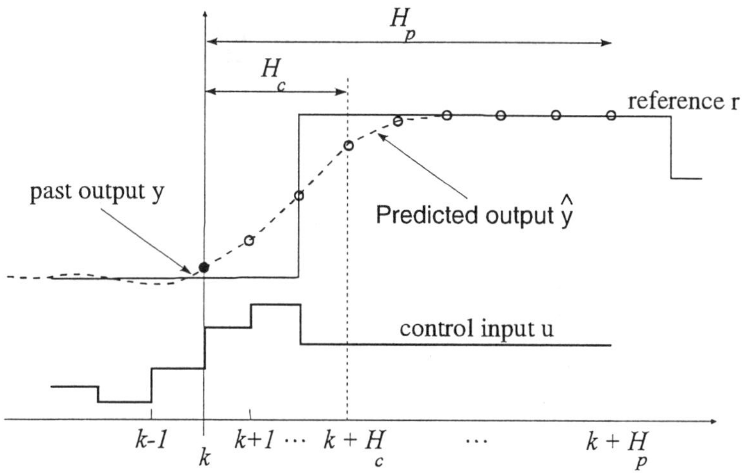

Figure 15 Basic principle of predictive control.

process outputs are as close as possible to the desired reference trajectories, under given operation constraints (optimisation process).

- Receding horizon strategy, so that at each sampling instant the optimisation process is repeated with the new measurements which has become available, and the first control action in the calculated control sequence is applied to the process (real-time receding horizon control).
- Sometimes an additional concept is introduced to compensate for model-plant mismatch and influences of disturbances.

Because of the explicit use of a process model and an optimisation approach, MBPC can be applied to complex systems, e.g. multivariable, non-minimum-phase, open-loop unstable, non-linear, or processes with a long delay time. Moreover, the method can efficiently deal with constraints on input and output variables of the process.

MBPC has been well accepted in industry due to the generality of the method and there exists already a large number of industrial applications (Richalet, 1993).

Extensive software packages are available which fit well in industrial instrumentation and process control systems. However, these packages are based on linear MIMO process models and on classical quadratic objective functions, including hard constraints. Extensive use is made of quadratic programming method in the optimisation process.

What could be the role of fuzzy techniques in MBPC?
A fuzzy model can be used to describe the non-linear behaviour of the process. This model can be used to predict the process behaviour in the future for a given control strategy.

Moreover, also the objective functions and constraints can be defined as fuzzy goals and constraints and the control problem is then translated to a fuzzy decision making problem. This is beneficial when some of the goals or constraints cannot be described analytically or should be fulfilled partially depending on other requirements. Sometimes a relaxation of some of the goals and constraints is possible.

Finally, a fuzzy model can be used in an IMC structure contained in the MBPC strategy to reduce process-model mismatch and the influence of disturbances.

Let us first describe the basic concepts of MBPC in more detail.

Control and prediction horizons The future plant outputs for an a priori prediction horizon H_p are predicted using a model of the process. The predicted output values $\hat{y}(k+j); j = 1...H_p$ depend on the state of the process at time k (e.g. given by past inputs and outputs) and on the future control signals $u(k+j)$ over a certain a priori determined control horizon H_c, assuming that $H_c \leq H_p$ and $u(k+j)$ remains constant for $j = H_c; ...., H_{p-1}$, see Fig. 15.

H_c is usually chosen to be equal to the order or the model or lower which is allowed when fuzzy objective functions are utilised. A low value of H_c reduces considerably the computational costs in the optimisation procedure.

H_p is usually related to the response time of the process (step response). For non-linear systems this is not clear because the time response can change considerably depending on the operating conditions. In a Takagi-Sugeno model this can be easily evaluated.

Objective function and reference trajectory The sequence of future control signals is obtained by the optimisation of an objective function which describes the goals of the control strategy. Constraints can be added as hard constraints separately to the objective function or should be included as soft constraints in the objective function.

In classical MBPC the objective function is usually expressed by the following quadratic function:

$$J(u) = \sum_{i=1}^{H_p} \alpha_i \left(w(k+i) - \hat{y}(k+i)\right)^2 + \sum_{i=1}^{H_c} \beta_i \Delta u(k+i-1)^2 \qquad (29)$$

The first term accounts for minimising the variance of the process output from the reference w, while the second term represents a penalty on the control effort. Sometimes u is used instead of Δu. The values α_i and β_i define the weightings of the different terms involved and indicate for instance that differences in the process out-

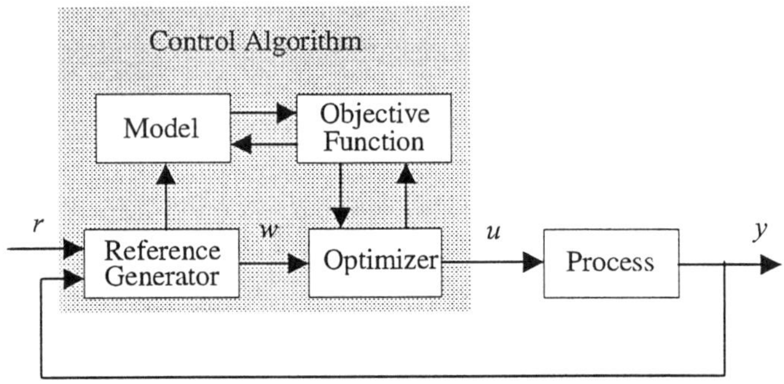

Figure 16 Classical model-based predictive scheme.

put should be weighted more severely when time evolves. When the process is linear and no constraints are involved, an analytical solution to the control problem can be obtained. Otherwise numerical (usually iterative) optimisation methods must be used.

In most cases the desired reference is known a priori and the system can react already before the change has actually been introduced. It is advisable to smooth the desired reference change to avoid sudden changes in the control actions at the cost of slower responses. This is obtained by introducing a reference generator.

Receding horizon principle and compensation of model-plant mismatch and disturbances MBPC is an open loop control strategy, but by using the receding horizon principle a certain feedback is introduced. It is, however, advantageous to use some additional measures to decrease the influence of disturbances and plant-model mismatch. This can be obtained by using an IMC scheme together with the MBPC strategy.

In Fig. 16 the complete classical MBPC scheme is presented. The optimiser delivers the control signal according to the receding horizon principle. The model is used for prediction and the objective function introduced by the user is optimised by the optimiser given the predicted process output and the reference values delivered by the reference generator.

Note that the MBPC can be regarded as a generalisation of IMC and that for $H_c = H_p = 1$, $\alpha_i = 1$ and $\beta_i = 0$ and requiring $y(k+1) = r(k+1)$ an inverse controller is obtained and the MBPC scheme becomes equivalent to the IMC scheme.

In Fig. 17 the MBPC controller is implemented in an IMC scheme to compensate for disturbances and modelling errors.

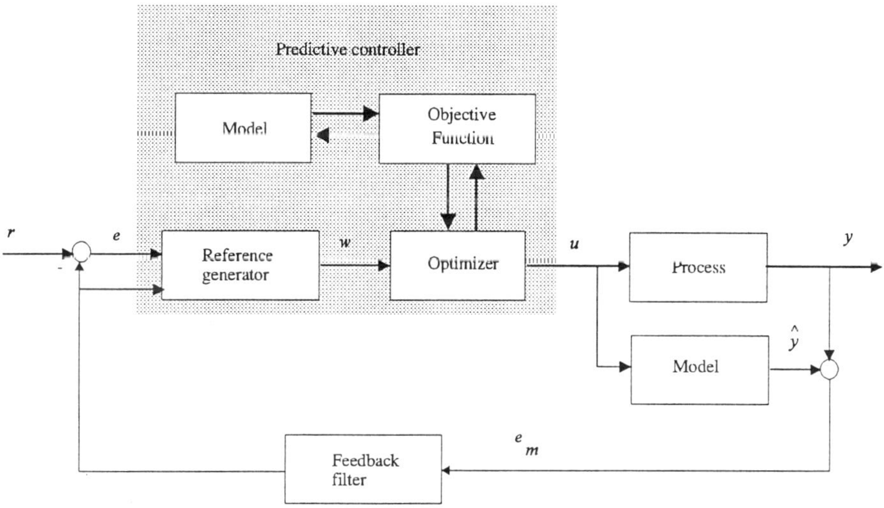

Figure 17 MBPC in an internal model control scheme.

It is clear that a fuzzy model can be included in this scheme for predicting the output in the IMC scheme as well as for providing the future outputs for the optimiser (MBPC strategy).

We have to keep in mind that using fuzzy models for predicting the output of a non-linear process means that a non-convex optimisation problem has to be solved. This is a time-consuming procedure which requires significant computer power and most of the optimisation methods do not guarantee a global optimum.

This hampers generally the application of MBPC for non-linear systems. However, this is a general problem not specifically related to a fuzzy approach.

When the number of linguistic terms is, however, small and we allow a restricted number of magnitudes of the control signal, it is possible to use standard techniques known from operations research such as dynamic programming and the branch-and-bound method. These techniques have been applied successfully in fuzzy MBPC (Sousa, 1998).

1.3.5 Fuzzy decision making in control

A different approach to fuzzy MBPC is to use fuzzy sets for defining fuzzy goals and constraints.

In most design procedures one assumes that goals and constraints are deterministic and well-defined. Although goals and constraints are expressed in an analytical way in an objective function, the outcome of the optimisation often results in changes of the parameters of the objective function when the designer is confronted with the results obtained, due to the high cost or the high energy consumption related to the solution. In many cases some of the goals or constraints can be relaxed, partly because the designer wants initially to be on the safe side when designing the control system.

Moreover, next to strict requirements and constraints, in many cases some desirable objectives play also an important role, but they do not have the same importance as the strict goals and constraints. They play a role in decision making when a number of alternatives should be compared which all satisfy the goals and constraints. In many cases a kind of tolerance is allowed with a preference for a certain value. It is clear that the concept of fuzzy sets fits well to this approach.

Moreover, when a system is non-linear, the advantages of having a quadratic cost function disappear. It is not longer possible to derive an analytical solution for the MBPC problem.

In fuzzy decision making, fuzzy goals and constraints are defined. It has been shown by Bellman and Zadeh (1970) that fuzzy goals and fuzzy constraints can be treated equally in fuzzy decision making.

Multiple goals and constraints can be defined in the same set or in different sets. All criteria can be defined in the same multidimensional set, which allows the use of aggregation operators between the different fuzzy criteria. Different accumulation operators have been considered between goals, between constraints and between goals and constraints. The predictive control problem is a multistage decision making problem which requires the definition of fuzzy criteria for each time step. The two main problems which emerge are the proper choice of the fuzzy goals and fuzzy constraints and the choice of the accumulation operators to combine these criteria.

Some standard membership functions can be selected for the requirements often met, such as a fast response without overshoot. Also hard constraints can be included requiring at the definition of a membership function that the support is restricted to the allowable region of the variable and the transient between the kernel and support is quite steep. Fig. 18 shows some membership functions satisfying decision criteria for the error, change in output and change in control action.

It is clear from this figure that only very few parameters are to be chosen, the K, H and S values, which have a physical meaning and are well suited to be chosen by the designer of the control system.

It has been shown (Sousa, 1998) that the choice of a t-norm operator is most straight forward in satisfying all the goals and constraints and is compulsory in order not to

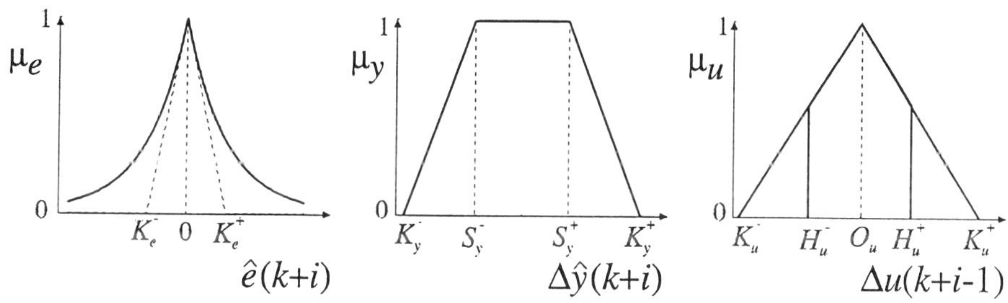

Figure 18 Membership functions that represent the satisfaction of decision criteria for the error, change in output, and change in the control action.

violate hard constraints.

The solution of the optimisation problem is of vital importance in any type of predictive control, especially because the solution should be obtained in real-time and the problem is non-convex.

It is in most cases sufficient to find a reasonably good solution as close as possible to the optimal one within the sampling period required for the dynamic system. It is shown (Sousa, 1998) that after discretisation of the possible control actions the branch-and-bound method provides a reasonable solution for the discrete optimisation problem. The method has been applied to a heating, ventilating and air condition problem in a real-time environment (Sousa, 1998). Also genetic algorithms have been applied successfully for this kind of optimisation problems. It is our expectation that the use of fuzzy goals and constraints methods will become increasingly popular in process control as long as the optimisation problem can be kept sufficiently small.

1.3.6 MIMO aspects of fuzzy control

Especially in the process industry we are confronted with MIMO systems with many interactions which means that it is often difficult for human operators to express their control strategies when several control inputs should be manipulated simultaneously.

A MIMO system can generally be described by a fuzzy inference system, with rules described by:

R_i: If x_1 is A_{1i} and x_2 is A_{2i} ... and x_p is A_{pi} then y_1 is B_{1i} ... and y_q is B_{qi} for $i = 1, 2...k$

Due to the function approximation approach followed, it is, however, difficult to partition the input space because the non-linearities of the different outputs have different dynamical variations.

By decomposing the MIMO system in q MISO systems the number of membership functions and rules are reduced considerably.

We still have, however, the problem that the complexity exponentially grows with the number of inputs; even for a simple SISO system the number of rules can already grow extensively when the order of the system is relatively high. In that case for a model with input vector

$$x(k) = [y(k), y(k-1)...y(k-n), u(k), u(k-1)...u(k-n)] \tag{30}$$

the number of rules k becomes

$$k = \prod_{i=1}^{2n} N_i \tag{31}$$

where N_i is the number of linguistic terms of the ith antecedent variable. For $N_i = 5$, for $i = 1...2n$ and $n = 2$ the number of rules becomes already: $5^4 = 625$.

For a MIMO system this number increases dramatically. For a system with m inputs and l outputs the number of rules becomes, when all single systems are of nth order,

$$k = \prod_{i=1}^{(m+l)n} N_i \tag{32}$$

So, even for a 2×2 MIMO system of the second order with $N_i = 3$ the number of rules is already: $3^8 = 6561$.

It is clear from these examples that we have to look for a compact description, preferably a state space model, and that a decomposition of a MIMO system could considerably reduce the number of linguistic rules.

Several approaches have been proposed in order to reduce the number of rules, either directly by reducing the number of state variables and linguistic variables or indirectly by changing the structure (aggregation, temporal or spatial decomposition).

To the aggregation type of approach belongs the numeric fusion of state variables. Two states, e.g., error signal and error derivative signal, are first added with a certain weighting and the resulting signal is fed to the rule base. The fusing has to be very well justified.

Removing less important rules is also a straight forward way to reduce the number of rules. This applies to rules which are either physically not realisable or have small firing probability. The reduction of membership functions and related to this also the reduction of rules can also be obtained using similarity measures (Setnes, Babuška, Kaymak and Lemke, 1998). The representation of a system as a chained hierarchical structure is based on the usage of a multilayer control structure where the first layer consists of the ones related to the most important state variables. The output of this MISO system is fed to the second layer and is combined with the second most important state variables, and so on. The number of rules in this case is a linear instead of an exponential function of the number of state variables. The reduction would be maximal if each layer contains only two state variables as input. When we compare the number of rules of this approach for the simple example of the SISO system with $N_i = 5$ we get $3 \times 5^2 = 75$ rules instead of 625 rules.

Another possibility is to divide the system into a number of different local domains each described by their own subrule bases. The consequent part of a meta rule base specifies the different local domains. As a result, only one subrule base may be active at a time and the number of rules is much smaller.

A very efficient method to define controllers for a MIMO system is to decouple the system first before applying a control structure. When the system can be fully decoupled a number of SISO controllers can be designed, each responsible for the control of one output variable. Research is going on to develop a unified approach for decoupling a fuzzy MIMO system, where also partly decoupling is possible. The result of this research will be very useful to design fuzzy controllers for MIMO systems in the near future.

1.4 HIGH LEVEL PROCESS CONTROL

The first industrial application of the theory of fuzzy sets was for high level process control of cement kilns, and it was developed by F. L. Smidth (FLS) in Denmark (Holmblad & Ostergaard, 1982). The inspiration came from Mamdani's early experiments with rule based control of a steam engine (Assilian & Mamdani, 1974) which, in combination with the rules of thumbs for manual control of a cement kiln (Peray & Wadell, 1972) resulted in the first high level process control strategy based on fuzzy logic. The obvious link between the rule based approach of fuzzy logic and high level control is illustrated by Table 3. The control instructions for the cement kiln operators use the everyday terms LOW, OK, SLIGHTLY etc., which also are some of the basic primary terms that are represented by membership functions in the theory of fuzzy sets. This section points out some of the practical considerations which have to be considered in connection with a project on development and installation of a high level process control strategy. *High level process control* may be defined as the *coordination of control loop set points, which normally is done by a human operator*.

A high level process control strategy works on the same level as the human operator.

Table 3 Extract from textbook (Peray & Wadell, 1972) for kiln operators.

Case	Condition	Action to be taken	Reason
10	BZ OK OX low BE low	a. Increase I.D. fan speed b. Increase fuel rate	To raise back-end temperature and increase oxygen percentage for action 'b' To maintain burning zone temperature
11	BZ OK OX low BE OK	a. Decrease fuel rate slightly	To raise percentage of oxygen
12	BZ OK OX low BE high	a. Reduce fuel rate b. Reduce I.D. fan speed	To increase percentage of oxygen for action 'b' To lower back-end temperature and maintain burning zone temperature
13	BZ OK OX OK BE low	a. Increase I.D. fan speed b. Increase fuel rate	To raise back-end temperature To maintain burning zone temperature
14	BZ OK OX OK BE OK	NONE. However, do not get overconfident, and keep all conditions under close observation.	

It takes over a part of, or all of the operator's job of controlling the process. High level control thus introduces an extra layer between the operator and the conventional control system in the process control pyramid as shown in Fig. 19.

The basic idea of high level process control based on fuzzy logic is to formulate sets of rules for automatic operation based on practical experience and knowledge about manual control of the process.

The terms of the rules, i.e. LOW, SMALL OK etc. are represented by membership functions, and the rules are evaluated by inference as described in the theory of fuzzy sets. Examples of fuzzy control rules are:

IF Temp is High and Pressure is Ok THEN Medium Flow

IF Temp is Ok and Pressure is Ok THEN Small Flow

IF .....

where High, Ok, Small and Medium are represented by membership functions, and where the result of the rules are calculated by inference, e.g. the Max. -Min. logical product. In fuzzy logic, the condition of a rule is fulfilled to a certain degree, and each rule will influence the result of the set of rules in accordance with its grade of fulfillment..

This heuristic approach for design of control strategies for automatic process control is useful, and probably the only practical solution, when the process is only partly known, difficult to describe by a mathematical model, if few measurements are avail-

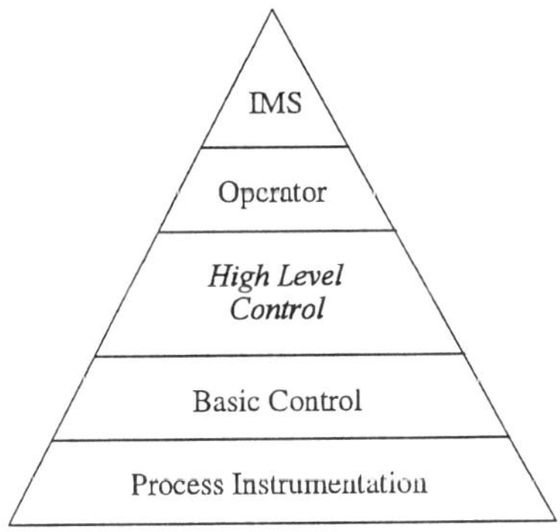

Figure 19 Process control pyramid.

able, or if the process is highly nonlinear.

1.4.1 High level control expectations

High level control in the process industry works on the same level as the human process operator, and the overall expectations to the benefits are that the automated operation will result in better process performance than produced by manual control. It is, however, not always obvious what the meaning of improved process performance is, and it may even differ within the same process industry depending on local conditions such as the present market situation, raw material costs and overall strategic goals. In some countries reduced fuel consumption, for instance, is not as attractive as it is in countries where the price of fuel is high. In general, the expectations to a high level control system will in be in comparison with manual control by an operator, or perhaps in comparison with an existing and more conventional control scheme.

In most cases, expectations will relate to profit in terms of reduced costs or increased productivity. Many industrial processes are very energy and raw material consuming. The glass industry, the cement industry, and power plants are examples of industries which consume large amounts of coal, oil, natural gas, or electrical energy. The plastic industry, for instance, and the processes for cleaning exhaust gases and waste water, use large amounts of raw materials which has a significant impact on the overall plant economy.

It is characteristic that the consumption of energy and raw materials depend very much on how the processes are being controlled, and improved operation through high level control may thus be an attractive possibility to increase profit. In general, improved operation can be defined as:

- More stable operation
- Running closer to the limits for acceptable product quality
- Running closer to the environmental emission limits

Stability, in fact, is the most important key to improve the operation of a process. Oscillations will always increase the consumption of energy and raw materials, and it will reduce the quality of the product. If the process is not running stable, the process measurements and the product quality parameters have to be on the very safe side to keep average values within the limits, and to prevent the process and the product quality from entering unacceptable or unsafe levels. Emission figures are very sensitive to unstable process operation, and in case of instability it is normally necessary to run far below the limits to ensure that the emission peaks are not unacceptably high.

A simple way to measure stability is through the standard deviation STD of the key measurements and quality parameters. Depending on the type of industrial process, the standard deviation may be calculated on a daily basis, and an average standard deviation is then calculated for a period which is representative for the performance evaluation. If the measurements and the quality parameters have target values, then a more feasible measure of stability is obtained by calculating the target value deviation TVD around the target value, instead of the variations around the average value AVR as done by STD. TVD may be calculated from the standard deviation and the average value by:

$$TVD = \sqrt{STD^2 + (SP - AVR)^2} \tag{33}$$

Improved control of the process is one way to make the process run more stable. It is realistic to expect reductions in STD and/or TVD of up to 50% or more as the result of a good working high level control system. The improved stability may be utilized to operate the process at a higher production level, and to run it closer to the limits for safe operation and also closer to the limits for emissions to the environment.

A waste incineration furnace is an example of a process where stability has a direct impact on profitability. In a waste incineration furnace, the temperature levels must not exceed certain values as that will cause build-up of material on the furnace walls, and result in unnecessary wear on the brick lining, and reduce the amount of burned waste. In case of severe build-up, the furnace has to be stopped for cleaning. However, the more stable the control strategy can keep the temperatures, the closer it is possible to run the temperatures at the upper limits. Higher temperatures result in more waste to be burned, which directly will increase profit.

1.4.2 High level control configurations

High level control systems may be integrated with the rest of the control system in different configurations as shown in Fig. 20, where PID stands for a conventional control scheme which in most cases consists of independent or coupled PID loops. Fuzzy in Fig. 20 refers to a high level control strategy. Normally, both the PID- and the Fuzzy box thus have more than one input and one output.

Fig. 20 *(a)*. In this configuration, the operator may select between a high level control strategy and conventional control loops. Very often the conventional loops represents an existing control scheme, which has been running the process before installation of the high level strategy. The operator thus has to decide which of the two alternatives are most likely to produce the best control performance. Waste incineration furnaces are examples of processes which are equipped with an existing system of coupled PID loops for control of the dosing of waste and the amount of combustion air, and here the high level control system by-passes the existing system when activated.

Fig. 20 *(b)*. This configuration represents the original high level control idea, where manual control carried out by a human operator was replaced by automatic control. Normally, the existing control loops are still active, and the high level control strategy makes adjustments of the controller set points in the same way as the operator used to do, before the automatic strategy was installed. Again it is up to the operator to decide whether manual or automatic control will result in the best possible operation of the process which, of course, may create conflict situations, which in general have to be addressed very seriously when working with high level process control.

Fig. 20 *(c)*. In this configuration, the high level strategy is used for adjustments of the parameters of the conventional control loops. A common problem with linear PID controllers used for control of highly nonlinear processes is that the set of controller parameters produces satisfactory performance only when the process is within a small operational window. Outside this window, other parameters or set points are necessary, and these adjustments may be done automatically by a high level strategy.

Fig. 20 *(d)*. Normally, conventional control systems which are based on PID controllers are capable of controlling the process when the operation is smooth and close to normal conditions. However, if sudden changes occur or if the process enters abnormal situations, then this configuration may be useful to bring the process back to normal operation as fast as possible. For normal operation, the additions to the output from the PID controllers are zero, whereas the PID outputs are modified in abnormal situations, which often is referred to as Abnormal Situation Management (ASM).

For all four configurations, the high level control system has an impact on the work done by the operator. Configuration (a) and (b) change directly the routines of the operator, which is a crucial point to take into account when the system is developed

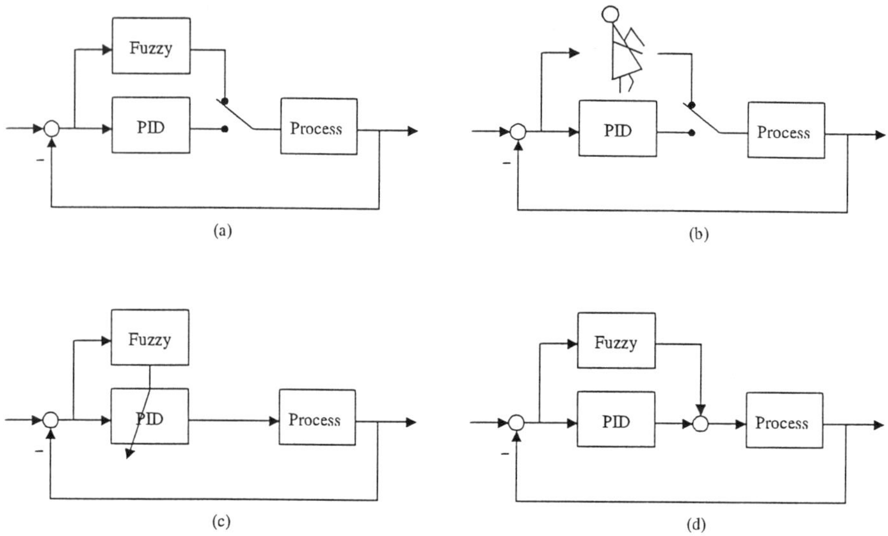

Figure 20 Fuzzy controller configurations. Fuzzy replaces PID (a), fuzzy replaces manual control (b), fuzzy adjusts PID parameters (c), and fuzzy adds to PID output.

and installed.

1.4.3 Design of high level control strategies

For the design or development of a high level process control strategy it is possible to point out some general subjects which should be considered by the project team. The design phase may by defined as the period from the start of the project until the system is being installed at the industrial plant. An example of a project schedule with four high level control design phases is

1. Process inspection
2. Design discussions
3. Preparation of design document
4. Programming and off-line test

The schedule emphasizes some of the important general elements of the development of a high level control strategy, as it is assumed that the system is going to be developed by an internal group at the industrial plant or by an external supplier in corporation with a project team from the plant. The project phases are basically the same for an internal project as for a system which is purchased for an external supplier.

As mentioned above, a high level control strategy has a more or less direct influence on the daily work of the operators, which has to be considered very carefully in the project plan.

Design phase 1. The process inspection phase consists of the following tasks:

- Verification of the status of the existing instrumentation.
- Interfacing of the high level controller to the existing control system

The existing instrumentation must be able to supply reliable signals, and the necessary process measurements must, of course, be available. It is crucial, that the inputs to and outputs from the high level control system are in a good condition.

In most cases, a high level control system is an add-on to an existing process control system, and it is thus necessary to establish a link between the two systems, which in a safe and reliable way can be used for transfer of data.

Design phase 2. The design discussions might be the most important phase of the project, as the following points have to be addressed very carefully:

- Introduction of the principles of high level control to the end users of the system, including the operators.
- Detailed discussions of the expectations to the high level control system, and definition of criteria for a successful project.
- Discussions of how to design a control strategy which can fulfil the expectations and the success criteria.

The ideas and principles behind the new system have to be explained to and discussed with the end user to eliminate misunderstandings, to reduce scepticism, and to create positive expectations. It is a natural human attitude to be sceptic and cautious about new things and ideas which are being forced upon you without proper explanations.

It is important to define as precise as possible the goals which should be attained. A project without well-defined objectives and specific expectations is very difficult to manage, and the project group is likely to function less efficient if goals and expectations have not been discussed and agreed upon.

Finally, the project group must design the control strategy including as many details as possible. Preferably, the project group should have a representative from the end user, so that the practical experience from manual control is taken as the start-

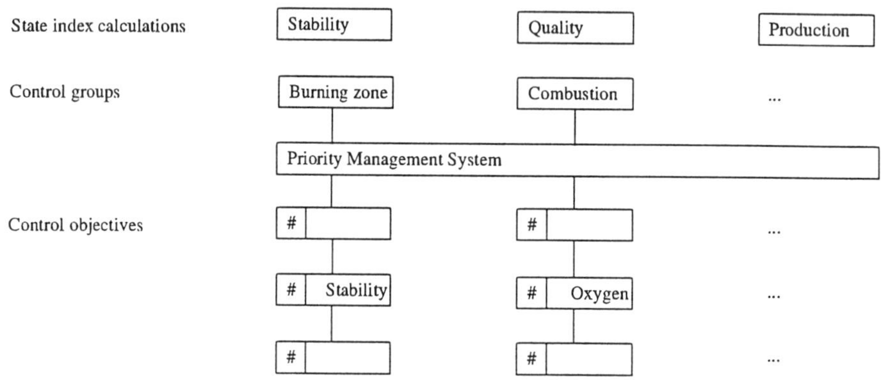

Figure 21 The FLS high level control strategy structure.

ing point, to be able to consider the common practice of the real process operation, and to ensure that the design decisions are transferred down through the organization. The design phase for a high level control strategy should focus on control objectives rather than discussing individual loops for maintaining single process measurements at set points. In the following an example is given of a design procedure for high level process control strategies, which has been developed by FLS. This procedure suggests a specific structure for a high level control strategy, and it also includes a scheme for implementation of the strategy in terms of functional tasks.

Design phase 3. The control strategy which has been designed by the project group must be described so that a document exists that can be referred to during the programming and implementation phase.

Design phase 4. Programming of the control strategy as it is described in the design document, followed by a test, where process simulators may be useful, if available. The test, of course, should be careful, but for high level control systems which are based on fuzzy logic and other uncertainty techniques it is against the philosophy to use lots of resources on model building and simulations.

The FLS design procedure For an operator, control of the process consists in achieving various goals, which are more or less precisely defined: maximum output, minimum consumption of raw materials and energy, high product quality, safe process operation, etc. Different processes have different control objectives but, in general, good process control may be defined through a list of control objectives which should be fulfilled as much as possible. The concept of control objectives is a key element in a high level control strategy.

For a cement kiln, typical control objectives are:

- stable operation,
- good cement clinker quality,
- high production
- complete combustion,
- low fuel consumption.

As control objectives are frequently in conflict, high level co-ordination means approaching optimal conditions in succession of importance. Priorities, in other words, have to be assigned to the various control objectives, specifying which objectives are considered the most important to fulfill. Control objectives with different priorities are thus the basic elements of the design of a high level control strategy.

The elements of the FLS design procedure for a process control strategy are the following:

- *State indices.* Calculations concerning the actual process condition.
- *Control groups.* Arrangements of the overall control strategy into groups of control objectives.
- *Priority management.* Determines the extent to which the control actions should be executed to fulfil the individual objectives.
- *Control objectives.* Specifications of the goals of the fuzzy control strategy.

Normally, a state index combines various measurements into a single figure. The degree of process stability, the product quality, the production level, etc., are all typical examples of state indices for a kiln control strategy. The state indices are important to the structure of the FLS design scheme for a high level control strategy, as they form the basis for dividing the overall strategy into control groups which can be treated independently. The state indices are used to co-ordinate control actions from the various control groups, see Fig. 21.

The *state index calculations* are calculations of index values by which process measurements are combined into state indices for the actual process stability, product quality, production level, etc. *Control groups* form a subdivision of the control strategy into groups of objectives which are related through priority numbers. A *priority management system* manages the scheduling of control actions in order of importance. *Control objectives* specify the individual goals that the fuzzy control strategy has to fulfill.

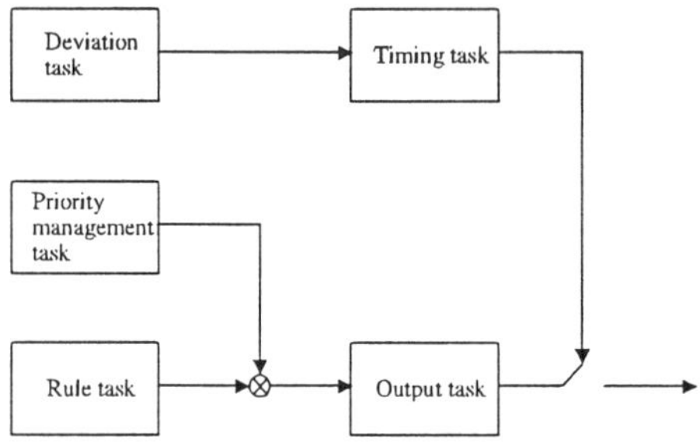

Figure 22 The control objective module.

Design of a control strategy consists of filling in the structure in Fig. 21 with specific state index calculations, control groups and control objectives, whereas the priority management system is a fixed module in the system. For a new process, it is perhaps not so clear which are the most appropriate indices and control groups. This is not a major problem, however, as the structure facilitates a stepwise implementation of control objectives and index calculations concurrent with increasing process control knowledge being available.

Every control objective is implemented in accordance with the so-called objective module, which consists of the following four tasks:

- deviation task
- rule task
- output task
- timing task

Figure 22 shows the objective module diagram, which is used for implementation of all types of control objectives.

Deviation task. This task calculates and evaluates the degree to which the objective is fulfilled. Normally, the calculation results in a "fuzzy value" between -1 and +1, which expresses how far the actual process situation is from fulfillment of the objective.
Rule task. This block holds the set of control rules for fulfillment of the specific objective. Normally, this block is formulated as a set of fuzzy control rules, but other techniques may also be used, like PID, neural nets, mathematical models, etc.
Priority management. The rule block for each control objective results in control actions which are multiplied by a weight factor between 0 and 1. The

smaller the weight factor, the more the control actions are suppressed. The weight factor for a given objective is a function of the deviation values for the higher priority objectives, which ensures that control actions to fulfill a given objective are reduced if higher priority objectives are not fulfilled.

Output program. The output program evaluates process constraints and selects among alternative control actions based upon the actual index values. The logic for selecting alternative adjustments may be fuzzy or non-fuzzy depending on whether a gradual or a hard switch is the most appropriate. In most cases, the fuzzy logic approach gives the best control performance, simply because no process operates with sudden changes between alternative control actions.

Timing calculation. The timing calculation determines when control actions are to be executed. It is just as important as the rule block for determining the proper function of the control strategy. Also the timing calculation is normally fuzzy in the sense that the time interval between control actions changes gradually as a function of the deviation value. The larger the deviation, the more frequent the control actions.

1.4.4 Installation of high level control strategies

The following list contains examples of installation phases which have to be carried out during installation of a high level control system.

Preparation for high level control. When the high level control system has been designed, programmed and tested it is ready for installation. A detailed installation schedule should be produced and discussed with the end users including the process operators. The plan should consider that the organization must have the necessary time to adapt to the new situation and to get familiar with the new control procedures.

Installation and test of signals. As a high level control system very often is an add-on to an existing control system, the installation normally starts with establishing a communication link between the two systems. The exchange of signals including transfer of new set point values which are calculated by the high level control strategy have to be tested. In general, the amount of data which is necessary for high level control is rather limited. For control of a cement kiln, the number of measurements is 25 - 40, and the number of set points which are adjusted is in the order of 10 - 15. In addition, the speed of the communication link is normally not critical, as high level control actions normally do not have to react very fast in process control applications.

Commissioning. During commissioning, where the designed control strategy is being tuned and refined, it is important to have regular status meetings for presentation of results and for discussion of future activities. Feedback from the end users is a very useful input for adjustment of the original installation schedule.

Evaluation. Having completed the installation, it is necessary to evaluate the performance of the high level control system so that it is documented whether the system results in improved process operation. If it is not clearly documented that the advanced control strategy yields measurable benefits compared with the conventional control scheme then it is likely that high level strategy will not be used in future applications.

1.4.5 Conclusion

High level control projects are challenging from both a technical and a project management point of view. The technical aspects include signal processing, control strategy design, test and installation, and these tasks, of course, have to addressed properly. The project management problems are at least as important and perhaps more difficult to solve than the technical, as they include selling the project to the plant organization including the process operators. Good technical projects may fail if the end users have not been properly involved in all the phases of the project. The project management has to ensure that the end users consider the new high level control system a help and not a threat.

1.5 References

Årzén, K.-E. (1993). Expert control: Intelligent tuning of PID controllers, *in* S. Tzafestas (ed.), *Applied Control: Current Trends and Modern Methodologies*, Marcel Dekker, New York, pp. 851–873.

Assilian, S. and Mamdani, E. (1974). Learning control algorithms in real dynamic systems, *Proc. Fourth Int. Conf. On Digital Computer Applications to Process Control, Zürich*, IFAC/IFIP, Springer, Berlin, pp. 13–20.

Babuška, R. (1998). *Fuzzy Modeling For Control*, Kluwer Academic Publishers.

Babuška, R. and Verbruggen, H. B. (1997). Fuzzy set methods for local modeling and identification, *in* R. Murray-Smith and T. A. Johansen (eds), *Multiple Model Approaches to Nonlinear Modeling and Control*, Taylor and Francis, London.

Bellmann, R. and Zadeh, L. (1970). Decision-making in a fuzzy environment, *Management science* **17**: B141–B164.

Cutler, C. and Ramaker, B. (1980). Dynamic matrix control - a computer control algorithm, *Proceedings JACC, San Francisco.*

Driankov, D., Hellendoorn, H. and Reinfrank, M. (1996). *An introduction to fuzzy control*, second edn, Springer-Verlag, Berlin.

Garcia, C. and Morari, M. (1982). Internal model control. 1. a unifying review and some new results, *Ind. Eng. Chem. Process Des. Dev.* **21**: 308–323.

Hill, G., Horstkotte, E. and Teichrow, J. (1990). *Fuzzy-C development system – userś manual*, Togai Infralogic, 30 Corporate Park, Irvine, CA 92714, USA.

Holmblad, L. P. and Østergaard, J.-J. (1982). Control of a cement kiln by fuzzy logic, *in* Gupta and Sanchez (eds), *Fuzzy Information and Decision Processes*, North-Holland, Amsterdam, pp. 389–399. (Reprint in: FLS Review No 67, FLS Automation A/S, Høffdingsvej 77, DK-2500 Valby, Copenhagen, Denmark).

IEC (1996). Programmable controllers: Part 7 fuzzy control programming, *Technical Report IEC 1131*, International Electrotechnical Commission. (Draft.).

Lee, C. C. (1990). Fuzzy logic in control systems: Fuzzy logic controller, *IEEE Trans. Systems, Man & Cybernetics* **20**(2): 404–435.

Li, M., Bruijn, P. and Verbruggen, H. (1994). Tuning cascade PID controllers using fuzzy logic, *Math. Comput. Simulation* **37**: 143–151.

MIT (1995). *CITE Literature and Products Database*, MIT GmbH / ELITE, Promenade 9, D-52076 Aachen, Germany.

Mizumoto, M. (1992). Realization of PID controls by fuzzy control methods, *in* IEEE (ed.), *First Int. Conf. On Fuzzy Systems*, number 92CH3073-4, The Institute of Electrical and Electronics Engineers, Inc, San Diego, pp. 709–715.

Passino, K. M. and Yurkovich, S. (1998). *Fuzzy Control*, Addison Wesley Longman, Inc, Menlo Park, CA, USA.

Pedrycz, W. (1993a). *Fuzzy control and fuzzy systems*, second edn, Wiley and Sons, New York.

Pedrycz, W. (1993b). S-t fuzzy relational equations, *Fuzzy Sets and Systems* **59**: 189–195.

Peray, K. E. and Wadell, J. J. (1972). *The Rotary Cement Kiln*, Chemical, New York.

Qiao, W. and Mizumoto, M. (1996). PID type fuzzy controller and parameters adaptive method, *Fuzzy Sets and Systems* **78**: 23–35.

Richalet, J. (1993). Industrial applications of model based predictive control, *Automatica* **29**: 1251–1274.

Richalet, J., Rault, A., Testud, J. and Papon, J. (1978). Model predictive heuristic control: Applications to industrial processes, *Automatica* **14**(5): 413–428.

Setnes, M., Babuška, R., Kaymak, U. and Lemke, H. V. N. (1998). Similarity measures in fuzzy rule base simplification, *IEEE Trans. Systems, Man & Cybernetics*.

Setnes, M., Fleer, R., Bruijn, P., Lemke, H. R. V. N. and Emck, F. (1997). Implementing fuzzy control in the manufacturing of washing powders, *in* EUFIT (ed.), *Proceedings Fifth European Congress on Fuzzy and Intelligent Technologies*, ELITE Foundation, Promenade 9, D-52076 Aachen, pp. 1360–1364.

Siler, W. and Ying, H. (1989). Fuzzy control theory: The linear case, *Fuzzy Sets and Systems* **33**: 275–290.

Sousa, J. M. (1998). *A Fuzzy Approach to Model-Based Control*, PhD thesis, Delft University of Technology, Delft, The Netherlands.

Sugeno, M., Murofushi, T., Nishino, J. and Miwa, H. (1992). Helicopter flight control based on fuzzy logic, *in* T. Terano, M. Sugeno, M. Mukaidono and K. Shigemasu (eds), *Fuzzy Engineering Toward Human Friendly Systems*, Ohmsha and IOS Press, Tokyo; Burke, VA, USA; Oxford, UK; Amsterdam, NL, pp. 1120–1121.

Takagi, T. and Sugeno, M. (1985). Fuzzy identification of systems and its applications to modeling and control, *IEEE Trans. Systems, Man & Cybernetics* **15**(1): 116–132.

Zhao, Z.-Y, Tomizuka, M. and Isaka, S. (1993). Fuzzy gain scheduling of PID controllers, *IEEE Trans. Systems, Man & Cybernetics* **23**(5): 1392–1398.

2 FUZZY SETS IN ENGINEERING DESIGN

Erik K. Antonsson

and

Hans-Jürgen Sebastian

"Fuzziness is more than the exception in engineering design problems: usually there is no well-defined best solution or design."

"L-Fuzzy Sets",
Journal Mathematical Analysis and Applications
1967, Volume 18, Page 146
—Joseph A. Goguen

Abstract: This chapter presents an introduction to several different methods of incorporating fuzzy information into engineering design.

2.1 INTRODUCTION

Innovative product development requires high quality and resource-efficient Engineering Design. Therefore, Intelligent Systems are needed to support the human designers to perform complex tasks, including trade-offs, negotiation, decision-making, design refinement, *etc.*, in all phases of the design process, including preliminary or conceptual design and detailed design. Such Intelligent Systems in Engineering Design are characterized by the combined use of different advanced design approaches and techniques, such as:

- Generic modeling of the design or configuration task by knowledge-based representation and processing of design objects (classes, individuals) and conceptual constraints.
- Modeling uncertainties and imprecision by Fuzzy Sets.
- Optimization or Multi-Criteria Analysis of the target system of the design task.
- Application of Metaheuristics, in particularly Evolutionary Strategies or Genetic Algorithms to generate a finite population of "good design alternatives".

In this section we focus on "Modeling of Imprecision by Fuzzy Sets". However, we briefly explore other techniques as well when trying to solve real world design applications. Therefore, this section starts with considering the design process from the information point of view.

Imprecision is an inherent characteristic of any engineering design. Imprecision is uncertainty that arises because of vague (*e.g.*, linguistic) or incomplete information. The design process begins with goals and requirements that are descriptive statements of which functions a product should perform. Such preliminary design information is almost always imprecise (especially in the preliminary design phase): specifications, goals and requirements are subjects to change, the design description is vague and incomplete. A major component of any design problem is how to model the transition from imprecise goals and requirements to precise specifications of the design objects. In other words, how to reduce imprecision during the design process with the aim to approach one or some "good design alternatives".

Traditionally, it is common for engineers to evaluate promising design alternatives one by one. Such an approach doesn't consider the nature of imprecision of the design process, and leads to expensive design computations. In Section 2.2 we will briefly present the main ideas of a selection of such methodological approaches to Engineering Design, which utilize Fuzzy Sets in order to realistically model imprecise design information and/or the imprecise design process. Several such approaches have been developed, most during the last decade, including:

- The Method of Imprecision (M_oI), discussed in Section 2.2.1, uses Fuzzy Sets to model the designers and the customers preferences. Aggregation of preferences by a family aggregation operators is used to get an overall evaluation of design alternatives. An extended Level Interval Algorithm (LIA) (extended for computational efficiency with many design variables and to handle nonmonotonic functions) is used in the M_oI.
- The Fuzzy Design Method (FDM), discussed in Section 2.2.2, is similar to the M_oI but without aggregation operators. Instead it uses linguistic variables to model user requirements and aspiration/reservation levels combined with the LIA to generate a finite set of feasible design alternatives. These alternatives are evaluated by Fuzzy MADM methods.

- A hybridization of Evolutionary Strategies, discussed in Section 2.2.3, to generate a finite population of design alternatives with a Fuzzy MADM to evaluate the fitness of the generated populations.
- Other interesting methods not covered in this chapter include
  - The Precision Convergence Approach, which models the design problem by a hierarchy of Fuzzy Constraint Networks and utilizes Fuzzy Constraint Propagation,
  - Fuzzy Multi-Criteria Decision Analysis and Optimization Approaches to Engineering Design Tasks (Rao et al., 1990; Dhingra, 1996).

In Section 2.3 we briefly describe two implementations of the Fuzzy Design Methods presented in Sections 2.2.1 and 2.2.2:

- The Imprecise Design Tool (IDT), discussed in Section 2.3.1, is an early implementation of the M_oI, which permits the designer to specify preferences on up to ten design variables and implements a simple trade-off (aggregation) strategy to evaluate the combined preference for design alternatives.
- The KONWERK workbench, discussed in Section 2.3.2, provides a user with a comfortable environment to add linguistic variables to objects and to apply Fuzzy MADM methods for dealing with a finite set of explicitly known design alternatives, which are described both by fuzzy and crisp attributes.

Section 2.4 presents three real world applications selected from different areas of Engineering Design to illustrate the breadth of application of these methods.

2.2 FUZZY SETS IN ENGINEERING DESIGN – METHODOLOGY

2.2.1 The Method of Imprecision

2.2.1.1 Introduction. *Imprecision* (the intrinsic vagueness of a preliminary, incomplete description) is an integral part of the engineering design process. Obtaining *precise* information upon which to base decisions is difficult or impossible, yet it is critical to make early engineering design decisions on a sound basis.

At the stage where technical solution concepts are being generated, the description of a design is largely vague or imprecise (fuzzy). This imprecision is reduced during the design process until ultimately the final description is precise (crisp), except for tolerances, which represent the allowable limits on stochastic manufacturing and material variations.

The need for a methodology to represent and manipulate imprecision is greatest in the early, preliminary phases of engineering design, where the designer is most unsure of the final dimensions and shape, materials and properties, and performance of the completed design. Recent studies have not only shown that imprecision is present throughout the engineering design process, but that it can effectively be represented using set-based methods (Ward et al., 1994). Additionally, the most important decisions, those with the greatest effect on overall cost, are made in these early stages (Holmes, 1984; Sullivan, 1986; Ulrich and Pearson, 1993; Whitney, 1988).

Work at Caltech over the past decade has focused on the development of methods for representing and manipulating imprecise descriptions of designs to permit the designer to compare alternatives during the preliminary design phase. Because design imprecision concerns the choice of design variable values used to describe an artifact or process, the designer's preference is used to quantify the imprecision with which design variables are known. Preference, as used here, denotes either subjective or objective information that may be quantified and included in the evaluation of design alternatives. This approach, built on the mathematics of fuzzy sets, is called the *Method of Imprecision* (M$_o$I) and has been the subject of a series of prior publications: Scott et al., 1998; Antonsson and Otto, 1997; Scott and Antonsson, 1998; Antonsson et al., 1996; Antonsson and Otto, 1995; Otto et al., 1993b; Otto et al., 1993a; Wood et al., 1992; Otto and Antonsson, 1994; Otto and Antonsson, 1991a; Wood et al., 1990; Wood and Antonsson, 1990; Wood and Antonsson, 1989.

Other researchers have also recently contributed to this area, including: Dhingra, 1996 (multiobjective fuzzy optimization techniques for engineering design); Diaz, 1988; Diaz, 1989 (fuzzy optimization methods); Djouad et al., 1996 (chemical process synthesis); Dubois, 1996 (fuzzy constraint propagation applied to manufacturing); Fargier, 1996 (fuzzy scheduling); Gerhart, 1997 (application of fuzzy methods to windturbine design); Grabot et al., 1996 (multiobjective scheduling); Hamburg and Hamburg, 1993 (management of uncertain knowledge in engineering design); Hsu et al., 1995 (engineering design optimization); Chen and Otto, 1995 (imprecise calculations in engineering design), Knosala and Pedrycz, 1992 (evaluation of design alternatives); Müller and Thärigen, 1994 (fuzzy MADM methods in system design); Posthoff and Schlosser, 1994 (fuzzy evaluations); Rao, 1987; Rao and Dhingra, 1989; Rao and Dhingra, 1991; Rao et al., 1990; Rao et al., 1992 (multiobjective fuzzy optimization techniques for engineering design); Sakawa and Kato, 1995; Shih and Wangsawidjaja, 1995 (multiobjective fuzzy optimization); Schleiffer, 1998 (fuzzy design with evolutionary algorithms); Thurston and Carnahan, 1990; Thurston and Carnahan, 1992 (fuzzy ratings and utility analysis in preliminary design evaluation of multiple attributes); Türkşen, 1996 (scheduling system design); and Zimmermann and Sebastian, 1993; Zimmermann and Sebastian, 1994; Zimmermann and Sebastian, 1995; Zimmermann, 1996 (fuzzy-multi-criteria decision making).

The following sections present a brief review of how imprecision is used to facilitate decision-making in engineering design using the M$_o$I, followed by a description of the calculation methods.

2.2.1.2 Overview of The Method of Imprecision. The M$_o$I begins with one or more design alternatives, at the concept level. The designer's preferences are then applied to each of the variables that (imprecisely) describe the design. Commonly, performance specifications will also be imprecisely described and can be elicited from customers through market surveys, *etc.* The imprecise design variables are then *induced* onto the performance space, by use of the algorithm described below. The design preferences induced onto the performance variables are compared to the specifications. At that point, trade-offs (using an aggregation function) can be made between the various competing aspects of the design, decisions can be made regarding

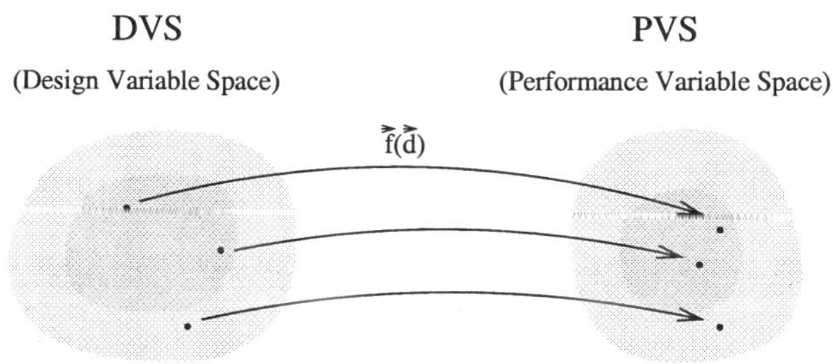

Figure 1 Evaluating individual designs one by one.

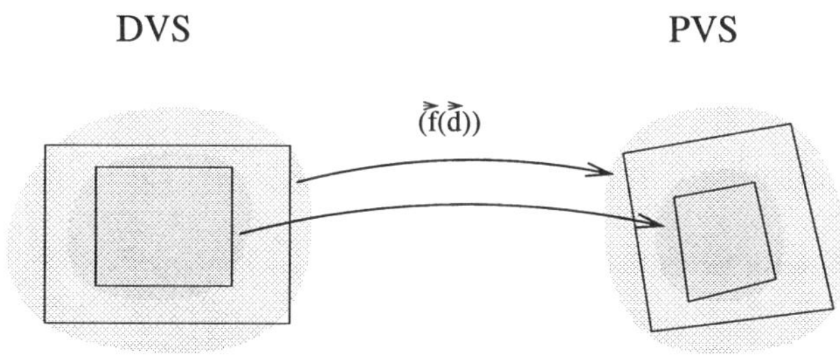

Figure 2 Evaluating sets of designs.

the feasibility of each alternative concept, and promising ranges (or sets) of design variables can be identified, as illustrated in Figure 2, in comparison with the more traditional engineering "point by point" approach illustrated in Figure 1.

2.2.1.3 Definitions. An imprecise variable is a variable that may potentially assume any value within a possible range because the designer does not know, *a priori*, the final value that will emerge from the design process. Even though the designer is uncertain about what value to specify, certain values will be preferred over others. This preference, which may arise objectively (*e.g.*, cost or availability of components) or subjectively (*e.g.*, from experience), is used to quantify the imprecision associated with a design variable. Thus the designer's experience and judgement are incorporated into the design evaluation.

Design Variables. Variables for which the designer is free to choose values (in an attempt to best meet the customer and performance requirements) are denoted d_i, where i ranges from 1 to n. The whole set of design variables for each design alternative is an n vector, $\vec{d}$, which is an element of the *design variable space* (*DVS*). The valid design variable values within the *DVS* form a subset $\mathcal{X}$. The set of valid values for d_i is denoted $\mathcal{X}_i$.

The preference that a designer has for values of d_i, the ith design variable, is represented by a preference function on $\mathcal{X}$, termed the *design preference*:

$$\mu_{d_i}(d_i) : \mathcal{X}_i \to [0, 1] \subset \mathrm{R}$$

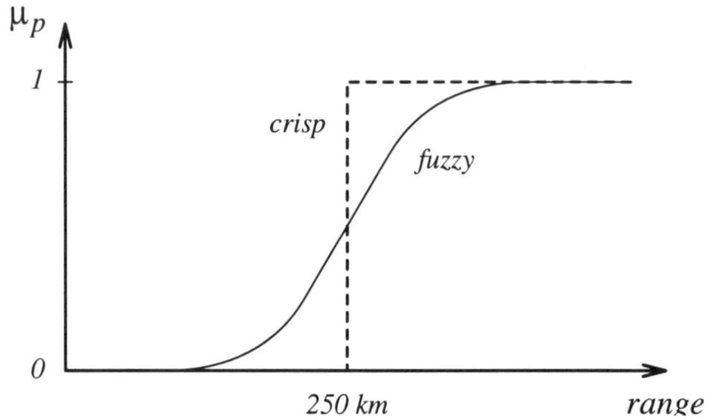

Figure 3 Example imprecise functional requirement.

where the $\mathcal{X}_i$ are assumed to be compact. $\mu_{d_i}(d_i)$ quantifies the designer's preference for values of d_i, and is distinct from the customary membership function in a fuzzy set, which quantifies the extent to which values belong to the set.

Specifications and requirements also embody design imprecision, even though most are written as if they were crisp, *e.g.*, "This device must have a range of at least 250 km." Such a requirement implies that given two designs arbitrarily close together, one with a range just above 250 km and one just below, the first would be acceptable but not the second, as shown by the dashed line in Figure 3. In contrast, specifications and requirements in the real world are commonly fuzzy. Often the designer must ask questions to distinguish the underlying fuzzy constraint so that the final design will satisfy the customer's actual requirements even though it may violate the crisp constraint initially given. The fuzziness of constraints and the fuzziness of preliminary design variables are both forms of design imprecision and can be represented in exactly the same way. The solid line in Figure 3 shows a fuzzy functional requirement or specification.

Performance Variables. Variables for which the customer establishes preferences are denoted p_j, where j ranges from 1 to q. Each performance variable p_j is defined by a mapping f_j such that $p_j = f_j(\vec{d})$. The mappings f_j can be any calculation or procedure to measure the performance of a design, including closed-form equations, iterative and heuristic methods, "black box" functions, experiments, and consumer evaluations. The set of performance variables for each design alternative is a q vector, $\vec{p} = \vec{f}(\vec{d})$, which is an element of the *performance variable space* (*PVS*). The subset of valid performance variable values $\mathcal{Y}$ is mapped from $\mathcal{X}$ and the set of valid values for p_j is denoted $\mathcal{Y}_j$.

The customer's preference for values of p_j, the jth performance variable, is represented by a preference function on $\mathcal{Y}$, termed the *functional requirement*:

$$\mu_{p_j}(p_j) : \mathcal{Y}_j \to [0,1] \subset \mathrm{R}.$$

2.2.1.4 Nomenclature. Symbols used in the MoI are summarized below.

d_i	design variable i
$\vec{d}$	vector of design variables $d_i, i = 1, \ldots, n$
DVS	design variable space
$\mathcal{X}$	valid design variable values, a subset of the *DVS*
p_j	performance variable j
$\vec{p}$	vector of performance variables $p_j, j = 1, \ldots, m$
PVS	performance variable space
$\mathcal{Y}$	valid performance variable values, a subset of the *PVS*
μ_d	design preference (designer)
$\mu_d(\vec{d})$	design preference on the *DVS*
$\mu_d(\vec{p})$	design preference induced onto the *PVS*
μ_p	functional requirement (customer)
$\mu_p(\vec{p})$	functional requirement on the *PVS*
$\mu_p(\vec{d})$	functional requirement mapped back to the *DVS*
μ_o	overall preference, $\mathcal{P}(\mu_d, \mu_p)$
$\mu_o(\vec{d})$	overall preference mapped onto the *DVS*
$\mu_o(\vec{p})$	overall preference mapped onto the *PVS*
μ_o^*	peak overall preference
$\mathcal{P}$	aggregation function
$\mathcal{P}_{\min}$	non-compensating aggregation function
$\mathcal{P}_{\Pi}$	compensating aggregation function

2.2.1.5 Aggregation. In order to evaluate a design $\vec{d} \in$ *DVS*, the various individual preferences must be combined or aggregated to give a single, overall measure. This aggregation, in practice, occurs in two stages. First, the individual design preferences μ_{d_i} are aggregated into the combined design preference μ_d with $\mathcal{P}_d$, and the customer's individual preferences (functional requirements) μ_{p_j} are aggregated into the combined functional requirement μ_p with $\mathcal{P}_p$. Second, once the imprecision on each design variable ($\mu_d(\vec{d})$) is induced onto the *PVS*, the induced preferences are combined with the functional requirements ($\mu_p(\vec{p})$) using $\mathcal{P}_c$, which combines all of the preferences to obtain an *overall preference* ($\mu_o(\vec{p})$) from Equation (2.1).

$$\begin{aligned}\mu_o &= \mathcal{P}_c\left[\mu_d, \mu_p\right] \\ &= \mathcal{P}_c\left[\mathcal{P}_d(\mu_{d_1}, \ldots, \mu_{d_n}), \mathcal{P}_p(\mu_{p_1}, \ldots, \mu_{p_q})\right]\end{aligned} \tag{2.1}$$

The *aggregation function* $\mathcal{P}$ must satisfy the following four axioms to be consistent with engineering design (Otto, 1992):

Monotonicity

$$\begin{aligned}&\mathcal{P}(\mu_1, ..., \mu_k, ..., \mu_{p+q}) \leq \mathcal{P}(\mu_1, ..., \mu_k', ..., \mu_{p+q}) \\ &\text{iff } \mu_k \leq \mu_k' \quad \forall k\end{aligned} \tag{2.2}$$

Continuity

$$\mathcal{P}(\mu_1, ..., \mu_k, ..., \mu_{p+q}) = \lim_{\mu'_k \to \mu_k} \mathcal{P}(\mu_1, ..., \mu'_k, ..., \mu_{p+q}) \qquad \forall k \tag{2.3}$$

Annihilation

$$\mathcal{P}(\mu_1, ..., 0, ..., \mu_{p+q}) = 0 \tag{2.4}$$

Idempotency

$$\mathcal{P}(\mu, ..., \mu) = \mu. \tag{2.5}$$

The *aggregation functions* $\mathcal{P}_c$, $\mathcal{P}_d$, and $\mathcal{P}_p$ reflect the trade-off strategies for each aggregation, formalizing how competing attributes of the design should be traded-off against each other, as discussed in Otto and Antonsson, 1991a; Otto and Antonsson, 1991b; Scott and Antonsson, 1995; Scott and Antonsson, 1998, and Scott and Antonsson, 1996. As the functional requirements and design preferences can be modeled as membership functions on fuzzy sets, the aggregation of preferences is a problem of the aggregation of fuzzy sets. The appropriate design strategy (aggregation) is usually dictated by the design problem or sub-problem and is not a choice the designer can freely make. A design problem will, in general, require a hierarchy of different trade-off strategies which successively aggregate design attributes.

Non-Compensating Aggregation. Consider a system of components, for example, where the failure of one component results in the failure of the entire assembly. A high preference corresponding to a long time to failure for one component cannot compensate for a low preference corresponding to a short time to failure for another component. This requires a *non-compensating* trade-off strategy for which the aggregation function is the minimum $\mathcal{P}_{\text{min}}$:

$$\mu_o(\vec{d}) = \min\left(\mu_{d_1}, ..., \mu_{d_n}, \mu_{p_1}, ..., \mu_{p_q}\right). \tag{2.6}$$

Compensating Aggregation. Alternatively, consider the performance variables energy storage and unit cost of an ordinary household battery. Low unit cost can partially compensate for short battery life and long battery life can partially compensate for high unit cost. These two attributes are appropriately combined with a *compensating* trade-off strategy, for which the aggregation function is the geometric weighted mean or product of powers $\mathcal{P}_\Pi$:

$$\mu_o(\vec{d}) = \left(\prod_{i=1}^{n} \mu_{d_i} \prod_{j=1}^{q} \mu_{p_j}\right)^{\frac{1}{n+q}} \tag{2.7}$$

The aggregation functions $\mathcal{P}_{\text{min}}$ and $\mathcal{P}_\Pi$ are just two examples of levels of trade-off that can be modeled by the MoI. Intermediate levels of aggregation are also possible, and importance weightings can be specified for each attribute (see Scott and Antonsson, 1995 for details). Within a single design problem, different groups of attributes may require different trade-off strategies. In general, preferences for individual attributes will need to be successively aggregated by a hierarchy of different trade-off strategies.

The point (or points) with the highest preference correspond to the performance of the overall most preferred design(s). The design problem is to find the corresponding set of design variables ($\vec{d}^*$) that produce the maximum overall preference ($\mu_o^*(\vec{d}^*)$). In the typical engineering design case, where the inverse mapping ($\vec{f}^{-1} : \mathcal{Y} \to \mathcal{X}$) doesn't exist, $\mu_o(\vec{d})$ can still be obtained point by point (Law and Antonsson, 1994). The peak overall preference in $\mathcal{X}$, μ_o^*, is equal to the peak overall preference in $\mathcal{Y}$ (Otto et al., 1993b).

Iteration. The method of imprecision is iterative. Preferences evolve through successive evaluations of imprecise design alternatives. Yet this is not traditional "point-by-point" design iteration: the methodology uses fuzzy sets representing preferences over many designs, providing more complete information earlier in the design process. This information can be propagated to downstream engineering groups, facilitating fuzzy set based concurrent design. Set-based concurrent design, introduced by Ward et al., 1994 is a powerful paradigm that is enhanced by the use of fuzzy sets.

2.2.1.6 Imprecision.

Quantifying Imprecision. Utility and risk-aversion are quantified in utility theory via the lottery method (Keeney and Raiffa, 1993). Unfortunately no such formal method exists for eliciting preference (Bandemer, 1993; Barrett et al., 1990; Dutta et al., 1986; Kacprzyk and Fedrizzi, 1988; Kacprzyk and Roubens, 1988). However, limits of acceptability for variable values, whether communicated formally or established informally by experience, are familiar to engineers in industry (Ward et al., 1994). Such acceptable limits correspond to intervals over which preference is greater than zero. This suggests that rather than determine the preference μ_d at each value of d, as shown in Figure 3, it may be more natural to determine the intervals in d, called α-cuts, over which μ_d equals or exceeds certain preference values α. The use of intervals encourages the passing of set-based design information between engineering groups early in the design process (Ward et al., 1994), and permits the early release of possible sets of design data from one engineering group to the next in advance of precise design information. This approach has many advantages over the traditional "point-by-point" design iteration. The MoI can extend set-based concurrent design by providing preference information over the possible range of design data.

Imprecision Calculations. After specifying preferences for the design variables μ_{d_i} on $\mathcal{X}_i$ and requirements or specifications for the performance variables $\mu_{p_j}(p_j)$ on $\mathcal{Y}_j$, (and identifying the appropriate aggregation function (Scott and Antonsson, 1995; Otto and Antonsson, 1991a)), the first step is to combine the individual μ_{d_i} to obtain μ_d, the combined design preference. Because design variables can be chosen independent of each other, these are combined with the *min*.

Next μ_{d_i} is induced onto $\mathcal{Y}$, using Zadeh's *extension principle* (Zadeh, 1965):

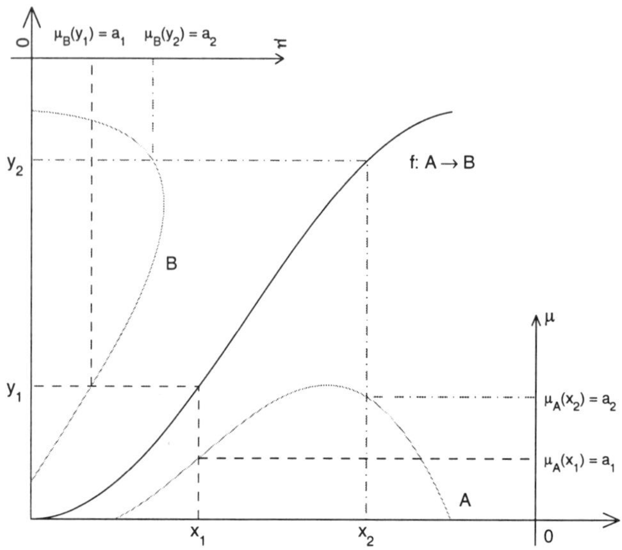

Figure 4 Zadeh's extension principle (Zadeh, 1965).

$$\mu_d(\vec{p}) = \sup_{\vec{d}:\vec{p}=\vec{f}(\vec{d})} \left[\mu_d(\vec{d})\right]$$

where sup over the null set is defined to be zero. $\mu_d(\vec{d})$ is the combined design preference on $\mathcal{X}$, as distinct from $\mu_d(\vec{p})$, the combined design preference induced onto $\mathcal{Y}$. $\mu_d(\vec{p})$ is obtained by mapping (or inducing) $\mu_d(\vec{d})$ onto the *PVS*.

A simple one-dimensional example of Zadeh's extension principle is shown in Figure 4. The performance p achieved for each value of the design variable d is given by the function f, which is a curve in this simple example. The corresponding $\mu_d(d)$ can be mapped onto p, producing $\mu_d(p)$, the design preference mapped onto the performance space (as illustrated by the dashed lines in Figure 4). For higher dimension design problems, each p will be a function of many d's, and each function f will be a hyper-surface.

An algorithm to compute Zadeh's extension principle (and thus to calculate $\mu_d(\vec{p})$) is the *Level Interval Algorithm* (LIA), first proposed in Dong and Shah, 1987 and Dong and Wong, 1987 as the "Fuzzy Weighted Average" algorithm and also called the "Vertex Method". Note that in the simple example above f is non-linear. Non-monotonic and discrete functions can also be used, as introduced in Otto et al., 1993a; Wood et al., 1992, and reviewed below.

2.2.1.7 Computations. As indicated above, one of the central elements of the procedure to represent and manipulate imprecision in engineering design, is the algorithm to induce preferences from independent variables to dependent ones. The following sections describe the algorithm as originally presented in the literature, and some of the extensions developed at Caltech.

The Level Interval Algorithm (LIA). Many discrete and analytical methods exist in the literature for carrying out extended operations with fuzzy sets (or fuzzy numbers). The Fuzzy Weighted Average (FWA) algorithm, as presented in Dong and Shah, 1987 and Dong and Wong, 1987, outlines a simple and efficient algorithm that is useful for carrying out engineering design calculations. This algorithm is extended below for generalized real functions of fuzzy variables, and the extended form is referred to as the *level interval algorithm* (LIA). Comparing the algorithm to the analytical method outlined in Kaufmann and Gupta, 1985, the LIA uses the interval analysis techniques as described; yet, the LIA simplifies the process extensively by discretizing the membership functions of the input fuzzy numbers into a prescribed number of α-cuts. Performing interval analysis for each α-cut and combining the resultant intervals, the output is a discretized fuzzy set, the performance variable output of input preference functions for the case of a design calculation. Dong and Wong also include a combinatorial interval analysis technique in order to avoid the problems of the multiple occurrence of variables for division and multiplication in an algebraic equation expression.

There are conditions which must be satisfied for application of the algorithm: the preference functions must satisfy the normality and convexity conditions and must be continuous over the design variables ($\tilde{d}$), no singularities of the functions can occur over $\tilde{d}$ (*i.e.*, no division by zero can occur, and no zero arguments can occur in $f_j(\vec{d})$ for each $\tilde{d}_i$ for the unary operations, such as the natural logarithm and the square root), and only monotonic regions of multi-valued functions, *e.g.*, sine and cosine, are computed for a given $\tilde{d}_i$.

Extending LIA for Internal Extrema. The LIA is valid only for real-valued functions f, and corresponding interval extensions $F(\mathcal{X}_0)$ that do not include internal bounded *extrema* for the intervals in question, $\mathcal{X} \in \mathcal{X}_0$. This is because only the endpoints (at a given α-cut) of the input variables d_i $i = 1, \ldots, n$ are used in the computation. An extension to the LIA was introduced in Antonsson and Otto, 1997 to determine the correct bounds $P^{\alpha_j}(\vec{D}^{\alpha_j})$ for a given α-cut α_j.

This extended LIA algorithm will not produce discontinuities in output preference that might arise from the multiplicity of points in the map's pre-image.

Optimization. One of the major limitations of the LIA stems from the assumption that the extreme values of f_j will occur at the corner points of the D_{α_k} n-cube (Law and Antonsson, 1995). The algorithm may be improved by relaxing this assumption. The problem restated is to find:

$$\begin{aligned} {p_j}^{\alpha_k}_{\min} &= \min\{p_j = f_j(\vec{d}) \mid \vec{d} \in D_{\alpha_k}\} \\ {p_j}^{\alpha_k}_{\max} &= \max\{p_j = f_j(\vec{d}) \mid \vec{d} \in D_{\alpha_k}\} \end{aligned} \tag{2.8}$$

Finding extrema within a subspace is a constrained optimization problem.

The traditional optimization algorithm utilized here is Powell's Method, which begins as a one-at-a-time search. After each iteration a heuristic determines whether to replace the direction of maximum decrease with the net direction moved during the last iteration. This allows minimization down valleys while avoiding linear dependence in the set of search directions (Adby and Dempster, 1974).

An important feature for a practical computational tool is a means to trade-off the number of function evaluations against accuracy. Such an adjustment enables the designer to use the same program to obtain quick estimates as well as precise evaluations. This is implemented as a user-specified fractional precision that defines termination criteria for the optimization algorithm.

Experimental design methods (Phadke, 1989) may also be used to identify design variables with near-linear effects, and remove those directions from the optimization search (Law and Antonsson, 1995).

2.2.1.8 Conclusion. The Method of Imprecision (M$_o$I), briefly described here, forms the basis for formally representing and manipulating imprecise information in engineering design while simultaneously incorporating engineering designers' experience and judgment into design decisions. The underlying methodology builds upon fuzzy mathematics by mapping fuzzy sets (design information) through crisp functions (engineering analyses), and utilizing a family of idempotent mixed connectives for aggregation. When the engineering analysis is expensive, traditional computation of the extension principle is made more efficient by incorporating design of experiments methods. The designer can adjust the precision of these computations in order to accommodate different stages of the design process (from the most imprecise preliminary stage to the precise end result).

The approach embodied in the M$_o$I encourages the designer and customer to specify preferences on design and performance variables (specifications), and thus promotes design communication to evolve from individual "point" designs to (fuzzy) sets of designs. Since a range of possible design variable values can be communicated to down-stream design processes earlier than a completed individual design, the M$_o$I can facilitate (fuzzy) set-based concurrent design.

2.2.2 A Fuzzy Design Method (FDM)

The M$_o$I briefly introduced in Section 2.2.1 not only formalizes imprecision in a conventional approach to design, but alters the whole design process by using Fuzzy Sets. These Fuzzy Sets, described by $\mu_{d_i}(d_i)$, $i = 1, \ldots, n$, and $\mu_{p_j}(p_j)$, $j = 1, \ldots, q$, provide the basis for determining an overall preference by use of an operator $\mathcal{P}$ (called aggregation function in Section 2.2.1). A computationally efficient implementation of Zadeh's extension principle (extended LIA) is used to perform the numerical calculations to induce designer preferences onto the performance space prior to aggregation. Now, we will use some of the main ideas of the M$_o$I, and we will add some others in order to propose another Fuzzy Design Method.

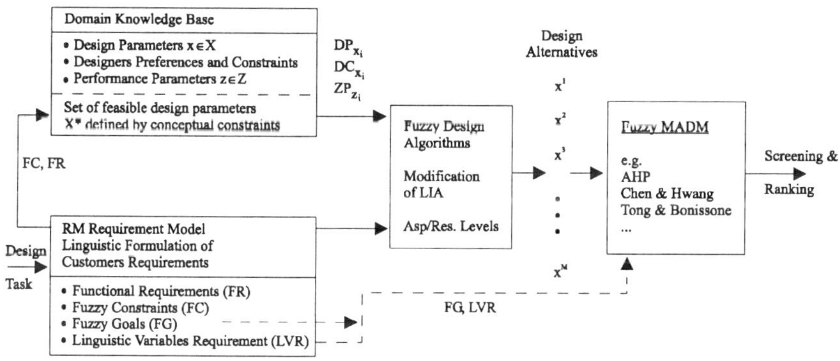

Figure 5 A Fuzzy Design Method

- The design variable space X will considered and additional constraints might be added in order to define a subset X^* of feasible design objects. This provides a mechanism to deal with fuzzy constraints, if desired.

- Designers and customers preferences respectively can be formulated using linguistic variables. This gives an imprecise linguistic user interface to formulate functional requirements, goals, and constraints.

- Instead of using a combination function $\mathcal{P}$ (operator) which realises aggregation of all preferences (introduced in the $M_{O}I$), aspiration and reservation levels of the performances are used in order to restrict the set of feasible design alternatives. These levels might be crisp numbers or linguistic terms defined by respective Fuzzy Sets. Using the terminology from the Intelligent Systems field, we can start with an overall architecture of the proposed Fuzzy Design Method (FDM) shown in Figure 5.

The knowledge base is divided into:

- A Domain Knowledge Base, and

- A User Requirements Knowledge Base.

Within the Domain Knowledge Base a generic model of the domain we are dealing with is represented.

For example, there is a Knowledge base of "Space Launch Systems" defined by taxonomies, design parameters, performance parameters and constraints. The Requirement Model is also a generic type model which predefines all possible customer requirements related to the domain which is considered. In particular, a closed world of Functional Requirements, Fuzzy Constraints, Fuzzy Goals and Linguistic Variables Requirements is predefined.

Now, a user formulates his or her design tasks by selecting requirements from the closed world. Then, the system is able to perform the design model of the task by generating instances of the domain knowledge model and the requirement model.

This model is used by the Fuzzy Design Algorithm which is a modification of the LIA approach and a combination with several types of aspiration and reservation levels which are derived from the requirement model instance. The output is a set of feasible design alternatives which will be transferred to a Fuzzy MADM in order to do screening and/or ranking. The Fuzzy MADM also gets input from the Fuzzy Goal and Linguistic Variables Requirements. Aggregation will be done by the Fuzzy MADM, but not by the Fuzzy Design Algorithm.

The following sections will present details of the:

1. Domain Knowledge Base,
2. Requirement Model,
3. Fuzzy Design Algorithm,
4. MADM Methods.

2.2.2.1 Domain Knowledge Base. Design objects are described by n design parameters (variables) $x_i, i = 1, \ldots, n$. $x = (x_1, x_2, \ldots, x_n)$ denotes the design vector where x_i describes either a real number (crisp case), a Fuzzy Set, or a linguistic Variable. It is assumed that $x_i \in X_i$ and X, $x \in X = X_1 \times X_2 \times \ldots \times X_n$, denotes the Design Variable Space (*DVS*). There is a hierarchy of design objects defined by specialization (*is - a*) and composition relations (*has-parts, is part of*).

For example, if none of the parameters x_i is specified, the x represents a generic object (a class), but, if all parameters x_i are instantiated each by a particular value of the sets X_i, then the object is an instance (individual). In addition, there might be constraints formulated in terms of design parameters x_i, which are called conceptual constraints (because they are defined on conceptual (generic) level). These constraints define a subset $X^* \subseteq X$ of feasible design objects. Then, each object has a degree of feasibility. (In this section we will not deal with fuzzy constraints of this type.) Up to this point we have not modeled designers preferences for the design parameters. Now, let's assume x_i to be either a crisp real parameter or a linguistic variable. Then, we are able to model preferences and constraints by Fuzzy Sets as follows:

- $DP_{x_i} = \{(x_i, \mu_{x_i}(x_i)) \mid x_i \in X_i\}$ is a Fuzzy Set, which defines the designers preference for x_i.
- $DC_{x_i} = \left\{\left(x_i, \mu^C_{x_i}(x_i)\right) \mid x_i \in X_i\right\}$ is a Fuzzy Set, which defines a constraint for x_i.

Figure 6 illustrates these two Fuzzy Sets in the case of a real-valued parameter x_i.

The requirements DP_{x_i} and DC_{x_i} have to be modeled using an appropriate "and" operator (Zimmermann and Sebastian, 1994).

Performance variables are denoted z_j, $j = 1, 2, \ldots, m$ where $z_j \in Z_j$ are either real numbers, Fuzzy Sets, or Linguistic variables. Z_j is a set of possible performance values. The z_j measures the performance (quality) of design objects. If x represents a design-object, then for each j there is a well-defined function f_j such that $z_j = f_j(x)$, f_j is a unique mapping from X into z_j. The overall performance of a design object is

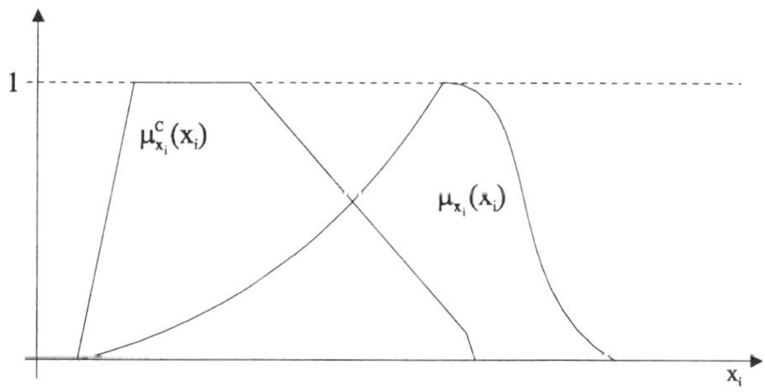

Figure 6 Fuzzy Preferences for Variable x_i

characterized by the vector $z = (z_1, z_2, \ldots, z_m) \in Z = Z_1 \times Z_2 \times \ldots \times Z_m$. Z is called Performance Variable Space (*PVS*).

As a simple illustrative example we use two real-valued performance parameters:

$z_1 = ms$	maximal speed of a car
$z_2 = t(0, 100)$	time a car needs to reach a speed of 100 km/h from 0

The design objects $x \in X$ are cars described by vectors x of design parameters belonging to a given set X of cars. We choose $Z_1 = [0, 300\text{km/h}]$ and $Z_2 = [0, 20\text{sec}]$. Of course, we can choose the parameters z_1, z_2 as performance parameters.

However, people who are not experts or who want to use imprecise terms while formulating their performance-requirements would prefer to ask for a fast car than to specify precisely z_1 and z_2. Therefore, we define a Fuzzy Set "Fast-Car" as:

$$\text{Fast} - \text{Car} = \{(x, \mu(x)) \mid x \in X, \mu(x) = \mu^*_{\text{FC}}(ms, t(0, 100))\}$$

μ^*_{FC} denotes a 2-dimensional membership function, which might be defined without knowing x, only by considering the two parameters ms and $t(0, 100)$. But, if x is given as an completely instantiated vector, then $ms(x) = f_1(x)$, $t(0, 100) = f_2(x)$ are known, and, therefore $\mu(x)$ is known. Figure 7 shows an example.

Analogously to the MoI, Fuzzy Sets representing the customers preferences with respect to a performance parameter are introduced:

$$ZP_{z_j} = \{(z_j, \mu_{z_j}(z_j)) \mid z_j \in Z_j\}$$

Such a definition particularly makes sense when z_j is a real-valued crisp parameter. Additionally, when z_j is linguistic variable, perhaps with terms low, medium, high (e.g., the price of a car), we can introduce the Fuzzy Set expressing the preference of a customer of low, medium or high priced car without considering a real scale of prices.

2.2.2.2 Requirement Model. We showed above how Fuzzy Sets might be used to derive imprecise linguistic requirements from precise real-valued performance parameters in order to approach a more human-like way to represent customer design

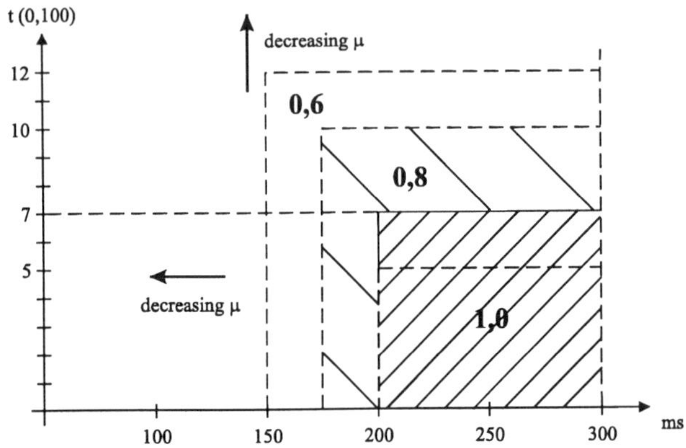

Figure 7 Two-Dimensional Membership Function

requirements. However, Fuzzy Sets such as: Fast-Car, Safe-Car, Energy-Efficient-House, *etc.*, might be used to obtain Functional Requirements (FR), Fuzzy Constraints (FC), Fuzzy Goals (FG) or Requirements which are formulated in terms of Linguistic Variables (LVR).

1. Functional Requirements

   It is not sufficient to describe a requirement (such as one related to a desired function of a design object) by defining a Fuzzy Set. For example "Fast-Car" or "Safe-Car" might be represented by Fuzzy Sets using real valued performance parameters to describe the crisp basic sets. However, the semantics of "Fast-Car" or "Safe-Car" are also related to determined components (or specializations of such components) that the design object (car) should have. For instance, "Safe-Car", in addition to the Fuzzy Set description (Fuzzy Set over X), means that the components "two independent brakes", "safety belts", "good tires", for example, *must belong* to the car, but one or two airbags *should belong* to the car with a "desirability measure of *e.g.*, 0.8". Also, ABS (anti-lock-brakes) could be a specialization-requirement for the brakes.

   We define the triple:

$$FR = (x \in FS, x \text{ has components belonging to } OC, X^{**} \in X^{*})$$

   to be a *Fuzzy Design Requirement*, where:

   - FS is a Fuzzy Set defined on real valued performance parameters or a Fuzzy Set which is equal to a performance parameter,
   - OC is a Fuzzy Set of optimal components where each has a degree of membership to the design object x, and
   - X^{**} describes a subset of the set of feasible design objects X^{*}.

Here x is used as the name of the considered design object, thus we don't consider the components of x.

The most important issue of Fuzzy Design Requirements introduced above is to model the dependencies between an imprecise linguistic expression such as "Safe-Car" and the component or specification structure which is required by this expression. However, such a Fuzzy Set might also be used to model the underlying requirement by a Fuzzy Constraint or a Fuzzy Goal.

2. Fuzzy Constraints and Fuzzy Goals

   We will choose again our simple illustrative example where ms denotes the performance parameter "maximal speed" and $ms(x)$ means the maximal speed of the object car which has the name x. Now we consider $ms(x)$ together with the imprecise linguistic expression "Fast-Car" in order to explain a "constraint".

   A *crisp constraint* could be formulated by $ms(x) \geq 190$ km/h.

   A *fuzzy constraint* is (according to the Fuzzy Set "Fast-Car") "x is feasible with the degree $\mu_{FC}(ms)$", where *e.g.*,

$$\mu_F C(ms) = \begin{cases} 0, & \text{if } ms(x) \leq 170 \\ \in [0,1], & \text{if } 170 \leq ms(x) \leq 190 \\ 1, & \text{if } ms(x) \geq 190 \end{cases}$$

   Using this kind of relaxation of the sharp boundary at 190 km/h between feasibility and infeasibility we get a smooth transition. We define a *Fuzzy Constraint* (FC) to be a Fuzzy Set FS where the membership function $\mu_{FC}(x)$ means the *degree of feasibility* of the design object x. Alternatively, the requirements could be modeled as Fuzzy Goals, where the designer wishes to maximize a real-valued function over X, which measures the utility.

Again, let's consider the imprecise requirements "Fast-Car", "Safe-Car", "Modern-Design-Car", and "Not-Expensive-Car" represented by Fuzzy Sets over the basic set of X of cars x. Several different approaches to use such Fuzzy Sets to model Fuzzy Goals are described below.

1. The Fuzzy Goal is described by a Fuzzy Set. Then, we introduce a specific defuzzification by: Find $x \in X$ such that $\mu_{FC_1}(x) \to \max$ with $\mu_{FC_1}(x)$ as the membership function of the Fuzzy Set.

$$FG_1 = \text{Fast-Car} \wedge \text{Safe-Car} \wedge \text{Modern-Style-Car} \wedge \text{Not-Expensive-Car}$$

   e.g., $\mu_{FC_1}(x) = \min\left(\mu_{\text{F-C}}(x), \mu_{\text{S-C}}(x), \mu_{\text{MS-C}}(x), \mu_{\text{NE-C}}(x)\right)$

   Another possibility is to consider an *aggregation tree* (see Figure 8). We aggregate "fast" with "modern style" to determine "attractiveness". "Attractiveness" and "in-expensiveness" are aggregated together to obtain "rationality". Finally, "rationality" and "safety" are aggregated to define the goal FG2. This aggregation uses a *compensatory operator* on the first two levels and a *non-compensatory operator* (min) on the final level.

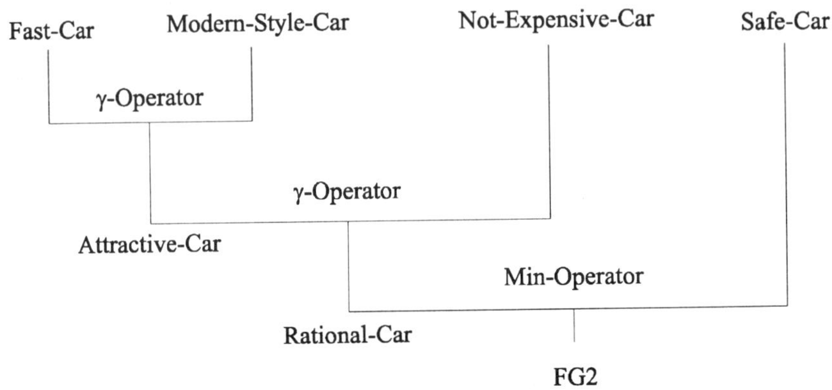

Figure 8 Example of an Aggregation Tree

A Fuzzy Goal (FG) is defined by a Fuzzy Set FG where the intention is to find $\hat{x} \in X^*$ *such that the membership function of FG becomes maximal:*

$$\mu_{FG}(\hat{x}) = \max_{x \in X^*} \mu_{FG}(x)$$

2. Another possibility is the *linguistic-variable* approach to define Fuzzy Goals (LVR in Figure 5). Here, we give an example. Let "Maximum-Speed" be a linguistic variable with the terms {slow, medium, high, very high}, then a Fuzzy Goal can be defined by "medium $\vee$ high". Then, again as a *defuzzification*, we are maximizing the membership degree of "medium $\vee$ high" over X using Fuzzy Sets "Maximum-Speed = medium" and "Maximum-Speed = high". A more detailed discussion, including the linguistic variable approach to Fuzzy Goal, is presented in Sebastian, 1996.

3. Specification of a Fuzzy Design Requirement model.

   In Figure 5 we introduced the architecture of the FDM, and briefly introduced the requirement types FR, FC, FG and LVR. Now, we address the question of obtaining the requirement model in a particular case. We denote:

   - FDR_j the *Fuzzy Design Requirement* j defined by a name, a Fuzzy Set, a set of additional required components and a set of specialized objects,
   - FC_k the *Fuzzy Constraint* k, where $k \in K$, K index set, if FC_k is a Fuzzy Set and μ_{FC_k} defines the "degree of feasibility" of an object represented by x,
   - FG_l the *Fuzzy Goal* l, where $l \in L$, L index set, if FG_l is a Fuzzy Set and μ_{FC_l} describes a "degree of desirability" of an object x.

In the sense of a closed world, we assume the sets K and L to be known. That means all possible candidates for Fuzzy Constraints or Fuzzy Goals have to be predefined in a knowledge base. Finally, a *Modeling Support System* helps to decide which of the candidates will be used as Fuzzy Constraints or Fuzzy Goals respectively for a particular case.

In the following an algorithm will be described which realises this concept.

Configuration of the Model Algorithm to build the RM Requirement Model

1. *Choose* the target (goal) system, (*e.g.*, car), from the conceptual hierarchy.

2. *Select* the Fuzzy Design Requirements FDR_j, in particular the Fuzzy Sets F_j, from the set of all attributes of the target system (object) in the conceptual hierarchy.

   - define J as an index set of the $FDR_j, j \in J$.
   - determine the set of additional components and specialized objects related to FDR_j for all $j \in J$. (Use the conceptual constraints, check inconsistencies, and solve conflicts).

3. *Select* such linguistic expressions from the set of all attributes which are candidates for Fuzzy Constraints or Fuzzy Goals. (Use the predefined sets K and L. The $F_j, j \in J$ are automatically candidates for Fuzzy Constraints or Fuzzy Goals as well.) For an initial-model define the Fuzzy Constraints and the Fuzzy Goals $FC_k, k \in K^* \subseteq K$, and $FG_l, l \in L^* \subseteq L$.

2.2.2.3 A Fuzzy Design Algorithm. According to Figure 5 there are several possibilities to formulate different versions of Fuzzy Design Algorithms depending on the input of the respective box in Figure 5.

1. Modifications of the LIA Algorithm by adding aspiration/reservation levels (input: Designer and Customer Preferences from the Knowledge Base, Aspiration/Reservation levels from the RM),

2. Bellman-Zadeh or Defuzzification Approaches (input: FC's and FG's from the RM and a finite set of design alternatives which are feasible with respect to X^*) (Bellman and Zadeh, 1970),

3. Fuzzy MADM (input: finite set of design alternatives from the Fuzzy Design Algorithm and FG's/LVR's from the RM. This approach assumes that only FC's are used to be input of the Fuzzy Design Algorithm while the FG's/LVR's are used within the Fuzzy MADM (Chen and Hwang, 1992).

Modifications of the LIA Algorithm. The modifications of the LIA proposed here rely on the detailed description of design objects x using design parameters x_i from the knowledge base, where:

1. - for all $i : x_i \in X_i, i = 1, 2, \ldots, n$,
   - $X^* \subseteq X = X_1 \times X_2 \times \ldots \times X_n$ is given,

- $\mu_{x_i}(x_i)$ denotes the designers preference with respect to the design parameters (x_i might be real-valued or a linguistic term-valued variable).

2. $z_j = f_j(x)$ are real valued performance parameters, $z_j \in Z_j \subseteq R^1$, $j = 1, 2, \ldots, m$ and $J', J' \subseteq \{1, 2, \ldots, m\} = J$, denotes a subset of J. f_j denote mappings from x into $Z_j \subseteq R$. If the idea is to maximize z_j, then there are given:

   - a reservation level r_j, and
   - an aspiration level a_j

   both formulated by customers. The levels are defined as follows:

   - reservation: if $z_j = f_j(x) < r_j$, then x is not feasible,
   - aspiration: if $z_j = f_j(x) \geq a_j$, then the customer is completely satisfied.

   If there is a Fuzzy Constraint defined for z_j in the RM, then it is easy to derive aspiration and reservation levels from such a constraint.

   Remarks:

   - There are design parameters which are not performances, however the customer might also have preferences with respect to such a design parameter, Therefore, we use "designers/customers preferences" with the respect to design parameters.
   - We do not consider Fuzzy Sets describing customers preferences with respect to performance parameters. Instead, in order to keep computation or effort low, we pick only two values, the r_j and a_j, for each $j \in J' \subseteq J$.

A Fuzzy Design Algorithm: Version 1. In the following section, we describe an algorithm which generates a finite set of design alternatives. Figure 9 shows an example with three design parameters illustrating the α-cut approach.

1. Initialization:

   $\alpha := 1$, choose an operator (min, product), choose a subset $I' \subseteq \{1, 2, \ldots, n\}$ of design parameters with highest priority, choose a subset $J \subseteq \{1, 2, \ldots, m\}$ of performance parameters that have aspiration and reservation levels a_j and r_j. Choose a number N of desired alternatives x.

2. Compute $S_{\alpha,i}$ for all $i \in I$.

3. Repeat until all $x'_i \in S_{\alpha,i'}$, $i' \in I'$ are considered or N is reached:

   For all $i' \in I'$ choose $x_{i'}, x'_i \in S_{\alpha,i'}$

   Compute x_i for all $i \in \{1, 2, \ldots, n\} \setminus I'$ such that:

   (a) $x_i \in S_{\alpha,i}$,

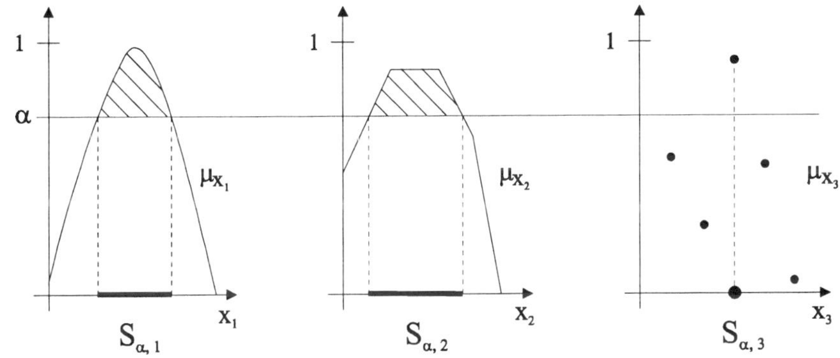

Figure 9 α-cuts on Three Design Parameters

(b) $x = (x_1, x_2, \ldots, x_n) \in X^*$ (feasible),

(c) $z_j = f_j(x)$ fulfills the aspiration/reservation levels for all $j \in J'$ by constraint propagation.

4. If $\sum_{\beta=1}^{\alpha} N_\beta < N$ and $\alpha > \min \alpha$ replace $\alpha := \alpha - \Delta\alpha$ go to step 2, otherwise "show results x".

All N_α solutions x found within the loop of step 3 are feasible, belonging to the α-cuts of the design parameters and fulfill the aspiration/reservation level constraints.

In this algorithm the f_j-functions are still considered to be crisp. We don't consider (in this first version) the induced design-parameter preferences into the performance space using the Extension Principle. This will be done in the second version.

Fuzzy Design Algorithm: Version 2. The idea is to construct the joint designer/customer preference function using a restricted LIA approach starting with the aspiration/reservation levels. This assumes that for each performance parameter z_j a customer preference with respect to z_j is specified by a Fuzzy Set. The first version algorithm will be modified as follows:

- Choose $I' = \{1, \ldots, n\}$
- Compute $S_{\alpha,1}, \ldots, S_{\alpha,n}$ for some of the α, $\alpha \in \{\alpha_0, \alpha_0 - \Delta\alpha, \ldots, \min \alpha\}$
- Before selecting x such that $x_i \in S_{\alpha,i}$, $i = 1, 2, \ldots, n$, $x \in X^*$ one has to compute the induced α-cuts for performance preferences from α-cuts of the designer preferences (with respect to design parameters) by the LIA or an extended LIA (see Section 2.2.1.7).

We consider a particular realization of this approach which keeps the computational complexity low.

1. We call

- α^{**} the aspiration level for designers preferences and
- α^{*} the reservation level for designers preferences.

$1 \geq \alpha^{**} > \alpha^{*} \geq \min \alpha$
(aspiration/reservation levels are chosen to be membership values instead of performance values we considered in the first algorithm)

We choose $S_{\alpha^{**},i}, S_{\alpha^{*},i}; i = 1, 2, \ldots, n$ where $S_{\alpha^{**},i} \supseteq S_{\alpha^{*},i}$.

2. Using the LIA we compute the induced α-cuts of performance preferences from the $S_{\alpha^{**},i}, S_{\alpha^{*},i}$. For each j, $j = 1, 2, \ldots, m$ and $z_j = f_j(x) \in Z_j, Z_j \subseteq R$, we get:

$$\tilde{P}_j^{\alpha^*} = [z_{j,\min}^{\alpha^*}, z_{j,\max}^{\alpha^*}]$$

$$\tilde{P}_j^{\alpha^{**}} = [z_{j,\min}^{\alpha^{**}}, z_{j,\max}^{\alpha^{**}}]$$

3. We assume: For each z_j there is a subinterval of Z_J;

$$r_{j,\alpha^*} \leq z_j \leq a_{j,\alpha^{**}}$$

where r_{j,α^*} denotes a reservation value on level α^* and $a_{j,\alpha^{**}}$ means an aspiration value on level α^{**}.

Then we get

- joint aspiration/reservation levels:

$$r_j := \max(r_{j,\alpha^*}, z_{j,\min}^{\alpha^*})$$

$$a_j := \min(z_{j,\max}^{\alpha^*}, a_{j,\alpha^{**}})$$

and an

- interval for the joint preference function:

$$r_j' := \min(r_{j,\alpha^*}, z_{j,\min}^{\alpha^*})$$

$$a_j' := max(z_{j,\max}^{\alpha^{**}}, a_{j,\alpha^{**}})$$

Now, there are two possible approaches:

1. Choose a subset $J' \subseteq J$
   Find x such that:
   - $x_i \in S_{\alpha^*,i}, i = 1, 2, \ldots, n$
   - $x \in X^*$
   - $r_j \leq f_j(x) \leq a_j$ for $j \in J'$
2. Compute the joint membership-functions $\mu_{z_j}^*(z_j)$ on $r_j' \leq z_j \leq a_j'$ for $j \in J'$ using an operator (*e.g.*, min-, prod-operators). (It holds $\alpha^* \leq \mu_{z_j}^*(z_j) \leq \alpha^{**}$ on

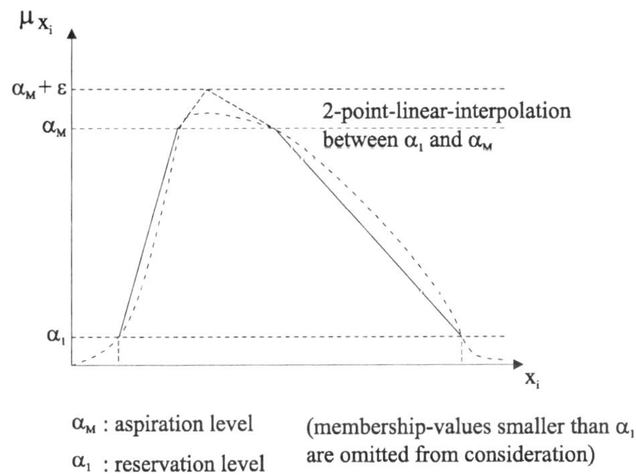

Figure 10 Two Point Interpolation

$[r'_j, a'_j]$.)
Find x such that:

- $x_i \in S_{\alpha^*,i}$,
- $x \in X^*$,
- $\mu_0^* = P\left(\mu^*_{z_{j_1}}(z_{j1}), \ldots, \mu^*_{z_{j_k}}(z_{jk})\right) \to \max, J' = \{j_1, j_2, \ldots, j_k\}$.

Approach (1) means one has to solve a crisp constraint problem. This problem has to be solved by constraint satisfaction or propagation with the goal to find a finite set of solutions x. Approach (2) is an optimization problem (after the computation of the joint membership functions $\mu^*_{z_j}(z_j)$ on $[r'_j, a'_j]$). However, the problem might be modified such that one has to compute a set of n-best solutions.

Fuzzy Design Algorithm: Version 3. Version 3, elaborated by R. Brehm, is not fully described here but illustrated by a sequence of pictures showing the main ideas. A detailed description is given in (Brehm, 1997). It is an iterative approach with the goal to construct a finite set of design alternatives. But, in contrast to the two versions above it does not generate either constraint satisfaction or optimization problems. A special type of membership function is used in order to permit the use of an inverse LIA-algorithm, LIA^{-1}. Figure 10 shows the type of membership function (2 point interpolations) of the designers preferences with respect to design variables. By introducing $\epsilon > 0$ the slightly manipulated membership function becomes inverse-unique.

Using the LIA-algorithm, the interpolation of the designers preferences will be mapped into the performance variable space (*PVS*). In addition, using the minimum operator, a joint induced preference function $\eta_j(z_j)$ will be determined (Figure 11).

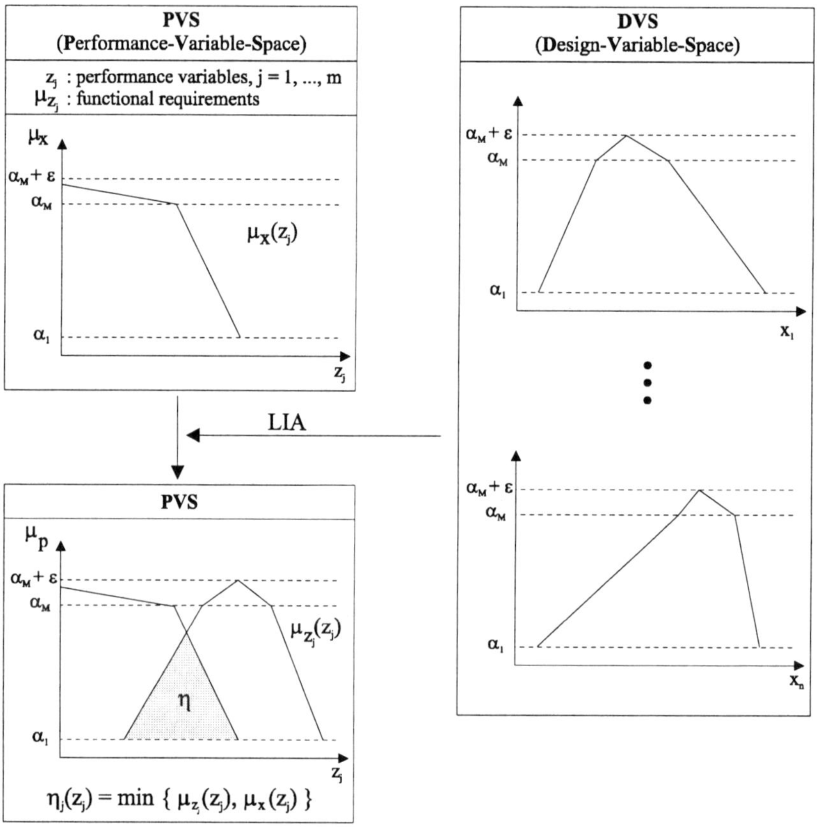

Figure 11 LIA Algorithm 1

After constructing the membership function $\eta_j(z_j), j = 1, \ldots, m$, we choose levels $\gamma_j \leq \alpha_M$ for each j, such that the γ_j cuts I_j related to $\eta_j(z_j)$ are not empty. Using the LIA^{-1} these sets are mapped back into the *DVS* and the intersections $S_i = S_i(\gamma_1, \ldots, \gamma_m)$ are determined (Figure 12).

If one of the sets S_i is empty, then the γ_j must be systematically reduced, until all $S_i \neq \emptyset$ (Figure 12) (Iteration $k \rightarrow k+1$). Then, design alternatives are selected from the sets $S_1, S_2, \ldots, S_n$ such that $x \in X^*$.

Bellman-Zadeh and Defuzzification Approaches. While we picked the case of real-valued performance and design variables in the previous sections, we now consider the case where requirements in the form of Fuzzy Constraints and Fuzzy Goals are considered. Let us propose the sets K^* and L^* of FC_k and FG_l to be known as a result of requirement modeling (see Section 8. Then, we can generate a "Fuzzy Multi Criteria Design Model" by choosing one of the following two approaches:

1. *Bellman-Zadeh-Model*
   (if $K^* \neq \emptyset \wedge L^* \neq \emptyset$)
   Define the Fuzzy Set F by:
   F = (AND FC_k) AND (AND FG_l), $k \in K^*, l \in L^*$

2. *Defuzzification approach*

   - Choose $x \in X'$ with $\max \mu_F(x), (F, \text{Bellman} - \text{Zadeh} - \text{Model}(1))$, or
   - $FC^* = \text{AND}FC_k, k \in K^*$, given by the Fuzzy Set over X'.

   Define an α-cut of FC^* and call $x \in X'$ α-feasible, if $\mu_{FC^*}(x) \geq \alpha$;

   $$X = \{x | x \in X' \wedge \alpha - \text{feasible}\}.$$

   Define FL^* by AND FG_l, $l \in L^*$ or by an *aggregation tree* (over several levels). Call x β-optimal, if x belongs to a β-cut of FL^* over X (particular case $\beta = 1.0$).

X' denotes a finite set of feasible (with respect to X^*) design alternatives which have to be generated from the domain knowledge base. Of course, if we choose the Bellman-Zadeh approach ($\mu_F(\hat{x}) = \max_{x \in X'} \mu_F(x)$), we get only one design alternative $\hat{x}$, if the max-operation is unique. If we get more than one maximizing $\hat{x} \in X'$ it makes sense to transfer them to the Fuzzy-MADM module. The particular α-cut oriented approach which deals with the FC's first and considers the FG's afterwards might easily be formalized in order to generate M design alternatives which will become inputs to the Fuzzy-MADM. This makes sense because aggregation of Fuzzy Sets is only partly used.

Fuzzy MADM. Many different methods for solving a MADM problem are known from the literature (Chen and Hwang, 1992; Müller and Sebastian, 1997), because on one hand the available data vary and on the other hand the type of desired solution can differ. The methods can be classified into two groups:

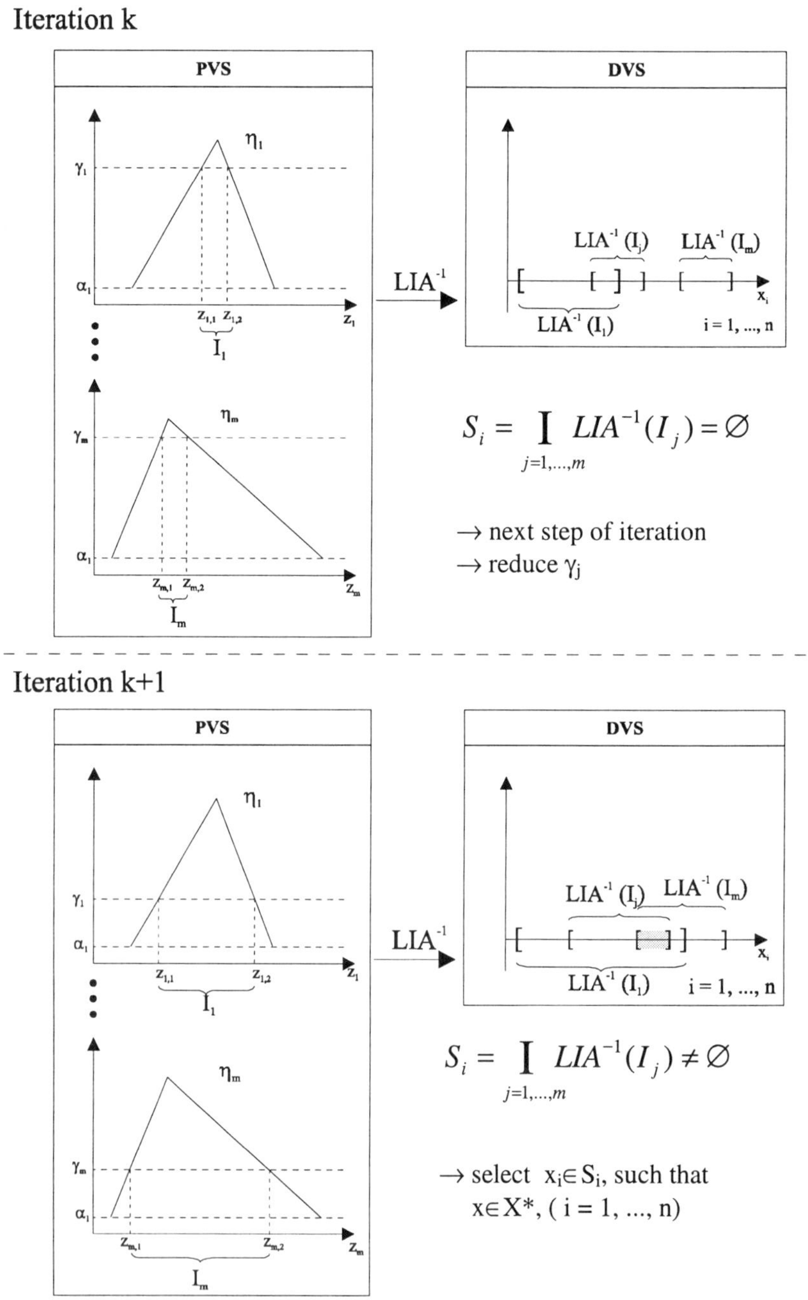

Figure 12 LIA^{-1} Algorithm

- the screening methods, and
- the ranking methods.

Screening methods eliminate alternatives that do not fulfill the conditions of being a feasible and desirable solution. A feasible alternative is considered to be not desirable if it is dominated by another alternative or if it not reaches a certain level for one or some of the attributes. Examples of screening methods are the Dominance Method, the Conjunctive Method or the Disjunctive Method.

Ranking methods create an order of all alternatives from the best to the worst. Examples are the:

- Simple Additive Weighting Method
- Weighted Product Method
- Distance from Target Method
- TOPSIS Method
- Saaty's AHP

The examples of MADM-methods cited above are crisp in the sense that the characterizing attributes of the finite set of given alternatives are crisp (real numbers, crisp linguistic characters). If all or some of the alternatives are imprecise, then Fuzzy-MADM methods are required. Again, there is a broad area of such kind of Fuzzy-MADM methods subdivided into screening and ranking methods. Some of the most prominent Fuzzy-MADM methods are

- Yager's Method
- Fuzzy AHP
- Chen and Hwang's Method
- Tong and Bonissone's Method

We will describe a workbench of software tools KONWERK in Section 2.3.2, where a MADM-toolbox is an essential part. In Sections 2.4.2 and 2.4.3 Chen and Hwang's Method, and Saaty's AHP will be used in order to solve a real world example.

2.2.3 Fuzzy Design by Evolutionary Strategies Combined with Fuzzy MADM

It seems to be a disadvantage of the FDM architecture presented in Figure 5 that there is no feedback from the Fuzzy MADM to the Fuzzy Design Algorithm (FDA). Of course, the FDA uses attributes and their aspiration/reservation levels, which are afterwards used within the Fuzzy MADM (on a more detailed level), but the procedure:

1. generate design alternatives $x^1, x^2, \ldots, x^m$

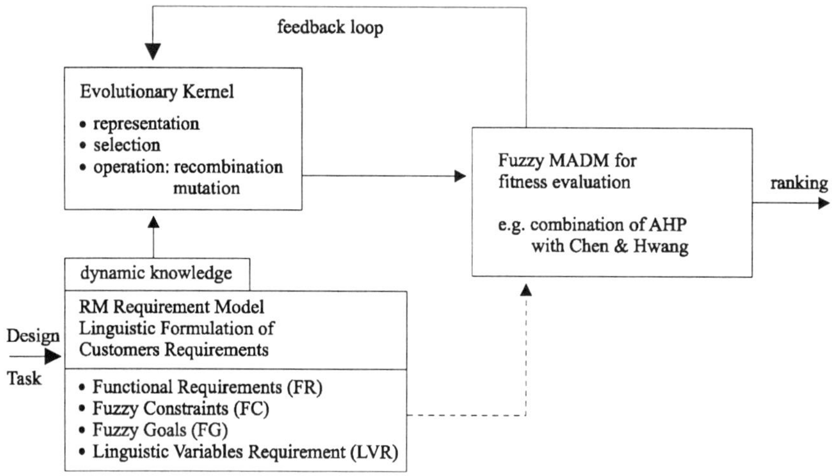

Figure 13 An Evolutionary Algorithm with a Fuzzy Fitness Evaluation

2. evaluate, screen and rank these alternatives

is strongly sequential.

Evolutionary strategies provide a more flexible approach. They also generate a finite set of alternatives (population), but this population is used (after evaluation and selection) to generate a new population using different genetic operators (mutation, recombination). The idea is, to achieve, after many such iterations, an improvement in the quality of the population with respect to the fitness evaluation. In other words, a feedback loop is introduced within the architecture of Figure 13.

The static knowledge base and the fuzzy design algorithm are thus replaced by the Evolutionary Strategy, which contains the representation of design objects, as well as operations with these representations. The feedback loop from the Fitness Evaluation to the Evolutionary Kernel (Figure 13), is most important (Brehm et al., 1997).

Of course, evolutionary strategies to solve complicated optimization problems are well known (*e.g.*, Schönberg et al., 1994), however, few applications to engineering design problems have been published. The reason seems to be that the representation of design objects using simple data structures is difficult in real world applications. Also, representing and maintaining of consistency of design alternatives during an evolutionary algorithm is necessary and difficult. Finally, modeling of customer requirements in design leads to Fuzzy Multi-Attribute fitness evaluation.

There are several attempts in the literature to realize concepts similar to those described above. For example, I.C. Parmee proposes population-based adaptive search techniques to engineering design problems (Parmee and Purchade, 1997). R. Schleiffer developed an application to an air tank design problem (Schleiffer, 1998). We will illustrate the approach described briefly above by the well-known example of personal computer configuration in Section 2.4.3.

2.3 IMPLEMENTATION OF FUZZY DESIGN METHODS – SOFTWARE TOOLS FOR FUZZY ENGINEERING DESIGN

2.3.1 The Imprecise Design Tool (IDT)

IDT is a C program that implements the Method of Imprecision (MoI) with a non-compensating design strategy ($\mathcal{P} = \min$) for a "black box" calculation $f : \mathcal{X} \rightarrow \mathcal{Y}$ Law and Antonsson, 1994. As presented in Section 2.2.1 the underlying mathematical capabilities incorporated in the MoI include considerable refinements and advancements not utilized in this early example. In our first test, this black box was the Engine Development Cost Estimator (EDCE) provided by General Electric Aircraft Engines, Cincinnati, Ohio. While EDCE requires crisp inputs and produces a single crisp output, IDT allows the designer to specify imprecise inputs and obtain an imprecise output, as illustrated in Figure 14. Eight of EDCE's inputs which represent the degree

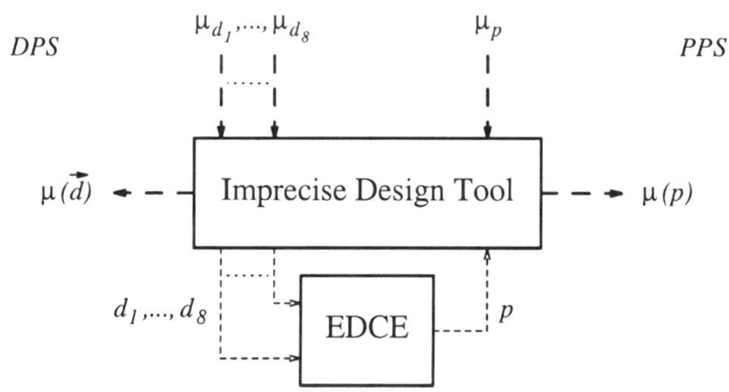

Figure 14 The Imprecise Design Tool and EDCE.

of innovation in eight subsystems of the new engine to be developed were chosen to be design parameters $d_1, \ldots, d_8$. EDCE produces a single output p: an estimate of the development cost for the new engine. $p = f(\vec{d})$ is continuous and monotonic: an incrementally larger percentage change in a subsystem results in an incrementally larger development cost. Recall that for $\mathcal{P} = \min$ and f continuous and monotonic, the α–cuts given by the LIA are exact. IDT uses the LIA to forward calculate $\mu_d(p)$ on $\mathcal{Y}$.

The user specifies designer preferences (μ_{d_i}) at each point in $\mathcal{X}_i$, $i = 1, \ldots, 8$, as an array of numbers. Since the LIA discretizes $\mu_{d_i}(\vec{d})$ into M levels $\alpha_1, \ldots, \alpha_M$, only these values should be specified. For IDT, $M = 10$ and $\alpha_k = 1.0, 0.9, \ldots, 0.1$: it is questionable whether a designer could (or should) distinguish more than ten levels of preference. The functional requirement on $\mathcal{Y}$ is specified as an ordered list of pairs (p, μ_p), which define a piecewise linear preference function. The first and last pairs in the list are extended to infinite values of p.

For two α–cuts D_{α_k} and $D_{\alpha_{k+1}}$ where $\alpha_k > \alpha_{k+1}$, $D_{\alpha_k} \subseteq D_{\alpha_{k+1}}$, since for all $\vec{d} \in D_{\alpha_k}$, $\mu_d(\vec{d}) \geq \alpha_k > \alpha_{k+1}$. But if $D_{\alpha_k} = D_{\alpha_{k+1}}$, the forward calculation need only be performed once, to obtain P_{α_k}, since $P_{\alpha_{k+1}}$ contains the same points but has lower preference. Hence $D_{\alpha_{k+1}}$ may be ignored. IDT uses a lookup table for $f(\vec{d})$

to avoid repeated evaluations of f for the current design calculation and subsequent iterations. The program evaluating f is called only once during the forward calculation for all the required points.

The calculated P_{α_k}, which represent $\mu_d(p)$, are combined with $\mu_p(p)$ to produce $\mu_o(p)$, which is saved as an ordered list of pairs (p, μ_o) defining a piecewise linear preference function on $\mathcal{Y}$. In the same step, the peak preference μ^* and the peak preference set of performance $\mathcal{Y}^*$ is also found.

After finding μ^*, IDT determines α^*, the largest $\alpha_k \leq \mu^*$, so that $\mathcal{X}^* \subseteq D_{\alpha^*}$. $\mu(\vec{d})$ is calculated at every $\vec{d} \in (D_{\alpha^*} \cap \mathcal{X}_D)$ and any $\mu(\vec{d}) = \mu^*$ are identified as peak preference design configurations. Where $\vec{d} \notin \mathcal{X}_D$ and $f(\vec{d})$ is not immediately available, $\mu_d(\vec{d})$ provides an upper bound since, for a non-compensating strategy, $\mu_o \leq \mu_d$. We now know $\mu(\vec{d})$ or an upper bound for $\mu(\vec{d})$ at every $\vec{d} \in D_{\alpha^*}$ that could potentially be a peak preference design configuration. A designer may also wish to visualize the variation of μ_o on the DPS, so IDT allows the user to specify points about which eight 1D cross–sections of μ_o in one design parameter, or four 2D cross–sections of μ_o in two design parameters, are generated.

2.3.1.1 Example. Consider estimating the cost of developing a new aircraft engine. Preference for cost is represented by a ramp rising linearly from $\mu_p = 0$ at $p = \$0$ million to $\mu_p = 1$ at $p = \$145$ million, with $\mu_p = 0$ elsewhere (representing a budgeted amount for new engine development of \$145 million, with a desire to spend the full development budget, see Figure 15). Given the requirements of the prospective customers, the design team has agreed on a set of designer preferences for the degree of innovation in each of its eight subsystems, shown in Figure 17. It is expected that only support-engineering will be required for the bearing and lubrication subsystem, and hence μ_{d_3} is non-zero only for a change value of 10%. Greater imprecision is associated with the high pressure turbine (μ_{d_5}). The design team has also decided to pursue a non-compensating design strategy.

$\mu_d(p)$ and $\mu_o(p)$ obtained from the forward calculation are shown in Figure 15. Note that the development costs shown are representative only, and do not reflect actual General Electric Aircraft Engines data. The peak preference in the PPS $\mu^* = 1.000$, at a development cost of \$145 million. The corresponding point in the DPS is not defined, and $\mu^{*\prime} = 0.998$, at $\vec{d^*} = (4, 2, 2, 3, 3, 4, 4, 3)$, the peak preference design. Other designs ($\vec{d} \in D_{\alpha^*}$) with near peak preference are listed in Table 1. Points indicated with asterisks are also on the 1D and 2D cross–sections in Figures 17 and 19. The ten points in Table 1 and $\vec{d^*}$ are the highest preference designs and may be considered to be the solution set. All other designs $\vec{d} \notin D_{\alpha^*}$ have $\mu(\vec{d}) \leq \mu_d(\vec{d}) \leq 0.9$, since $\alpha^* = 1.0$ and the next highest value of μ_{d_i} specified is 0.9. The design team may now introduce other considerations to reduce the number of design solutions, or adjust the designer preferences and functional requirement preferences and run IDT again, or both.

2.3.1.2 Conclusion. This section summarized the MoI calculation procedures used by IDT, and presented a contemporary commercial engineering design example. IDT applies this formal method for representing and manipulating imprecision to a real

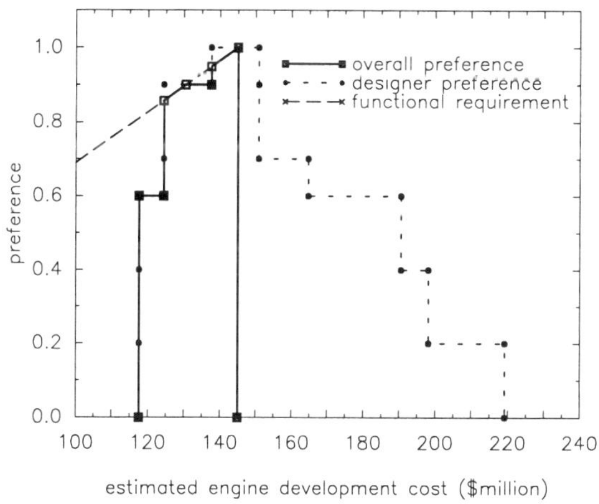

Figure 15 Preferences μ_o, μ_d, and μ_p on the PPS.

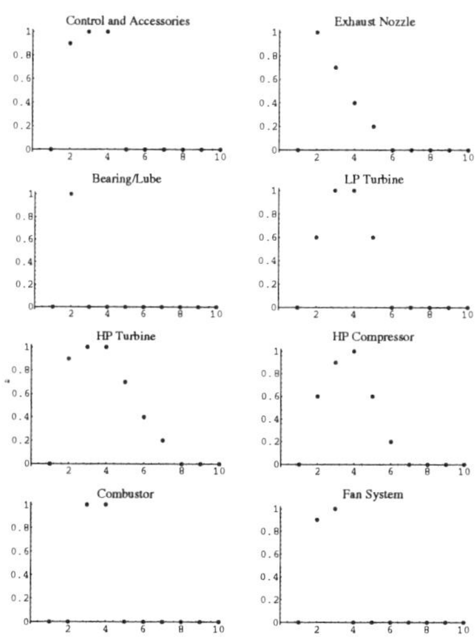

Figure 16 1D cross–sections of μ_d on the DPS.

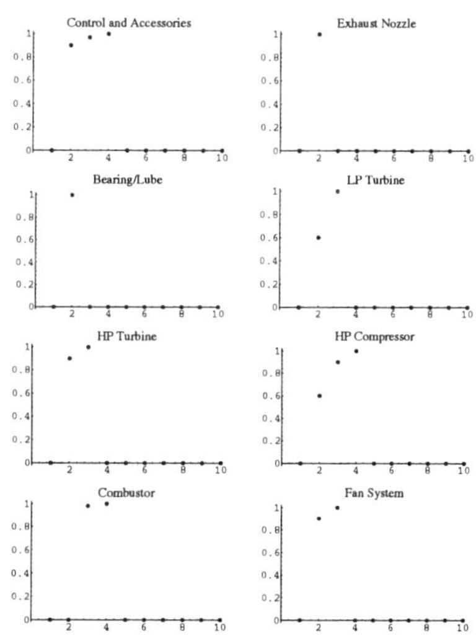

Figure 17 1D cross–sections of μ_o on the DPS.

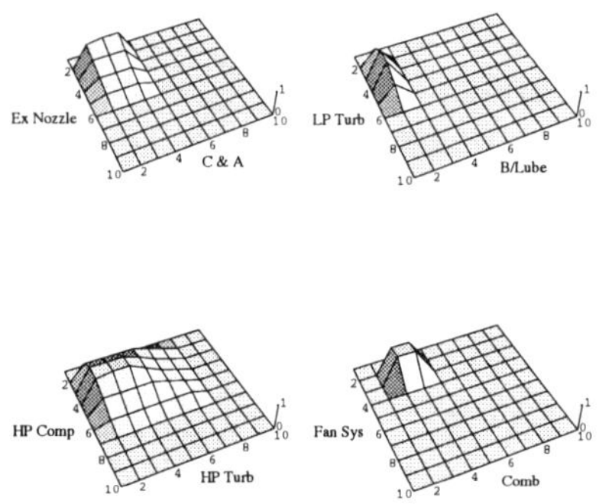

Figure 18 2D cross–sections of μ_d on the DPS.

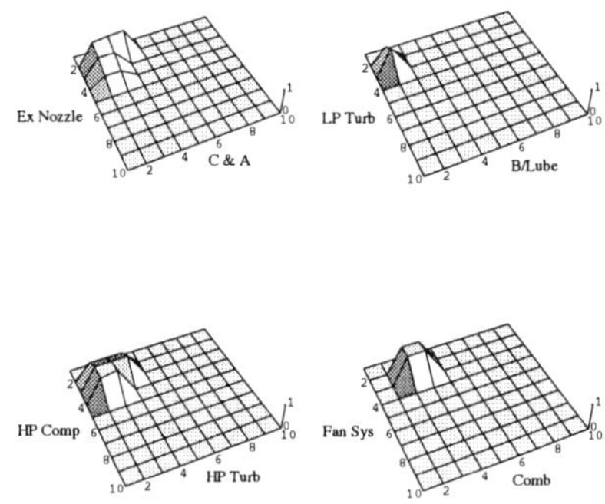

Figure 19 2D cross–sections of μ_o on the DPS.

$\vec{d}$	$\mu(\vec{d})$	$\vec{d}$	$\mu(\vec{d})$
$(3,2,2,3,3,4,3,3)$	0.949	$(3,2,2,4,3,4,4,3)$	0.985
$(3,2,2,3,3,4,4,3)$	0.968∗	$(3,2,2,4,4,4,3,3)$	0.983
$(3,2,2,3,4,4,3,3)$	0.966	$(4,2,2,3,3,4,3,3)$	0.979∗
$(3,2,2,3,4,4,4,3)$	0.985	$(4,2,2,3,4,4,3,3)$	0.996
$(3,2,2,4,3,4,3,3)$	0.966	$(4,2,2,4,3,4,3,3)$	0.996

Table 1 Points in D_{α^*} with μ_o close to $\mu^* = 0.988$.

engineering design problem, allowing the designer to explore and evaluate a design while it is still imprecisely described.

2.3.2 The KONWERK Workbench

2.3.2.1 General Introduction. KONWERK is a toolkit for conceptual design and configuration of complex systems, such as engineering and socio-economic systems. Its development was part of the PROKON (Günter, 1995), which was sponsored by the German Ministry of Research and Technology. A central requirement for tools like KONWERK is applicability in different domains. Thus its concepts must be universal. Yet every domain has special properties which should be used to solve problems. Therefore, the modules are divided into basic modules, and extension modules which either enhance the abilities of basic modules or provide new abilities.

The basic modules cover the following general tasks in configuration and design systems:

- representation of domain objects,
- representation and processing of relations, constraints and heuristics,
- formulation of the configuration task,
- control of the configuration.

We will now introduce the modules (basic and extension modules) which form the kernel of KONWERK. The basic modules are necessary in most applications. They offer standard methods for the main tasks in configuration systems. If required by the domain, they may be complemented by extension modules. The basic modules of KONWERK are:

- *Conceptual* hierarchy for the representation of domain object
  Domain objects and their properties are represented in a conceptual hierarchy which comprises a taxonomic (is-a) and a compositional (part-of) hierarchy. The possible values for the properties are predefined with help of so-called object descriptors.
- *Constraints* for the representation of relations
  For the representation of configuration restrictions and their processing we use a constraint system. By the use of constraint propagation it is possible to restrict values and to discover inconsistencies.
- *Control*
  The control is based on a depth-first search and supports chronological backtracking. The control knowledge is represented in several *strategies* which control the sequence of configuration steps chosen from an agenda.
- Specification of the configuration task
  A configuration task in KONWERK is specified by the selection of a target object which has to be configured. It is possible to select further objects (components of the target object) and to specify their properties (slots).

Extension modules enlarge the functionality of KONWERK in several ways. KONWERK is be seen as a toolbox with a variety of modules which the user can select to build his tool in a flexible way. Conceptually, extension modules extend the basic modules. On the implementation level, they are often complex modules which contain all the functionality of a basic module in addition to many functional extensions of this basic module.

In Table 2 a survey of the basic and extension modules of KONWERK is given. This table shows that KONWERK is able to deal with imprecision in design (fuzzy related modules), and contains a MADM toolbox. We will explain in the following some ideas of the implementation of fuzzy sets in KONWERK. For more details see (Müller and Sebastian, 1997; Sebastian and Antonsson, 1996).

2.3.2.2 Fuzziness in KONWERK. Section 2.3.2.1 briefly mentioned several possibilities for including fuzziness in KONWERK. Here more details are presented of how imprecise properties of objects might be modeled and processed within KONWERK. The user is able to:

- use a fuzzy set instead of a crisp number while parameterizing an attribute (restricting the slot value of a parameter),
- generate imprecise classes (concepts) on the basis of imprecise properties of the objects,
- solve so-called "selection problems" by fuzzy MADM-methods (in case one has a finite number of explicitly given alternatives and imprecise attributes).

The first of these possibilities will be explained in the following, using the KONWERK syntax as necessary.

Imprecise Attributes. The objects defined in the knowledge-base of a KONWERK application have "normally" crisp properties. With the extension"*linguistic variable*" it is possible to use "linguistic terms" (terms of the linguistic variable) instead of crisp values. This concept is introduced with a simple example shown in Table 3.

The syntax used is: *def-do.name car* means definition of a domain-object which has the name *car*. To be able to find the object in an object hierarchy, we declare that *car* is a domain-object using the slot descriptor *superconcept*.

Also the slot-names *performance*, *speed*, *etc.*, their ranges and their type descriptions, are self-explanatory (PS means horse power, m/h miles per hour, *etc.*). Let's now consider the linguistic variable *style* in more detail. It might be defined by one of three terms: bad, medium, good. In Table 4 a possible graphical definition of the fuzzy sets "bad", "medium", and "good" is represented. Of course, one has to define this linguistic variable in KONWERK. We do this using the syntax shown in Table 4.

Mapping describes the three membership functions in Figure 20 which define the linguistic terms. The user of KONWERK can choose for the variable *style* either any value in [0100], a subinterval of this interval, or one of the linguistic values *bad*, *medium*, or *good*.

The above has illustrated only the most basic concepts of fuzziness in KONWERK. More details can be found in (Müller and Sebastian, 1997; Günter, 1995).

Task	basic modules	extension modules
• Representation of domain objects	• conceptual hierarchy	• views and mixins in the conceptual hierarchy • fuzzy properties in the conceptual hierarchy • linguistic values in the conceptual hierarchy • spatial properties (2D and 3D) • fuzzy is-a hierarchy • measures
• representation and processing of relations	• constraints	• uni-directional functions • rules for the representation of heuristics • constraints with fuzzy variables • spatial relations
• control	• depth-first search with backtracking	• case based configuration • exploration based configuration • basic optimization module • optimization module MADM • linear optimization • non-linear optimization with constraints • branch-and-bound optimization • knowledge based backtracking • knowledge based backtracking with data acceptance • repair instructions • complete search in an explicit solution space • ATMS-based configuration
• task specification	• target object, components and their properties	• requirement modelling (functional task specification, fuzzy task specification, goals and wishes for the optimization)
• special extensions	•	• integration of simulation methods • compilation of constraints • interface to data base systems • interface to CAD systems

Table 2 Survey of the basic and extension modules

```
(def-do  : name car
         : superconcept domain-object
         : parameters ( (performance      [0PS 800PS]      :basicunit "PS")
                        (speed            [0m/h 300m/h]    :outputunit "m/h")
                        (price            [0$ 300000$]     :basicunit "$")
                        (style            [0 100]          :ling-var ´style)
                        (luggage-space    [0.0 1.0]        :ling-var ´size-ls)
                        (safety           [0.0 1.0]        :ling-var ´safety))
         : relations    ((self: nop: optimization-problem ´selection-car)))
```

Table 3 KONWERK example

```
(def-linguistic-variable: name style
         : mapping      ((#!bad         {<0 10 1.0 10 40>})
                        (#!medium       {<10 40 1.0 60 90>})
                        (#!good         {<60 90 1.0 90 100>})))
```

Table 4 KONWERK syntax

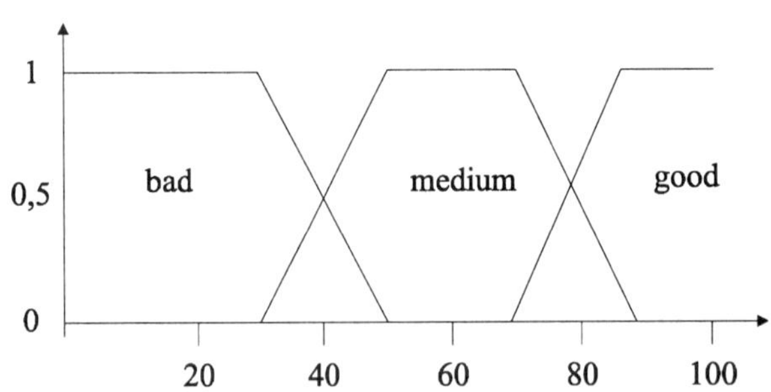

Figure 20 Linguistic variable "Style" (of object car)

2.4 FUZZY SETS IN ENGINEERING DESIGN – APPLICATIONS

2.4.1 Preliminary Vehicle Structure Design

The general vehicle structure design problem is the engineering of a *body-in-white*, which consists of the (usually metal) frame to which components and exterior panels are fastened. While there are interesting alternative solutions such as space frames and monocoques, this paper is concerned with the welded metal structure typical of automobiles of the present day (see Figure 22). The vehicle structure engineers must design a body-in-white that meets certain measurable engineering targets such as stiffnesses, stress levels under load, and weight. In addition, they must satisfy many performance targets associated with less easily measured concepts such as style, manufacturability, and requirements of other engineering groups involved in the design process. These unmeasured performances are handled informally, often by negotiation between groups working on the same vehicle. The Method of Imprecision, or M_oI (see Section 2.2.1) was developed to allow for a formal approach to the incorporation of this imprecise information.

An application of the M_oI to vehicle structure design was demonstrated, using a 1980 VW Rabbit (see Figure 21) as a benchmark for the design. The vehicle was stripped to the structural body-in-white, and torsional and bending stiffnesses were measured. The intact body-in-white was found to have a torsional stiffness of approximately 4900 N-m/degree and a bending stiffness of approximately 2500 N/mm. In addition, geometric data were gathered and used to create a solid model (Figure 22). The solid model and the structural stiffness information together were used to create a finite element model (Figure 23).

The finite element model was parameterized with five design variables, listed here with their units and design ranges:

1. A-pillar thickness (0.7–1.1 mm)
2. B-pillar thickness (0.8–1.5 mm)
3. floor pan thickness (1.0–1.4 mm)
4. floor rail thickness (0.8–1.2 mm)
5. B-pillar location (-50–200 mm aft of a nominal point chosen by stylists)

and the performance was assessed with three measures:

1. Bending stiffness (N/mm)
2. Torsional stiffness (N-m/deg)
3. Weight (kg)

The demonstration here was conducted using a subset of the design variables; the method can be applied directly to a larger set of variables. The stated design problem was to achieve 10% improvements over the reference model in the three measured performances. In addition, it was understood that that the design must not be difficult to manufacture, and that this year's model should have a somewhat longer and sleeker look.

Figure 21 1980 VW Rabbit in stiffness testing

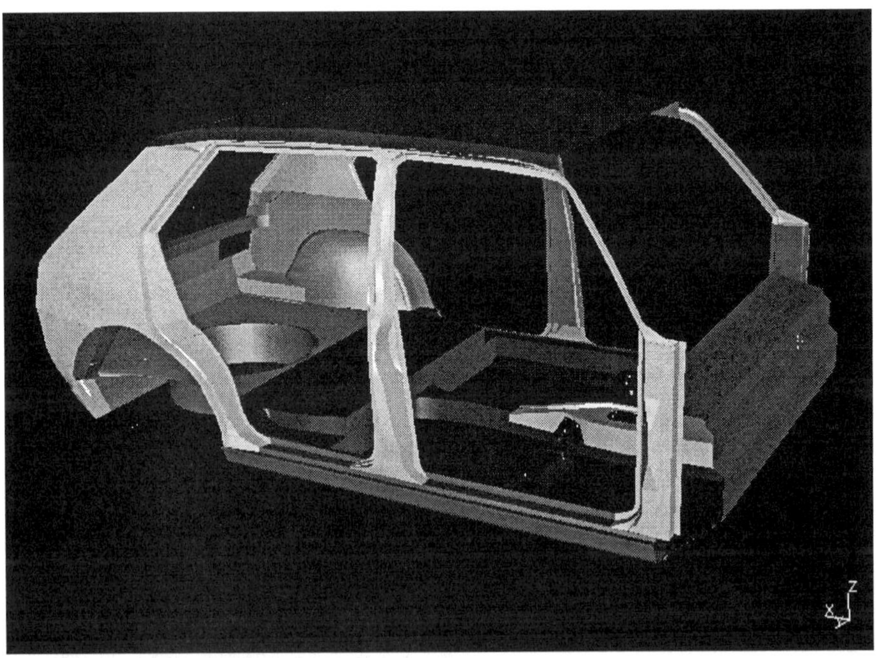

Figure 22 Geometric model of body-in-white in SDRC I-DEAS

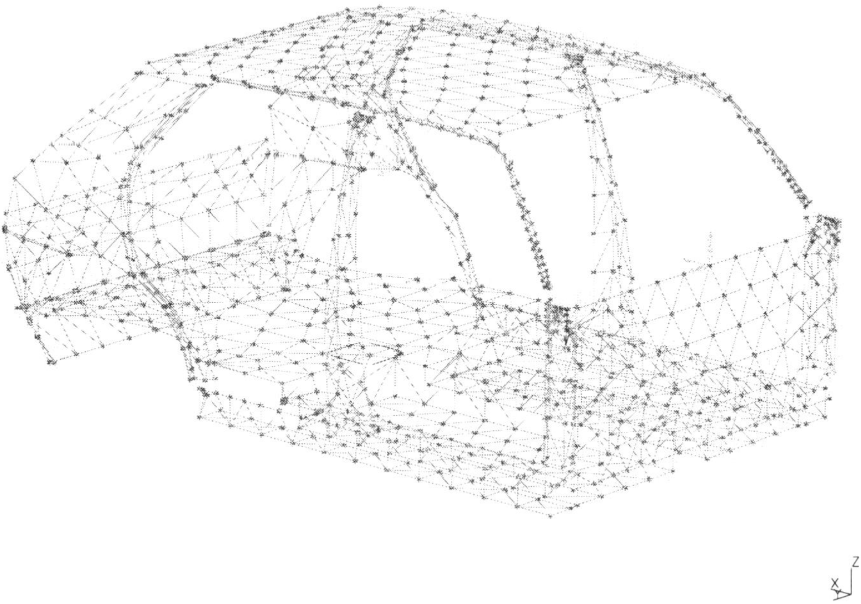

Figure 23 Finite element model of body-in-white

2.4.1.1 Applying the MoI to Include Imprecise Information. While standard optimization methods could be used to attempt to determine the highest achievable bending stiffness, the highest achievable torsional stiffness, or the lowest achievable weight for this analysis model, such an optimization would not tell the designer which designs are the most promising when other relevant considerations are taken into account. On the one hand, there is a necessary trade-off between the stiffnesses and the weight; it is impossible to optimize both simultaneously. Additionally, there is other (imprecise) information to consider when making the decision, such as manufacturing and styling concerns. The application of the MoI to this problem involves constructing a different "optimization" problem that includes the imprecise information that would be left to the negotiation stage in traditional design.

The calculated performance requirements on bending stiffness, torsional stiffness, and weight were originally expressed as targets of 10% improvements over the reference model. The MoI was developed to avoid dependence on such unrealistically, and indeed unproductively, precise targets. In place of these hard targets, imprecise performance requirements were specified with a linear interpolation between two points. In the implementation of the MoI, it is common to name the customer as the source of the performance preferences; in fact, it is more likely to be a manager, perhaps informed by market research, serving as the customer's proxy. To specify these imprecise requirements, the manager must answer two simple questions: "What is the lowest performance you can live with (where is $\mu = 0$)? What performance would satisfy you completely (where is $\mu = 1$)?" These bounds are clearly dependent on a number of factors, including the target market and the performance of competitors'

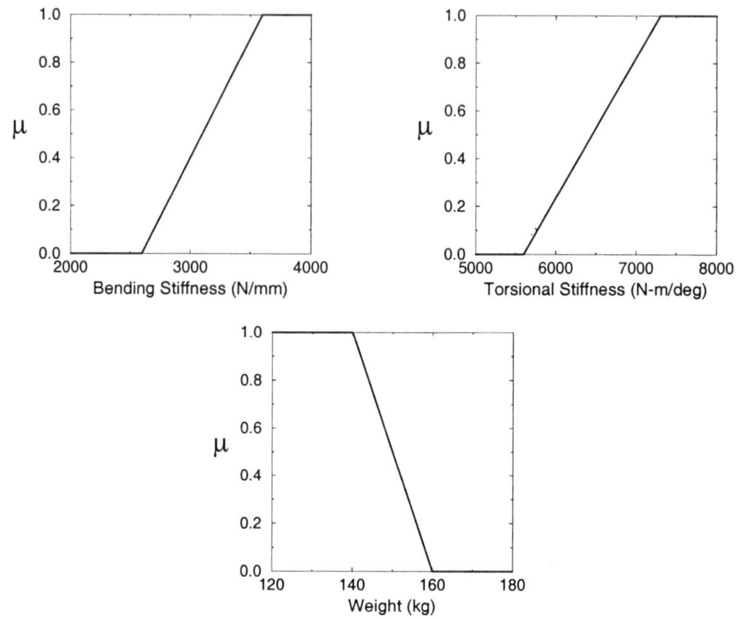

Figure 24 Imprecise Performance Requirements

products; we have found that managers can answer these two questions with little more effort than is needed to settle on the initial crisp target. Figure 24 shows the imprecise requirements on stiffnesses and weight.

To include requirements on manufacturing, availability, style, and other things which are not calculated in the finite element analysis, designer preferences are specified on the design variables. As with the imprecise performance requirements, they range from $\mu = 0$ at the unacceptable limit to $\mu = 1$ at the most preferred. A preference is defined on each of the five design variables. Each preference is representative of imprecise information that can be incorporated using the MoI:

1. The sheet steel for stamping the A-pillar is only available in 0.1 mm increments, so this preference is discrete rather than continuous. The manufacturing engineer has a higher preference for thinner sheets (from $\mu = 1$ at 0.7 mm down to $\mu = 0.4$ at 1.1 mm), since they are easier to form; this is a design preference for manufacturability.

2. The B-pillar thickness is continuous and decreasing over the range. Such a preference does *not* indicate that the physical B-pillar might be 1.113 or 1.114 mm thick; rather it means that the designer knows that the finite element model is simplified, and that a high number for B-pillar thickness means that more reinforcing features will need to be added to the B-pillar. The designer would like to keep the B-pillar as simple as possible.

3. The floor pan thickness is preferred thicker by the designer for ease of attachments and for durability.

4. The floor rail thickness preference is an example of a sourcing, or availability preference; it states that some thicknesses are more easily obtained than others.

5. The design preference for B-pillar location comes from the stylists, and captures the directive for a longer, sleeker look for this year's model. It has been specified using *α-cuts* (refer to Section 2.2.1.6). The stylists have indicated a range of perfectly acceptable values (25–75 mm), a range of barely acceptable values (-50–200 mm), and a range of values that fall in the middle (0–150 mm). The α-cut method of specifying preferences has computational advantages.

In addition to these preferences, each attribute is assigned a weight indicating its relative importance, and the way in which attributes trade-off against each other must also be specified. In this test example, it was determined that bending and torsional stiffness traded-off in a non-compensating manner — the lowest preference is maximized. Together they traded-off with weight in a compensating manner, so that high performance on stiffness could partly make up for low performance on weight, and vice versa. The designer preferences all traded-off in a compensating manner as well, with the styling preference assigned a relatively high weight to reflect the importance of styling considerations in automobile design. The preferences for the computed performance variables are also weighted heavily.

Through the application of the MoI, the design problem has been reformulated to be the maximization of the (scalar) overall preference $\mu_o(\vec{d})$. The computation of $\mu_o(\vec{d})$ for a single design point $\vec{d}$ is limited by the finite element stiffness calculation, which takes about a minute on a Sun Ultra1-170MHz workstation; the calculations of weight and preference aggregation are of negligible cost. Even in this relatively modest problem, where there are only five design dimensions, an exhaustive calculation of preferences over the design space is prohibitively expensive. The MoI exploits the structure of the problem to speed the search for preferred solutions.

2.4.1.2 Results. The design problem, including all imprecise information, was solved in two different ways. First, in order to demonstrate the method, the finite element analysis was run 3125 times to provide a coarse but complete check of the entire design space. The point of peak overall preference of $\mu_o = 0.50$ was found at $\vec{d} = (1.0, 0.9, 0.9, 1.0, 50)$, where the design preferences μ_d are $(0.6, 1.0, 1.0, 0.5, 1.0)$; the stiffnesses and weight at this point were $\vec{p} = (2832, 5836, 147)$, with preferences $(0.23, 0.14, 0.62)$. The maximum achievable stiffnesses are 3365 N/mm ($\mu_p = 0.77$) in bending and 6029 N-m/degree ($\mu_p = 0.25$) in torsion, but the corresponding weight of 170 kg is unacceptable. Similarly, a weight of 144 kg ($\mu_p = 0.78$) is achievable, but stiffnesses drop to 2803 N/mm ($\mu_p = 0.20$) and 5730 N-m/degree ($\mu_p = 0.08$). The combined overall preference μ_o also takes into account the design preferences on style, manufacturability, and the like.

The power of the method lies not in an ability to find a single overall "best" point, but in the information it contains of how the total combined preference μ_o varies with each of the design variables. Although it is impossible to display all five dimensions varying at once, a tool was written that uses a commercial package (Matlab) to display results interactively. Using the tool, the designer can see the change in preference that would occur by varying each design variable independently from a chosen beginning point. Results can be seen on five simultaneous plots in two dimensions (see Figures 25), or on a three-dimensional surface plot (see Figure 26) with the remaining design variables set to nominal values.

Approximations. Naturally, the exhaustive evaluation of points in the design space would not be performed on a real design problem. It was performed here *only* for comparison purposes. An approach that utilizes Design of Experiments (DOE) (Barker, 1985) to approximate the finite element calculations for bending and torsional stiffnesses reached substantially similar results in only 21 runs (or approximately 20 minutes). The average difference (from the exhaustive evaluation) in bending stiffness was approximately 1%, with a maximum difference of less than 4%, while the average and maximum differences for torsional stiffness were both less than 1%.

In some cases, the nonlinearities of the analysis function f will defeat a linear or even polynomial approximation, but in many cases, such as the example presented here, these simple approximations can drastically reduce the required computation. Since precise answers are not required for preliminary design, it is sensible to exploit approximation tools when possible. If more computation can be justified, a more thorough calculation can always be made.

2.4.1.3 Discussion. Engineering analysis usually requires some judgment on the part of the designer. Unless a full-scale exact prototype is to be built and tested, the accuracy of any calculated performance measure depends on the fidelity of the model employed. Even when exact data are available for some attributes, final decisions about a design incorporate other, unmodeled concerns, such as manufacturing and styling.

The M_oI constructs a model of the entire decision process, expressing the calculated overall performance $\mu_o(\vec{d})$ as a function of the design variables. It depends on many factors: the function f for calculating measurable performances, the specification of design preferences μ_d and performance preferences μ_p, the weighting of these preferences, and the specification of trade-off aggregations between attributes. A change in any of these will affect the shape of the function μ_o in design space, and thus affect the decision. The analysis f is here relatively expensive to compute, and changes in the finite element model are costly to propagate to overall preference. Changes in the other factors, on the other hand, are easily incorporated, as finite element results $f(\vec{d})$ are stored so that the same design point need never be analyzed more than once. This allows the M_oI to support an iterative decision process, when the information from the first round of calculations inspires a change in the preference structure.

In the example shown in this paper, the shape of the function $\mu_o(\vec{d})$ is sensitive to changes in the styling preference $\mu_d(d_5)$, which is not surprising, since this preference is accorded a large weight. This and other features of the design problem can be seen in the advanced interface shown in Figure 25. The vertical dashed (red) lines indicate the selected values of the design variables. In this figure, $\vec{d} = (0.8, 1, 0.9, 1.2, 50)$. The solid (blue) lines indicate how the overall preference μ_o would change by varying that

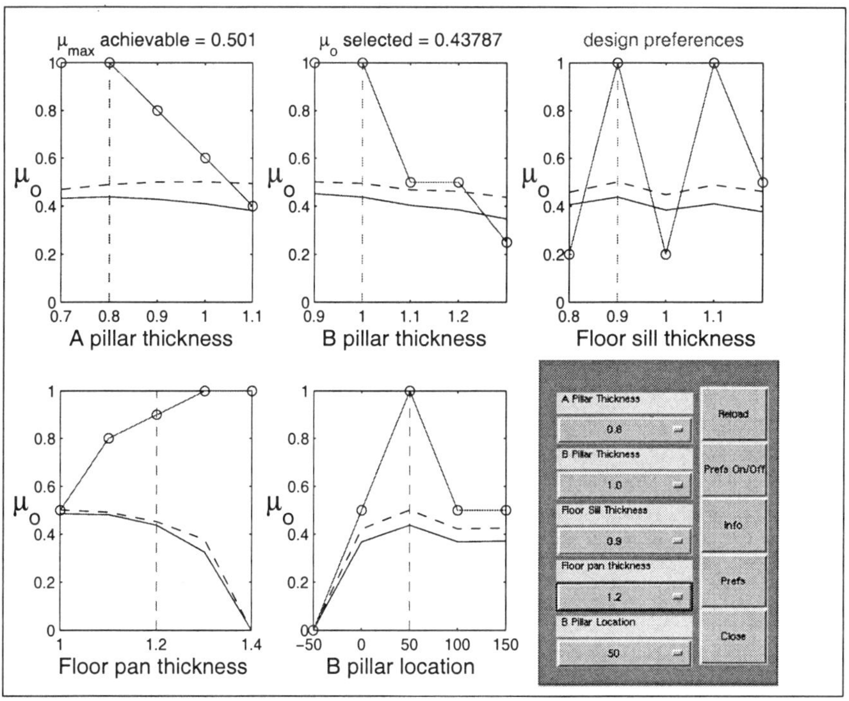

Figure 25 Graphical User Interface for Preference Display

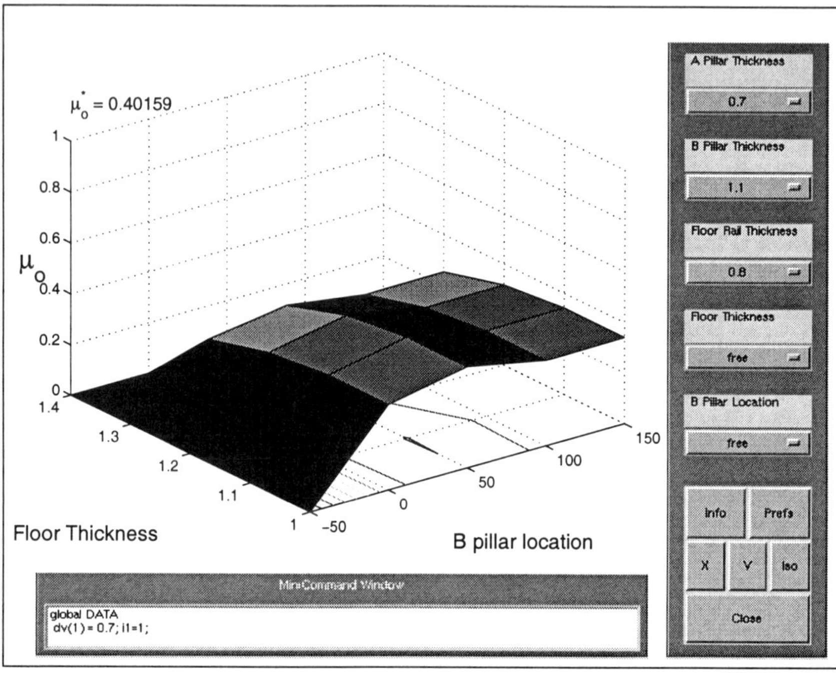

Figure 26 3-D Graphical User Interface for Preference Display

design variable while holding the other four fixed at their current values. The dashed (black) lines show the maximum achievable μ_o for each value of each design variable. Finally, the solid (red) lines joining the circles are the specified preferences on design variables. Thus Figure 25 shows that the overall preference μ_o varies qualitatively, though not quantitatively, as four of the five design variables; the exception is d_4, floor pan thickness, where the overall preference μ_o tends to decrease as the designer preference $\mu_d(d_4)$ increases. The designer can interact with the preference display to examine trends in the structure of the overall preference.

2.4.2 Preliminary Design and Configuration of Future Space Launch Systems

In space flight a change of paradigm has taken place in recent years. While in former times national prestige and the "cold war" were the main reasons for space activities, today economical aspects gain more importance. Furthermore, environmental compatibility and other requirements act not only as aims but as restricting arguments. The task of space launch systems (SLS) is to provide the transportation demand for orbital applications such as positioning satellites in orbit, and planetary missions. SLS are the key technology for all space activities. Today a large number of different systems and a large body of experience are available (Figure 27). Future space launch system

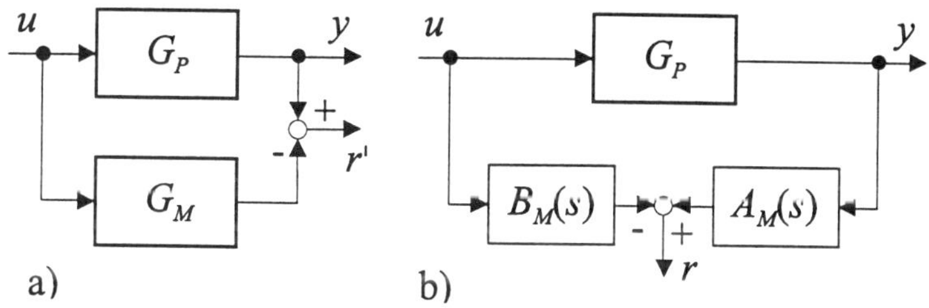

Figure 14. Parity equation methods
a) output error b) equation error

However, for SISO processes only one residual can be generated, and it is therefore not easy to distinguish between different faults.

More freedom in the design of parity equations can be obtained if for SISO processes intermediate signals can be measured (Fig. 4), or for MIMO processes. Then a state-space model is appropriate, as shown by Gertler (1995) for discrete-time models. For a continuous-time model

$$\dot{\boldsymbol{x}}(t) = \boldsymbol{A}\boldsymbol{x}(t) + \boldsymbol{B}\boldsymbol{u}(t) \tag{46}$$

$$\boldsymbol{y}(t) = \boldsymbol{C}\boldsymbol{x}(t). \tag{47}$$

Eq. (47) is inserted in eq. (46) and differentiated p times, resulting in the equation system

$$\begin{bmatrix} y(t) \\ \dot{y}(t) \\ \ddot{y}(t) \\ \vdots \end{bmatrix} = \begin{bmatrix} C \\ CA \\ CA^2 \\ \vdots \end{bmatrix} \boldsymbol{x}(t) + \begin{bmatrix} 0 & 0 & 0 & \cdots \\ CB & 0 & 0 & \\ CAB & CB & 0 & \\ \vdots & & & \end{bmatrix} \begin{bmatrix} u(t) \\ \dot{u}(t) \\ \ddot{u}(t) \\ \vdots \end{bmatrix} \tag{48}$$

$$Y_F(t) = \boldsymbol{T}\boldsymbol{x}(t) + \boldsymbol{Q}\boldsymbol{U}_F(t).$$

In order to remove the non-measurable states $\boldsymbol{x}(t)$, eq. (48) is multiplied by $\boldsymbol{W}$, such that

$$\boldsymbol{W}\boldsymbol{T} = \boldsymbol{0}. \tag{49}$$

This leads to residuals

$$\boldsymbol{r}(t) = \boldsymbol{W}\boldsymbol{Y}_F(t) - \boldsymbol{W}\boldsymbol{Q}\boldsymbol{U}_F(t) \tag{50}$$

compare Fig. 15.

The derivatives in $\boldsymbol{Y}_F(t)$ and $\boldsymbol{U}_F(t)$ can be obtained by digital state variable filters for orders $n \leq 3$. The design of the matrix $\boldsymbol{W}$ gives some freedom to generate a structured set of residuals. One possibility is to select the elements of $\boldsymbol{W}$, such that one measured variable has no impact on a specific residual. Then this residual remains small in the case of an additive fault on this variable, and the other residuals increase.

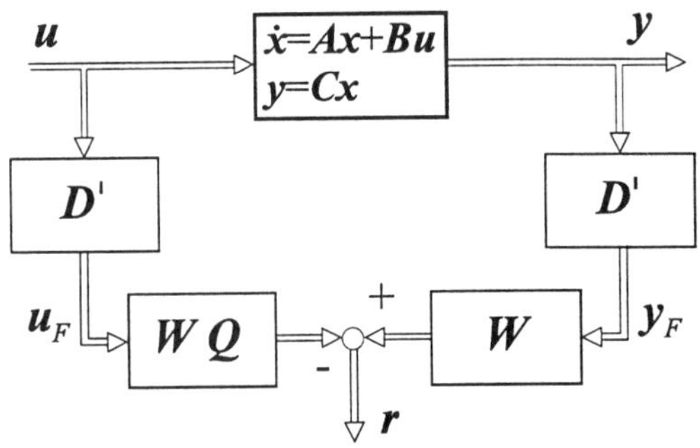

Figure 15. Parity equation methods for a MIMO state-space model; ***D'***: differentiation filter

3.3.5 Fault detection with signal models

Many measured signals $y(t)$ show oscillations that are of either harmonic or stochastic nature, or both. If changes in these signals are related to faults in the process, actuator or sensor, a signal analysis is a further source of information. Especially for machine vibration, sensors for position, speed or acceleration are used to detect, for example, imbalance and bearing faults (turbo machines), knocking (Diesel engines) or chattering (metal-grinding machines). But also signals from many other sensors, like electrical current, position, speed, force, flow and pressure, may show oscillations with a variety of frequencies higher than the usual process dynamic responses. The extraction of fault-relevant signal characteristics can in many cases be restricted to the amplitudes $y_0(\omega)$ or amplitude densities $|y(i\omega)|$ within a certain bandwidth $\omega_{min} \leq \omega \leq \omega_{max}$ of the signal by using bandpass filters , Stearns (1975).

Other possibilities are the determination of autocorrelation functions or the Fourier Transform (especially in the form of the Fast Fourier Transform, FFT) or the spectral density, Barschdorff (1987). Correlation functions and spectral densities are especially suited to separate stochastic and periodic signal components. If the frequencies are known, cross-correlation with, e.g., the known basic and higher-order harmonics can be performed.

However, a correlation function or a frequency spectrum forms a nonparametric representation of the signal, and is (both for harmonic and stochastic signals) not well-suited to direct automatic fault detection if the frequencies are unknown. Therefore, parametric signal models should be preferred, which allow the main frequencies and their amplitudes to be directly estimated, and which are especially sensitive to small frequency changes. This is possible by modelling the signals as a superposition of damped sinusoidals in the form of discrete-time ARMA (autoregressive moving average) models, Burg (1968). By means of a two-step maximum-entropy estimator via correlation functions, a selected number of distinct, unknown frequencies ω_j can be estimated as well as their amplitudes y_{0j}, Neumann (1991), Janik (1993). Symptoms are then generated by changes of estimated frequencies and amplitudes.

3.3.6 Change detection and symptom generation

The measured or estimated quantities, such as signals, parameters, state variables or residuals are usually stochastic variables $S_i(t)$, with mean value and variance

$$\bar{S}_i = E\{S_i(t)\} \; ; \; \bar{\sigma}_i^2 = E\{[S_i(t) - \bar{S}_i]^2\} \tag{51}$$

as normal values for the non-faulty process. *Analytic symptoms* are then obtained as changes

$$\Delta S_i = E\{S_i(t) - \bar{S}_i\}; \; \Delta\sigma_i = E\{\sigma_i(t) - \bar{\sigma}_i\} \quad t > T_F \tag{52}$$

with reference to the normal values. Usually, the time instants T_F of fault occurrences are unknown.

To separate normal from faulty behaviour, usually a fixed threshold

$$\Delta S_{tol} = \epsilon \bar{\sigma}_S \qquad \epsilon \geq 2$$

has to be selected. By this means, a compromise has to be made between the detection of small faults, and false alarms because of short time violations, Fig. 16a). Methods of change detection, e.g., a likelihood-ratio test or Bayes decision, a run-sum or a two-probe t-test, Basseville (1985), may improve the (binary) decision.

Another possibility is to represent the feature as fuzzy set $\mu_S(S)$ (e.g. a triangle, with the center as the mean and lower width $\Delta = \kappa\sigma_S$, $\kappa = 2,3,\ldots$, Fig. 16b). By matching the current value $\mu_{S'}(S)$ with the feature´s membership function, the "increased" $\mu_{S+}(S)$,

$$\mu_S = \max_S [\min(\mu_{S'}(S), \mu_{S^+}(S))], \tag{53}$$

one obtains a gradual measure for exceeding a threshold, a *fuzzy threshold*, Isermann (1995).

The generation of analytic symptoms concludes the task of model-based fault detection, Fig. 2.

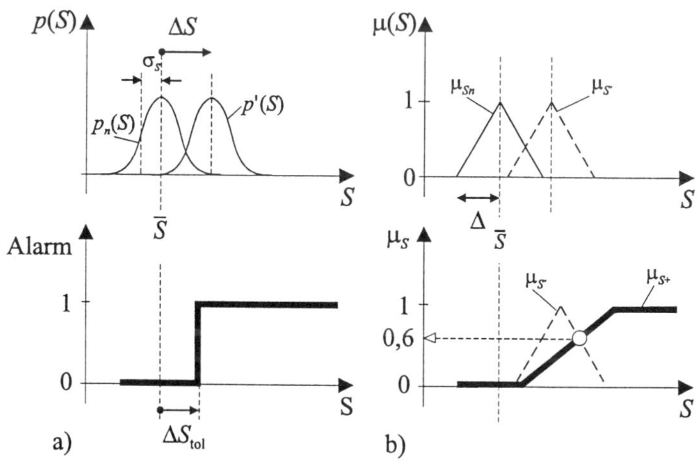

Figure 16. Symptom detection of simple features
a) Fixed threshold b) Fuzzy threshold

3.3.7 Comparison of fault-detection methods

The different approaches for model-based fault detection result in different properties with regard to the detection of specific faults. For the applicability the following aspects have to be considered.

- mapping of real faults into the generated residuals
- speed of fault development (abrupt, incipient)
- a priori knowledge of the model (structure, parameters)
- excitation through input signals (static, dynamic)
- extent of information for a (deep) fault diagnosis

Parameter estimation methods are especially suitable for the detection of multiplicative faults in the form of parameter changes. These changes happen mainly in processes and actuators. Parameter estimation yields also several symptoms for single-input single-output processes. They can be directly used for some classes of nonlinear processes (e.g. Hammerstein-structure) and also for static nonlinear characteristics. However, they require an excitation of the process input with a certain frequency spectrum. This is no problem for many processes with input changes in normal operation, as e.g. servo systems, actuators, machine tools, vehicles, etc. For processes with mainly steady state operation, as e.g. chemical and industrial processes, a small excitation signal may be permitted if a fault is suspected. Parameter estimation methods show a unified, modular structure and can therefore easily be configured. Only the calculation of process coefficients depends on the very processes (if required).

Parity equations are especially suited for the detection of additive faults. They appear, e.g., for sensors and actuators and also processes, e.g., through leaks and short-cuts. However, the process model used must describe the process relatively precise. An advantage is the small computational effort and that for a part of the faults no input excitation is required. The parity equations are very sensitive to nonmeasurable process disturbances, which are not decoupled during design. Multiple residuals are only obtained for multi-measurable output signals.

State observers have similar properties as parity equations. They are especially suitable for additive faults. The design effort is relatively small for fault sensitive filters and relatively large and process dependent for the others.

A basic difference between the methods is, that the parameter estimation is designed to determine constant values out of disturbed signals and a state observer (and parity equations) to determine time variant values from undisturbed signals. Therefore the response of a parameter estimator for noisy signals is slower, but more precise. State observers react faster but with unreliable values (large variance). If a parameter estimation is designed for time variant parameters, e.g. by exponential forgetting or inclusion of a dynamic parameters model, [9], also parameter estimates follow faster on cost of noise suppression. Hence, the ability of fast following abrupt changes depends, as well for parameter estimation as for state estimation, on the design. For further detailed comparison of parity equations and parameter estimation see Höfling (1996).

3.3.8 Combination of different detection methods

As shown above parameter estimation on one hand and state estimation and parity equations on the other hand show advantages and disadvantages with regard to the detection of the various types of faults. Therefore, if all faults should be detectable, different detection methods should be integrated properly in order to mainly use their advantages. As in most cases the model parameters are unknown, it is quite natural to apply parameter estimation first. Then, the following combinations of model-based detection methods result, compare Fig. 17.

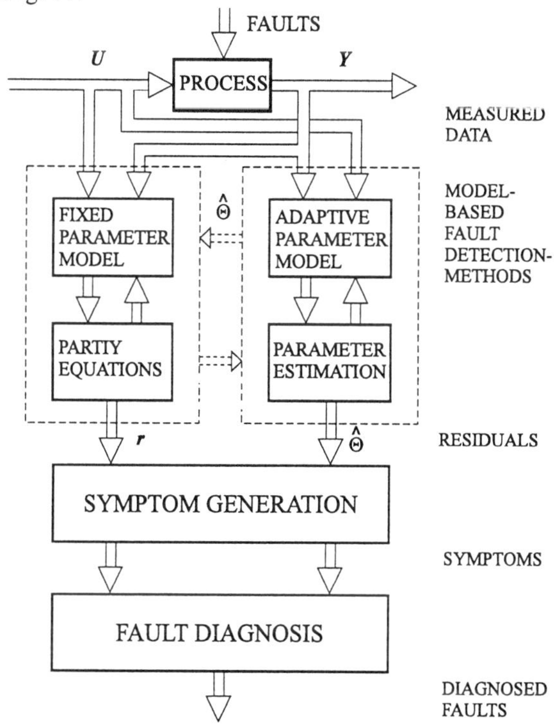

Figure 17. Combination of different model-based fault-detection methods for symptom generation (example)

I) Sequential parameter and state estimation

- parameter estimation to obtain the model
- state estimation for fast change detection
- parameter estimation (on request) for deep fault diagnosis

II) Sequential parameter estimation and parity equations, Höfling and Pfeufer (1994), Höfling (1996), compare Fig. 17.

- parameter estimation to obtain the model
- parity equations for change detection with less computations
- parameter estimation (on request) for deep fault diagnosis

III) Parallel parameter and state estimation

- for multiplicative and additive faults
- depending on input excitation

The way of integration depends very much on the process, the faults to be detected and the allowable computational effort. In some cases also the integration of process model-based and signal model-based detection methods gives a good overall information:

IV) Parameter estimation and vibration analysis

- parameter estimation for parameter mapping faults
- vibration analysis for other type of faults like unbalance, knocking, chattering
  (This is especially attractive for rotating machines)

3.4 FAULT-DIAGNOSIS METHODS

The task of fault diagnosis consists of the determination of the type of fault with as many details as possible, such as the fault size, location and time of detection. The diagnostic procedure is based on the observed analytic and heuristic symptoms and the heuristic knowledge of the process, as shown in Fig. 1.

In this section the heuristic part of the knowledge and inference mechanisms for diagnosis are described, in order to build on-line expert systems for fault diagnósis.

3.4.1 Symptom representation

The inputs to the knowledge-based fault-inferencing mechanism are all available symptoms as facts, and the fault-relevant knowledge about the process, mostly in heuristic form.

a) Analytic symptoms

The analytic symptoms S_{ai} are the results of the limit value checking of measurable signals, signal or process-model fault-detection methods and change detection, as described in Section 3.3.

b) Heuristic symptoms

Heuristic symptoms S_{hi} are the observations of the operating personnel in the form of acoustic noise, oscillations or optical impressions like colours or smoke, obtained by inspection. These empirical facts can usually only be represented in form of qualitative measures, e.g., as linguistic expressions like "little", "medium" or "much".

c) Process history and fault statistics

A third category of facts depends on the general status, based on the history (past life) of the process. This process history includes the information of previous running time, load measures, last maintenance or repair. If fault statistics exist (e.g., from "statistical process control") they describe the frequency of certain faults for the same or similar processes. Depending on the quality of these measures, they can be used as analytic or heuristic symptoms. However, the information on the process history in general is vague and their facts have to be taken as heuristic symptoms. The symptoms based on process history will be called S_{pi}.

The knowledge about the symptoms can be represented, e.g., in the form of data strings, and can include, for example: number, name, numerical value, reference value, calculated confidence or membership value, time of detection, explanatory text, Freyermuth (1991).

d) Unified symptom representation

For the processing of all symptoms in the inference mechanism, it is advantageous to use a unified representation. One possibility is to present the analytic and heuristic symptoms with confidence numbers $0 \leq c(s_i) \leq 1$ and treatment in the sense of probabilistic approaches known from reliability theory, Freyermuth (1991). Another possibility is the representation as membership functions $0 \leq \mu(s_i) \leq 1$ of fuzzy sets, Isermann and Ulieru (1993). Fig. 18 shows some examples for cases where the symptom ΔS_i, eq. (52), either increases or decreases. Fig. 18a),b) has the advantage that only one membership function has to be processed, in contrast to Fig. 18b) where 5 membership functions for linguistically expressed changes have to be processed.

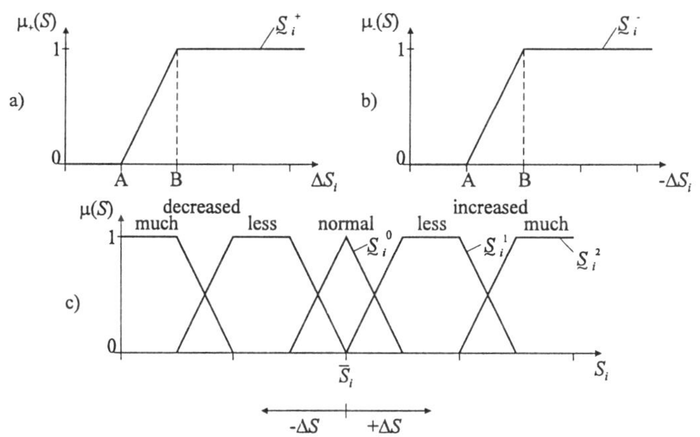

Figure 18. Membership functions of symptoms $S_i = \Delta S_i$ represented as fuzzy sets

By these kinds of fuzzy sets and corresponding membership functions, all the analytic and heuristic symptoms can be represented in a unified way within the range $0 \leq \mu(s_i) \leq 1$. These integrated symptom representations are then the inputs for the inference mechanism, Fig. 1.

e) Fault-symptom relationships

The propagation of faults to observable features or symptoms in general follows physical cause-effect relationships, where physical properties and variables are connected to each other quantitatively and also as functions of time. However, the underlying physical laws are frequently not known in analytical form, or are too complicated for calculations. If no information is available on the fault-symptom causalities, experimentally trained *classification methods* can be applied for fault diagnosis. This leads to an unstructured knowledge-base. In the case that the fault-symptom causalities can be expressed in the form of if-then rules *reasoning methods* can be used. In the following subsections both diagnostic procedures will be briefly described. As the timely behaviour (dynamics) between causes and effects is not known in most cases, the relations are considered to be static.

3.4.2 Diagnosis using classification methods

In Section 3.3 it was shown how analytic and heuristic features S (or symptoms ΔS_i) can be generated. These features are now represented in a feature vector

$$\boldsymbol{S}^T = [S_1, S_2, \dots, S_n]. \tag{54}$$

The corresponding faults are assumed to be known, also in vector form

$$\boldsymbol{F}^T = [F_1, F_2, \dots, F_m]. \tag{55}$$

The elements of $\boldsymbol{F}$ may be binary $F_j \in [0,1]$ expressing the faults as having either "occurred" or "not occurred". They may also represent gradual measures for the size of the faults $F_j \in [0 \dots 1]$.

If no further knowledge is available for the relations between features and faults, classification or pattern recognition methods can be used, Fig. 19. Here, reference vectors $\boldsymbol{S}_n$ are determined for the normal behaviour. Then the corresponding input vectors S of the features are determined experimentally for certain faults F_j. The relationship between $\boldsymbol{F}$ and $\boldsymbol{S}$ is therefore learned (or trained) experimentally and stored, forming an *explicit knowledge base*. By comparison of the observed $\boldsymbol{S}$ with the normal reference $\boldsymbol{S}_n$, faults $\boldsymbol{F}$ can be concluded.

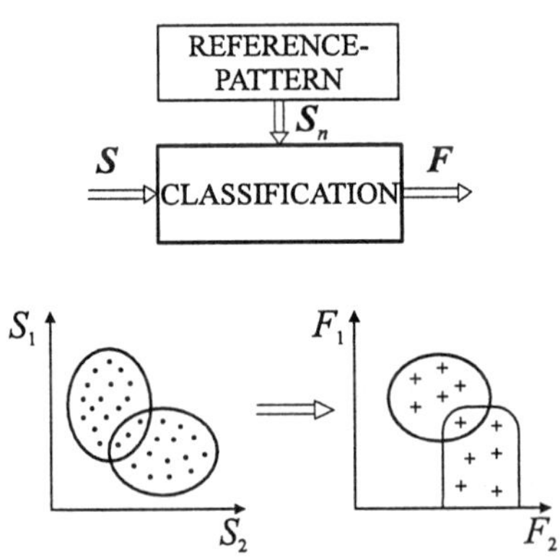

Figure 19. Fault diagnosis using classification methods

One distinguishes between *statistical or geometrical classification methods*, with or without certain probability functions, Tou and Gonzalez (1974). A further possibility is the use of *neural networks*, because of their ability to approximate nonlinear relations and to determine flexible decision regions for $\boldsymbol{F}$ in continuous or discrete form, Barschdorff and Becker (1990), Leonhardt (1996). By *fuzzy clustering* the use of fuzzy separation areas is possible, Halgamunge (1996). (If features $\boldsymbol{S}$ are used for classification according to Fig. 19, the generation of symptoms ΔS_i is not necessary because this is included in the classification procedure, together with the reference S_n)

3.4.3 Diagnosis using reasoning methods

For some technical processes the basic relationships between faults and symptoms are at least partially known. Then this a-priori-knowledge can be represented in causal relations

$$\text{fault} \rightarrow \text{events} \rightarrow \text{symptoms.}$$

Fig. 20a) shows a corresponding causal network, with the nodes as states and edges as relations. The establishment of these causalities follows the fault-tree analysis (FTA), proceeding from faults through intermediate events to symptoms (the physical causalities) or the event-tree analysis (ETA), proceeding from the symptoms to the faults (the diagnostic forward-chaining causalities), see e.g. Lee et al (1985). To perform a diagnosis, this qualitative knowledge can now be expressed in form of rules

IF<condition>THEN<conclusion>.

The condition part (premise) contains facts in the form of symptoms S_i as inputs and the conclusion part includes events E_k and faults F_j as a logical cause of the facts. If several symptoms indicate an event or fault, the facts are associated by AND and OR connectives, leading to rules in the form

$$\text{IF} < S_1 \text{ AND } S_2 > \text{THEN} < E_1 >$$
$$\text{IF} < E_1 \text{ OR } E_2 > \text{THEN} < F_1 >$$

compare Fig. 20b).

For the establishment of this heuristic knowledge several approaches exist, see Frost (1986), Torasso and Console (1989).

In the classical fault-tree analysis the symptoms and events are considered as binary variables, and the condition part of the rules can be calculated by Boolean equations for parallel-serial-connection, see e.g. Barlow and Proschan (1975), Freyermuth (1993). However, this procedure has not proved to be successful because of the continuous nature of faults and symptoms.

For the fault diagnosis of technical processes, methods of *approximate reasoning* are more appropriate.

Based on the available heuristic knowledge in form of *heuristic process models* and *weighting of effects* different diagnostic forward and backward *reasoning strategies* can be applied. Finally the diagnostic goal is achieved by a *fault decision* which specifies the type, size and location of the fault as well as its time of detection.

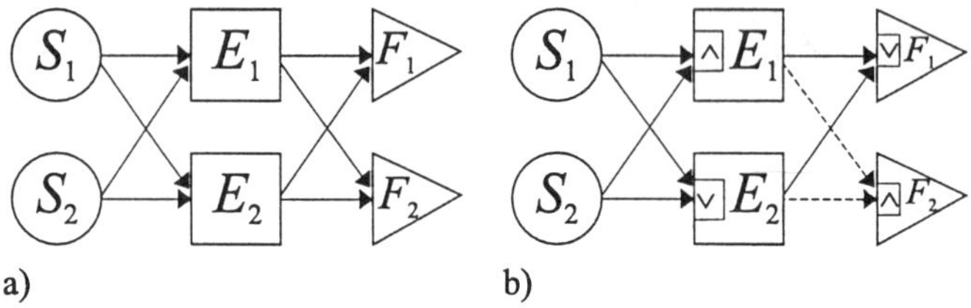

Figure 20. Fault diagnosis using inference methods
a) causal networks b) fault symptom trees

A review of recent developments in reasoning oriented approaches for *diagnosis* was given in Isermann and Ulieru (1993). As major areas of interest medicine, Shortliffe (1976), Miller et al. (1979), Sanchez (1979), and engineering, Fink and Lusth (1987), Milne (1987a), can be observed. Engineering research especially with regard to reliability analysis of nuclear power stations and aero space systems and electrical equipment started much earlier and followed in many cases the concept of *fault tree analysis*, DIN 40042 (1970), IEC 271, DIN 25424 (1977), DIN 25419 (1979), Barlow and Proschan (1975), Himmelblau (1978), Lee et al. (1985), Freyermuth (1991), Cayrac et al. (1996). On the other side *artificial intelligence* (AI) offered new methods for the treatment of cause-effect relations, Pearl (1988), and for diagnostic problem solving, Milne (1987b), Peng and Reggia (1987, 1990), Torasso and Console (1989). The development of AI approaches was oriented initially to medical diagnosis and then extended to technical processes. Therefore, also for technical diagnosis only heuristic symptoms were considered. Then more sophisticated diagnostic reasoning strategies were developed by increasing the level of abstraction see e.g. Reiter (1987) and de Kleer (1986) using the causalities as deep logic interdepencies. The fault-symptom trees known from engineering and the AI strategies for treating causalities can be brought together for fault diagnosis. Especially the analytical symptoms generated by the model-based fault detection then allow to perform a deep fault diagnosis by pinpointing on the possible physical fault origins (roots).

a) Forward chaining

By using the strategy of forward chaining a rule, the facts are matched with the premise and the conclusion is drawn based on the logical consequence (Modus ponens). Therefore with the symptoms $S_i \triangleq \Delta S_i$ as inputs the possible faults F_j are determined using the heuristic causalities.

In general the symptoms have to be considered as uncertain facts. Therefore a representation of all observed symptoms as confidence functions $c(S_i)$ or membership functions $\mu(S_i)$ of fuzzy sets in the interval [0,1] is feasible, especially in unified form as described in section 3.4.1)

Approximate reasoning with fuzzy logic

With the structure of a fault-symptom tree, obtained by knowledge acquisition via the event-tree analysis (ETA) of the process, a fuzzy rule based system with multiple levels of rules can be established, as shown in Fig 21, see also Vaija and Järveläinen (1985).

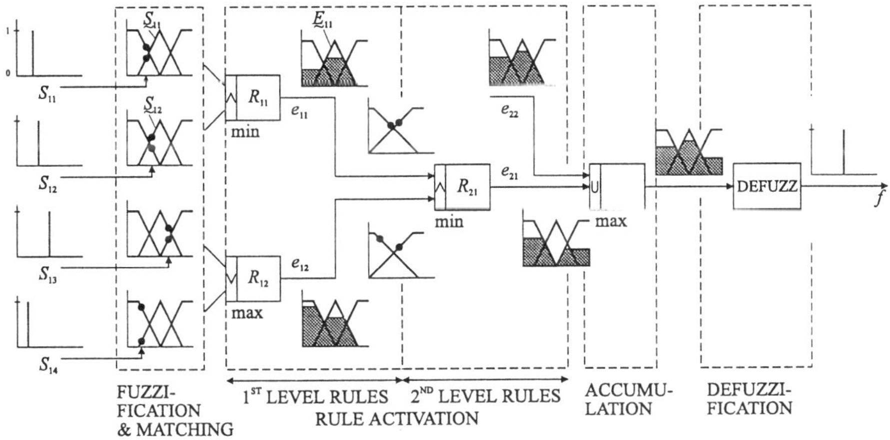

Figure 21. Signal flow in a fuzzy rule based system with two levels of rules and max/min operations and singletons as inputs

The symptoms ΔS_i of the features S_i are now reprenseted by fuzzy sets S_i^0, S_i^1, S_i^2,... with linguistic meanings like "normal", "less increased", "much increased", etc. The general procedure within the fuzzy IF-THEN rule based system, Mamdani and Assilian (1975), Zimmermann (1991), Driankov, Hellendoorn, Reinfrank (1993), then follows:

- *inference*
  (The fuzzy IF-THEN rule expresses a fuzzy implication relation between the fuzzy sets of the premise and the fuzzy sets of the conclusion. If the rule is interpreted by the Mamdani-implication and the compositional rule of inference by Zadeh (1973) is applied, following steps can be distinguished:
  - matching of the facts with the rule premises (determination of the degrees of fulfillment of the facts to the rule premises → fit values)
  - if the rule contains logic connections between several premises by fuzzy AND or fuzzy OR the evaluation is performed by t-norms or t-conorms (The result gives then the degree of fulfillment of the facts to the complete premise)
  - the evaluation of the resulting conclusion follows the max-min composition, Zadeh (1973)
    (the degree of fulfillment of the premise restricts the fuzzy set of the conclusion))
- *accumulation* of all conclusion fuzzy sets if several rules contribute to the output (A union operation yields a global fuzzy set of the output)
- *defuzzification* to obtain crisp outputs (various defuzzification methods can be used, as e.g. center of gravity, maximum-height, and mean of maximum, to obtain a crisp numerical output value). For a fault diagnosis, however, the

defuzzification is often replaced with a simple maximum operation.
For the conditional part of the IF-THEN rules relatively simple fuzzy-logic operations are obtained by max-min composition.

$$\textit{Fuzzy-AND}:\ \mu(\eta)=\min[\mu(\xi_1)\ ,\ldots,\ \mu(\xi_\nu)] \tag{56}$$

$$\textit{Fuzzy-OR}:\ \mu(\eta)=\max[\mu(\xi_1)\ ,\ldots,\ \mu(\xi_{nu})] \tag{57}$$

These operations agree with corresponding operations of Boolean logic. However, some information is not used as only the minimal or maximal value is taken. Another possibility is to use the prod-sum-operation

$$\textit{Fuzzy-AND}:\ \mu(\eta)=\mu(\xi_1)\mu(\xi_2)\ \ldots\ \mu(\xi_\nu) \tag{58}$$

$$\textit{Fuzzy-OR}:\ \mu(\eta)\ =\ 1-\prod_{i=1}^{\nu}\ (1-\mu(\xi_i)) \tag{59}$$

In this case all values are represented in the result. Note also the similarity to equations of probability theory.

The NOT-Operation is in both cases

$$\textit{NOT}:\ \mu(\eta)\ =\ 1-\mu(\xi_i) \tag{60}$$

The dimensions of the fuzzy rule based system is given by

- number of symptoms $j = 1, 2, \ldots, M$
- number of levels $i = 1, 2, \ldots, P$
- number of rules per level $\upsilon = 1, 2, \ldots, N$
- number of faults $l = 1, 2, \ldots, L$

The overall dimension may therefore blow up strongly even for small components or processes. Therefore the software implementation is important. Mainly two procedures to perform the reasoning are known:

- *Sequential rule activation* from level i to level $i + 1$ (horizontal procedure)
  - matching of symptoms $\underset{\sim}{S}'_{ij}$ with the premises of rules of level $i = 1$ (including their logical operations) to determine the events $\underset{\sim}{E}'_{1\upsilon}$
  - matching of events $\underset{\sim}{E}'_{i\upsilon}$,i =2,3, ... ,P,with the premises of rules of level i to determine the events $\underset{\sim}{E}'_{(i+1)\upsilon}$
  - matching of events $\underset{\sim}{E}'_{(i+1)\upsilon}$ with the fault´s fuzzy subsets $\underset{\sim}{F}'_l$
  - accumulation of fault fuzzy subsets to obtain the global possibility for each fault
  - fault decision by maximal possibility
  - defuzzification to obtain a crisp size of the most possible fault.
- *Multiple rule activation* from all symptoms to all faults (vertical procedure), by using the fuzzy relational equation

$$\boldsymbol{F} = \boldsymbol{S} \circ \boldsymbol{R} \tag{61}$$

proposed by Ulieru (1993, 1996).

  - chaining $\underset{\sim}{S}_{1j} \rightarrow \underset{\sim}{E}_{1\upsilon} \rightarrow \ldots\ \underset{\sim}{E}_{i\upsilon}\ \rightarrow \ldots \rightarrow\ \underset{\sim}{F}_l$ by multiple composition to obtain overall relations $\underset{\sim}{R}_{S1jxFl}$ in form of a matrix $\boldsymbol{R}$
  - matching of $\boldsymbol{R}$ with all current symptoms $\boldsymbol{S}'$ by using the fuzzy relational equation $\boldsymbol{F}' = \boldsymbol{S}' \circ \boldsymbol{R}'$
  - accumulation to obtain the global possibility for each fault

- fault decision by maximal possibility
- defuzzification to obtain a crisp size of the most possible fault.

In both approaches a defuzzification can be performed after each level or not.

A similar approach using fuzzy relation equations was proposed earlier by Kitowski and Bargiel (1987) using derivations of directly measurable variables as symptoms and stating the fuzzy relation equation

$$\boldsymbol{F} = \boldsymbol{S} \circ \boldsymbol{R}^* \tag{62}$$

The knowledge for the establishment of this equation stems from a fault-tree analysis (FTA). Then the inverse problem of eq. (61) has to be solved, by calculating the possibilites of the faults $\boldsymbol{F}$, e.g. by using Sanchez´s α operator, Sanchez (1979). See also Gmytrasiewicz et al. (1990) uses the relational fuzzy eq. (61), for a diagnosis after a failure has happened (post mortem diagnosis) and searching for more or less binary events (symptoms) as causes. Also in this case the inverse problem is solved.

Indeed, in many diagnosis application cases the knowledge acquisition first follows a fault-tree-analysis (FTA), by proceeding from assumed faults through events to symptoms by physical causality inspection according to Fig. 20b). If the knowledge is complete, a graphical representation like Fig. 21 can be obtained directly, from which eq. (61) can be established. By this way the solution of the inverse problem can be circumvented. Then, also basic fuzzy-logic software tools can be directly used.

To reduce the *computational effort*, simplifying assumptions may help. Following simplifications are possible:

- singletons for the symptoms: matching is reduced to direct fuzzification of crisp values.
- singletons for the events and faults: the conclusions are singletons with a height of the matching degree between facts and rule premise.
- universal fuzzy sets for the events and faults: the conclusion is a universal set with height of the matching degree between facts and rule premise. But then the size of the event or fault cannot be determined.

In the cases of singletons or universal sets for events and faults the conclusions are identical with the membership degree of the facts with the logical operation in the premise. This means the whole procedure reduces to an aggregation of the facts with appropriate *t*-norms and *t*-conorms, and chaining of the results, Isermann (1994). Fuzzy fault diagnosis can be expanded to time dependent faults and symptoms, including incipient and intermittent faults and multiple faults, Ulieru (1996).

Probabilistic fault-symptom reasoning

Another possibility to cope with uncertain facts is to assign probabilities $P(\xi_i)$ to the symptoms (events) and $P(\eta_k)$ to the events (faults). Based on the causal tree a simplified Bayesian network can be assigned, including conditional probabilities $P(\xi_i|\eta_k)$, Pearl (1988).

With the assumption that the symptoms ξ_i are statistically independent among each other and the symptoms ξ_i are statistically independent of the events η it holds for an AND connection, Ulieru (1996)

$$\begin{aligned} P(\eta,\xi_1 AND \xi_2) &= P(\xi_1 AND \xi_2|\eta)P(\eta) \\ &= P(\xi_1)P(\xi_2)P(\eta) \end{aligned} \tag{63}$$

Similarly one obtains for an *OR*-connection

$$P(\eta, \xi_1 \, OR \, \xi_2) = [P(\xi_1) + P(\xi_2) - P(\xi_1) P(\xi_2)] P(\eta) \tag{64}$$

If in addition the events are assumed to have happened surely, $P(\eta)=1$, one obtains for the AND-connection

$$P(\eta, \xi_1 \, AND \, \xi_2 \, AND \ldots \xi_\nu) = P(\xi_1) P(\xi_2) \ldots P(\xi_\nu) \tag{65}$$

and the *OR*-connection

$$P(\eta, \xi_1 \, OR \, \xi_2 \, OR \ldots \xi_\nu) = 1 - \prod_{i=1}^{\nu} (1 - P(\xi_i)) \tag{66}$$

The similarity of the formulas to fuzzy-logic operations with the prod-sum operation, e.g. eqs. (58), (59), is obvious. However, the assumptions on statistical independence do not take into account the existing causalities of the fault-symptom trees.

b) Backward chaining

The strategy of backward chaining assumes the conclusion as known and searches for all relevant premises (modus tollens). This is especially of interest if the symptoms are not complete. Therefore the concluded events and faults are displayed to the operator after forward chaining with all known symptoms. A refinement of the diagnosis can then be achieved by selecting the most plausible events and faults as hypothesis and applying backward chaining by asking for missing symptoms. Then forward chaining is restarted. This procedure is implemented best within an interactive dialogue and repeated until terminated by the operator (Freyermuth, 1991, 1993).

Other chaining strategies are Establish-Refine, Hypothesize and Test, Depth and Width searching, etc. see Harmon and King (1985).

If all symptoms are already taken into account during forward reasoning, backward reasoning can be applied for the validation of diagnosed faults, Ulieru (1993).

The fault diagnosis considered until now assumed that all symptoms appear simultaneously and do not change with time. Further cases are therefore dynamically developing faults and symptoms (e.g. either incipient or intermittent faults) and therefore dynamic fault trees, Nold (1991), Ulieru (1996) and also multiple faults.

c) Self-learning fuzzy diagnosis system

Following the success in other applications, several researchers propose self-learning fuzzy systems for diagnosis and supervision tasks. These systems have the ability to learn a fuzzy diagnosis system based on example measurements (faulty/not faulty) of the process under consideration. Successful applications include automotive systems, Ayoubi (1996), chemical processes, Zhang et al. (1994), and electrical drives, Füssel and Isermann (1998), Goode and Chow (1994).

Next to fuzzy derivatives of machine learning approaches, most publications have used combinations of neural network's learning features with fuzzy logic systems.

Sometimes, however, simple clustering techniques are used. In those approaches, algorithms like the Fuzzy-C-Means (FCM) clustering algorithm, Bezdek (1993), are applied to the given symptom data and the area of support of the individual clusters is declared a multi-dimensional membership function, see Fig. 22. This membership function is then modelled e.g. with a multi-dimensional Gaussian

$$\mu(s) = k \cdot e^{-\frac{1}{2}(s-s_0)^T T^{-1}(s-s_0)} \tag{67}$$

Here, $\boldsymbol{T}$ determines the shape and orientation of the Gaussian, while s_0 carries the information about the location of this multi-dimensional fuzzy attribute.

Each cluster defines the premise of a fuzzy rule. The conclusion of such a rule is typically given by the majority of the fault classes of the data points belonging to the same cluster. The usage of this approach, however, should be limited to systems of two or three input dimensions, because membership functions with more than three dimensions are not comprehensible.

More oriented along the standard fuzzy logic procedure with defuzzification, inference and one-dimensional membership functions are *hybrid* approaches embedding the fuzzy logic in a neural network structure. These *neuro-fuzzy* structures employ a fuzzification layer with membership functions (mostly Gaussian and/or sigmoidals) as the first active layer of a neural network. This is followed by one or two layers, implementing the inference engine, see Fig. 23. These layers implement linear or sigmoidal neurons

$$\mu = \sum_{i=1}^{m} w_i \mu_i \qquad \text{or} \qquad \mu = sig\left(\sum_{i=1}^{m} w_i \mu_i\right) \tag{68}$$

where m denotes the number of connections leading to that neuron. Alternatively, fuzzy logic operators like the minimum for AND and maximum for OR-operators as in (48) and (49) are used. Usually, there is either a single AND-layer or and AND followed by a layer implementing an OR.

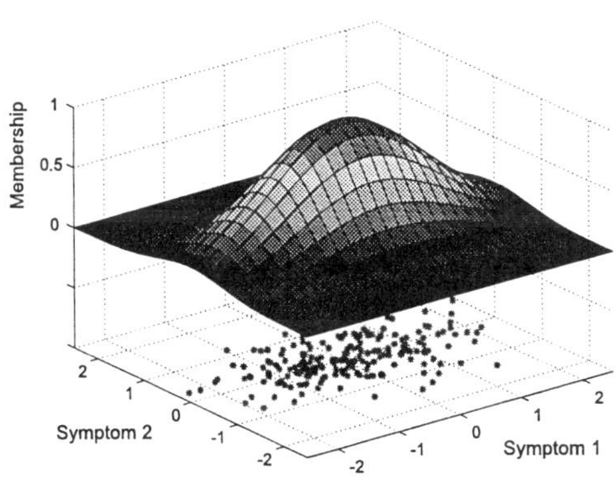

Figure 22. Example of a multi-dimensional membership function. The distribution of the underlying data points determines the shape and location of the function. This approach assumes that clusters of data belong to the same fault and can be diagnosed by simple membership functions like the one shown here

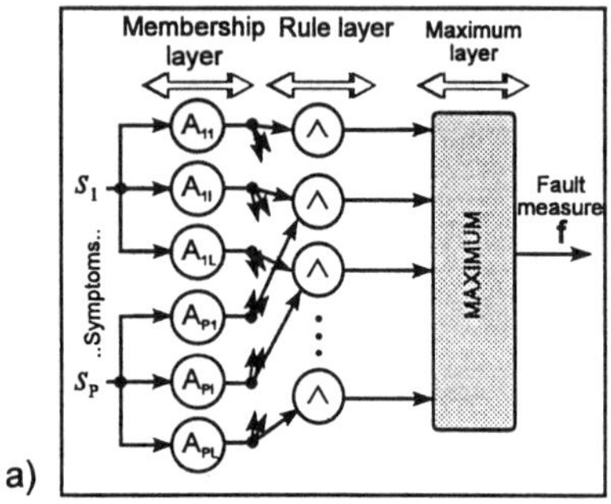

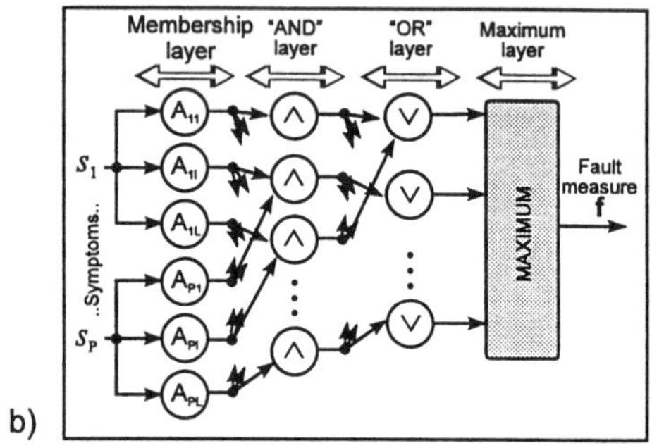

Figure 23. Two typical neuro-fuzzy schemes used for fault diagnosis. Implementation with a single layer of operator neurons, a), or two layers, b)

The design procedure is usually split in two stages: A *structural initialisation* followed by a *parameter optimization.* The first step fixes the rule base. Connections are made and weights initialized. It requires the determination of fuzzy membership functions. This is either based on the data with clustering techniques, Ayoubi (1996), Füssel and Isermann (1998), see Fig. 24, self-organizing maps, Pistauer and Steger (1995), or heuristically by hand, Zhang et al. (1994). Once the first layer is determined, the set of connections between the network layers has to be found. This is equivalent with finding the rules of the rule base. The data set is presented to the first layer of the network and connections are established between those fuzzy membership neurons which yield the strongest output and the corresponding node in the inference layer which corresponds to the desired output. Connections repeatedly activated by the data are enforced and those seldom used might be dropped. This is achieved either through Hebbian learning, Ayoubi (1996), competative learning, Pistauer and Steger (1995) or similar schemes.

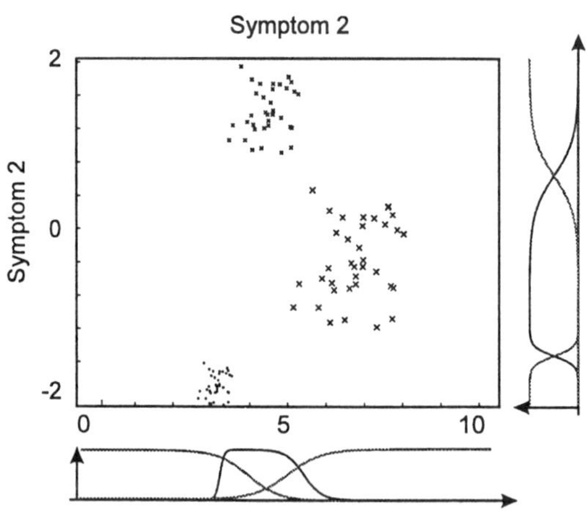

Figure 24. Creation of fuzzy membership functions through clustering. Before clustering, a projection of the data on the individual axes was performed

In the second phase, the parameter optimization, both, the connection weights as well as the parameters of the fuzzy attributes and even the operators of the network can be subject to an optimization. The majority of the publications reports the use of backpropagation optimization. The interpretability of the resulting structures, however, is not always taken care of. The neurons used in the inference engine are sometimes not properly designed to reflect the AND and OR-operators of fuzzy logic systems, which in connection with arbitrarily changing connection weights hinders a linguistic representation. Only sometimes, measures are taken to prevent fuzzy attributes to be modified too rigorously, Goode and Chow (1994), Füssel and Isermann (1998). An example is given in Füssel and Isermann (1998) where the implementation of a special structure even allows an interpretation as a fault tree.

Overall, the applications show that the neuro-fuzzy approaches can facilitate the problem of missing expert knowledge in designing fault diagnosis systems. The optimization schemes involved lead to high performance fuzzy rule bases that can also more easily be adapted to varying conditions than completely hand-made fuzzy systems.

3.5 APPLICATIONS

3.5.1 Fault Diagnosis of a D.C. Motor

As one of the most common electrical devices, drives play an important role in industrial applications. In the following, a diagnosis scheme will be presented which performs the supervision of a d.c. motor and is able to differentiate between 15 different faults based on a fuzzy logic diagnosis system without costly sensor usage, see Table 1 for the faults considered. In this application, a real-time fault detection scheme is implemented on a digital signal processor hardware to supervise the motor. In case of any irregularities, the fuzzy logic diagnosis system is initiated and reveals the exact fault location and cause.

The fault detection scheme is based on the mathematical model of the motor, Fig. 25. The model for a permanently excited d.c. motor from literature was augmented with some nonlinear terms to account for the effects at the brushes. Three signals can be measured: the rotor current, voltage and speed. The rotor voltage U_A^* and external load M_L^* are inputs, the rotor current I_A and speed ω are outputs of the model which can conveniently be written in state space form as

$$\dot{x} = \begin{pmatrix} \dot{I}_A \\ \dot{\omega} \end{pmatrix} = \begin{bmatrix} -\dfrac{R_A}{L_A} & -\dfrac{\Psi}{L_A} \\ \dfrac{\Psi}{J} & -\dfrac{M_{R1}}{J} \end{bmatrix} \begin{pmatrix} I_A \\ \omega \end{pmatrix} + \begin{bmatrix} \dfrac{1}{L_A} & 0 \\ -\dfrac{1}{J} & 0 \end{bmatrix} \begin{pmatrix} U_A \\ M_L \end{pmatrix} \tag{69}$$

$$\dot{y} = \begin{pmatrix} I_A \\ \omega \end{pmatrix} = \begin{pmatrix} 1 & 0 \\ 0 & 1 \end{pmatrix} x$$

The nonlinear terms from Fig. 25a) are included in the corresponding input signals, so that the model with U_A and M_L becomes linear. In the experimental setup, Fig.

25b), the external load can be influenced and is assumed to be known. In other applications, a load model would have to be included. The model allows the design of four continuous-time parity residuals $r_1 \ldots r_4$, each being independent of a different input or output of the process. Table 2 gives an overview over the four residuals. The first two are computed in real-time and form the base of the fault detection scheme.

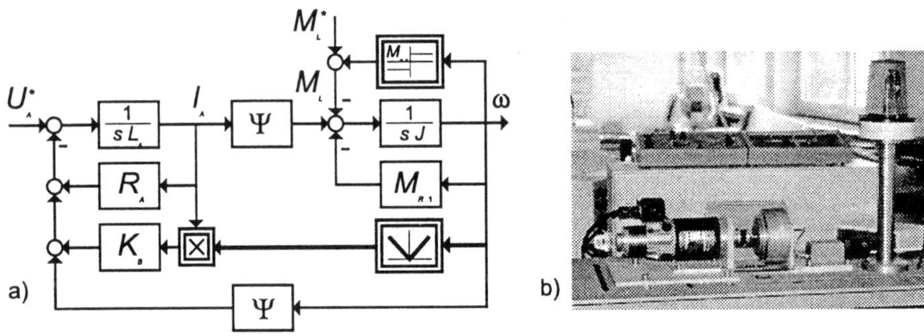

Figure 25. D.c. motor model, a), and experimental setup, b). The signals U_A^*, I_A, and ω are measurable

Table 1. Overview over faults considered

Location	Fault
wiring	resistance change
wiring	inductance change
wiring	broken
speed sensor	offset fault / gain fault / total
current sensor	offset fault / gain fault / total
voltage sensor	offset fault / gain fault / total
rotor	intertia change

Table 2. Four decoupled residuals

Residual	De-coupled from
$r_1(t)=L_A\dot{I}_A(t)+R_AI_A(t)+\Psi\omega(t)-U_A(t)$	M_L
$r_2(t)=J\dot{\omega}(t)-\Psi I_A(t)+M_{RI}\omega(t)+M_L(t)$	U_A
$r_3(t)=JL_A\ddot{I}_A(t)+(L_AM_{RI}+JR_A)\dot{I}_A(t)+$ $+(\Psi^2+R_AM_{RI})I_A(t)-J\dot{U}_A(t)-$ $M_{RI}U_A(t)-\Psi M_L(t)$	ω
$r_4(t)=JL_A\dot{\omega}(t)+(L_AM_{RI}+JR_A)\dot{I}_A(t)+$ $+(\Psi^2+R_AM_{RI})\omega(t)-\Psi U_A(t)+$ $L_A\dot{M}_L(t)+R_AM_L(t)$	I_A

The overall fault detection and diagnosis scheme is pictured in Fig. 26. It can be seen that, additionally to the residuals $r_1 \ldots r_4$, also the results of a parameter estimation and further symptoms (like mean values of the signals) are exploited for the fuzzy diagnosis scheme.The diagnosis itself is usually performed off-line. It can be accompanied by a special test cycle delivering the necessary excitation for a parameter estimation.

tion.

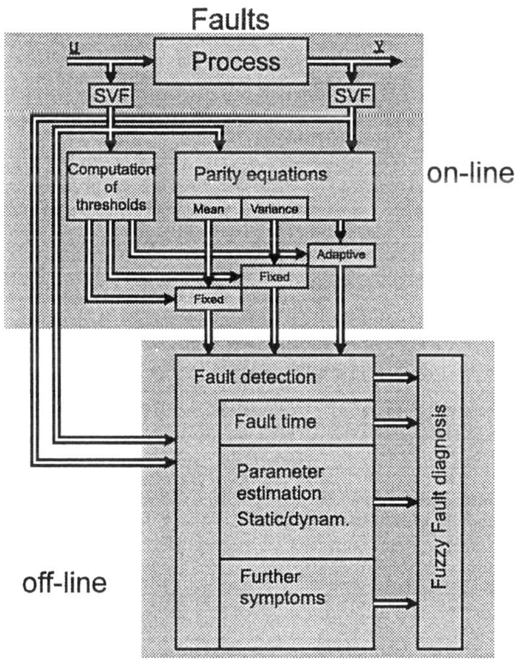

Figure 26. Supervision and fault-diagnosis concept

The structure of the fuzzy diagnosis scheme, that was designed from expert knowledge, is depicted in Fig. 27. First, a fuzzy residual evaluation determines the residual which has not been deflected by the fault. Since each of the residuals is decoupled from a different signal, this information will already be very useful for the determination of the fault cause. This is done by evaluating both the means as well as standard deviations of the residuals. Since it does not yield a hard 0/1 decision like normal thresholds would, the information about possible other minimal residuals is not lost.

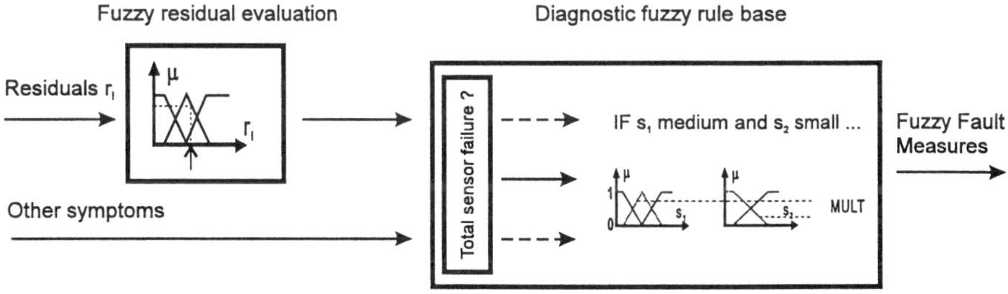

Figure 27. Fuzzy diagnosis scheme for the d.c. motor

The result of the residual evaluation is an intermediate step and serves in connection with the other symptoms as inputs for the rule base determining the exact fault. That rule base consists of 15 rules, one for every fault. The first of the rules identifies total failures of the sensors as they would appear when a cable was broken. The ouput of this first rule is also used in the remaining rule base. As an example, the rule for the diagnosis of a brush fault of the motor is given by

IF	Standard_deviation(residual 1) is very small
AND	Standard_deviation(residual 2) is large
AND	Standard_deviation(residual 3) is very small
AND	Estimation(inductivity) is normal
AND	Estimation(resistance) is small
AND	Both estimations(resistance) equal
AND	Estimation(flux linkage) normal
THEN	Fault at the brushes

It can be seen that the individual rules employ only a subset of all available symptoms. That is due to a relatively high degree of redundancy of the symptoms. Fig. 28 displays some of the membership functions used. For the conjunction in the rule premises a product operator is applied.

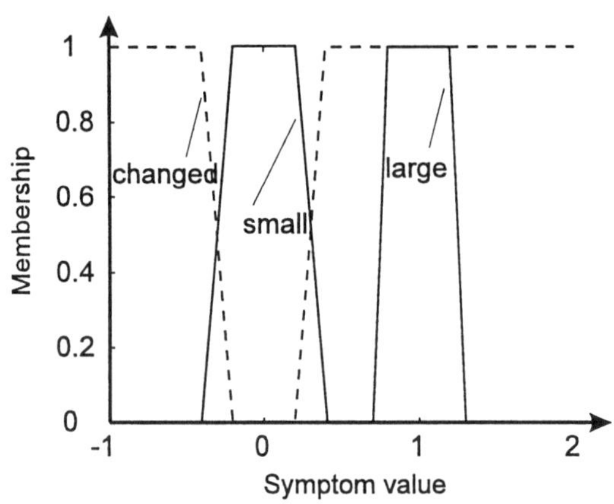

Figure 28. Example of membership functions used. In order to reflect the expert knowledge properly, it was necessary to use not only grid-based membership functions but also non-standard shapes like "changed".

The results of the fuzzy rule base are presented to the operator in order to assist him for maintenance and repair, see Fig. 29. The fault with the highest fault measure is additionally quantitatively evaluated.

The use of fuzzy logic proved to be especially suitable in this application as expert knowledge could be exploited. Measurement noise is accounted for since no hard thresholds are used. Furthermore, the presentation of the fuzzy results allows the operator to consider different fault causes.

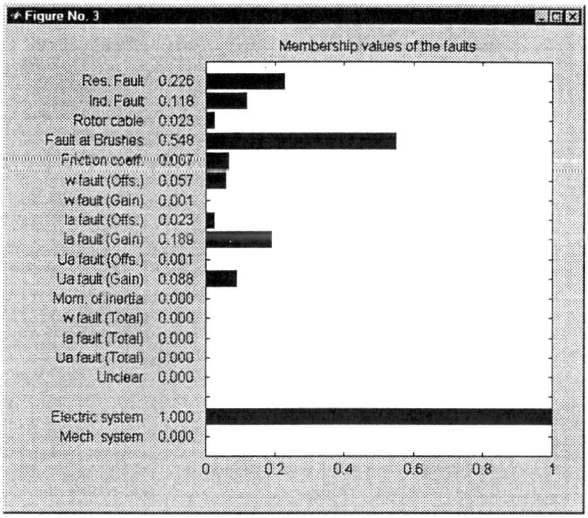

Figure 29. Example of the output from the diagnosis system presented to the operator. Since all results are given, it is possible to consider not only the most likely fault, a brush failure, but also other alternatives.

3.5.2 Fault diagnosis of a machine tool feed drive with fuzzy reasoning

The feed drive shown in Fig. 30 consists of a table which is moved on the slide ways by a ball screw spindle. The spindle is driven by a speed controlled d.c. motor through a toothed belt. With a tension screw at the casing the tension of the belt can be influenced. The measured variables are only the voltage $U_A(t)$, the armature current $I_A(t)$, the speed $\omega_M(t)$ of the d.c. motor and $\omega_s(t)$ of the spindle. By discrete square root filtering 10 continuous-time parameters were estimated for sinusoidal $U_A(t)$ and among them following process coefficients were calculated: moment of inertia J_{sm}, Couloumb friction coefficient M_{FC}, belt stiffness c_{TR} and damping d_{TR}. For details, see He (1993) and Isermann (1993).

Fig. 31 shows the fault-symptom tree for a belt defect with analytic and heuristic symptoms for the case that one thread of the belt was cut. The values of the analytic symptoms were obtained through parameter estimation and δ_A=1; δ_B = 5 ; 5 ; 30 ; 10 for S_{a1} ; S_{a5} ; S_{a7} ; S_{a9}.

As fuzzy relations between symptoms and effects are difficult to specify, simplified fuzzy logic is applied. Forward chaining of the analytic symptoms with prod-sum (max-min) operation yields:

$$\mu(E_{31}) = 1.0.7 = 0.7 \ (0.7)$$

$$\mu(E_4) = 1.1 \ = 1 \ (1)$$

The heuristic symptoms from operator's observation are

$$\mu(E_7) = 1\text{-}(1\text{-}0.5)(1\text{-}0.3)(1\text{-}0.2) = 0.72 \ \ (0.5)$$

Hence,

$$\mu(E_3) = 1\text{-}(1\text{-}0.7)(1\text{-}0.72)=0.92 \qquad (0.7)$$

The fault F_6 is now supposed and by backward chaining the operator is asked about the life time relevant symptoms of the belt history. This leads to

$$\mu(E_6) = 1\text{-}(1\text{-}0.7\ 0.8)(1\text{-}0.8\ 0.6)(1\text{-}0.2) = 0.82\ (0.7)$$

(The symptom S_{10} refers to an increased axis distance between the pulleys by adjusting the belt tension screws. This is an indication of a too large belt elongation).

By forward chaining the fault F_6 can be diagnosed with the joint possibility

$$\mu(F_6) = 1\text{-}(1\text{-}0.92)(1\text{-}0.82) = 0.99\ (0.7)$$

Hence, possibly a defect of the belt has occurred, for example by breaking of cord threads.

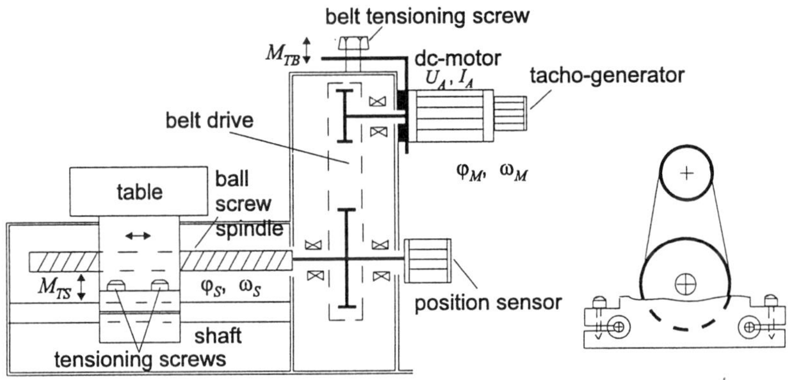

Figure 30. Scheme of the feed drive test rig

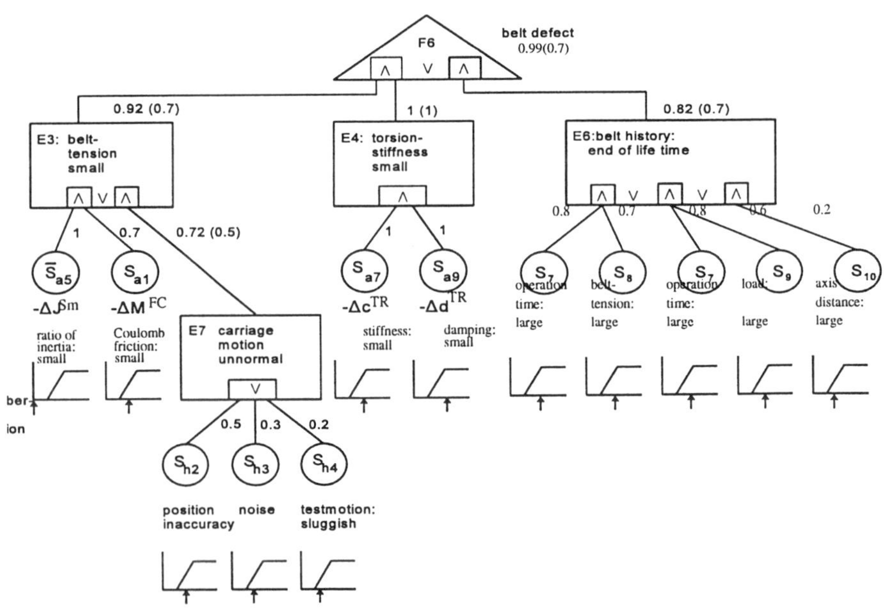

Figure 31. Fault-Symptom Tree for the belt status of a feed drive. Membership values with prod-sum (max-min) fuzzy logic

REFERENCES

Ayoubi, M. (1996), Fuzzy systems design based on hybrid neural structure and application to the fault diagnosis of technical processes, *CEP Control Engineering Practice* **4** (1) 35-42.

Barlow, R.E., F. Proschan (1975), *Statistical theory of reliability and life testing*. Rinehart & Winston Inc., Holt.

Barschdorff, D. (1987), *Gearbox failure diagnostics*. VDI-Berichte Nr. 644, 241-248. VDI-Verlag, Düsseldorf.

Barschdorff, D., D. Becker (1990), Neuronale Netze als Signal- und Musterklassifikatioren, *Technisches Messen* **57**, 437-444.

Basseville, M., A. Benveniste (eds.) (1985), Detection of abrupt changes in signals and dynamical systems, in *Lecture notes in Control and Information Sciences*, **77**, Springer-Verlag, Berlin.

Beard, R.V. (1971), Failure accommodation in linear systems through self-reorganization. *Rept. MVT-71-1*. Man Vehicle Laboratory, Cambridge.

Bezdek, J. (1993), Fuzzy models - what are they, and why?, Editorial at the first issue of *IEEE Transactions on Fuzzy Systems* **1** (1).

Burg, J.P. (1968), A new analysis technique for time series data, in *NATO Advanced Study Institute on Signal Processing with Emphasis on Underwater Acoustics* (12-23. August 1968).

Caminhas, W.M., H. Tavares, F. Gomide (1996), A Neurofuzzy Approach for Fault Diagnosis in Dynamic Systems, *Proceedings of the Fifth IEEE International Conference on Fuzzy Systems*, FUZZ-IEEE '96, 2032-2037.

Cayrac, D., D. Dubois, H. Prade (1996). Handling uncertainty with possibility theory and fuzzy sets in a satellite fault diagnosis application. *IEEE Transactions of Fuzzy Systems* **4** (3) 251-269.

Clark, R.N. (1978a). A simplified instrument detection scheme, *IEEE Transactions on Aerospace Electron. Systems* **14**, 558-563.

Clark, R.N. (1978b). Instrument fault detection, *IEEE Transactions on Aerospace Electron. Systems* **14**, 456-465.

Chow, M.Y. (1997), Motor Fault Detection and Diagnosis, *IEEE Industrial Electronics Society Newsletter* **44** (4) 4-11.

DIN 40042 (1970). *Zuverlässigkeit elektrischer Geräte, Anlagen und Systeme-Begriffe* (reliability of electrical equipment-definition in 4 languages). Deutsche Industrie Norm. Beuth-Verlag, Berlin.

DIN 25424 (1977). *Fehlerbaumanalyse* (fault tree analysis). Deutsche Industrie Norm. Beuth-Verlag, Berlin.

DIN 25419 (1979). *Störfallablaufanalyse* (incident sequence analysis). Deutsche Industrie Norm. Beuth-Verlag, Berlin.

Driankov, D., H. Hellendoorn, M. Reinfrank (1993). An Introduction to Fuzzy Control. Springer-Verlag, Berlin.

Fink, P.M., J.C. Lusth (1987), Epxert systems and diagnostic expertises in the mechanical and electrical domains. *IEEE Transactions on Systems Man and Cybernetics* **17**, 340-349.

Frank, P.M. (1987), Advanced fault detection and isolation schemes using nonlinear and robust observers, in *Proceedings of the 10th IFAC-Congress*, München, **3**, 63-68.

Frank, P.M. (1990), Fault diagnosis in dynamic systems using analytical and knowledge-based redundancy, *Automatica* **26**, 459-474.

Frank, P.M., J. Wünnenberg (1989), Robust fault diagnosis using unknown input observer schemes, chapter 3 in R.J. Patton, P.M. Frank, R.N. Clark (eds.), *Fault diagnosis in dynamic systems - theory and applications*, Prentice Hall, London.

Freyermuth, B. (1991), Knowledge based incipient fault diagnosis of industrial robots, in *Proceedings of the IFAC-Symposium SAFEPROCESS* (Baden-Baden 1991), Pergamon Press, Oxford, **2**, 31-37.

Freyermuth, B. (1993), *Wissensbasierte Fehlerdiagnose am Beispiel eines Industrierobotors.* Dissertation TU Darmstadt. Fortschr. Ber. VDI Reihe 8 Nr. 315, VDI-Verlag, Düsseldorf.

Frost, R.A. (1986), *Introduction to knowledge base systems*, Collins, London.

Füssel, D., R. Isermann (1998), Hierarchical Motor Diagnosis Utilizing Structural Knowledge and a Self-Learning Neuro-Fuzzy Scheme", Proceedings of the 24th Annual Conference of the IEEE Industrial Electronics Society, IECON 98, Aachen, Germany.

Gertler, J. (1988), Survey of model-based failure detection and isolation in complex plants, *IEEE Control Systems Magazine*.

Gertler, J. (1991), Analytical redundancy methods in fault detection and isolation, in *Proceedings of the IFAC-Symposium SAFEPROCESS* (Baden-Baden 1991), Pergamon Press, Oxford, **1**, 9-21.

Gertler, J. (1995), Generating directional residuals with dynamic parity equations, *Automatica* **31** (4) 627-635.

Goode, P.V., M.Y. Chow (1994), A Hybrid Fuzzy/Neural System Used to Extract Heuristic Knowledge from a Fault Detection Problem, *FUZZ-IEEE 94*, Orlando, 1731-1736.

Gmytrasiewicz, P., J.A. Hassberger, J.C. Lee (1990), Fault tree based diagnostics using fuzzy logic. *IEEE Transactions on Pattern Analysis and Machine Intelligence* **12** (11) 1115-1119.

Harmon, P., D. King (1986), *Expertensysteme in der Praxis - Perspektiven, Werkzeuge, Erfahrungen*, Oldenbourg-Verlag, München.

He, X. (1993). Modellgestützte Fehlererkennung mittels Parameterschätzung zur wissensbasierten Fehlerdiagnose an einem Vorschubantrieb. Dissertation TH Darmstadt. VDI-Fortschrittsberichte, Reihe 8, Bd. 354. VDI-Verlag, Düsseldorf.

Himmelblau, D.M. (1978), Fault detection and diagnosis in chemical and petrochemical processes, Elzevier, Amsterdam, 342-393.

Höfling, T., T. Pfeufer (1994), Detection of additive and multiplicative faults - parity space vs. parameter estimation, in *Proceedings of the IFAC Symposiom SAFEPROCESS* (Espoo 1994), 539-544.

Höfling, T. (1996), *Methoden zur Fehlererkennung mit Parameterschätzung und Paritätsgleichungen.* Dissertation TU Darmstadt Fortschritt-Ber. VDI Reihe 8 Nr. 546, VDI-Verlag, Düsseldorf.

Isermann, R. (1984), Process fault detection based on modelling and estimation methods - a survey. *Automatica* **20** (4) 387-404.

Isermann, R. (1992a), *Identifikation dynamischer Prozesse.* **Vol. 1 & 2**. 2nd edition. Springer-Verlag, Berlin.

Isermann, R. (1992b) Estimation of physical parameters for dynamic processes with application to an industrial robot, *International Journal of Control* **55**, 1287-1298.

Isermann, R. (1993), Fault diagnosis of machines via parameter estimation and knowledge processing, *Automatica* **29** (4) 815-835.

Isermann, R. (ed.) (1994), Überwachung und Fehlerdiagnose - Moderne Methoden und ihre Anwendungen bei technischen Systemen., VDI-Verlag, Düsseldorf.

Isermann, R. (1995), On fuzzy logic application for automatic control, supervision and fault diagnosis, in *Proceedings of the Third European Congress on Fuzzy and Intelligent Technologies EUFIT'95*, Aachen, **2**, 738-753.

Isermann, R., K.-H. Lachmann, D. Matko (1992), *Adaptive Control Systems*, Prentice Hall, New York.

Isermann, R., M. Ulieru (1993), Integrated fault detection and diagnosis, in *Proceedings of the* IEEE/SMC Conference "*Systems Engineering in the Service of Humans*", Le Touquet, **1**, 743-748.

Isermann, R., P. Ballé (1997), Trends in the application of model based fault detection and diagnosis of technical processes. *CEP Control Engineering Practice* **5** (5) 709-719.

Janik, W. (1993), *Fehlerdiagnose des Aussenrund-Einstechschleifens mit Prozess- und Signalmodellen.* Dissertation TU Darmstadt. Fortschr. Ber. VDI Reihe 2 Nr. 288, VDI-Verlag, Düsseldorf.

Jones, H.L. (1973), *Failure detection in linear systems.* Ph.D. Thesis, Dept. of Aeronautics, M.I.T., Cambridge.

Kitowski, J., M. Bargiel (1997), Diagnostics of faulty states in complex physical systems using fuzzy relational equations. In Sanchez, E., L.A. Zadeh (1997): *Approximate Reasoing in Intelligent Systems, Decision and Control.* Oxford, Pergamon.

Kleer, de (1986), Reasoning about multiple faults. *Proceedings of the AAAI-86*, Philadelphia, 132-139.

Lee, W.S., D.L. Grosh, F.A. Tillmann, C.H. Lie (1985), Fault tree analysis, methods and applications - a review, *IEEE Transactions on Reliability* **R-34** (3) 194-202.

Leonhardt, S. (1996), *Modellgestützte Fehlererkennung mit neuronalen Netzen - Überwachung von Radaufhängungen und Diesel-Einspritzanlagen.* Dissertation TU Darmstadt. Fortschr. Ber. VDI Reihe 12 Nr 295, VDI-Verlag, Düsseldorf.

Mamdani, E.H., S. Assilian (1975). An experiment in linguistic systesis with a fuzzy logic controler. *International Journal of Man-Machine-Studies* **7**, 1-13.

Marcu, T., M. Voici (1995), A Neuro-Fuzzy Approach to Pattern Recognistion for Process Change Detection, *Proc. of 3rd Euopean Control Conference*, Rome, 1999-2004.

Mehra, R.K., J. Peschon (1971), An innovations approach to fault detection and diagnosis in dynamic systems, *Automatica* **7**, 637-643.

Milne, R. (1987a), Strategies for Diagnosis. *IEEE Transactions on System Man and Cybernetics* **17**, 333-339.

Milne, R. (1987b), Artificial intelligence for ON-LINE diagnosis. In IEE Proceedings, **134** (4).

Neumann, D. (1991), Fault diagnosis of machine-tools by estimation of signal spectra, in *Proceedings of the IFAC-Symposium SAFEPROCESS* (Baden-Baden 1991), Pergamon Press, Oxford, **1**, 73-78.

Nold, S. (1991). Knowledge based real time fault diagnosis with EFTAS 1. *IFAC-Symposium SAFEPROCESS*, Baden-Baden. Proc. Pergamon Press, Oxford.

Patton, R.J., J. Chen (1991), A review of parity space approaches to fault diagnosis, in *Proceedings of the IFAC-Symposium SAFEPROCESS* (Baden-Baden 1991), Pergamon Press, Oxford, **1**, 239-255.

Pearl, J. (1988), *Probabilistic reasoning in intelligent systems: networks of plausible inference*, Morgan Kaufmann, San Mateo.

Peng, Y., J. Reggia (1990), *Abductive inference models for diagnostic problem-solving.* Springer, New York.

Pistauer, M., Ch. Steger (1995), Multistep Parameterleraning in a Neural Network Based Fuzzy Diagnosis Module, *IEEE Int. Conf. On System, Man and Cybernetics*, Vancouver, 3249-3254.

Reiter, R. (1987), A theory of diagnosis from first principles. *Artificial Intelligence* **32**, 57-95.

Sanchez, E. (1978), Inverses of fuzzy relations. Applications to possibility distributions and medical diagnosis. *Fuzzy sets and systems*, **2**, 75-86.

Schneider, H., P.M. Frank (1996). Observer-based supervision and fault detection in robots using nonlinear and fuzzy logic residual evaluation. *IEEE Transactions on Control Systems Technology* **4** (3) 274-282.

Shortliffe, E.H. (1976). Computer-based medical consultations, MYCIN. *Artificial Intelligence Series 2*, Elsevier, Amsterdam.

Stearns, S.D. (1975), *Digital signal analysis*, Hayden Book Company, New Jersey.

Torasso,P., L. Console (1989), *Diagnostic problem solving*. North Oxford Academic, U.K.

Tou, J.T., R.C. Gonzalez (1974), *Pattern recognition principles*, Addison-Wesley Publishing, Reading.

Tsui, C.-C. (1993), A general failure detection, isolation and accomodation system with model uncertainty and measurement noise, in *Proceedings of the 12th IFAC World Congress*, Sydney, **6**, 231-238.

Ulieru, M. (1993). A fuzzy-logic based computer assisted fault diagnosis system. *TOOLDIAG Int. Conf. on Fault Diagnosis*, Toulouse, France.

Ulieru, M. (1996), *Fuzzy logic in diagnostic decision: possibilistic networks*. Dissertation TU Darmstadt.

Vaija, P., M. Järveläinen (1995), Multilevel failure detection systems. *Computers in Industry* **6**, 253-263.

Willsky, A.S. (1976), A survey of design methods for failure detection systems, *Automatica* **12**, 601-611.

Yen, G.G.(1995), Autonomous Neural Control in Flexible Space Structures, *CEP Control Eng. Practice* **3** (4) 472-483.

Zadeh, L.A. (1978). Fuzzy sets as a basis for a theory of possibility. *International Journal of Fuzzy Sets and Systems*, **1**, 3-28.

Zhang, J., A.J. Morris, G.A. Montague (1994), Fault Diagnosis of a CSTR using Fuzzy Neural Networks, *IFAC Artificial Intelligende in Real-Time Control,* AIRTC 94, Valencia, 191-196.

Zimmermann, H.J. (1991). *Fuzzy Set Theory and Its Applications*. 2nd ed. Kluwer Acadamic Publishers, Boston, Dordrecht, London.

4 QUALITY CONTROL AND MAINTENANCE

Jens Strackeljan
Richard Weber

Abstract: Quality control and maintenance are areas where fuzzy sets have been applied very intensively. Various industrial solutions using fuzzy approaches emphasize their benefits in the respective fields. This chapter gives an overview of fuzzy set applications in quality control and maintenance.

4.1 INTRODUCTION

Quality control and maintenance are two areas which have recently become more and more important in order to ensure a company's success. Customers take for granted the functions of products they are searching for. Differences between competing products exist mainly in their prices and quality. Quality control techniques assure high quality standards whereas efficient maintenance of processes could lead to low production costs resulting in competitive prices.

Many approaches from the areas of statistical quality control (SQC) and statistical process control (SPC) have been suggested to support quality control and maintenance. Intelligent technologies, such as fuzzy logic and neural networks offer many potentials to extend the set of techniques to solve the respective tasks. These potentials have been realized in several projects and are described in various publications. This chapter gives an overview of different applications of fuzzy logic for quality control and maintenance. Thus showing the advantages and potentials of soft computing for the respective realizations.

Section 2 gives an overview of various applications where soft computing techniques and especially fuzzy logic have been used for quality control and maintenance. In section 3 realizations in quality control are described in more detail. Section 4 presents applications of fuzzy logic for maintenance. Conclusions and future prospectives are given in section 5. To enhance the exchange of knowledge about the respective techniques, a competition on some test cases has been initiated. Goals, task descriptions, and results of this competition are also presented in chapter 5.

4.2 OVERVIEW OF INDUSTRIAL APPLICATIONS OF FUZZY TECHNIQUES IN QUALITY CONTROL AND MAINTENANCE

A general definition of quality control is given by Ishikawa:

"Quality control consists of developing, designing, producing, marketing, and servicing products and services with optimum cost-effectiveness and usefulness, which customers will purchase with satisfaction. To achieve these aims, all the separate parts of a company (...) must work together. All the company's departments must strive to create cooperation-facilitating systems, and to prepare and implement standards faithfully. This can only be achieved through full use of a variety of techniques such as statistical and technical methods, standards and regulations, computer methods, automatic control, facility control, measurement control, operations research, industrial engineering, and market research." [Ishikawa 1990, p.2].

As mentioned in this reference, a variety of techniques is needed to reach the goals of quality control. Although all cited approaches are important for the respective tasks, it has become more and more evident that also new technologies from the area of soft computing (e.g. fuzzy logic, neural networks) could offer advantages when it comes to solve complex problems due to steadily increasing quality standards. An introduction to fuzzy logic and most of the techniques mentioned in this chapter is given e.g. in [Zimmermann 1996].

This section gives an overview on different realizations where soft computing techniques and especially fuzzy logic have been used for quality control. These applications show the broad range of such approaches concerning

- different fuzzy methods (e.g. fuzzy clustering, rule-based fuzzy systems),
- various application areas (inspection of machines, testing of software, analysis of chemicals), and
- products (software, hardware).

A hardware realization of fuzzy systems for air quality control has been suggested in [Breiden 1994]. Different sensor signals like humidity, temperature, and air speed are processed on the presented fuzzy micro controller in order to control different devices such as air conditioners which assure good air quality. There are further publications dealing with the efficiency of fuzzy methods for the assessment of environmental issues, see e.g. [Chen 1993].

A software solution for quality control of software is described in [Ebert 1994] where a rule-based fuzzy system has been proposed to classify different software components as having "a high error rate" or "needing high development effort" [Ebert 1994, pp. 349]. Since such classes of software components are often not well separated, fuzzy logic turned out to be a well suited approach for this kind of quality control.

Rule-based fuzzy systems have also been used to support quality control in electronics production where solder joints in printed circuit are inspected [Feldmann, Sturm 1994]. X-ray sensing data are acquired and analyzed to find the most appropriate quality class for each inspected item.

Quality control of diesel fuels has been suggested by interpreting the respective gas chromatograms [Meier 1994]. Characteristic peaks from such chromatograms have been taken as input values for fuzzy clustering by fuzzy c-means [Bezdek 1981]. This way, a separation between good and bad samples of diesel fuels could be realized. A similar application of spectra analysis is presented in [Kasper, Weber 1995] where fuzzy clustering and neural networks have been used for acoustic quality control of ceramic products, e.g. tiles. This application as well as the following realizations have been performed using the software tool DataEngine [MIT 1998].

At the production line of washing machines a system for quality control of the final products based on laser vibration measurements has been developed in a European research project called MEDEA [Fochem et al. 1997b]. Automatic diagnosis of antifriction bearings with fuzzy logic is described in [Fochem et al. 1997a] and [Strackeljan, Behr 1996]. With an advanced monitoring system for machinery with variable rotor speed inadmissible rotor deflections, surface caking, crack initiation in turbine blades, instabilities, and damage to journal and roller bearings can be reliably detected [Strackeljan, Behr 1998].

Fuzzy clustering has been applied to detect inductive motors characterized by an undesired noise, due to rotor shaft vibrations [Fogliardi et al. 1997]. Motor vibrations are measured using two orthogonal accelerometers. A set of features related to these vibration characteristics is defined. Using these features, fuzzy clusters of *good* (noiseless) and *bad* (noisy) motors are generated. These classes are used for an on-line application, in order to classify motors as *good* or *bad*. The results indicate that fuzzy c-means is a suitable method for vibration evaluations in industry.

Quality control of tape deck chassis has been performed based on sound instead of vibration data. Here, a Kohonen network [Kohonen 1995] is applied to the problem of securing quality standards and fault detection of tape deck chassis [Hofmeier et al. 1997]. The acoustic behavior of the chassis is measured using an acoustic sensor. A set of features related to the acoustic characteristics of the chassis is defined. Four classes of faulty chassis and the „good“ class are predefined by experienced specialists. The results indicate that the trained Kohonen network is able to classify the tape deck chassis online on the production line.

Quality control of gearboxes has been performed successfully with techniques from dynamic fuzzy data analysis, see e.g. [Joentgen et al. 1999b]. In dynamic data analysis objects are described by trajectories rather than by points in the feature space. Clustering of these trajectories considers therefore the history of all objects instead of just looking at the current feature values. To cluster such trajectories the functional fuzzy c-means (FFCM) has been developed [Joentgen et al. 1999a].

The area of image processing also offers various possibilities for quality control. Analyzing the information contained in an image of a product to be tested can give valuable information about its quality. The analysis of the break structure of windshields has been performed by image processing combined with fuzzy technologies and neural networks [Fochem 1998].

Newer applications of electronic noses demonstrate the suitability of neural networks and fuzzy classifiers for the detection and identification of different odors [Ping, Jun 1996], [Fochem et al. 1998]. These systems are in general composed of two interacting subsystems: an array of olfactive sensors and a pattern recognition algorithm.

4.3 PRESENTATION OF APPLICATIONS FOR QUALITY CONTROL

Various applications of soft computing techniques show their high potential for quality control. Next, three realizations are presented in more detail. In the first one, neural networks are used for acoustic quality control of saw blades. The second deals with fuzzy classification for non-destructive testing of metal-bonded-to-rubber components. Finally, an optical color measuring system based on a fuzzy classifier is presented.

4.3.1 Acoustic Quality Control of Saw Blades

For the following application, neural networks have been used to build a system for quality control of saw blades [Brandt et al. 1996]. The quality of such saw blades is established by assigning each saw blade to one of four possible classes. A blade can be unrolled, weakly rolled, accurately rolled, and strongly rolled. Based on vibrations which are measured by laser techniques each saw blade is assigned to one of these four classes.

The different ways how the saw blades are rolled influence the measured parameters so that the analysis of the respective vibration measurements allows a classification into the four quality classes.

Various techniques for data acquisition have been used, for example a microphone to record the time signal. This technique has the disadvantage of background noise influencing the results. Therefore laser vibration measurements were performed which showed superior results as only the vibration of the saw blade was detected, opposed to any sound. In order to have statistically reliable data, eight measurements have been taken for each saw blade.

The measured time signals are pre-processed by the appropriate signal processing techniques, e.g. [Famili et al. 1997]. By applying a Fourier transform the time signal is transferred into the frequency spectrum. For good and bad saw blades these frequency spectra show different characteristics which lead to the extraction of the features separating them into different quality classes. These features are input for a neural network which has been trained to perform the separation. In this particular case, backpropagation algorithm (see e.g. [Rumelhart, McClelland 1988]) and a self organizing feature map proposed by Kohonen [Kohonen 1995] have been used. Both algorithms have been applied using the software tool DataEngine [MIT 1998].

The following figure shows the described process of vibration analysis.

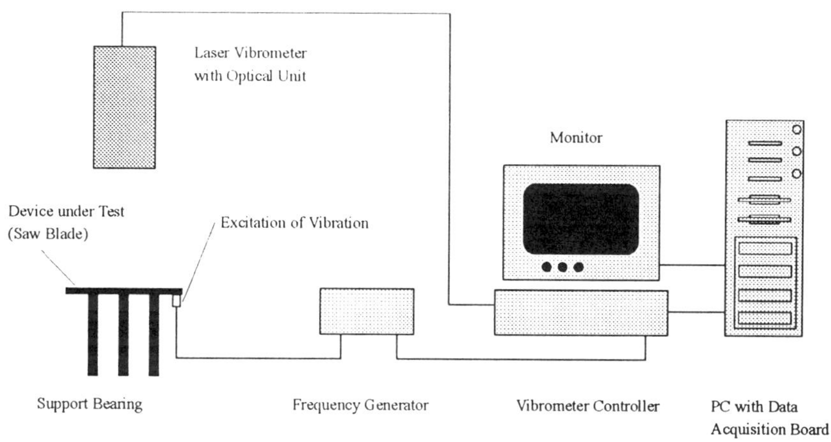

Figure 1. Vibration analysis of saw blades

The first approach for feature selection is based on features which are computed from the frequency spectrum by dividing this spectrum into 96 intervals of equal size. In the second approach, characteristic features have been selected from the frequency spectrum in order to separate between the different classes.

By combining both algorithms with both approaches for feature selection we arrive at four different combinations. In all these combinations 120 examples have been used for training and 30 new examples have been classified. By comparing the classification given by the neural networks with the known classes for each saw blade, one could compute the rate of misclassification which is given below.

Table 1. Misclassification rate (in %)

	Backpropagation Algorithm	*Kohonen Feature Maps*
First approach (intervals of equal size)	36.7	27.7
Second approach (Features based on feature selection)	10	6.7

In both approaches Kohonen feature maps lead to better results than the back-propagation algorithm. In the application described here a misclassification rate of 6.7% was satisfactory.

4.3.2 Non-Destructive Testing of Metal-Bonded-to-Rubber Components

Metal-bonded-to-rubber components are important elements in many mechanical constructions where safety aspects must have absolute priority. For the isolation of machinery vibration these elements are used as vibration dampers. Most tires contain valves which are typical combinations of rubber and metal components. Unfortunately there are only a few non-destructive methods, like X-ray, to detect faults in the bending between the two components. Therefore, a method which will be able to make a sure classification of a test piece during the production process will be of great interest.

Essentially the quality of a vulcanization process is influenced by the parameters of temperature, state of the metal surface and the correctness of the rubber mixture. So it is evident that different production batches are not exactly identical and for the development of a non-destructive quality control system it is important that the variation of production does not cover the fault induced components.

The presented method concerns the evaluation of the state of a technical system capable of vibration. In such cases, it is totally immaterial at first whether the object in question is an active system capable of displaying characteristically natural behavior during operation or whether a passive system is present which offers a system response only as a result of a forced external excitation. To the first group belong, for example, all rotating machine components for which the vibrations of the system can be registered by measuring the body or air sound with suitable sensing devices. The evaluation of such objects is ordinarily subsumed under the heading of vibro-acoustic diagnosis. For the technical diagnosis of passive systems, conversely, an excitation system is required which, for example, introduces a very high energy and extremely short shock wave in order to generate a broad frequency spectrum into the test object. To be sure, for this purpose a very precise and arbitrarily reproducible trigger must be at hand in order to obtain a phase relationship for different measurements. Upon the triggering of unconditioned signals vibrating with frequencies in the kHz range these conditions are technically only rarely realized so that time signals should be used only under certain reservations. The transformation of the time signal into the frequency range allows these problems to be almost totally avoided. This is due to the calculation of a power density spectrum the phase information of the signal is of no significance. The omission of the phase spectrum naturally means a 50% loss of information. Figure 2 illustrates the testing equipment for the non-destructive testing of metal-bonded-to-rubber components.

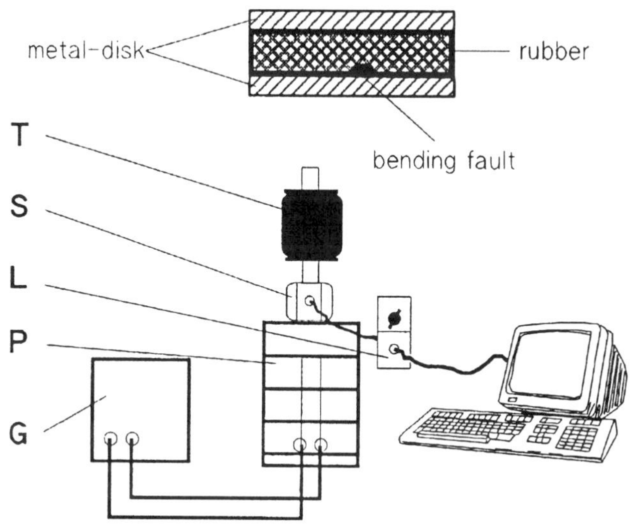

Figure 2. Test appliance for non-destructive testing of metal-bonded-to-rubber components

The test piece (T), the piezo-ceramic exciting system (P) and the mounting is excited to oscillations with a short pulse duration (G) of 10^{-5} s. A force sensor (S) measures the reaction to this excitation, which depends only on the variations of the test piece, when all other components of the test appliance keep sufficiently constant.

After the charge amplification, the time signal is transformed to the frequency domain by calculating the power spectra of the digitized data. Now a total of 400 spectral amplitudes are available as potential features for a classification algorithm. The vibration signals of a tire valve with a bending fault varies from a valve without production mistake because there are small variations in the natural frequencies and the damping behavior.

During a first test phase the present system was able to classify a test set of 70 valves without any mistakes.

4.3.3 An Optical Color Measuring System based on a Fuzzy Classifier

Expanding on the previously presented application for the classification of mechanical vibrations, the functional capacity of fuzzy classification techniques will be demonstrated by the example of a realized optical measurement system. In doing so, we make use of the fact that it is completely a matter of indifference for the classifier whether the problem in question involves a mechanical or an electromagnetic vibration signal.

The automatic color determination has great importance within the context of quality assurance as the product quality is directly related to the product color. Besides the actual production of the color and its further processing, e.g. in the case of the enameling of body parts in the automobile industry, which presumes the highest degree of reproducibility possible, the use of automated color measuring systems is also advisable wherever the color of a product must be established by color comparisons. Thus, for example, the color determination of prosthetic material in the dental sector is generally done by subjective evaluation by means of a visual spectrum with the naked eye. In developing an automatic color measuring system, the foreground is not occupied by the ability to improve color distinction by human beings who can differentiate effortlessly more than 3*E5 colors. It is rather to quantify the color perception which depends strongly on the radiation distribution of the light sources by an objective measurement procedure.

Difference reflectometry provides the principle for a measurement process in which the characteristic, i.e. the directly reflected radiation intensity for the object in question, $R_p(\lambda)$ is measured relative to a $R_s(\lambda)$ standard in the wavelength range between 380 and 780 nm. In this case the test surface to be examined is irradiated by a photoconductor on an area of approximately 2 by 2 mm with white light. The reflected light passes through a second photoconductor into a spectrometer in which, after breaking up according to wavelength dispersion on a grating, the light intensities assigned to the individual wavelengths are simultaneously measured by a photodiode array line. The diode voltages which are proportional to the light intensity are digitized, transmitted to a computer and processed further.

In practical use of this system, however, problems arise in the reproducibility when plotting the surface-specific reflection spectra. The measurement head must be guided without trueing on an unpolished and therefore uneven surface. Due to angular variations with respect to the normal to the surface, as a rule extremely varying intensities of the reflected light are registered and a comparison with previously recorded reference patterns is no longer possible.

As a brief demonstration of the system's capacity, three colors were distinguished which are classified by the basic shade of the color "green" which differ quite clearly in optical impression.

In the first experiment, the photoconductor was guided by hand but positioned essentially perpendicular to the sample surface. Figure 3 shows for the classes $\Omega_{1,2,3}$ as typical representatives of the learning set in each case 8 curves which are selected randomly from the total set of all 300 random samples. An unequivocal separation of the reflection spectra on the basis of nearly every intensity amplitude in the wavelength range from 400 to 780 nm is possible without any problems so that the use of a classification system on this measurement series may not be explained.

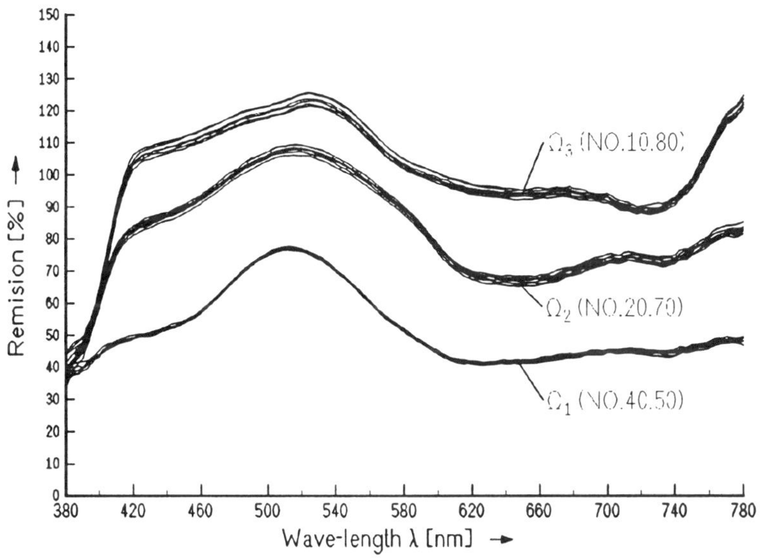

Figure 3. Reflection curves of 3 shades of green with a constant position of the measurement head

The distinguishability, however, becomes more difficult if the photoconductor is placed at different angles on the sample surface because then the intensity of the back-radiated light changes upon each measurement.

From the absolute position of the reflection curve now no conclusions can be drawn regarding the pertinent colors. Figure 4 shows that a separation of the spectra is no longer possible by analyzing a single feature.

With a wavelength distance $\Delta \lambda = 2.5$ nm of the diodes, in the wavelength range in question there are 160 amplitudes which are available basically as input features of the classification process.

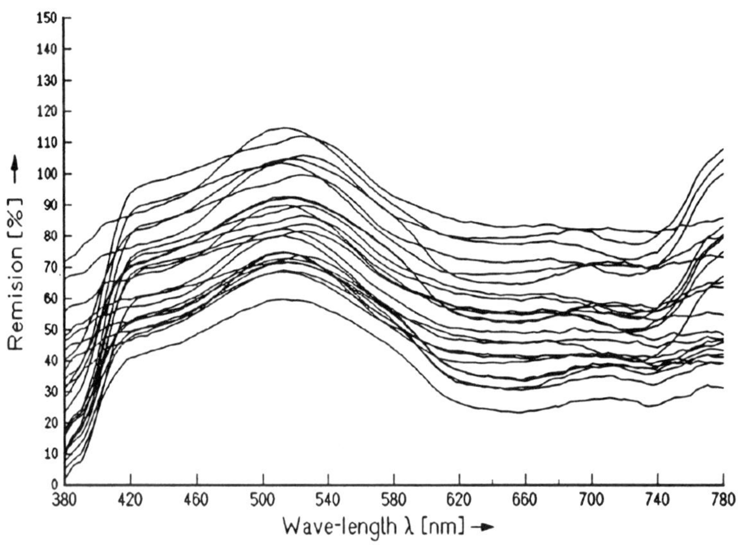

Figure 4. Reflection curves of 3 shades of green with the measurement head in variable position.

Table 2 shows the classification results for the most suitable features. Already with just two features the classification problem can be solved for a learning set with 30 samples for each class without reclassification errors.

Table 2. Results of the classification process for distinguishing different shades of green

Number of features	*Wave-length [nm]*	*Recall [%]*	*Test [%]*
2	460, 503	100%	93%
3	395, 450, 503	100%	95%
4	395, 423, 460, 503	100%	98%
5	395, 423, 460, 503, 620	100%	100%

A 100% correct reclassification of the recorded test set of 50 data sets was accomplished with 5 features.

Since these results can also be confirmed in online operation of the classifier even under altered incidences of light, the selection of features can be considered as a demonstration of the functional/performance capacity of the classification algorithms used.

4.4 APPLICATIONS OF FUZZY LOGIC FOR MAINTENANCE

This section describes applications of fuzzy logic for condition monitoring of rotating machinery (4.4.1) as well as diagnosis systems for maintenance management (4.4.2) and alarm management in petro-chemical plants (4.4.3). Other realizations for maintenance management have been suggested e.g. for power plants where the best from a given set of inspection alternatives has to be selected by considering multiple criteria [Jovanovic et al. 1993].

4.4.1 Condition Monitoring of Rotating Machinery

Safety- or production-relevant plant components are already subject to monitoring at present and the measurement of vibration signals is a suited diagnosis parameter. However, most commercially available machine monitoring systems are not sufficiently well suited for performing diagnostic functions. The following shortcomings are especially worth mentioning:

- insufficient stability towards noise load and interference, such as the effect of cavitation;
- utilization of only one, or very few diagnostic parameters;
- susceptibility to severe fluctuations in measured signals;
- limit-value-oriented monitoring of a restricted frequency range by mask techniques;
- methodically inadequate method for classification of specific industrial process malfunctions.

For example, there could be a significant change in the vibration behavior of a pump, but the high background level makes it impossible to detect changes in a signal by simply measuring the overall level of a signal or some other obvious characteristics.

Additionally, machine malfunctions often display similar symptoms and can be interrelated. If only vibration amplitude and frequency are observed many malfunctions can appear as imbalance.

A decided improvement in diagnostic reliability can be achieved by simultaneous consideration of several vibrational features and supplementary processing of as many process parameters as possible, such as reactor pressure, temperature, power etc. In this conjunction, no competitive alternative is currently available for the application of classifying pattern recognition methods for automatic appraisal of a technical object. As a result, a multi-dimensional feature vector which includes all measured data as components must be projected as a scalar value to describe a class membership. This parameter is then employed for deciding whether the actual feature vector is included in one of the previously defined damage classes, and in which of these. This task can be performed continuously and completely automatically by a computer, without the need of a human expert for conducting the diagnosis. This fact could be important because experts are generally not available 24 hours a day and the human decision may be affected by some environmental factors.

For the task of machine monitoring, it is clear from the start that a model - and thus mathematically describable - relationship between the cause of damage and vibration signal is available only in rare cases for such complex combinations of many machine components under the influence of external fluctuation parameters. The efficiency of the associated classification algorithm is thus determined essentially by a 'learning process' which precedes the diagnostic process itself. In this learning process, certain free parameters of the classifier must be specified on the basis of unambiguous, pre-classified representatives of various classes. During the subsequent working phase, unknown objects which 'resemble' the previously recorded learning example only to an extent yet to be specified are then independently assigned to one of the previously defined significance classes. An essential aid for designing a classifier of this kind is the mathematical formulation of fuzzy sets.

As an application of the diagnostic system, we describe here the early detection of defects on roller bearings. The initial stage of bearing damage is characterized by dislodging of very small material particles from the running surface, by the appearance of fatigue cracks near the surface, or by rupture or bursting of bearing components. The geometrical alteration of the race is revealed by the generation of a short shock by the roller during passage over the site of damage. This shock propagates in the form of an approximately spherical wave front from the site of origin and excites nearly all bearing components capable of vibration to characteristic vibrations in the corresponding frequency range.

For the detection of bearing faults we calculate 24 secondary features from the vibration signal by a computer-aided data acquisition system. The implemented classifier calculates the membership values and provides a complete appraisal of the current state of the bearing at intervals of 1 min. For the convenience of the operating personnel, the membership values are determined only for the four global classes, 'shut-down', 'intact state', 'change in condition', and 'defect'. In figure 5 the evolution of the membership values is plotted over a period of 19 hours.

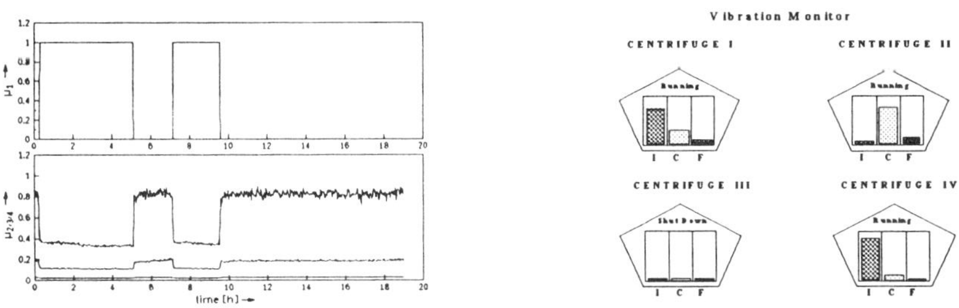

Figure 5. Membership values for the classes Ω_1 to Ω_4

The membership μ_1 for class Ω_1 is plotted in the upper section of figure 5. This membership is determined in a fuzzy manner, but only the values 0 and 1 are employed for indicating membership in the 'shut-down'-class. The lower section of figure 5 shows the class membership values for the classes Ω_2 to Ω_4. The advantage of projecting 24 characteristic values onto a single value, which can be followed without difficulty on a trend plot, is evident. If the feature vector is considered more exactly, of course, the system can provide information which extends beyond the pure representation of class membership.

Figure 6 illustrates the learning set for the four defined classes for the sole consideration of the Kurtosis features and the KPF- value. As a ratio between the fourth statistical moment and the square of the standard deviation (second moment), the Kurtosis value represents a measure which describes the convexity of a distribution. KPF is a characteristic damage frequency which can be calculated from the bearing geometry and rotational speed of the shaft and represents the associated, referred amplitude in the power density spectrum.

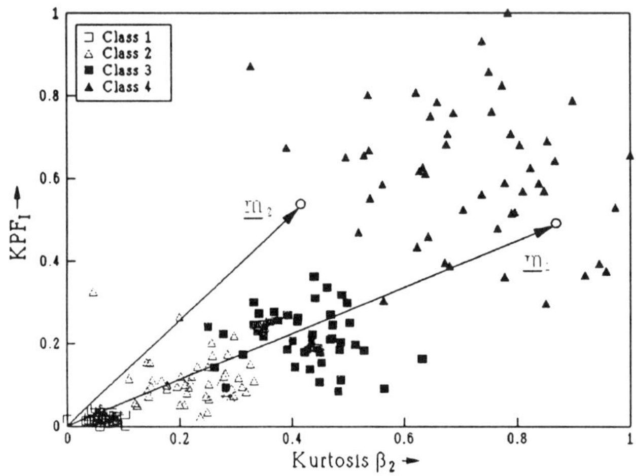

Figure 6. Learning samples for the diagnosis of roller bearings

Only a very slight scatter of the feature vectors occurs for the exactly defined shut-down class, whereas the scatter increases considerably for the 'change' and 'defect' classes. An example of the importance of recording more than one process parameter during diagnosis is presented in figures 7 and 8. Figure 7 shows a typical time signal for a roller bearing with an outer race damage. In figure 8 a decided variation in the damage signal for the same roller bearing and defect as a result of cavitation effects is evident.

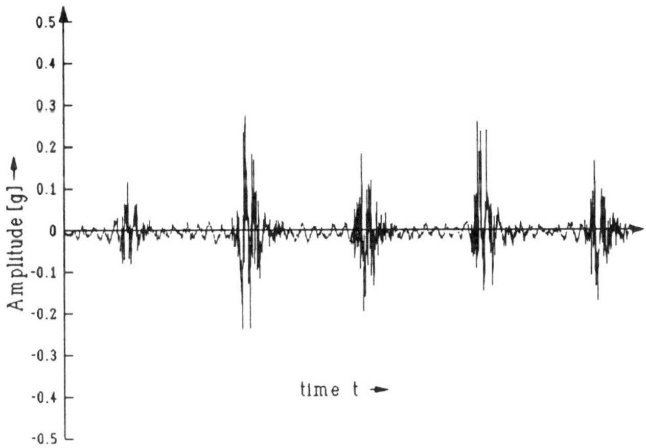

Figure 7. Time record showing typical fault signature

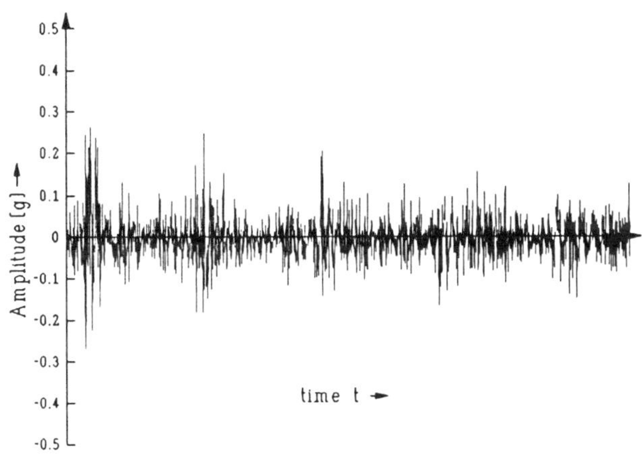

Figure 8. Time record of a defect roller bearing with cavitation interference

The Kurtosis factor decreases to the value for a nearly intact bearing because of the pronounced superposition of the additional excitation as a result of cavitation. In this case, it would not permit any conclusion concerning the machine status. In combination with the KPF-values, however, the diagnostic system provides a correct appraisal which is largely independent of external effects.

The example presented here indicates the suitability of a fuzzy pattern recognition approach, precisely for designing a monitoring system for rotating components for which an analytical description of the relation between cause and effect cannot be provided. At a plant of the primary chemical industry, the described diagnostic system has been in operation for more than two years to monitor several continuously operating centrifuges, and the effectivity has been proved repeatedly by early detection of various faults [Strackeljan, Behr 1996].

4.4.2 Diagnosis Systems and their Integration into the Site-Wide Production and Maintenance Management

The implementation of diagnosis systems can be achieved on different production management levels, although goals and assumptions will be very different. Decision support systems proposing appropriate strategic actions (off-line) or classifiers proposing low level control actions are applied to the production, plant or enterprise management level.

An operator's point of view could be different to the one of plant managers who are more concerned with plant wide handling and control of energy, raw material, intermediate and final products. The use of diagnosis systems in this context requires the application of various information and automation technologies and a certain strategy to fit them together to one entity (Fig. 9). The diagnosis results should not only prevent production loss or environmental exposure, but also adapt specific maintenance strategies to the production schedules on the different management levels (unit, plant, integrated plant environment) and their product strategies.

While the monitoring system described in section 4.1 allows only diagnosis of roller bearings the following example shows that similar techniques could be integrated on a higher level in a decision making system.

To achieve a comprehensive company strategy some targeted decisions and appropriate actions on the plant management level are necessary in operations, maintenance and safety standpoints (Fig. 9). The present diagnosis system technology which had been installed in a chemical company is not thought to automatically shutdown units or plants. These actions are taken in dedicated systems as single sources of process controllability. The input from diagnosis systems will be valued as decision support in the control system. So a consistent strategy is possible without duplication of functions. The significant function of human experts is so far incorporated into the control program and manual action requests.

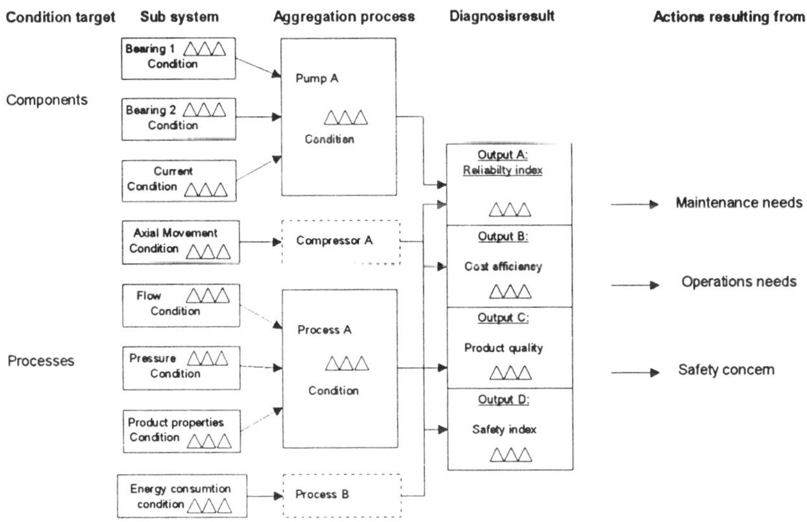

Figure 9. Hierarchical structure of fuzzy evaluations on the plant management level

The need exists to merge single, uncertain judgements into one consolidated result on different management levels of the production process. Different methodologies are selected in accordance to the appropriate levels. It is necessary to condense various single components and sub-process valuations, e.g. motor bearings as part of a compressor to one judgement, how the machine or process as a whole runs, whereby the fuzzy methodology contributes to the input tolerances and uncertain human decision making process. Variables like motor current, bearing oil pressure and flows deliver additional information that do not influence the complexity of the system, since a transparent process exists to gain the resulting machine or process valuations out of class related single component evaluations. The triangles in figure 10 demonstrate the general fuzzy approach of the hierarchical evaluation system.

Figure 10 [MIT 1998] shows an example of the fuzzy input processing the variable 'Safety Index' and three rules out of 65 in the rule base. The results are high level indicators supporting human decisions regarding process safety, plant reliability, process quality and cost efficiency through output versus input evaluations. These linguistic variables could be processed further in interlinking support systems. Knowing the business strategy the decision maker can set priorities and conduct further the processing of indexes to final strategies to run whole entities [Schubert et al. 1998].

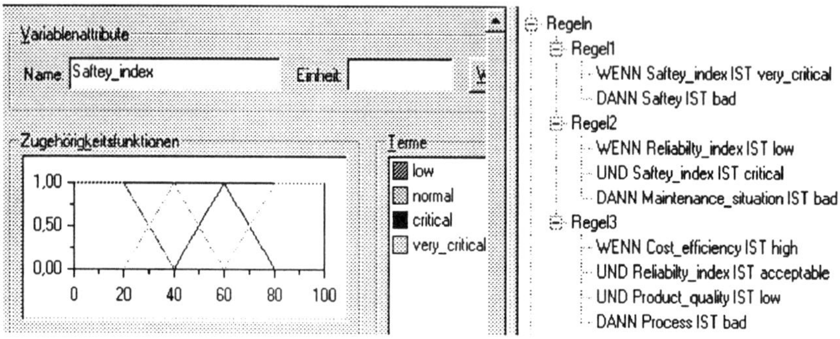

Figure 10. Using a fuzzy rule base for the diagnosis of process, safety and maintenance on the level of site management

4.4.3 Maintenance of petro-chemical plants

Commercial thermal cracking plants produce about 360,000 t ethylene a year. The process of thermal cracking of petroleum hydrocarbons with steam is commonly called pyrolysis or steam cracking [Meier et al. 1994]. Naphtha, which is obtained by distillation of crude oil, is the principal raw ethylene material. Boiling ranges, densities, and compositions of Naphtha depend on crude oil quality.

Naphtha is heated in cracking furnaces up to 820 ^{o}C - 840 ^{o}C, where the chemical reaction starts. During the cracking process also acetylenic, diolefenic and aromatic compounds are produced, which are known to deposit coke on the inside surfaces of the furnace tubes. This coke layer inhibits heat transfer from the tube to the process gas, so that at some time the furnace must be shut down to remove the coke. To guarantee a continuous run of the whole plant, several furnaces are integrated into the production process. Figure 11 shows the described scheme of thermal cracking.

By the process data alone, it is not possible for the operator to decide whether the furnace is coked or not. His experience and the running time of the regarded furnace is the basis for his decision. Next, a method is described, which is suitable to determine the degree of coking based on on-line measured process data.

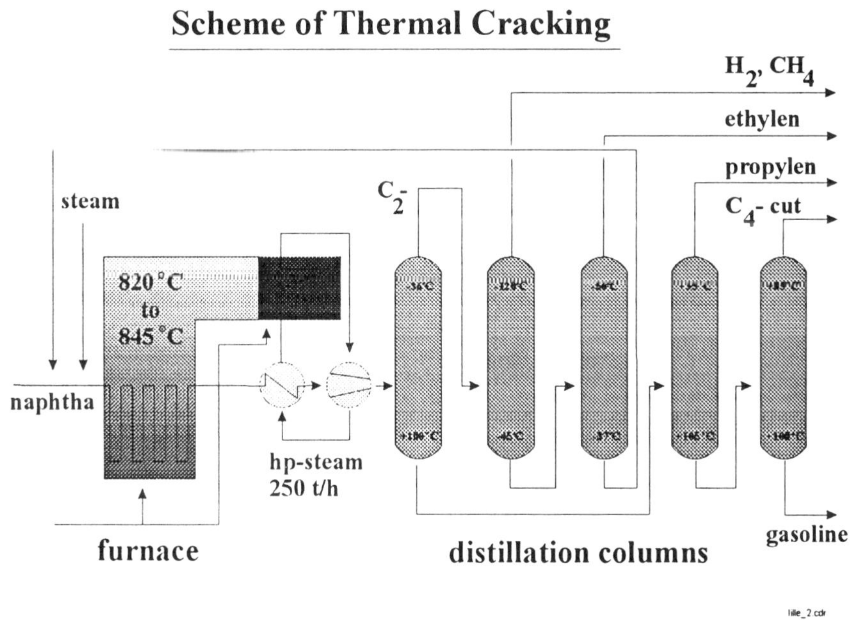

Figure 11. Scheme of thermal cracking

Solution by data analysis

Modern process control systems collect process data and archive them. Based on this data set, the presented classifier is designed by clustering. Clustering methods compress information of data sets by finding classes, which can be used for classification. Similar objects are assigned to the same class. In our case, objects are different states of a cracking furnace during a production period. Objects are described by different features. Features are the on-line measured quantities like temperatures etc.. The problem is to find the right features for the regarded problem. There are some mathematical methods like principal component analysis to reduce the number of features down to three or two. Alternative approaches for automatic feature selection are presented e.g. in [Behr et al. 1997].

For the task of feature selection the support of experts is also required. Each feature leads to one dimension of the feature space. Clustering algorithms find accumulations of objects in that space. These accumulations are the different classes. For this application, fuzzy clustering based on fuzzy c-means has been used [Bezdek 1981]. It is important, that the so found classes can be interpreted by the practitioner. He may recognize that one class contains good process states, the other class bad ones. The state of the furnace is described by several features. Figure 12 shows a brief sketch of the furnace and the measured features.

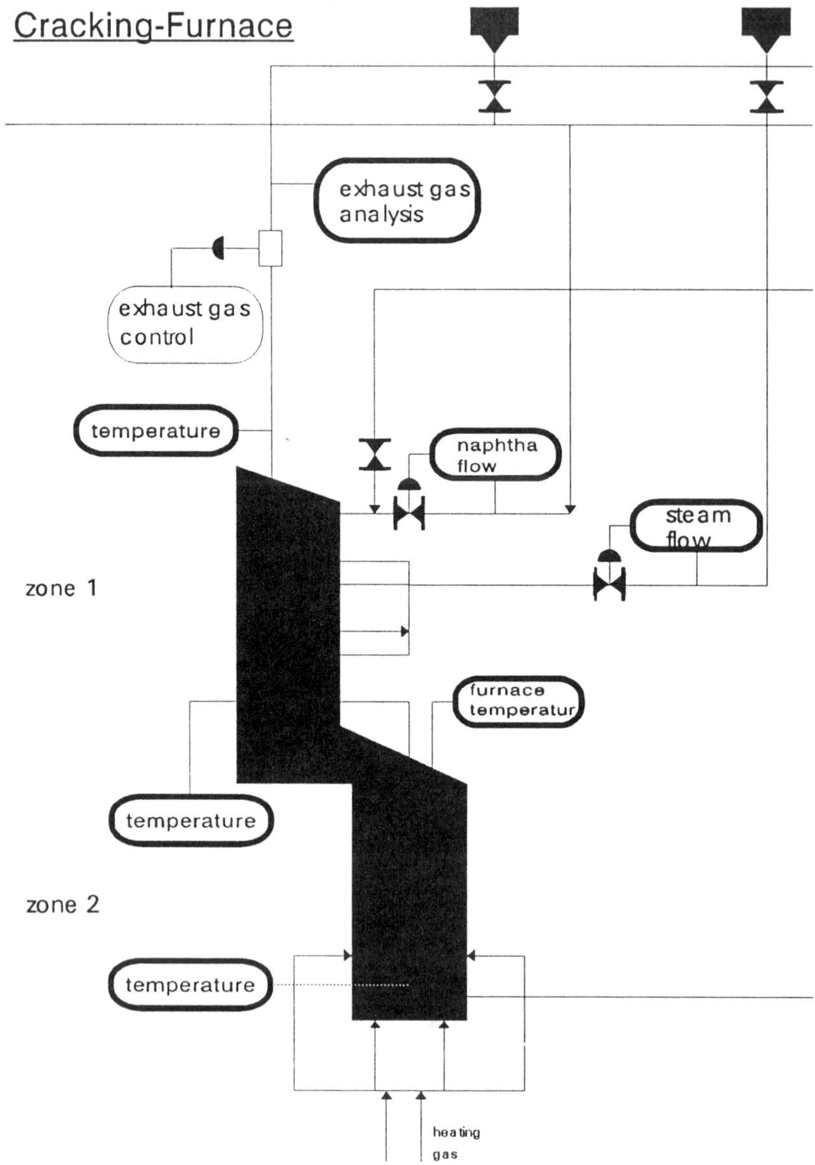

Figure 12. Cracking furnace

About 20 measurements of different indicators, such as, temperatures, pressures, or flows are taken each minute. So the information, which is hidden in a large data set, can be compressed by finding classes and designing a classifier. With fuzzy classification, processes can be studied, which continuously move from one state to another. One of these processes is the above described coking of cracking furnaces. After the production period the furnace is shut down and the coke is burned out with a mixture of steam and air. After that the furnace is reintegrated into the production process until it has to be decoked again. During a production period the furnace state starts at the decoked class and runs continuously towards the coked class as is shown in figure 13.

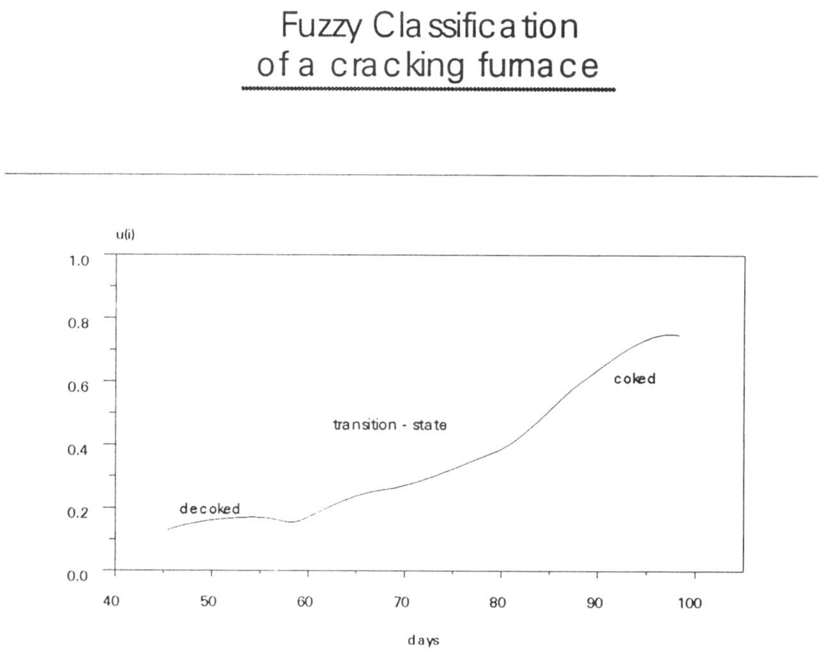

Figure 13. Membership of process state to the class 'coked process state'

The membership value shows the degree to which the current process state belongs to the class of coked situations and hence the feature which describes coking of cracking furnaces.

The classifier works on-line and classifies the current furnace state concerning the coking problem. The operator can use this information to check how long the regarded furnace will be able to run until it has to be decoked. Now it is easier to make arrangements concerning logistical questions like ordering the right amounts of raw material or not to be understaffed at certain times.

4.5 CONCLUSION AND FUTURE PROSPECTIVES

This chapter has given an overview of applications of soft computing and especially fuzzy logic in the areas of quality control and maintenance. Section 2 mentions several brief references to relevant publications, whereas sections 3 and 4 describe realizations in the two investigated areas in more detail. Thus the high importance of fuzzy logic in the respective application fields is underlined and possibilities to transfer the described applications to other similar tasks are given.

In the tasks described in this chapter, signals have been analyzed and the information contained in these signals could be used for quality control or maintenance. Signal analysis is a very broad area where different techniques, methods, and theories have been employed. In order to build a basis for reliable decisions on such approaches, an "International Competition on Signal Analysis" has been initiated in the framework of the European network of excellence ERUDIT. Problem formulation and data from now four test cases have been published and everybody could apply her or his approach to solve the problem. An independent board of referees has judged on the submitted solutions and the first three winners have been awarded during the conferences EUFIT '96 and EUFIT '98 in Aachen (Germany). Their solutions are published in the ERUDIT newsletters 3/96 and 3/98 which are available on request from both authors.

4.6 REFERENCES

Bezdek, James C. (1981). *Pattern Recognition with Fuzzy Objective Function Algorithms.* New York London: Plenum Press

Brandt, H., Lahmann, H.-W., Weber, R. (1996). "Quality control of saw blades based on neural networks and laser vibration measurement." *Second International Conference on Vibration Measurements by Laser Techniques: Advances and Applications.* SPIE - The International Society for Optical Engineering, Vol. 2868, 119 – 124.

Breiden, O. (1994). "Fuzzy Micro Controller and Gas-Sensors in Air-Quality Control Applications." *Proceedings EUFIT '94 - Second European Congress on Intelligent Techniques and Soft Computing*, September 1994, Aachen, Germany, 1746-1748.

Chen, G. N. (1993). "Assessment of environmental water with fuzzy cluster analysis and fuzzy recognition." *Analytica Chiminca Acta* 271, 115 – 124.

Ebert, C. (1994). "Rule-based fuzzy classification for software quality control." *Fuzzy Sets and Systems* 63, 349-358.

Famili, A., Shen, W.-M., Weber, R., Simoudis, E. (1997). "Data Preprocessing and Intelligent Data Analysis." *Intelligent Data Analysis* 1 (1), http://www.elsevier.com/locate/ida

Feldmann, K., Sturm, J. (1994). "Closed Loop Quality Control in electronics production based on fuzzy logic and neural nets." *Proceedings EUFIT '94 - Second European Congress on Intelligent Techniques and Soft Computing*, September 1994, Aachen, Germany, 775-781.

Fochem, M., Geropp, B., Joentgen, A. (1997a). "Automatic Antifriction Bearing Diagnosis with Fuzzy Logic." *Proceedings of ESIT - European Symposium on Applications of Intelligent Technologies*, September 9-10, 1997, Aachen, Germany

Fochem, M., Weber, R., Wischnewski, P. (1997b). "Classifier Design Using Intelligent Technologies and Integration into the Production Line of Household Appliances." *Proceedings of EUFIT '97 - 5th European Congress on Intelligent Techniques & Soft Computing*, September 8-11, 1997, Aachen, Germany, 1749 – 1754

Fochem, M. (1998). "Automatic Break Pattern Analysis of Automobile Glass." *NIweek 98*, Proceedings CD, August 25 – 27, 1998, Austin, Texas

Fochem, M., Schrötter, T., Tselentis G., Wischnewski, P. (1998). "Automatic feature selection applied to electronic nose data for classifying odours." *Proceedings of EUFIT '98 - 6th European Congress on Intelligent Techniques & Soft Computing*, September 1998, Aachen, Germany, 1205-1210.

Fogliardi, R., Wischnewski, P. Fochem, M. (1997). "Online Fuzzy Identification of Noisy Electric Motors on the Production Line with LabVIEW® and DataEngine V.i." *NIweek 97*, User Symposium Proceedings, August 18-21, 1997, Austin, Texas

Hofmeier, R., Fochem, M., Wischnewski, P. (1997). "Online Quality Control System on the Production Line of Tape Deck Chassis using Kohonen Networks." *NIweek 97*, User Symposium Proceedings, August 18-21, 1997, Austin, Texas

Ishikawa, K. (1990). *Introduction to Quality Control*, Tokyo: 3A Corporation

Joentgen, A., Mikenina, L., Weber, R., Zimmermann, H.-J. (1999a). "Fuzzy Dynamic Data Analysis Based on Similarity Between Functions." *Fuzzy Sets and Systems*, accepted for publication

Joentgen, A., Mikenina, L., Weber, R., Zeugner, A., Zimmermann, H.-J. (1999b). "Automatic Fault Detection in Gearboxes by Dynamic Fuzzy Data Analysis." *Fuzzy Sets and Systems*, accepted for publication

Jovanovic, A., Psomas, S., De Witte, M., Bath, U. (1993). "Development of an Integrated System for Decision Optimization in Power Plants Maintenance." *Proceedings EUFIT '93 - First European Congress on Fuzzy and Intelligent Technologies*, Aachen, September 1993, 83-86.

Kasper, C., Weber, R. (1995). "Quality Control with Fuzzy Logic and Neural Networks in LabVIEW®." *NIweek 95*, User Symposium Proceedings, July 26-28, 1995, Austin, Texas, pp. 4-1 ,..., 4-5.

Kohonen, T. (1995). *Self-Organizing Maps*, Berlin, Heidelberg: Springer-Verlag

Meier, W. (1994). "Quality control of diesel fuel with fuzzy methods." *Proceedings of EUFIT' 94 - Second European Congress on Intelligent Techniques and Soft Computing*, Aachen, Germany, 111-115.

Meier, W., Weber, R., Zimmermann, H.-J. (1994). "Fuzzy Data Analysis - Methods and Industrial Applications." *Fuzzy Sets and Systems* 61, 1994, 19-28.

MIT (1998). *DataEngine 3.0 - Manual*. Aachen

Ping W., Jun, X. (1996). "A novel recognition method for electronic nose using artificial neural networks and fuzzy recognition." *Sensors and actuators* 37 (3), 169-174.

Rumelhart, D.E., McClelland, J.L. (1988). "Parallel Distributed Processing". 8th ed., MIT Press

Strackeljan, J., Behr, D. (1996). "Condition monitoring of rotating machinery using a fuzzy pattern recognition algorithm." 6th *International Conference on Vibrations in Rotating Machinery*, Oxford, Sept. 9-12, 1996, 507 – 516.

Strackeljan J., Behr, D., Kocher, T., (1997). "Fuzzy Pattern Recognition for Automatic Detection of Different Teeth Substances." *Fuzzy Sets and Systems* 85, 275-286.

Strackeljan, J., Behr, D. (1998). "Vibration Monitoring of Non-Stationary Rotor Systems." *Proceedings of the* 7th *ISCOMAC Conference* Feb. 1998, Hawaii, 135-144.

Schubert, A., Strackeljan, J. (1998). "Fuzzy Diagnosis Methods for production optimization and maintenance." *Proceedings of EUFIT '98 - 6th European Congress on Intelligent Techniques & Soft Computing*, September 1998, Aachen, Germany, 1289 – 1293.

Zimmermann, H.-J. (1996). *Fuzzy Set Theory - and Its Applications*. Third Edition. Boston, Dordrecht London: Kluwer Academic Publishers

5 USING FUZZY LOGIC FOR MOBILE ROBOT CONTROL

Alessandro Saffiotti
Enrique H. Ruspini
Kurt Konolige

Abstract: The development of techniques for autonomous operation in real-world, unstructured environments constitutes one of the major trends in the current research on mobile robotics. In spite of recent advances, a number of fundamental difficulties remain. In this chapter, we discuss how fuzzy logic techniques can be used to address some of these difficulties. To illustrate the discussion, we describe the fuzzy-logic solutions developed on Flakey, the mobile robot of SRI International.

5.1 INTRODUCTION

The operation of an autonomous mobile robot in a real-world unstructured environment requires consideration of multiple issues.

First, the controller must be able to operate under conditions of imprecision and uncertainty. For example, prior knowledge about the environment is, in general, incomplete, uncertain, and approximate. Perceptually acquired information is also typically noisy and incomplete. Furthermore, the execution of control commands is not completely reliable while the dynamics of real-world environments is complex and unpredictable. To cope with these difficulties, the controller must be able to respond *reactively* to unforeseen events as soon as they are perceived.

Second, the controller must be able to reach explicit goals such as the objectives specified by users or the subgoals generated by the high-level deliberation processes of automated intelligent planners. The controller must be able to incorporate these goals into *purposive* procedures that promote their attainment.

Finally, the controller must be able to consider multiple concurrent requirements. For example, a mobile robot may need to reach the end of a hallway, while keeping its load well balanced, and avoiding static and moving obstacles. It is important to note that these requirements may demand the activation of both reactive and goal-achieving procedures. A major problem in the design of an effective controller for mobile autonomous devices is the *combination* of goal-specific strategies by resolution of conflicts between multiple objectives.

The autonomous robotics literature contains numerous examples of approaches that seek to deal with these problems by decomposition of the overall control function into a number of simple units, called *behaviors*, (Brooks, 1986). Each behavior is responsible for producing certain actions (typically, motion and perception actions), aimed at achieving or maintaining a particular goal such as staying away from obstacles.

Behavior-based approaches to mobile robot control have gained increasing popularity, being supported by considerations arising from the study of animal behavior (Arbib, 1981), by architectural concerns (Brooks, 1986), and by their implementation convenience (Gat, 1992). These approaches face, however, two major problems: the combination of simple behaviors to form more complex ones, and the integration of behaviors with higher-level processes, like a strategic planner.

Fuzzy logic provides tools that are of potential interest to mobile robot control. Most applications of fuzzy logic in this field concern the use of fuzzy control techniques to implement individual behavior units. Fuzzy controllers are a convenient choice when an analytical linear model of the system to be controlled cannot be easily obtained; they have shown a good degree of robustness in face of large variability and uncertainty in the parameters; and they lend themselves to efficient implementations, including hardware solutions. These characteristics fit well the needs of autonomous navigation where: (i) a mathematical model of the environment is usually not available; (ii) sensor data is uncertain and imprecise; and (iii) real-time operation is of essence. It is no surprise, then, if fuzzy control has been the first application of fuzzy logic techniques in this domain since Sugeno and Nishida's model car (Sugeno and Nishida, 1985), and it is still the most common one. (See (Saffiotti, 1997b) for a critical survey.)

Although less noticed until recently, fuzzy logic can also provide useful tools to address another important difficulty in behavior-based approaches to autonomous robotics: how to coordinate the concurrent execution of several behaviors aimed at the achievement of different, possibly conflicting objectives; and how to link this coordinated activity to the global analysis performed by the higher-level reasoning processes. This chapter presents a high-level discussion of a fuzzy-logic based approach to autonomous navigation that addresses these fundamental difficulties.

We will describe techniques to control an autonomous mobile robot based on the use of fuzzy logic for trading off conflicting goals. The formal bases for these techniques have been set forth by Ruspini (Ruspini, 1990; Ruspini, 1991a) after the seminal works by Zadeh (e.g., (Zadeh, 1978)). In a nutshell, each goal is associated with a function that maps each perceived situation to a measure of desirability of possible actions from the point of view of that goal. Behaviors induced by many simultaneous goals can be smoothly blended into a dynamic sequence of control actions. In particular, reactive and goal-oriented behaviors are blended into a single sequence of control actions.

These techniques have been first implemented on Flakey, the mobile robot platform of the Artificial Intelligence Center of SRI International (Saffiotti et al., 1993; Konolige et al., 1997). The illustrative examples that we present in this paper are drawn from our experience with this robot. The scope of these techniques extends, however, beyond this particular test-bed, as they effectively deal with basic problems in the development of intelligent autonomous systems such as the attainment of multiple, possibly conflicting, objectives; the integration of numerical control and symbolic planning; and the construction and utilization of approximate and incomplete models of the world.

5.2 FUZZY BEHAVIORS

In our approach to robot control, we express desirable behavioral traits as quantitative *preferences*, defined over the set possible control actions, from the perspective of the goal associated with that behavior. For example, a behavior for avoiding obstacles could map configurations of sonar readings that correspond to the presence of an obstacle on the left of the robot into a function that prefers actions that steer the robot to the right.

Following the formal semantic characterization of Ruspini (Ruspini, 1991a; Ruspini, 1991b), we describe each behavior B in terms of a desirability function

$$Des_B : \text{State} \times \text{Control} \rightarrow [0,1]\,,$$

that measures, for each state vector x and control vector c, the desirability $Des_B(x,c)$ of applying the control c when the state is x *from the point of view of attaining goals associated with B*. Equivalently, we can say that Des_B associates each situation x with the fuzzy set $\mathbf{C}$ of control values characterized by the membership function $\mu_C(c) = Des_B(x,c)$.

In general, c is an n-dimensional vector of control values (the nature of the input vector x is further characterized in Sect. 5.4). In the case of our test-bed robot, Flakey, c is a pair of set points (v,θ), where v is a linear velocity and θ is a turning angle. An internal conventional motor controller is employed to achieve the set-points recommended by the fuzzy controller.

Desirability functions are less committing than classical control functions or similar formal objects employed in other approaches to robot control, like Khatib's pseudo-potential fields (Khatib, 1986) or Arkin's motor schemas (Arkin, 1990a). In fact, desirability functions do not map input states to controls but to a measure quantifying preferences over the space of these controls. This

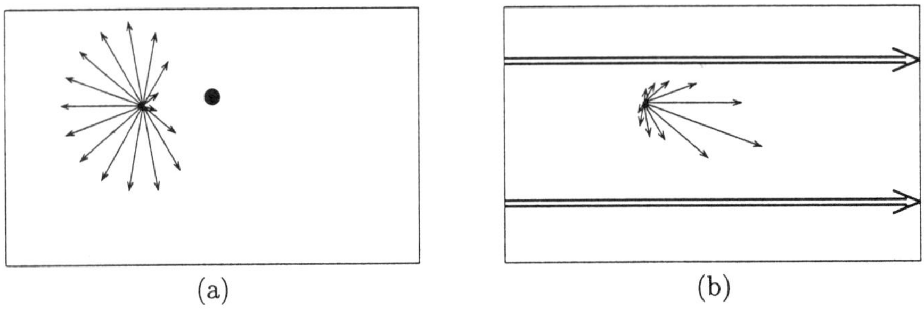

Figure 5.1 Desirability functions for two behaviors: (a) Keep-Off; (b) Follow.

approach recognizes that many alternative controls can generate, to a greater or lesser extent, the same *type* of behavior. The control that should actually be applied in a given situation depends not only on that situation but also on the interaction with other concurrent behaviors. We will see in Sect. 5.7 that the major difference between desirability functions and potential-field approaches lays in the way that behaviors are merged.

To better understand the role of desirability functions, consider a behavior named KeepOff seeking to keep the robot away from a point P with coordinates (P_x, P_y) (e.g., P may correspond to a sensed obstacle). We may define a desirability function for KeepOff by considering, for each control $c = (v, \theta)$, how close the robot would come to P after the next control cycle, if c were to be actually applied.[1] If the new position of the robot after application of the control c is denoted by $\hat{R} = (\hat{R}_x, \hat{R}_y)$, then the desirability of employing that control might be evaluated by means of a ramp function defined on the distance between the predicted position and the obstacle:

$$Des_{\mathsf{KeepOff}}(x, c) = \min\left(1, \frac{1}{d_0}\sqrt{(P_x - \hat{R}_x)^2 + (P_y - \hat{R}_y)^2}\right)$$

where d_0 is a threshold denoting the safe distance.[2]

Figure 5.1 (a) plots this desirability function for a fixed value of v and different values of θ. In this figure, the position of the robot is indicated by the smaller dot while that of the obstacle is shown by the larger one. Each vector indicates a possible turning control θ with its length proportional to the corresponding value of the desirability function. Figure 5.1 (b) shows a possible desirability function for the Follow behavior, which seeks to proceed along a given lane (indicated by the double lines).

Simple behaviors may be composed into more complex sets of actions by combining their desirability functions by means of the logical operations of fuzzy logic. In particular, we utilize the min, max, and complement to one operations to compute the truth value of logical conjunction ($\wedge$), disjunction ($\vee$), and negation ($\neg$), respectively.[3] We will also utilize $\oslash$, the quasi-inverse

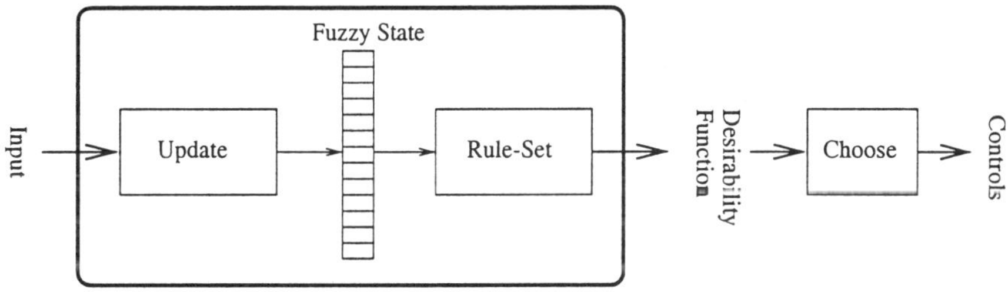

Figure 5.2 Implementation of a behavior in the fuzzy controller.

of $\wedge$, to compute the truth value of a logical implication from those of its antecedent and consequent.

5.3 IMPLEMENTATION OF A FUZZY BEHAVIOR

The definition and implementation of desirability functions is a major aspect of our approach. Proper characterization of these functions is important to assure efficient performance and to facilitate effective combination of multiple behaviors.

We have already suggested a procedure to define such functions in the context of the KeepOff behavior. This type of procedure starts from an analytical specification of the intended behavior in terms of desirable states building then the desirability function from this specification and from a model of the robot. In general this is a complex procedure, which we are currently seeking to simplify and systematize. Currently, however, we follow a more empirical path, and implement desirability functions by a set of fuzzy rules that are experimentally tuned until the produced behavior is judged to be a satisfactory approximation of the intended one.

We implement each behavior by a fuzzy machine having an architecture illustrated in Figure 5.2. The **Fuzzy State** is a vector of fuzzy variables, each representing the truth value, on the $[0, 1]$ scale, of a fuzzy proposition of interest (e.g., "obstacle-close-on-left"). At every cycle, the **Update** module produces a new fuzzy state based on the current input. The **Rule-Set** module contains a set R of fuzzy rules

$$\text{IF } A_i \text{ THEN } C_i, \quad i = 1, \ldots, n,$$

where A_i is a propositional formula in fuzzy logic built from the fuzzy propositions in **Fuzzy State**, plus the logical connectives $\wedge$, $\vee$ and $\neg$, evaluated as explained above, and where C_i is a fuzzy set of control values. For each possible control value c, $C_i(c)$ quantifies the extent by which c is a good instance of C_i. From these fuzzy rules, a desirability function Des_R is computed by

$$Des_R(x, c) = [A_1(x) \wedge C_1(c)] \vee \cdots \vee [A_n(x) \wedge C_n(c)]. \tag{5.1}$$

IF (lane-too-right $\wedge$ $\neg$lane-angled-left) THEN turn-medium-right
IF (lane-too-left $\wedge$ $\neg$lane-angled-right) THEN turn-medium-left
IF (lane-angled-right $\wedge$ $\neg$centerline-on-left) THEN turn-smooth-right
IF (lane-angled-left $\wedge$ $\neg$centerline-on-right) THEN turn-smooth-left

Figure 5.3 Fuzzy rules for the Follow behavior.

Intuitively, the equation (5.1) characterizes a control c as being desirable in the state x , if there is some rule in R that supports c and whose antecedent is true in x. This interpretation of a fuzzy rule set is that of a classical (Mamdami type) fuzzy controller, generalized so as to allow each antecedent A_i to be an arbitrary fuzzy-logic formula.

Figure 5.3 shows the rules that implement the previously discussed Follow behavior in our robot test-bed, Flakey. These rules keep the robot within the lane boundaries (first two rules) and lined up with them (last two). The **Fuzzy State** is composed of six propositional variables, whose truth value depends on the the current position of the lane with respect to the robot. Flakey's controller incorporates a dozen of behaviors such as this one—including behaviors for avoiding obstacles, for facing a given object, for crossing a door, and for reaching a near location. Each such behavior typically consists of four to eight rules involving up to a dozen fuzzy predicates.

Regardless of its implementation approach, a desirability function specifies, for each input variable value, a ranking over possible controls rather than a unique control value to apply in that situation. The robot eventually employs this ranking to *choose* one specific control $\hat{c}$ that is sent to the plant. A possible mechanism to accomplish that selection is centroid defuzzification:

$$\hat{c} = \frac{\int c \, Des_R(x,c) \, dc}{\int Des_R(x,c) \, dc}, \tag{5.2}$$

which computes the mean of possible control values, weighted by their degree of desirability.

Centroid defuzzification has been found satisfactory in our experiments whenever the rules in a rule set do not suggest dramatically opposite actions. In those cases, centroid defuzzification obviously does not work since averaging of multi-modal desirability measures might result in selection of a very undesirable choice (e.g., the best trade-off between avoiding an incoming train by jumping to the left or to the right is hardly to stay on the track!). In our development of mobile robot controllers, our empirical strategy has been to design the rule sets so as to avoid production of multi-modal desirability functions—roughly, we insist that rules that propose opposite controls have mutually exclusive antecedents. Other authors (Yen and Pfluger, 1992) have relied, however, on alternative defuzzification functions.

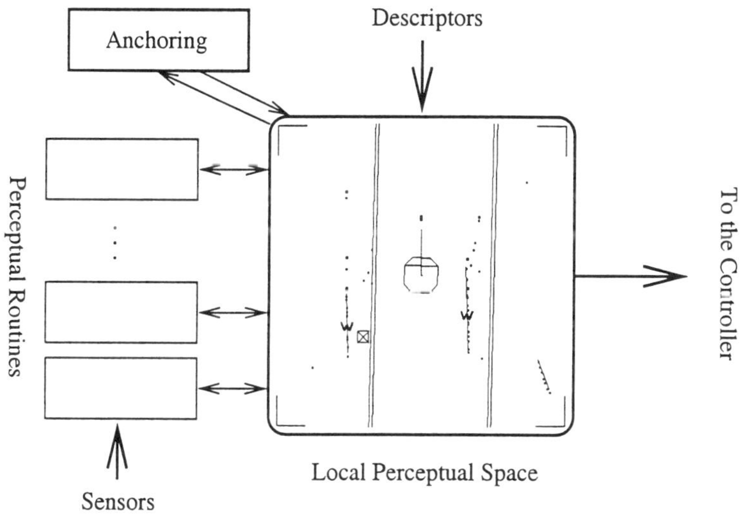

Figure 5.4 Local perceptual space and controller input.

5.4 TAKING GOALS INTO CONSIDERATION

We have been, so far, rather vague about the nature of the input to the controller's behaviors. The simplest form that might be taken by such input is direct sensor observations, e.g., readings from the sonar sensors indicating the presence of obstacles. Behaviors that only use the current perceptual data, possibly after some perceptual interpretation, are often called *reactive*. The amount of perceptual interpretation that is processed by such behaviors is usually small, in the interest of short response time, but it need not be.

When explicit *goals* are sought, however, it is often necessary to resort to some internal representation that does not originate in the sensor data. For example, suppose that the robot is trying to reach a landmark that is out of the range of its sensors. The mobile device must then rely on assumptions about the landmark position based, for example, on a map of the environment. Consequently, goal-achieving behaviors generally require some non-perceptual input to relate relevant properties of the goal with the adequacy of control actions.

Our strategy to give explicit goals to purposive behaviors is based on the utilization, as the input source for the controller, of a global data structure, called the *Local Perceptual Space* (LPS), which fuses both sensor and prior information. This approach is illustrated in Figure 5.4, where the LPS is represented in a Cartesian plane centered on the robot, itself drawn in the center of the figure as seen from above. The LPS includes sensor data at different levels of abstraction and interpretation. The small dots represent, for example, readings from the sonar sensors while the long segments labeled "W" are

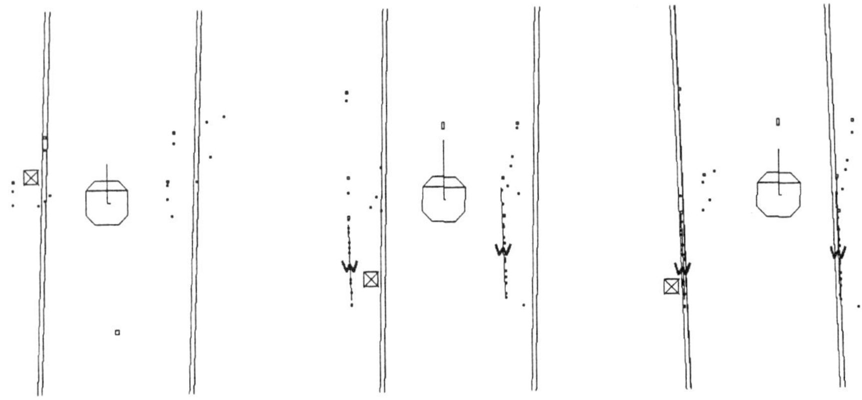

(a) (b) (c)

Figure 5.5 Anchoring a corridor descriptor to sensor readings.

linear features extracted from these readings, assumed to be walls. The LPS also includes *descriptors* of various objects used by the purposeful behaviors. For example, the two double lines in the picture indicate the descriptor of a corridor to follow.

Descriptors are a general, flexible mean of representing the abstract data required by goal-achieving behaviors. Descriptors are typically specified by a planner or some other higher-level deliberation process on the basis of prior information such as that contained on a map of the environment. For example, a lane descriptor might be employed to specify the expected position of a specific corridor that we want to follow.

Descriptors permit the implementation of goal-achieving behaviors using an approach similar to that employed to specify reactive behaviors (see Fig. 5.2). In general, however, the **Fuzzy State** specification for the former class of behaviors depends on properties of a descriptor (e.g., its position) rather than on direct perceptual inputs. The Follow behavior above is an example of a goal-achieving behavior.

Descriptors do not necessarily correspond to physical objects in the operational environment of the robot. For example, we may use an abstract lane descriptor to implement a "foray" behavior that takes the robot out a few meters from a given home position and then back again (this behavior was actually implemented in the controller employed to guide Flakey in the 1992 AAAI Robotics Competition (Congdon et al., 1993)). Whenever the descriptor corresponds to an actual environmental feature, however, it is important to maintain an accurate correspondence between this feature and its descriptor since control rules are based solely on the properties of the descriptor while the robot must perform the intended behavior with respect to the real feature.

The process employed in our approach to maintain the correspondence between descriptors and environmental features is called *anchoring* (Saffiotti,

1994). Figure 5.5 shows an example of anchoring during the execution of the Follow behavior. Initially (a), the robot is given a descriptor—produced from prior map information—of a corridor to follow (double lines). After a while (b), enough sonar readings are gathered to allow the perceptual routines to recognize the existence of two parallel walls, marked by "W." Note that the descriptor does not match precisely the position of the real corridor as there may be errors either in the robot's estimate of its own location or in the map of the environment. Finally (c), the perceived position of the walls is used to update the descriptor, i.e., the descriptor is *anchored* to the sensor data. As the rules in Follow are based on the properties of the descriptor, this anchoring entails that the robot will now follow the actual (sensed) corridor.

In control theory terms, we could say that the controller in the example above initially operates in an open-loop fashion with respect to prior information, switching then to a closed-loop, sensor-based regime when the walls are perceived. The corridor descriptor is normally re-anchored at each control cycle whenever the walls are visible. The ability to recover from vague or imprecise assumptions is particularly important in practice since our maps typically contain only approximate metric information. The descriptor serves as a source of credible assumptions when perceptual data is not available, as is the case when first engaging a new corridor, or when the walls are momentarily occluded by obstacles.

5.5 BLENDING OF BEHAVIORS

In previous sections we have discussed the implementation of reactive and purposive behaviors by means of fuzzy rules. As it was noted in the Introduction, several behaviors of either type may be active at any time during the operation of the intelligent mobile agent. Correspondingly, several instances of the fuzzy machine shown in Fig. 5.2 are simultaneously active in the fuzzy controller. At any given moment, the overall behavior of the robot is the result of the combination of all the behaviors that are active at that moment. In this section, we discuss the combination or *blending* of fuzzy behaviors.

The simplest type of behavioral combination is that where the desirability functions are simply conjoined to indicate that, regardless of context, we aim to attain all the objectives that they represent. In this case, for every state x, the combined desirability of a control value is given by the minimum of the degrees of desirability assigned by each behavior to that value:

$$(Des_1 \wedge Des_2)(x, c) = Des_1(x, c) \wedge Des_2(x, c) \; . \tag{5.3}$$

We call the expression $(Des_1 \wedge Des_2)$ the *conjunctive combination* of Des_1 and Des_2. A disjunctive combination can be defined in a similar way using the operator $\vee$.

Note that we could use triangular norms different from min in Equation (5.3). The nature of the triangular norm being employed essentially characterizes the the tradeoff between goals that underlies their combination (Dubois and Prade, 1985). Utilization of the min triangular norm means that, in order for the conjunction to be deemed desirable to the level α, each of the conjuncts must be also desirable at least to the level α. An alternative conjunction operator, the product T-norm, results in a different notion of tradeoff: a decrease in the level of desirability of one conjunct can be compensated by a corresponding increase in the other.

Conjunctive combination works well in situations where the behaviors being combined do not conflict—that is, for every state there is some control that is desirable from the perspective of each behavior. When behaviors conflict, however, the combined desirability function would be identically zero. Unfortunately, this is situation is often the norm rather than the exception.

We may want to combine, for example, a behavior for traveling along the center of a corridor and another one for avoiding obstacles. When the robot is approaching an obstacle ahead, the first behavior would prefer controls that go forward while the second would favor controls that turn the robot, say, right. The key observation here is that each behavior has its own *context* of applicability and that each desirability function should be considered only when that context is appropriate. In our example, the first behavior can be sensibly applied only in situations where the space in front of the robot is free. When the obstacle is detected, this behavior is outside its area of competence and we should (at least partially) disregard its preferences.

We represent contexts of applicability as fuzzy sets of the set of possible states. We define the *context restriction* of a desirability function *Des* to a context *Cxt* by the expression

$$Des^{\downarrow Cxt}(x,c) = Des(x,c) \oslash Cxt(x)\,,$$

where *Cxt* is a fuzzy subset of state space.

Utilization of the quasi-inverse $\oslash$ to compute truth values of implications might be interpreted as follows. When the state x is in the core of the context *Cxt*, i.e., $Cxt(x) = 1$, then the desirability of a control c is measured by $Des(x,c)$. When the state x is fully outside of the context of applicability, i.e., $Cxt(x) = 0$, then any control is admissible (as far as the particular goal associated with *Des* is concerned). Intermediate values of *Cxt* make fully admissible those control values c for which $Des(x,c) \geq Cxt(x)$.[4] In other words, context restriction leads to a less committed desirability measure that only expresses a definite preference when the agent is in one of the context states, but (partially) allows other possibilities when it is (partially) outside these states.

We can combine context restriction and conjunctive combination into a general combination pattern, called *context-dependent blending*, where each behavior is associated with a context, which identifies the states where the behavior is competent. The blended desirability function is defined by

$$Des = Des_1^{\downarrow Cxt_1} \wedge Des_2^{\downarrow Cxt_2}\,. \tag{5.4}$$

The previous definition characterizes context-dependent blending in terms of a desirability function that agrees with Des_1 to the extent that the current

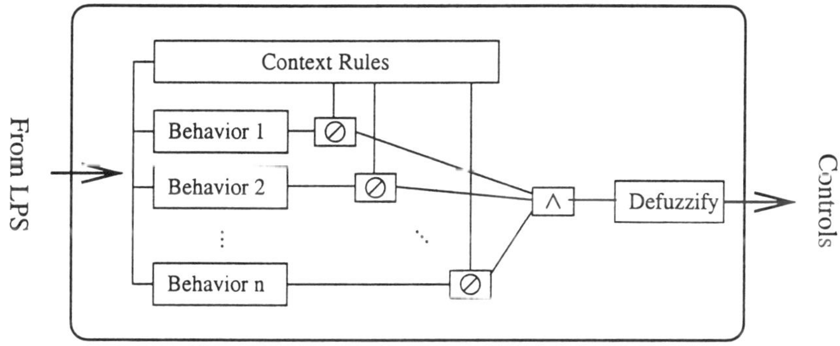

Figure 5.6 A hierarchical fuzzy controller for context-dependent blending.

situation satisfies Cxt_1, and that agrees with Des_2 to the extent that it satisfies Cxt_2. Multiple behaviors may be combined by context-dependent blending. The commutativity and associativity of the $\wedge$ operator assure that the generalization of the right hand side of Equation 5.4 does not depend on the order of the behaviors being combined.

Context-dependent blending may be implemented in a hierarchical fuzzy controller as shown in Fig. 5.6. In this configuration, each behavioral module is a fuzzy machine as illustrated in Fig. 5.2. The context rules are fuzzy meta-rules of the form

$$\text{IF } A_j \text{ THEN } \textit{apply } B_j, \quad j = 1, \ldots, m, \tag{5.5}$$

where A_j is a fuzzy-logic formula describing a context, and B_j is a behavior. The truth values, given the current state, of the formulas defining the contexts are used to discount the desirability function produced by the corresponding behavior. The discounted values are then merged together according to (5.4). Finally, the resulting blended desirability function is defuzzified by (5.2) to produce a crisp control. As we shall see in Section 5.7, the fact that defuzzification is performed *after* combination is essential.[5]

Figure 5.7 shows a simple example of context-dependent blending where FOLLOW is employed to follow a corridor, and KEEP-OFF is activated to avoid obstacles. The meta-rules used to encode the contexts in this example are:

```
IF approaching_obstacle THEN apply Keep-Off
IF NOT(approaching_obstacle) THEN apply Follow
```

The bars (a) and (b) show the activation level and the preferred controls (turn and acceleration) for each behavior at two instants while the bottom line in each graph shows the result of the blending. In (a) an obstacle has been detected and the preferences of Keep-Off dominate, thus making the robot abandon the mid-line of the corridor. Later, when the path is clear, the goal-oriented preferences expressed by Follow regain importance (b), and the robot recovers its mid-corridor course. The graph at the bottom of the picture (c)

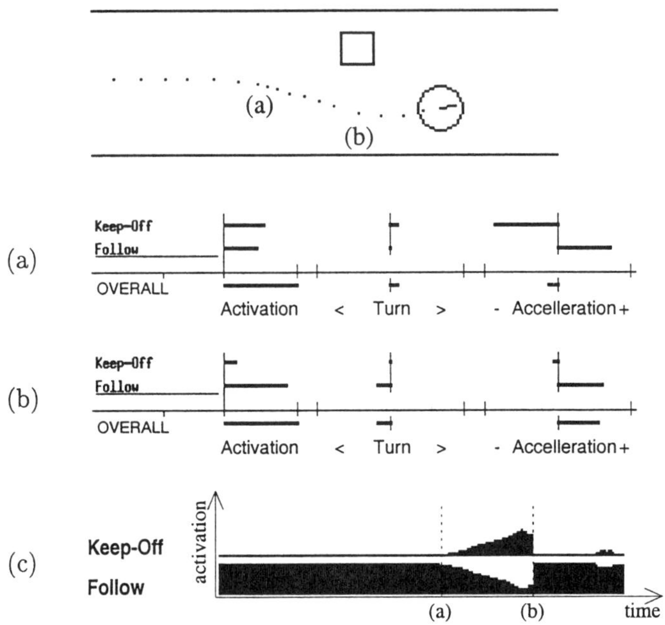

Figure 5.7 Blending corridor following and obstacle avoidance.

shows the evolution of the activation level of each behavior (i.e., the truth value of the corresponding context) over time.

5.6 EXPERIMENTS

The approach presented in this paper has been implemented on the SRI International mobile robot Flakey having been tested extensively in indoor robot navigation experiments. In this section, we present two experimental runs that illustrate the performance of Flakey.

Flakey is a custom-built mobile robot, with a height of 1 meter and a diameter of 0.6 meters, driven by two independent wheels at maximum linear velocity of about 0.5 meters/sec. Sensors include a ring of 12 sonars, bumpers, wheel encoders, and video and audio equipment (not used in the experiments reported here). On-board computation capabilities include a 2-processor Sparc 10-51. The control software is distributed over several processes that are daisy-chained with a basic cycle time of 100 msecs. Most of the interpretation and control routines have execution times ranging from 1 to 10 msecs on the Sparc 10 while the fuzzy controller requires about 2 msecs with one behavior active and approximately 10 msecs when 8 behaviors are active (on average, 2–3 behaviors are active at any given time during a typical navigation task). Al-

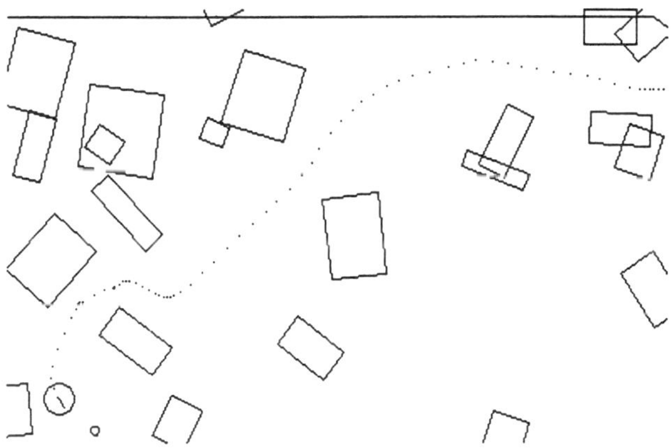

Figure 5.8 A run of the Wander behavior in an unknown environment.

though Flakey is capable of running all required computational processes on its on-board computers, we typically execute, for programming convenience, most functions remotely, communicating with Flakey via a radio-link.

Our first experimental result was obtained while the robot wandered around in an unknown environment avoiding static and moving obstacles. This task was performed by Flakey at the First AAAI Robotics Competition (Congdon et al., 1993). In this competition Flakey placed second in overall rankings and was particularly commended by the judges for its reliable reactivity. Flakey did not make use of maps of the environment, simply roaming around at random while trying to stay away from obstacles as they were perceived. Figure 5.8 shows a run of the Wander behavior in an environment similar to that of the competition. The smoothness of motion, both in turning and acceleration, can be seen in the wake of small dots that indicate Flakey's trajectory (one dot per second).

The Wander behavior is built by blending three basic behaviors:

1. The Go-Forward behavior keeps a constant cruising speed *when* the way is clear;

2. The Keep-Off behavior keeps the robot a safe distance away from obstacles *when* an obstacle is detected; and

3. The Avoid-Collisions behavior activates emergency maneuvers *when* an immediate collision danger is sensed.

All these behaviors are purely reactive, simply monitoring sonar readings to determine the presence of obstacles. The operation of these behaviors, however, does not require object descriptors.

Our second example, shown in Fig. 5.9, illustrates how context-dependent blending of behaviors may be employed to perform full navigation plans. Each

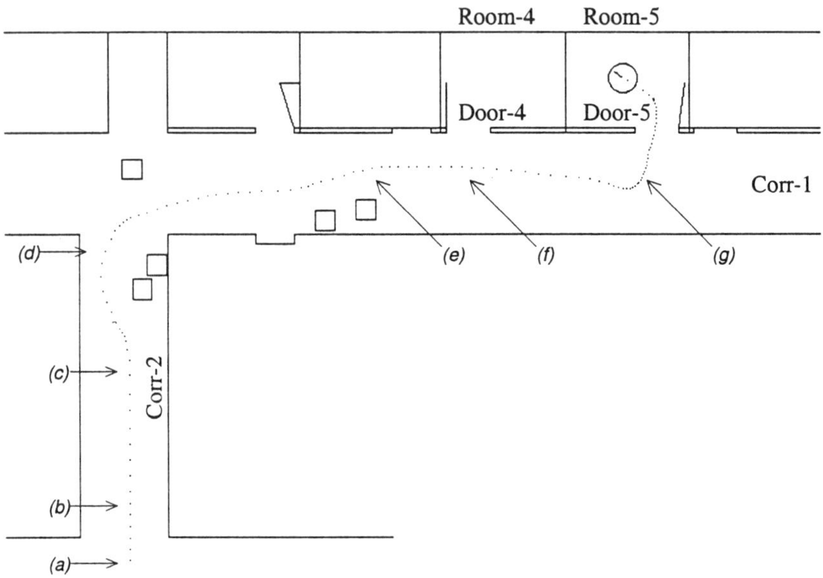

Figure 5.9 Execution of a full navigational plan.

action in the navigation plan is represented as a behavior having a context of applicability. This view of plans as sets of "situation $\rightarrow$ action" rules has been advocated by several authors as a most useful perspective when dealing with situated agents (Suchman, 1987), (Schoppers, 1987), (Payton et al., 1990). In our experiment, Flakey was initially at the lower left intersection in the environment shown in the picture. The ultimate navigation goal was to guide Flakey to Room-5.

The following behaviors and contexts constitute a plan to reach this goal:

IF $near(Obstacle)$ THEN Keep-Off
IF $\neg near(Obstacle) \wedge at(Corr2) \wedge \neg at(Corr1)$ THEN Follow(Corr2)
IF $\neg near(Obstacle) \wedge at(Corr1) \wedge \neg near(Door5)$ THEN Follow(Corr1)
IF $\neg near(Obstacle) \wedge near(Door5)$ THEN Cross(Door5)

where we have explicitly indicated the descriptors employed by the goal-achieving behaviors, like in Follow(Corr1). Intuitively, the robot must Follow Corr-2 until it reaches Corr-1; then Follow Corr-1 until it is in the neighborhood of Door-5; and finally Cross this door. The Keep-Off behavior may be activated at any moment to avoid unexpected obstacles (the Og descriptor is an 'occupancy grid' used for sensor-based obstacle avoidance). Note that Follow(Corr1) and Follow(Corr2) denote two distinct behaviors that share the same rules but differ in the descriptor they use.

We have shown elsewhere (Saffiotti et al., 1995) that plans of this type can be generated by classical AI planning techniques. For example, the above plan has

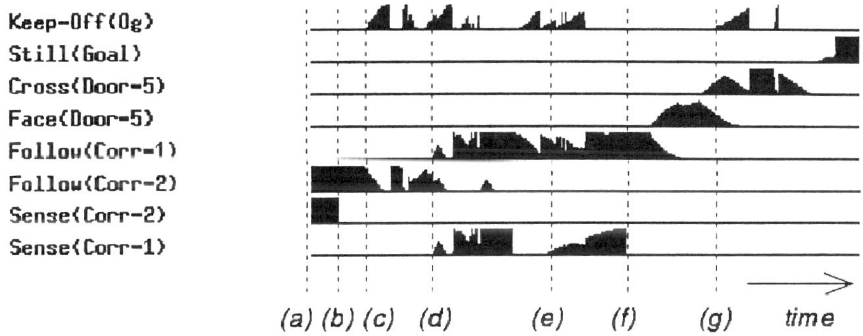

Figure 5.10 Behavior activations during the execution shown in Figure 5.9.

been automatically generated by Flakey from knowledge of the navigation goal of the plan by a simple goal-regression planner employing a sparse topological map annotated with approximate metric information as the source of prior environmental information.[6]

Figure 5.9 shows an actual execution of this plan, while Figure 5.10 plots the temporal evolution of the level of activation of each behavior—that is, of the truth level of the corresponding context. Total execution time was approximately 120 sec at maximum speeds of 400 mm/sec. Each behavior is activated when and to the extent by which its contextual conditions are verified. Although the plan does not specify a pre-determined order of behavior execution, this sequencing may be implicit in the definitions of the contexts of applicability as execution of one behavior may be necessary to reach situations where another might be activated as is the case in the vicinity of (d) and (g).

It is important to remark that behaviors do not "terminate" in the usual sense, being simply deactivated when their context become false and being possibly reactivated if this context becomes true again. This particular aspect of our approach is exemplified by the Sense behavior, which helps Flakey to detect walls by preferring straight and slow motions. This behavior is activated when Flakey enters a new corridor, and remains operational until the walls are anchored (a, d); the Sense behavior may be reactivated later on if anchoring is lost, e.g., if the walls are occluded by obstacles for too long (e).

Finally, note the interaction between purposeful behaviors and reactive obstacle avoidance after (c) and around (d), (e), and (g). In particular, the interaction immediately following (g) deserves further comment. The Cross behavior relies on prior information about the position of the door. In our experiment, this prior knowledge was rather imprecise and the robot was required to activate the Keep-Off behavior to avoid colliding with the edge of the wall. The combined effect of the blending of these two behaviors was to lead Flakey into the opening that is "more or less at the estimated position," correcting its motion to find the true entrance. The elasticity of fuzzy rules thus guarantees a smooth degradation of performance when the prior knowledge employed to characterize the descriptors is partly incorrect.

5.7 DISCUSSION

The majority of the current implementations of behavior-based architectures for mobile robot control use on-off schemas to combine behaviors with behaviors being enabled or inhibited based on some a priori hierarchy. Furthermore, only one behavior at a time has control over effectors. The limitations of this approach have been noticed by several authors who have advocated a weighted form of combination where the outputs of different behaviors are merged and the overall control is the result of a tradeoff computation (Maes, 1989; Payton et al., 1990; Arkin, 1998).

Our approach offers a solution to the behavior combination problem that is based on two distinguishing principles:

1. Behaviors are viewed as modules that express goal-specific preferences as to the suitability of control actions rather than as low-level procedures that compete for the control of the robot's effectors.

2. Behavior blending relies on a general context-dependent mechanism based on sound interpretations (Ruspini, 1991a) of fuzzy-logic concepts and structures rather than on ad-hoc forms of combination.

One of the major advantages of weighted combination with respect to crisp activation-inhibition is a better integration between reactive and goal-achieving behaviors (Payton et al., 1990; Saffiotti, 1997a). Weighted combination permits to consider the preferences of purposive behaviors during reactive maneuvers thus biasing the control choices toward the achievement of explicit goals.

It is interesting to compare context-dependent blending with the *artificial potential field* methods, first introduced by Khatib (Khatib, 1986) and now extensively employed in motion planning and plan execution (Latombe, 1991; Arkin, 1998). In this approach, vector fields are employed to model the nature of desired interactions between robots and environmental features.

For example, obstacles are represented as generators of repulsive fields that keep the robot away from collisions. Correspondingly, attractive fields are utilized to steer the robot towards desirable locations in its workspace. At each point, the robot responds to a pseudo-force proportional to the gradient of the field. Multiple potential fields are combined by linear superposition of these vectors thus leading to a result based on blending of the "most desirable" motion choice for each goal given the current situation. In contrast, when combining two desirability functions, we *first* combine the component desirability functions, effectively forming a full preference function, choosing *then* the preferred control by defuzzification of the blended function (see Fig. 5.6). In this respect, our combination schema is close to the one proposed by Payton and Rosenblatt (Payton et al., 1990). That schema, however, is an ad-hoc mechanism which is not grounded in any mathematical formalism.

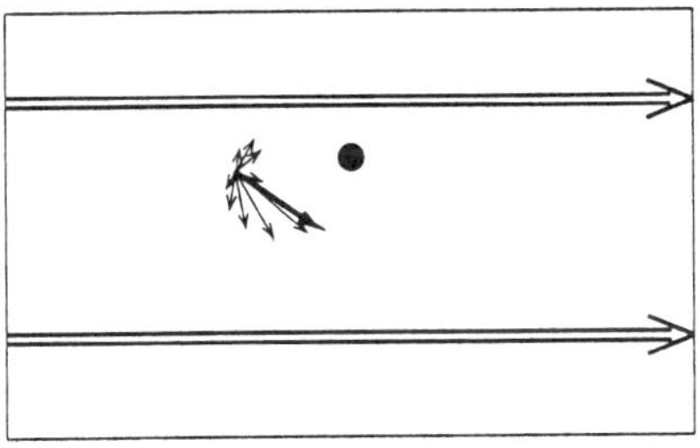

(a)

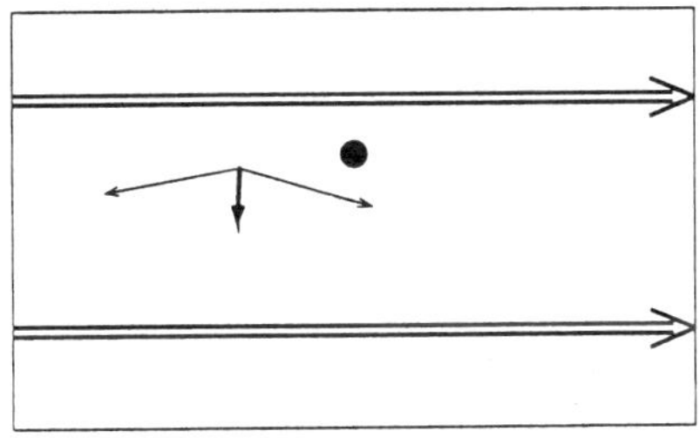

(b)

Figure 5.11 Combination of desirability functions (a) and of pseudo-forces (b).

The difference between potential-field and desirability-function combination is illustrated in Fig. 5.11. The left side (a) of the figure illustrates the result of blending, with equal weights, the two desirability functions shown in Fig. 5.1 above; the thick arrow indicates the preferred control obtained by applying centroid defuzzification to the blended function. On the right side (b), it is shown the veçtor composition of the two preferred controls that would be individually selected by separate consideration of each desirability function. These vectors may also be thought of as pseudo-forces generated by the Follow-Corridor and by the Keep-off behaviors, respectively. As it appears, the two methods of composition may lead to different results. It is intuitively clear that desirability functions carry more information than pseudo-forces since they also measure the desirability of suboptimal controls.

The second important problem that we have also addressed is that of relating control-level processing with the high-level abstract deliberative tasks required by plan production and evaluation. Our approach in this regard relies on two major conceptual tools:

1. *Object descriptors* employed to guide the activity of basic behaviors towards attainment of abstract goals, specified by higher-level modules and anchored to low-level perceptual data.

2. *Context-dependent activation*, in which the conditions that render a particular behavior applicable are expressed by formulae of fuzzy logic.

Fuzzy-logic methods have been of essential importance to take advantage of these conceptual devices. The elasticity inherent in fuzzy formalisms facilitates, for instance, the utilization of object descriptors even when there is not an exact correspondence between the position of the descriptor and the one of the object that it describes. Reliance on the language of fuzzy logic permits, on the other hand, to define complex activation conditions more expressively than is the case for alternative approaches. Furthermore, the ability to integrate abstract controller descriptions and to specify patterns of behavior combination as logical formulae facilitates integration with classical symbolic planning techniques.

Our fuzzy-logic based approach to behavior combination has been shown to be ameanable to a formal description and analysis (Saffiotti et al., 1995). We have been able to derive formal results, for example, showing that, under certain assumptions, two behaviors that promote independent goals may be combined by context-dependent techniques to produce a blended behavior that promotes a combination of those goals. Our approach leads, therefore, to sound formalisms for the study of controller properties.

It is also clear that our approach, like similar treatments, has its share of methodological problems, which we are currently seeking to understand and overcome. Fuzzy behaviors, for example, have been so far based primarily on local "greedy" descent gradient methods, which are highly reactive and simple to compute. As is the case with other local techniques this approach to behavior definition cannot guarantee that the device will not be trapped on local minima or infinite loops. More elaborate formulations should emphasize integration with techniques such as dynamic programming that are based on a more thorough analysis of the eventual consequences of control actions. Our treatment of this type of problems has relied, so far, on higher-level deliberation processes only to limit behavior instantiation and combination to those cases where local techniques are appropriate (e.g., on the basis of a coarser global analysis). We have also relied on execution-monitoring approaches to detect local minima and failure conditions.

Our experience has also shown that the process of tuning the parameters of fuzzy rules may be rather difficult. Although the ability to decompose complex behaviors into simpler ones facilitates the identification of rules, we have found that some behaviors—such as obstacle avoidance—demanded several days of experimental debugging. We are currently studying methods for the automatic synthesis of behaviors from specifications and the possible role of learning techniques to generate behaviors as well as to improve behavioral performance.

Finally, much still remains to be done to develop effective analytical formalisms that permit to determine the degree by which composite behaviors attain a variety of control objectives.

5.8 CONCLUSIONS

Fuzzy logic methods have been proved to be effective tools to design highly responsive controllers for autonomous mobile agents. These controllers are capable of implementing motion and perception behaviors so as to attain multiple, possibly conflicting, goals. Fuzzy-logic approaches have also been useful as the bases of formal results that facilitate the analysis of two important problems inherent in behavior-based approaches: the coordination of multiple behaviors and the incorporation of prior knowledge into the controller.

Our experiments have shown that fuzzy behaviors are able to operate relying only on approximate maps and imprecise sensing: a most important requirement for an autonomous vehicle intended for use in unstructured environments. In related work (Saffiotti and Wesley, 1996), we have also shown how fuzzy logic may be used to address the self-localization problem in robot navigation. The major problems that we have found with our approach concern the empirical nature of the definition and tuning of the fuzzy rules. We have hinted above at some of the directions that we are exploring to mitigate these problems.

The focus of this chapter was on the application of fuzzy logic techniques to the design and implementation of basic navigation behaviors, and to the combination of basic behaviors to form complex behaviors to execute full navigational plans. While these issues are pivotal to any autonomous navigation capability, there are many other imporatant problems in autonomous robotics for which solutions based on fuzzy logic can be, and have been sought. These include the perception and modelling of the robot's environment; the ability to self-localize with respect to a given (partial) map; the ability to recognize and recover from failures due to unexpected environmental changes or hardware failures; and the ability to learn from experience. The interested reader is addressed to (Saffiotti, 1997b; Hoffmann, 1998; Driankov and Saffiotti, 1999) for some up-to-date surveys of these applications.

Acknowledgments. Work by the first author was partly supported by the BELON project founded by the *Communauté Française de Belgique.* Enrique H. Ruspini was partially supported by the U.S. Air Force Office of Scientific Research under Contract No. F49620-91-C-0060. Additional support was provided by SRI International. The authors benefitted from discussions with H. Berenji, P. Bonissone, D. Driankov, N. Helft, O. Khatib, J. Lowrance, K. Myers, D. Ruspini, L. Valverde, and L. Zadeh.

Notes

1. This example clearly oversimplifies matters for the sake of clarity. A more realistic treatment should consider more factors, e.g., the velocity obtained by applying c.
2. This approach to the definition of a desirability measure is known as *predictive fuzzy control* (Yasunoby and Miyamoto, 1985).
3. These operators are those actually used in our experiments on the robot Flakey. See (Saffiotti et al., 1995) for a generalization of our treatment to include arbitrary continuous triangular norms, triangular co-norms, and generalized negation operators (Weber, 1983).
4. This characterization is valid for the inverse of the minimum T-norm. Other T-norms have inverses that lead to a more gradual relaxation of the inadmissibility of control values outside the core of the context.
5. A similar approach to the treatment of multi-objective control problems utilizing fuzzy-logic methods has been previously proposed by (Berenji et al., 1990). Context-dependent blending generalizes and extends that approach by allowing dynamic modification of the degrees of importance of each goal.
6. The actual plan has more behaviors (Saffiotti et al., 1995), including some for perceptual actions such as the Sense behavior discussed below, and more complex contexts than the simplified account given in the text.

References

Arbib, M. A. (1981). Perceptual structures and distributed motor control. In Brooks, V., editor, *Handbook of Physiology – The Nervous System II*, pages 1449–1465. American Physiological Society, Bethesda, MD.

Arkin, R. C. (1990a). The impact of cybernetics on the design of a mobile robot system: a case study. *IEEE Trans. on Systems, Man, and Cybernetics*, 20(6):1245–1257.

Arkin, R. C. (1998). *Behavior-Based Robotics.* MIT Press, Cambridge, MA.

Berenji, H., Chen, Y.-Y., Lee, C.-C., Jang, J.-S., and Murugesan, S. (1990). A hierarchical approach to designing approximate reasoning-based controllers for dynamic physical systems. In *Proc. of the Conf. on Uncertainty in Artif. Intell.*, pages 362–369, Cambridge, MA.

Brooks, R. A. (1986). A robust layered control system for a mobile robot. *IEEE Journal of Robotics and Automation*, RA-2(1):14–23.

Congdon, C., Huber, M., Kortenkamp, D., Konolige, K., Myers, K., Ruspini, E. H., and Saffiotti, A. (1993). CARMEL vs. Flakey: A comparison of two winners. *AI Magazine*, 14(1):49–57.

Driankov, D. and Saffiotti, A., editors (1999). *Fuzzy logic techniques for autonomous vehicle navigation.* LNCS. Springer, Berlin, DE. Forthcoming.

Dubois, D. and Prade, H. (1985). A review of fuzzy set aggregation connectives. *Information Sciences*, 36:85–121.

Gat, E. (1992). Integrating planning and reacting in a heterogeneous asynchronous architecture for controlling real-world mobile robots. In *Proc. of the AAAI Conf.*, pages 809–815, San Jose, CA.

Hoffmann, F. (1998). Soft computing techniques for the design of mobile robot behaviours. *Information Sciences.* To appear.

Khatib, O. (1986). Real-time obstacle avoidance for manipulators and mobile robots. *The International Journal of Robotics Research*, 5(1):90–98.

Konolige, K., Myers, K., Ruspini, E., and Saffiotti, A. (1997). The Saphira architecture: A design for autonomy. *Journal of Experimental and Theoretical Artificial Intelligence*, 9(1):215–235.

Latombe, J. C. (1991). *Robot Motion Planning.* Kluver Academic, Boston, MA.

Maes, P. (1989). The dynamics of action selection. In *Proc. of the Int. Joint Conf. on Artificial Intelligence*, pages 991–997, Detroit, MI.

Payton, D. W., Rosenblatt, J. K., and Keirsey, D. M. (1990). Plan guided reaction. *IEEE Trans. on Systems, Man, and Cybernetics*, 20(6):1370–1382.

Ruspini, E. H. (1990). Fuzzy logic in the Flakey robot. In *Proc. of the Int. Conf. on Fuzzy Logic and Neural Networks (IIZUKA)*, pages 767–770, Iizuka, JP.

Ruspini, E. H. (1991a). On the semantics of fuzzy logic. *Int. J. of Approximate Reasoning*, 5:45–88.

Ruspini, E. H. (1991b). Truth as utility: A conceptual synthesis. In *Proc. of the Conf. on Uncertainty in Artif. Intell.*, pages 316–322, Los Angeles, CA.

Saffiotti, A. (1994). Pick-up what? In Bäckström, C. and Sandewall, E., editors, *Current Trends in AI Planning — Proc. of EWSP '93*, pages 166–177. IOS Press, Amsterdam, NL.

Saffiotti, A. (1997a). Fuzzy logic in autonomous robotics: behavior coordination. In *Proc. of the 6th IEEE Int. Conf. on Fuzzy Systems*, pages 573–578, Barcelona, Spain. IEEE Press.

Saffiotti, A. (1997b). The uses of fuzzy logic for autonomous robot navigation. *Soft Computing*, 1(4):180–197. On-line at http://iridia.ulb.ac.be/FLAR/.

Saffiotti, A., Konolige, K., and Ruspini, E. H. (1995). A multivalued-logic approach to integrating planning and control. *Artificial Intelligence*, 76(1-2):481–526.

Saffiotti, A., Ruspini, E. H., and Konolige, K. (1993). Blending reactivity and goal-directedness in a fuzzy controller. In *Proc. of the 2nd IEEE Int. Conf. on Fuzzy Systems*, pages 134–139, San Francisco, California. IEEE Press.

Saffiotti, A. and Wesley, L. P. (1996). Perception-based self-localization using fuzzy locations. In Dorst, L., van Lambalgen, M., and Voorbraak, F., editors, *Reasoning with Uncertainty in Robotics*, number 1093 in LNAI, pages 368–385. Springer-Verlag, Berlin, DE.

Schoppers, M. J. (1987). Universal plans for reactive robots in unpredictable environments. In *Proc. of the Int. Joint Conf. on Art. Int*, pages 1039–1046.

Suchman, L. (1987). *Plans and situated actions: the problem of human machine communication.* Cambridge University Press, Cambridge, MA.

Sugeno, M. and Nishida, M. (1985). Fuzzy control of model car. *Fuzzy Sets and Systems*, 16:103–113.

Weber, S. (1983). A general concept of fuzzy connectives, negations and implications based on t-norms and t-conorms. *Fuzzy sets and systems*, 11:115–134.

Yasunoby, S. and Miyamoto, S. (1985). Automatic train operation by predictive fuzzy control. In Sugeno, M., editor, *Industrial Applications of Fuzzy Control*, pages 1–18. North-Holland, Amsterdam, NL.

Yen, J. and Pfluger, N. (1992). A fuzzy logic based robot navigation system. In *Proc. of the AAAI Fall Symp. on Mobile Robot Navigation*, pages 195–199, Boston, MA.

Zadeh, L. A. (1978). Fuzzy sets as a basis for a theory of possibility. *Fuzzy Sets and Systems*, 1:3–28.

6 CIVIL ENGINEERING
Including Earthquake Engineering

Felix Wong
Karen Chou
James Yao

Abstract: This chapter gives a summary of some applications of fuzzy sets, fuzzy logic and fuzzy control in civil and earthquake engineering that are available in the open literature. Due to the language limitation of the authors, the review is based primarily on publication in English, and even at that, it cannot be complete due to the large number of papers, books and studies that are in the literature. The discussion focuses on (civil engineering) applications; mathematical and other theoretical developments are covered elsewhere in this handbook. The general evolution of civil engineering involvement in fuzzy sets is first reviewed. From a humble beginning in the early 1970's, fuzzy sets technology is now embraced by all disciplines of civil engineering. The presentation describes how the technology is used, but not why, as the use of fuzzy sets remains highly controversial in civil engineering and "why" is best left to other forums. In general, applications in civil and earthquake engineering fall into one of several standard protocols, which include structured processing, unstructured processing, expert systems and intelligent systems. These application groups are illustrated with selected examples, while details are left to the extensive (but certainly not exhaustive) list of references cited. Prognosis for continued growth is excellent in view of the trends of the profession/industry towards efficient maintenance and management of existing infrastructure, and improved functionality, performance and durability of new constructed facilities.

6.1 HISTORIC PERSPECTIVE

Civil engineering is noted for its dealings with uncertainties. Recognizing the limits of our knowledge of the real world and the dire socio-economic consequences of any miscalculation, civil engineers have long used linguistic descriptors to describe uncertain and fuzzy information, and conservative factors to ensure design safety. In fact, in modern information science context, civil engineering may be viewed as mainly the aggregation of uncertain and fuzzy information by analytical and judgmental means, into an understandable conclusion - which may, nevertheless, remain imprecise. As an example, even in the most recent probability-based LRFD (Load and Resistance Factor Design) specification, the values of many factors are based on experience and intuition rather than analytical methods.

Civil engineers had long practiced the art of fuzzy thinking and processing even prior to Lofti Zadeh's paper in 1965 (Zadeh [1965]) that formally introduced fuzzy sets to engineers. Witness the prevalent use of terms such as clayey sand (a cross between clay and sand), hard rocks (such as granite or limestone), moderate damage (which is between light and severe damage), etc., in the profession. Zadeh's paper gave these terms a firm, mathematical foundation, and explicit procedures to process them. However, this quantitative advance would have gone unnoticed, if not for naught, had it not been for the work of two engineers and educators: Professors Colin Brown and David Blockley. In an ironic twist, the profession had difficulty accepting the formalism initially even though the technology reflects the essence of the profession. But the works of Brown and Blockley and others such as James Yao, David Elms, Paul Jowitt, etc., began to turn the attitude around. Already esteemed for their work in civil engineering, Brown, Blockley and others published in the leading professional journals a number of papers in the late 1970's and early 1980's which have become the watershed events in civil engineering for fuzzy sets applications (see Brown [1979], Brown and Yao [1983], Yao [1980a], Brown et. al. [1984], and Blockley [1979]). In particular, the paper by Blockley [1977] in the Institution of Civil Engineers has been credited by Brown as a complete eye-opener, without which he would have no idea how to use fuzzy sets in engineering. Note that it took over 10 years from Zadeh's introduction for fuzzy sets to trickle down to civil engineering.

Fuzzy sets appeals to Brown because of his interest in (civil engineering) systems science, to Yao because of his long involvement in assessing damage in existing structures, and to Blockley because of his interest in structural design and safety. All are involved with structures. It is not surprising then that the first centers of fuzzy sets research in the US are the University of Washington, Purdue University and, in the UK, the University of Bristol where Brown, Yao and Blockley worked, respectively. Through their published works and teaching (and preaching), fuzzy sets began to spread from structural engineering to other civil engineering disciplines at the three universities, and later to other institutions and the profession at large. By mid-1980's, other centers appeared in Stanford, Ohio State, Syracuse, North Carolina, Georgia Tech and Alabama. The books authored by Blockley [1980] and Yao [1985] around that time remain must-reads for structural engineers.

Development in other countries and continents follows a similar pattern: Munroe at Imperial College, Shibata at Tokyo University, Shiraishi and Furuta at Kyoto University, Kawamura at Kobe University, Elms at University of Canterbury, Feng and Liu at the Institute of Engineering Mechanics (State Seismological Bureau, China), and Wang at the Harbin Architecture Institute (China). However, the spread of fuzzy sets into civil engineering in Asia and Europe had a different flavor than in the US. The movement was spearheaded by fuzzy sets inroads into mathematics circles; civil engineering benefited only indirectly. Fuzzy (gray) mathematics societies were formed in these countries and became established much earlier than their counterparts in civil engineering. In these continents, mathematics activity was more robust, and, civil engineering became affected because of it. This situation remains true at present even though activities in all areas and in every continent have intensified to date.

6.2 SCOPE OF APPLICATIONS

Since the early 1980's, applications of fuzzy sets have extended to all disciplines of civil engineering including structural, earthquake and geotechnical engineering, architecture, project management, water resources, environmental engineering, transportation and construction. As in other fields of engineering and science, the number of civil engineering applications as reported in the literature has exploded to the point that a summary within limited space can neither be comprehensive nor satisfactory. For example, a search of the ASCE web site (American Society of Civil Engineering, *http://www.ascepub.infor.com*) led to over 150 records with fuzzy sets as the keyword term, covering the period 1983-1996. Excluding applications in environmental engineering which are addressed in a separate chapter of this handbook, the remaining records can be organized into major groups as shown in **Table** 1. General, cross-discipline applications are organized as in **Table** 2.

Table 1. Civil engineering applications from the ASCE Web site.

CE Discipline	Application Area	Sub-Area	Phenomenon Addressed
Building structures	Architectural design	Construction planning	Location and capacity of storage facilities
		Space decision	Preference for height of room
		Facility management	Office zoning and layout
		Urban design	City layouts as function of time period
	Structural design	Shape optimization	Least weight

		Architecture reasoning	Henry's problem
		Preliminary design	Uncertainty modeling
		Plastic design	
	Structural reliability	Ship structures	Uncertain loads and parameters
		Building	Qualitative uncertainties
			Inspection frequency
			Safety assessment
		R/C structures	Frame structures
		Methodology	
	Structural control	Base isolation	Sliding structures
		Active control	Multi-degree, nonlinear systems
		Adaptive systems	Tuned mass dampers
		Structural reliability	Effect of active control on
		Intelligent Infrastructures	
	Environmental control		Air-conditioning and heating
	Structural analysis	Finite element	
		Mechanics Computation	Uncertain loads
		Modal vibration	Classification of data
		Matrix analysis	Qualitative loads
		Beam on Winkler foundation	
		Aseismic structures	
		Plasticity theory	
		Random vibration	
		Aging stochastic systems	
	Building retrofit		Masonry buildings
	Damage assessment		Judgment of importance
		During construction	Pre-stressed concrete piles
		Protective structures	
Earthquake Engineering	Earthquake damage assessment	Modified Mercalli intensity	
		Fragility	
		Vulnerability	Masonry buildings
			Underground lifelines
			City-wide gas systems
			General buildings
			Damaged buildings
	Structural design	Optimization	
		Planning	Capacity of reinforced-concrete structures
	Structural degradation	Response uncertainties	Structure dynamics

		Forensic investigation of damage	Damage in reinforced concrete buildings
	Architectural design	Refuge planning	Emergency routes and exits
Civil infrastructure	Damage, reliability and maintenance research		
	Highway bridges	Inspection and rating	Damage assessment
		Rehabilitation	Strategies for
			Priority for
Transportation	Traffic control	Traffic surveillance	Detecting freeway incidents
		Accident prone location	Accident factors, road conditions
	Intelligent vehicle/highway		Vehicle control & traffic congestion
			Vehicle platoon control system
	Bus routing		Preference analysis
	Vehicle scheduling		Many-to-many algorithm
	Pavement	Highway pavement maintenance	Skid resistance and roughness
			Pavement condition and treatment
			Distress-index
Construction	Constructability		Work-zone traffic control
	Scheduling and planning	Network	Duration estimates
		Decision support	Duration estimates
		CPM/PERT	Resource allocation
		Risk management	Linguistic
	Project control		Evaluating alternatives
	Bids & Tendering		Safety and risk analysis
			Pricing risk and consequences
		Design/build methodology	Selection of winning proposal
			Tender evaluation
	Job site evaluation		Record interpretation and management
		Automated corrective actions	Interpretation of daily site reports
	Performance Evaluation	Construction systems/services	Value engineering
		Constructed facilities	
		Lunar & space structures	
	Construction strategy		
	Bridge construction	Cable tension adjustment	Efficiency
	Intelligent mining	Excavator control	Autonomous lunar mining
	Safety assessment	Job site accidents on operation	
		During construction	
		Failure of temporary structures	
	Contracting	Legal expert	

Geotechnical	Cone penetration tests		Sand compressibility
	Slope stability	pile bearing capacities	
		Failure potential	
		Failure potential	Failure potential maps
	Foundation	Soil mechanics	
	Pile driving		Concrete piles
	Capacity of piles	Piles in sand	

Table 2. Examples of general, cross-discipline applications.

CE Discipline	Application Area	Sub-Area
Risk analysis	Fault trees and event trees	
	Safety of existing structures	
Uncertain Systems Analysis	Structural design	Reliability
	Engineering decision	
	Knowledge-based inference	
	Expert opinions	
Engineering education	Intelligent teaching system	Student module

Other helpful databases are available. For example, a quick scan over the National Information Service Center located at the University of California at Berkeley *http://www.nisee.ce.berkeley.edu/html/nisc.html*, or the QUAKELINE at the National Center for Earthquake Engineering (NCEER) at *http://www.nceer.eng.buffalo.edu*, will reveal the breadth and scope of fuzzy sets applications in earthquake and structural engineering, which may not be fully conveyed here due to space limitation.

The nature of the uncertainty involved in civil engineering decision is complex and varied, and is the subject of intense research. The form of the uncertainties, the methods of assessment, and the evidence to support the use of one methodology or another are being studied. Most civil engineers are familiar with probability theory, but fuzzy set theory contains ways of utilizing semantic information and has a self-contained logic that is absent from probability. However, the reader should be aware that fuzzy set theory is but one candidate for the expression of ambiguity. As yet no overwhelming evidence is available to sustain its case as an expression of human cognition; neither has contrary evidence been provided (see Brown and Louie [1984] for an excellent discussion of the subject, and, more recently, Natke and Ben-Haim [1997]). Be that as it may, the theory has been successfully applied to practical decision and assessment problems as attested by the contents of the following sections.

6.3 HOW FUZZY SETS ARE APPLIED

In essence, all civil engineering applications use fuzzy sets in one of three ways:

- To represent fuzzy information, so that it can be quantified;
- To process such information appropriate to the application; and
- To draw conclusions which are readily understandable.

Other types of information and uncertainties may be involved in these steps, notably crisp and probabilistic information, as depicted in **Figure** 1. Representation of fuzzy as well as other types of information may be processed in myriad ways, fuzzy or otherwise, depending on the application needs. Likewise, the conclusion may be a combination of fuzzy, probabilistic and crisp statements as appropriate, but always in a form that is readily understood by members of the community.

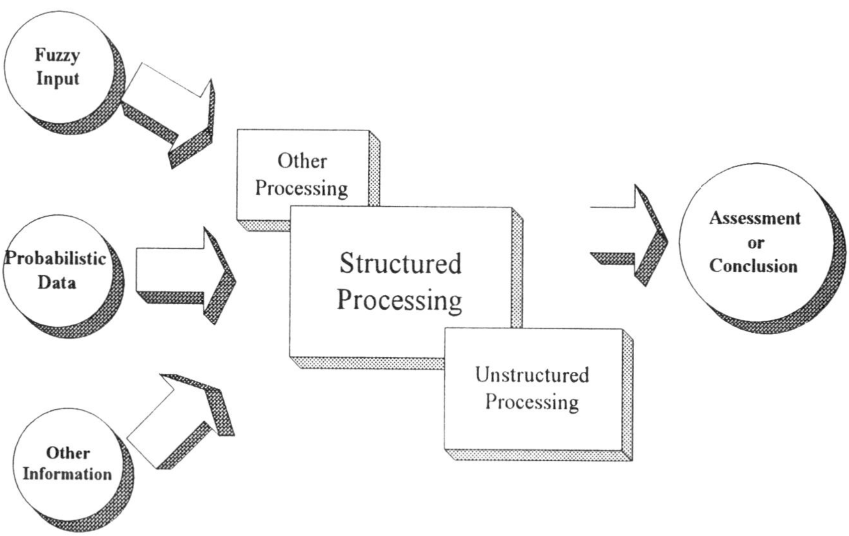

Figure 1. Elements of a fuzzy sets application in civil engineering.

6.4 APPLICATIONS PROTOCOLS

While the number of applications and disciplines of interest may be overwhelming, fuzzy sets applications in civil engineering follow only a finite number of processing protocols. Hence, by understand these protocols, one grasps the essence of how fuzzy sets have been utilized in civil engineering. These protocols are described in the following in an order that reflects their evolution. As one would readily observe, the evolution in civil engineering application is influenced strongly by the evolution of fuzzy sets as a technological tool. It starts as a way of handling imprecise information; information is then expanded to include knowledge. Knowledge processing, in turn, evolves into expert systems, fuzzy logic, and, finally, intelligent systems.

6.4.1 Structured processing

Most of modern civil engineering is based on mathematical models using the laws of physics, called first-principle models. The implicit assumption is made that these models represent the real-world behavior of structures, soils, earth movement, traffic flow, etc. Hence, the system description is crisp. Only the information feeding the models may be imprecise, uncertain, etc., thereby leading to imprecise and uncertain conclusions. Such applications may also be viewed as the propagation of uncertainties from one end of the analysis chain to the other, or from one level of an hierarchy to the next.

A simple example of structured processing is the work by Dong et. al. [1990]. The "system" is represented by algebraic equations relating the dynamic amplification factor of a vibrating structure to the mass, stiffness and damping of the structure. Fuzziness in mass, stiffness and damping leads to fuzziness in the response. The system representation can be more complex, as in Chao and Ayyub [1996] where a static finite-element model of the structure is used; or very complex, as in the random vibration model investigated by Chiang et al [1987]; and extremely complex as, as in Wadia-Fascetti and Smith [1996]. Other interesting applications include Murlidharan [1992] where the model analyzed corresponds to a beam on Winkler foundation, and Klisinski [1988] who extends the theory of plasticity with fuzzy sets. However, regardless of the complexity, the processing requirement is the same: the fuzzy variables must be manipulated through algebraic, differential or integral equations using the Extension Principle. As is well known, the Extension Principle can be extremely difficult to implement computationally despite its simple appearance, and linear programming (Dubois and Prade [1980]), fuzzy arithmetic (Kauffmann and Gupta [1985]) or numerical methods such as the Vertex Method (Dong and Shah [1987]) and Monte Carlo (Juang et. al. [1992]) must be used.

Compared with probability applications, fuzzy sets applications have the advantage of being procedurally consistent. The Extension Principle applies to problems of all complexity; only the computation implementation is affected. That is not the case with the probabilistic approach where, for example, the uncertain parameters of the governing differential equations such as those considered by Chiang et. al. in said reference elevate the problem into an extremely difficult class called parametric vibration, which currently has no general solution.

All fuzzy sets applications based on first principle models can make use of this processing protocol. In particular, for buildings and structures, the applications include structural analysis and reliability (Dong [1986]), and for geotechnical engineering, they include slope and foundation stability (Juang and Elton [1996]). However, a traditional, deterministic engineering model is not a prerequisite. Other "models" can be readily extended. For example, Chou and Yuan [1993] have enriched the Bayesian method of assessing structural reliability by incorporating the effects of fuzzy and subjective information; the base model in this case is the standard Bayesian (purely probabilistic) methodology (see also Itoh and Itagaki [1990]). In a similar vein, Shiraishi and Furuta [1983] generalized classical reliability theory with the use of fuzzy probability to account for subjective uncertain factors such as omissions, mistake and incorrect modeling.

Other broad classes of models that can be fitted under this protocol are those encountered in design and optimization. These applications usually involve structured models as well as constraints (see Furuta et. al. [1989], Munroe and Chuang [1986]), but they sometimes also contain unstructured models. Another hybrid class is that associated with (event, fault) trees and networks, used mostly in association with project review, construction management and risk analysis. Because these classes often involve elements of structured and unstructured processing, they do not, strictly speaking, fall under one discussion or another, and will be addressed in subsequent sections. As will also be seen, depending on the scope of problem addressed, these applications may also fall under the domain of expert systems.

Because even the most rigorous physical models are inadequate in predicting nature with absolute accuracy, civil engineers are also aware of the presence of the so-called "modeling uncertainty". However, a methodology for quantifying this uncertainty has remained elusive, mainly because the real-world reference is difficult to establish. Fuzzy sets have been used to represent this uncertainty in analysis as an alternative to the probabilistic approach (see Wong et al. [1987], Juang and Wey [1991]), and enjoy the same implementation and interpretation advantages mentioned previously. However, this area of application remains to be explored.

6.4.2 *Unstructured processing*

Unstructured processing refers to systems which have less founding in the laws of physics than structured systems described previously, and can range from slightly unstructured systems (e.g., empirical laws based on data) to highly unstructured systems (e.g., purely judgmental reasoning). Most applications in assessment, ranking and decision involve unstructured processing. However, that is not to say that assessment, ranking and decision exclude structured processing; they in fact often include some form of structured processing. But the outcome of structured processing is often only part of the information considered in those applications. Other parts are often unstructured, which is why fuzzy sets are called upon in the first place.

In unstructured processing, reasoning and/or logic may be used in lieu of physical laws. The reasoning can be simple, qualitative substitute of the physics (Bobrow [1985]), or subjective and judgmental as derived from experience. It can be a fault diagram, event tree, and algorithmic or conditional rules. However, very often, the basis is less rigorous than would be desired. Unstructured processing can be further subdivided into two kinds: bottom-up or top-and-bottom. In the former, processing begins with the basic information inputs and moves through the chain (or up the hierarchy), which can be functional or relational (logical), until a conclusion is reached. Examples are fuzzy aggregation, fuzzy inference, fuzzy reasoning and fuzzy classification. In the latter, the engineers have some causal knowledge but lack the quantitative underpinning of that knowledge. Fuzzy sets enable it to be formalized. Examples are fuzzy relations and fuzzy identification. More often than not, the same protocol can be used in both bottom-up and top-and bottom fashion, e.g., fuzzy relations. Applications may also include many unstructured processing elements as well as structured processing elements.

6.4.2.1 Fuzzy aggregation. The simplest form of this application is as follows: given the relevant (fuzzy) attributes, synthesize them into a (fuzzy) index, which can then be used for further processing. Aggregation into a single-index, though not always necessary, is desirable in civil engineering because it is easy to perceive and facilitates (Pareto) ranking and decision-making. Among the many viable aggregation methods (see Dong et. al. [1988]), the weighted average method has found most acceptance in civil engineering. Variants such as max-min weighted, min-geometrically weighted and delimited, weighted sum have seen less application.

Note that weighted sum is an algebraic operation, and if it is used, can be implemented according to the structured processing protocols discussed previously. Such aggregation is considered unstructured nevertheless because the weights are not derivable from laws of physics. They define the unstructured system, and may be deduced by a direct approach in which expert inputs are shaped into a consistent set (such as by Saaty's eigenvector method [1978]). Methods of aggregation that expand the weighted sum concept and combine algebraic and logical operations (Bellman and Zadeh [1970]), such as the confluence method (Yager [1980]; Weber [1983]), the γ-interpolation method (Zimmermann and Zysno [1980]), and fuzzy confluence rule (Kawamura et. al. [1988]), have seen little application in civil engineering. Evidence aggregation methods gleaned from Dempster and Shafer's theory (Ishizuka et al [1983]) are also not widely practiced in civil engineering.

Applications with fuzzy aggregation are omnipresent in civil engineering as assessment is a major function of the profession. Sample applications include assessing seismic risk based on hazards and building vulnerability (Miyasato et. al. [1986]), evaluating earthquake damage in light of urban development (Song et. al. [1996], Pandey and Barai [1994]), determining soil density from penetrator tests (Juang et. al. [1996]), interpreting job site records (Russell [1995]), treatment of highway pavement (Prechaverakul et. al. [1994]), Juang and Amirkhanian [1992]), pricing of construction taking into account the associated risk and consequence (Paek et. al. [1993]), evaluating the failure potential of soil slopes (Lee and Juang [1992], Juang et. al. [1992]), assessing safety of construction (Ayyub and Eldukair [1990]), evaluating the performance of constructed or dream facilities (Hadipriono

[1988]), Hadipriono and Ross [1988]), evaluation of bridge condition (Tee et. al. [1988]), simple analysis of risk in project planning (Kangari and Bakheet [1994]), water resources and management (Tulbure [1997] and Natke [1999]), and in a much grander scale, the extension of project-network analysis to account for fuzzy task duration and available resources (see, e.g., Lorterapong and Moselhi [1996], Chang and Crandall [1990]) or in the scheduling transportation vehicles (Kikuchi and Donnelly [1992]).

Less frequently, one may wish to apply unstructured processing protocols to a structured-process when only an order of magnitude answer suffices. An example is that of the work of Schwartz and Chen [1993]. A novel way of aggregating fuzzy and objective information is proposed by Brown (see Brown et. al. [1984], Kam and Brown [1983, 1984], and Mar and Brown [1984]). The updated probability distribution is determined through the constructions of objective (probabilistic) and fuzzy entropy equations. Since the fuzzy entropy function is not unique, calibration on the actual probability distribution is needed. The methodology is used to assess the seismic safety of existing and aging buildings.

6.4.2.2 Fuzzy ranking and ordering. A companion to fuzzy aggregation is fuzzy ranking or ordering. Selecting from among a group of fuzzy indices is not as straight forward as for crisp numbers and herein lies a dilemma: the rich information content in fuzzy sets preclude them from unequivocal ranking and decision.

One way out of the dilemma is, as in the case of statistical decision, to convert the indices into crisp values so comparison among them can be readily made. Many criteria for "defuzzification" have been used in application, including: pseudo-expectation, gravity center, most possible, pessimistic/optimistic, average mean, nearest to the ideal optimum, dominance, the four point average, and template classification. Each has its computation advantages and disadvantages and decision rationale, but like aggregation there is no unique choice. Sample applications include the rating of bridge conditions (Melhem and Aturaliya [1994], Sobanjo [1993], Sobanjo et. al. [1994]), pavement maintenance priority (Chen and Gupta [1995]), bus route selection (Ogawa and Koike [1994]), tender evaluation (Nguyen [1985]), and the evaluation of performance in geotechnical engineering (Chameau and Gunaratne [1984]).

Ranking of fuzzy indices does not need to be preceded always by defuzzification. Ranking can be handled as a classification procedure (see discussion to follow), but this approach is not popular because extra efforts are required while benefits are unclear; the unstructured nature of the solution procedure is not any less.

6.4.2.3 Fuzzy relations. Fuzzy relations constitute one of the first fuzzy sets technologies adopted and applied by civil engineers, and continue to play several important roles in reasoning and information synthesis. An unstructured mechanism, one whose inner workings are fuzzy, is best represented by a fuzzy relation. Hence, any input to this system, fuzzy or otherwise, can be processed in a manner similar to structured processing, producing the corresponding fuzzy outputs. Conversely, given the (fuzzy) input-output pairs, the mechanism may be inferred by the corresponding fuzzy relation; the fuzzy system is said to have been identified. These early ways of usage have been extended and evolved to envelope expert rules, fuzzy inference, and knowledge-based and intelligent systems as will be described in subsequent sections.

Applications that use fuzzy relations as the system engine include: the evaluation of earthquake intensity (Feng et. al. [1982a, 1982b]), seismic safety assessment (Brown et. al. [1984]), and estimating project duration (Ayyub and Haldar [1984]). Brown has given an excellent expose on fuzzy identification (inverse fuzzy relations) in contrast with probability (Brown [1987]). The backing-out of a fuzzy relation, given the input-output quantities, is often referred to as system identification or the identification of a fuzzy system, to be exact. Methods used by civil engineers most include Mandani [1974, 1977a, 1977b], Pedrycz [1983, 1984], and a modified version of Pedrycz's (Boissonnade et. al. [1985]). Examples of applications involving identification of fuzzy relations include: the identification of seismic intensity (Boissonnade et. al. [1983, 1985]) and the correlation between intensity and peak ground acceleration (Liu [1982]).

A close relative of fuzzy relations when used in the identification context is fuzzy linear regression. In both cases, the input-output pairs are given, and their relationship is to be estimated in the form of a matrix, namely, a fuzzy relation matrix and an algebraic matrix with fuzzy elements, respectively. Kaneyoshi et al. [1990] show that fuzzy linear regression can be recast into a *non-fuzzy* linear programming problem. Using this simple technique, very sophisticated engineering systems have been built for the design and erection control of cable-stayed bridges (see Kaneyoshi et. al. [1993]). This technique should be contrasted with the non-linear hill-climbing method (e.g., Pedrycz's method mentioned previously) that is usually required in identifying fuzzy relations.

6.4.2.4 Fuzzy modus ponens, fuzzy logic and fuzzy reasoning. Fuzzy logic and fuzzy reasoning denote the processing of fuzzy propositions, e.g., "the stress is large"; "the wind is strong", etc. The fuzziness arises because it is difficult to assess the truth of the proposition. The processing parallels that of fuzzy sets, with the truth function of the proposition corresponding to the membership function in fuzzy sets.

Fuzzy modus ponens and logic may take one of many forms, one of which is fuzzy relation and composition. In particular, inference rules $If\, A_i\ Then\ B_i$ may be synthesized into a relation R, and reasoning performed on new antecedent A_i^* by composition $A^* \circ R$ (see Wong et. al. [1986], Dong and Wong [1987a], and Wong and Dong [1986]). However, the companion to modus ponens, viz., modus tollens, cannot be effected by fuzzy relations operations because of non-uniqueness of the inverse operation. Consequently, and with the additional advantage of ease of computation, most fuzzy logic applications in civil engineering are based on extension of binary logic, most notably the union and intersection operation for OR and AND combinations. However, nearly all logical applications use the maximum and minimum operators for disjunction (OR) and conjunction (AND), respectively; rarely are other forms of the general families of T-conorms and T-norms used in a civil engineering application (Boinissone [1987]).

In general, inference with fuzzy sets may be broken out into two general categories consistent with the philosophy of this chapter: inference with algorithm rules and inference with conditional rules (see Dong and Wong [1987b, 1987c]). Algorithmic rules are rules which can be represented as a mathematical algorithm (e.g., hierarchical network) or equation. Logically, algorithmic rules are equality relations, viz., $y \leftrightarrow (x)$ when the mapping is one-to-one; y is true when x is true and vice versa. When x is uncertain, y is as well. Uncertainty can also arise because of uncertainty in the validity of the equation relation, but that is seldom addressed.

Conditional rules are more general than algorithmic rules and constitute relations that are less well defined, such as common forms of knowledge representation in expert systems. In contrast to the mapping representation, conditional rules may be denoted symbolically as $(y|x)$, i.e., the knowledge of y is depends somehow on the knowledge of x. Hence, even when x is known with certainty, y can be uncertain due to uncertainty in the rule. Naturally, when x is uncertain, y is uncertain due to the uncertainty in x and, possibly, the uncertainty in the rule. Logically, $(y|x)$ is an implication relation and can be denoted by $x \rightarrow y$, i.e., x implies y. It is also referred to as x entails y, or x is contained in y. The two modes of deduction associated with implication propositions are modus pollens and modus tollens, and have been described previously.

Examples of fuzzy logic applications include: seismic risk evaluation (Dong et. al. [1986], Juang and Elton [1986]), detecting metal fatigue (Bowman et. al. [1987], Hinkle and Yao [1988]), fault trees and event trees (Ross and Sunil [1994]), and damage assessment of lifeline systems (Katayama et. al. [1991]). With time, fuzzy logic applications have evolved along with the general technology trend to become fuzzy reasoning in expert and intelligent systems. More will be said about these applications later, but a good overview can be found in Yao [1988]. Blockley and Baldwin [1987] portent the modern-day systems approach where various forms of uncertain reasoning such as fuzzy logic, probabilistic inference, and the theory of evidence are integrated for civil engineering applications.

6.4.2.5 Fuzzy control. Fuzzy control is a special kind of fuzzy information processing in which control actions are the end results. Fuzzy control may be embodied as simply as a fuzzy relation, and is probably the most visible among all fuzzy sets technology because of its association with successful engineering applications such as industrial process, robotics, navigation, power plants, traffic control, etc. In the early days, fuzzy control was practiced in two directions: the dynamic programming approach which is analogous to classical optimal control based on a performance criterion, and the linguistic synthesis approach that is akin to fuzzy reasoning. The latter has become by far the more popular approach (see review of early applications by Tong [1977] or Mamdani [1977b]), and the trend is the same in civil engineering.

Fuzzy control has become perhaps the most robust field of application today. Recent applications in construction and transportation include: Hybrid control of sliding structures (Subramaniam et. al. [1996]), intelligent cruise control (Chang and Choi [1966]), bridge construction (Furuta et. al. [1995]), and automatic highway (Liu and Frank [1991]). Applications to structures include: Control of sliding structures (Reinhorn et. al. [1996]), general multi-degree-of-freedom systems (Casciati and Faravelli [1996]), active tuned mass dampers (Abe [1996]), active control with multi-objective optimization (Rao [1994]), and control and reliability (Yao and Natke [1993]). Frank [1994] is representative of applications in process monitoring and fault diagnosis.

We shall use structural control to illustrate the typical application. For vibration control of buildings and structures, available methods of control include those that control the rigidity, mass and damping of a structure and those that control a combination of these three properties. The energy required with rigidity and mass methods is quite large, while damping control uses a relatively small amount of energy. For extreme environments, attempts are made in general to control the excitation (e.g., base isolation) and/or the response (e.g., actuator control).

Table 3. A fuzzy control table.

		ΔE, Rate of Change of Position Error						
		NL	**NM**	**NS**	**Z**	**PS**	**PM**	**PL**
E, Position Error	**NL**	PL	PL	PL	PL	PM	Z	Z
	NM	PL	PL	PL	PM	PS	Z	Z
	NS	PL	PL	PM	PS	Z	NS	NS
	Z	PM	PM	PS	Z	NS	NM	NM
	PS	PS	PS	Z	NS	NM	NL	NL
	PM	Z	Z	NS	NM	NL	NL	NL
	PL	Z	Z	NM	NL	NL	NL	NL

Fuzzy control in civil engineering is similar to that in other fields. Its core consists of a set of fuzzy rules that reflect intuitively on what should be done when the structure is found to deviate from its nominal state., e.g., lateral displacement, velocity and acceleration which should all be zero in the ideal situation. For instance, in excursion control, a typical rule of thumb may be "If the position deviation is large positive, PL, and the rate of change in position is zero, Z, then the counter force should be zero", i.e., *IF E=PL and ΔE=Z, THEN F=Z* where E denotes position error, ΔE the rate of change of E, and F the control force, because the structural motion is steering itself back towards the equilibrium position; there is no need for external interference. On the other hand, "If the position deviation is zero and the rate of change in position is large negative, then the control force should be large positive", i.e., *IF E=Z and ΔE=NL, THEN F=PL* in order to apply the brakes, so to speak. These rules of thumb are usually collected in a table, such as shown in **Table** 3, which is the crux of a fuzzy control strategy.

The control strategy in **Table** 3 corresponds to a 2-to-1 mapping; the two monitored parameters, position and its rate of change (velocity), are mapped to the control force which may correspond to an actuator output, damping force, or additional mass. Acceleration may also be involved in a control strategy, either in place of or in addition to position and velocity. The control table may be 1-to-1, 3-to-1, many-to-many, etc., and the variations are endless just in structural application alone (see references cited previously and Yamada et. al. [1994]; Shingu and Funamoto [1992].

As with other fuzzy sets technologies, fuzzy logic control has become an integral part of many modern intelligent systems. Fuzzy control is practiced not in isolation but as an element of a more general framework. The interested reader can refer to Casciati et. al. [1995] or Thint et. al. [1997]. Compared with conventional algorithms such as classical optimal control (Abdel-Rohman et al. [1980]) or instantaneous optimal control (Yang and Akbarpour [1987]), fuzzy control has several practical advantages:

- Its control algorithm is extremely simple, as can be seen from **Table** 3.
- It is suitable for real-time control by virtue of its simplicity
- It does not require accurate information on structural and vibration characteristics of the system being controlled
- System is more robust in terms of performance and (software) implementation

The last point should be expanded on. Although fuzzy control does not enjoy the theoretical proofs of controllability, stability and optimality that classical control methodologies enjoy, civil engineering researchers have found from experience in application that fuzzy control is especially effective when the system response is nonlinear, the measurement is polluted by noise, and the feedback signals are weak. The effectiveness can be attributed to the ability of fuzzy control to discriminate and treat irregular input by sets according to their significance, e.g., extreme control forces or uncertain input near zero. The fuzzification into membership functions, the output aggregation and defuzzification steps all contribute to reduce the influence of input noise and computer imprecision (see Subramaniam [1994]).

A novel approach described by Kawamura and his colleagues (see Fujitani et. al. [1996], Tani and Kawamura [1995], Tani et. al. [1992]) is worth noting because it differs from the mainstream fuzzy control approach in how fuzzy sets are used. Called optimal, adaptive and predictive control, the method in essence has three main elements that are, respectively, optimal, adaptive and predictive (**Figure 2**).

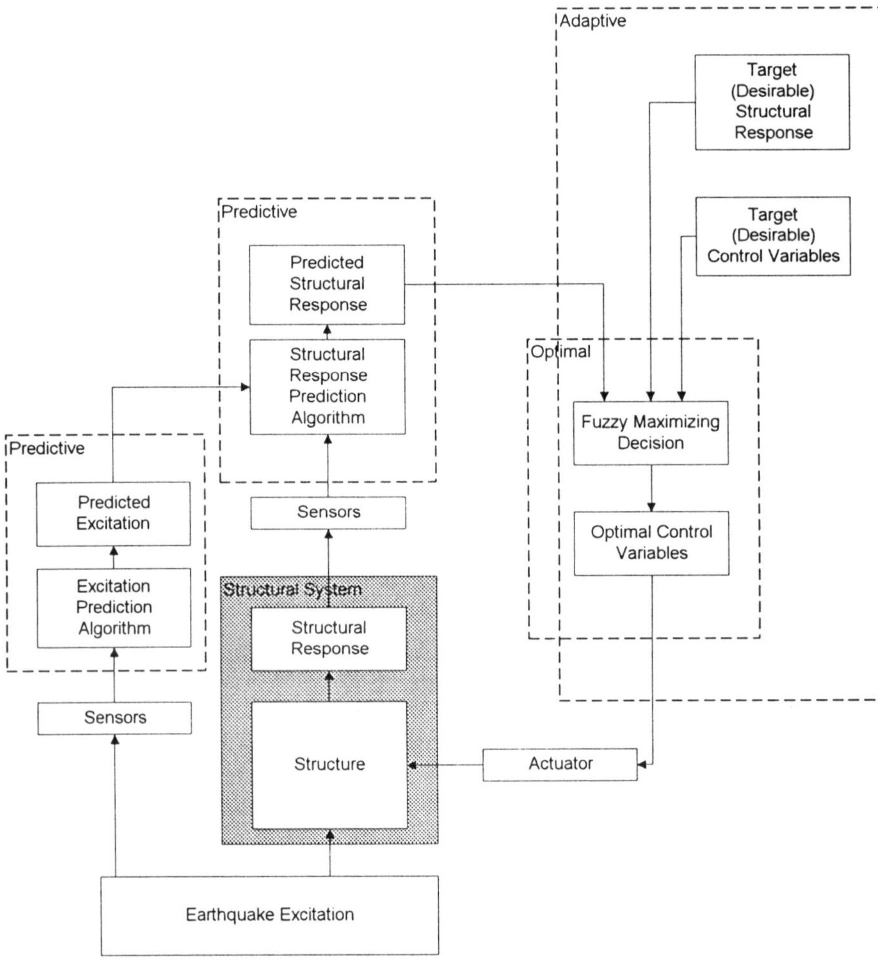

Figure 2. Optimal, adaptive and predictive control (from Tani et. al. [1992]).

At any time in the process, future excitation is predicted based on past values. Likewise, structural response is predicted based on past values, the predicted excitation and a candidate control force. The predicted response is then compared with the desired response envelope (in the form of a fuzzy sets membership function which reflects comfort and structural safety) and the candidate control force compared with its corresponding desired envelope (a membership function which reflects economy and actuator limitation). The latter is adjusted (hence, adaptive) so that both its own desirability and the desirability of the prospective structural response are acceptable; the tradeoff is effected by a fuzzy maximizing (hence, optimal) decision algorithm, after Bellman and Zadeh [1970].

The prediction function is initially mechanized by using fuzzy rule sets (see Kawamura and Yao [1990]) which is similar to the fuzzy control table used in conventional fuzzy control (e.g., **Table** 3 above), except that the inputs are past quantities of excitation (or response) and the output is future quantities of excitation (or response). As in many other civil engineering applications, the predictive task is taken over later by neural networks. Consequently, the maximizing decision part of the methodology is the only part that remains fuzzy-sets related.

6.4.2.6 Fuzzy classification and pattern recognition. Two other major areas of applications that stand out by themselves among unstructured processes are fuzzy classification and optimization. Although classification or optimization can be considered to follow a processing chain of events, such as input-processing-output discussed so far, the processing is more operational and less physical. In classification, an operation is performed on a group of objects with the goal of dividing them. In optimization, many attributes are considered with the goal of picking the combination that meets certain criteria subject to certain constraints. There is usually no relation between the operational mechanism and a physical mechanism, structured or otherwise.

Loosely speaking, the process of classification is to divide a number of objects into several groups or clusters according to some criteria. The process of pattern recognition is to use the resulting groups as templates and identify new data as being part of one of those patterns. Both are important steps prior to any selection of decision process. The advantage of fuzzy sets over conventional approaches is that the class boundary is not crisp; the partition is soft. An object may belong to several classes with different degrees of belonging-ness. Hence, the information contents of the data are retained through the processing, up until the time selection or decision is made.

The key to classification is coming up with a criterion or classifier. Generally speaking, the graph-theoretic method is well suited to data with chaining characteristics, and the objective function criterion is applicable to data that are basically hyper-spherical. Most civil engineering applications use the latter approach. The partition sought is one that optimizes a certain objective function, and the more popular objective functions are the least-square function and Ruspini's function (Ruspini [1977]). As in fuzzy ranking, the information content of a fuzzy partition is inherently richer than a crisp partition, but in practice, the partition may need to be "hardened" before a decision can be made. The defuzzification may proceed according to the maximum membership or nearest center.

Recognizing data with fuzzy features as belonging to one group or another requires a measure of distance: similarity or dissimilarity. Conventional distance measures such as the Hamming distance of Euclidean distance, generalized to fuzzy sets, have been used. In evaluating earthquake intensity and predicting earthquakes, Feng et. al. ([1982a], [1982c]) use the "approaching-degree" metric proposed by Wang [1983]. Recent applications include: Classification of modal vibration data (Ross and Hasselman [1995]), and identifying accident-prone locations on roadways (Sayed et. al. [1995]).

6.4.3 Expert Systems

Applications of fuzzy sets in civil engineering follow closely the evolution of the technology, usually after a warm-up period to allow for discovery and assimilation. The structured or unstructured processing methods discussed above are brought over to civil engineering as their usefulness becomes known. However, the transfer of technology is initially piece-meal; the initial applications are significant and important, but limited in scope nevertheless. An application usually involves using one or at most two of the techniques discussed. This mode of application continues today, as can be seen from the references cited in this chapter. However, with the maturation of the expert system and intelligent system technologies, and as the synergistic power of these technologies with fuzzy sets becomes obvious, civil engineering applications begin to evolve in that pattern as well. Today in civil engineering, it is seldom that fuzzy sets are not mentioned in the context of a decision-support or knowledge-based systems.

That expert systems technology and civil engineering prove a good match should not come as a surprise, considering the abundant use of expert judgment in the profession. Take assessment of damage in existing structures, for example. Efforts have been made over the years to standardize personal judgment in that profession, but neither the OEDC committee (OEDC [1976]) that addressed bridge inspection nor the ASCE committee (ASCE [1991]) that addressed existing buildings was able to accomplish its goal. Members of the committee agreed that because evaluation by its nature implies the use of personal judgment by those qualified as experts, the requirements for evaluation could not be standardized. Until a better approach comes along, expert systems are at present the most effective and practical tool for judgment and professional experience to be explicitly incorporated in the decision process.

Expert systems applications in civil engineering may entail only the basic rule-based methodology, as many earlier systems do, or just about everything. It is advantageous today to look at expert systems as the re-structuring of formerly unstructured and structured processing tools that have been applied fragmentarily. The system technology provides the necessary framework to bind previously uncoupled and unconnected pieces so that the whole becomes a powerful structure.

With the popularization of experts systems, rule-based inference becomes much more firmly established in civil engineering than previously possible with fuzzy modus ponens and fuzzy logic alone. Fuzzy sets evolve more and more into an uncertainty attribute under the general category of approximate reasoning, alongside other measures such as probability, evidence, etc., and less so as a self-contained reasoning schema. Fuzzy reasoning in the form of fuzzy relations discussed previously all but disappears from the literature. IF-THEN rules can be implemented explicitly, and there is no need for representation in fuzzy matrices.

An early but significant milestone in civil engineering is the publication of SPERIL (Ishizuka et. al. [1980], Ogawa et. al. [1984], Zhang and Yao [1986]), which addresses damage assessment of built structures. Subsequent expert systems for damage assessment include Brown et. al. [1984], Tee [1988], Ross et. al. [1986] and Dong et. al. [1986]. In its essence, these expert systems extend the core idea of conventional methods (Yao [1980]) to other observable quantities, whether they can be measured or not, e.g., energy absorbing potential, seismic behavior, external look, amount of cracking and spall, and so on. The relations between these observable quantities and the sought-after damage are stipulated in the form of analytic algorithms where they are appropriate, such as in the conventional approach, and in the form of rules where analytic algorithms are not possible. The rules are based on aggregated expert knowledge and opinions.

Figure 3 is an overview diagram of the inference network used in SPERIL-1. SPERIL is unusual in that it not only represents an extremely prompt transfer of the expert system/fuzzy sets technology to civil engineering, but also is noteworthy for its rich content. Although dated more than 17 years ago and superseded by two important revisions, the basic ideas of SPERIL are as good today as when they were first introduced. Note that changes in natural frequency and damping are included as part of the "evidence" derived from analysis of measurement records and are used in the assessment of global and local damage, just as in conventional methods. Drift, fatigue and creep records can also be included as major elements of the data analysis. Hence, the expert system approach can be viewed as a generalized framework that considers evidence from a multitude of sources, including the conventional ones. Given today's technology, the scope of **Figure** 3 could be properly updated to enhance its capability but the framework remains valid. One obvious enhancement is to include evidence from acoustic, ultrasonic, magnetic testing to augment visual inspection. Another is to leverage the pattern learning power of artificial neural networks to improve the cognitive power of the system as will be discussed in a subsequent section.

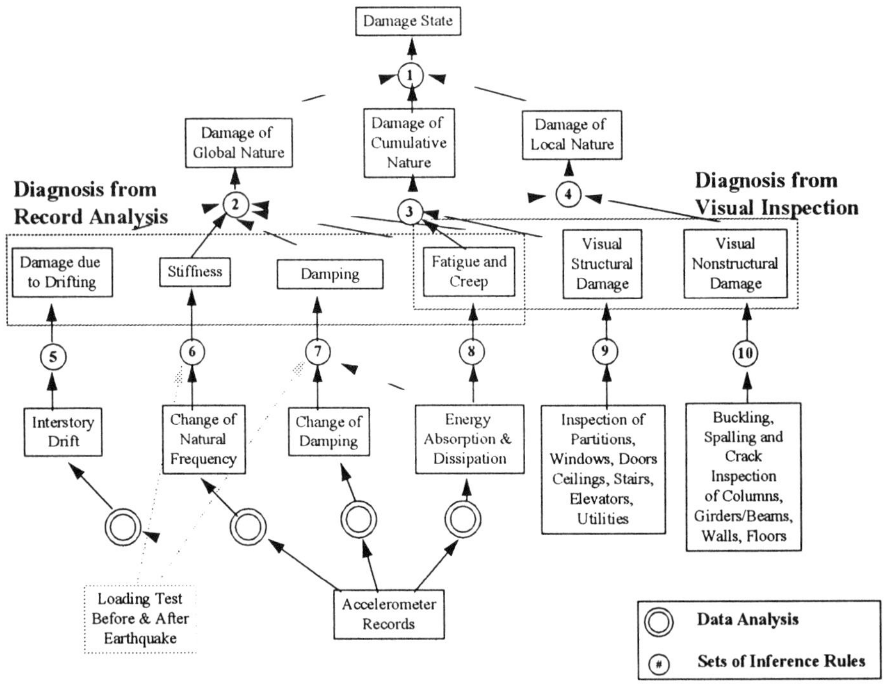

Figure 3. Inference network of SPERIL-1 (from Ishizuka et. al. [1980]).

Yao and Fu [1985] and Furuta et. al. [1985] are two earlier reviews of expert systems applications and potentials in civil engineering. Among the first expert systems or prototypes developed for other civil engineering applications, are: Fault diagnosis (Lai et. al. [1986]), water quality management (Jowitt and Lumbers [1982]), and characterization of soil data (Mullarkey et. al. [1985]). More recent expert system applications involving fuzzy sets are: Investigation of seismic damage (Castaneda and Brown [1994]), damage to concrete piles (Yeh et. al. [1991]), durability assessment of bridge decks (Furuta et. al. [1990]), evaluation of design/build proposals (Paek et. al. [1992]), aseismic structural planning (Kawamura et. al. [1992]), earthquake proofing of underground lifelines (Takada et. al. [1991]), planning of work-zone traffic control (Fisher and Rajan [1996]), automatic project correction based on daily job site reports (Russell and Fayek [1994]), legal advice for the construction industry (Diekmann and Kraiem [1990]), bridge condition assessment (Natke et. al. [1995]), and fault diagnosis and detection (Issermann [1994]).

6.4.4 Intelligent Systems

The evolution of computer and modeling sciences into intelligent systems is readily embraced by civil engineers. Artificial intelligence technologies such as neural networks and genetic algorithms are applied singly or jointly under an expert system or generalized fuzzy inference framework. Artificial neural networks are powerful because many functions of the human brain, such as memory, learning, recognition, prediction and adaptive reasoning, can be simulated by neural networks with basically the same architecture. Genetic algorithms are useful for combinatorial and optimization problems, as they are based on heredity and evolution of a living organism. Compared with conventional methods, genetic algorithms are unusual in that they can provide efficient solutions even if the evaluation function is discontinuous. Other new technologies such as geographic information systems, image recognition, database management have also been incorporated to facilitate the handling and management of data in various media and formats, thus enhancing the intelligence of the system. *In this new setting, fuzzy sets become just one of many methodologies for modeling and processing uncertainties.*

Several examples are given below to illustrate the state of advancement, breadth and scope in application, and the complementary role of fuzzy sets in modern intelligent systems. A recent issue of the Journal of Microcomputers in Civil Engineering (Volume 11, 1996) is devoted to works in intelligent systems; it describes ingenuous uses of neural networks and genetic algorithms in conjunction with fuzzy sets, and should be of interest to civil engineers.

6.4.4.1 Neural network with fuzzy parameters (Ni et. al. [1996]). A neural network is trained to encapsulate expert knowledge relating geologic and environmental conditions to failure potential of slopes. The application is noteworthy because the slope attributes are described in linguistic terms, necessitating representation by fuzzy sets. Hence, conventional neural network must be extended to process fuzzy numbers as not only the data set but also the resultant weights and biases of the network are fuzzy. The required processing is implemented using the vertex method (see Dong and Wong [1987d] and also earlier discussion on structured processing in this chapter).

One critical problem in training a network with fuzzy numbers is that as the input-output data sets are processed one by one, local minimum conditions may be encountered which will adversely affected the correctness of the membership function so established. Therefore, a two-stage training method is used wherein the first stage corresponds to preliminary training and the second stage corresponds to fine-tuning (**Figure** 4). In the preliminary training stage, all available training data sets are used and a larger relative error is allowed to expedite convergence. In the fine-tuning stage, each data set is regarded as a unit, and only one unit is processed at a time. The error criterion is also customized to meet specific requirements in the training, so that one set of network parameters is obtained for each data set. Repeating this process for *n* data sets result in *n* sets of network parameters, which

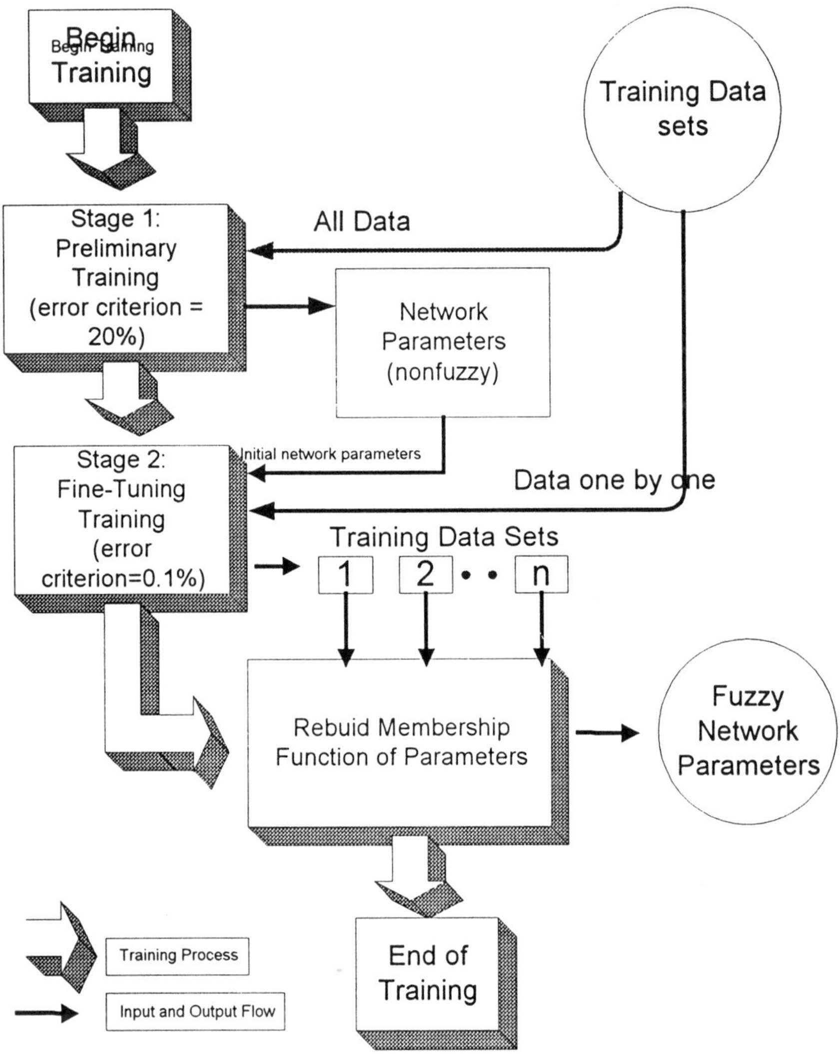

Figure 4. Schematics of a two-stage training algorithm, neural network for fuzzy training patterns (from Ni et. al. [1996]).

are then used to establish the membership functions of the parameters. Based on a study with a database of 50 training patterns, Ni et. al. found that topographic factors (such as gradient, profile, location and height) to be important to slope failure whereas environmental factors (such as vegetation and land use) were unimportant.

6.4.4.2 Adaptive network-based fuzzy inference system (Faravelli and Yao [1996]). A neuro-fuzzy system combines the strength of fuzzy systems in dealing with uncertainties with the advantage of neural nets in learning and recalling patterns. In an earlier discussion on fuzzy control, it is mentioned that the task of predicting future excitation (or response), that is previously handled by fuzzy conditional rules, can be taken over by trained neural networks. Faravelli and Yao use neural network to take over the complete fuzzy inference process. In particular, the fuzzy (rule) inference scheme of Takagi and Sugeno [1983] of the form:

Rule i: If x is A_i and y is B_i, Then $f_i = p_i x + q_i y + r_i$

is implemented within an ANFIS (Adaptive Network-Based Fuzzy Inference System, by Jang [1992]) of the form in **Figure** 5.

The square boxes indicate nodes that are adaptive, i.e., they contain parameters which are changed upon learning. The first set of parameters denote those that define the membership functions for linguistic labels such as Negative (NE), Positive (PO), etc., that govern the response variables x (displacement), y (velocity), etc.; that is, the antecedent part of the rule. The second set of parameters corresponds to those that define the consequent part of the rule such as p_i, q_i and r_i in the equation above. The neuro-fuzzy system learns fuzzy rules and terms by using training sets of data which are time histories of controlled response which are either prescribed or generated by others methods such as conventional fuzzy control. In application, the membership functions of the linguistic labels are adjusted so that the time integral of the squared control-force is minimized. Based

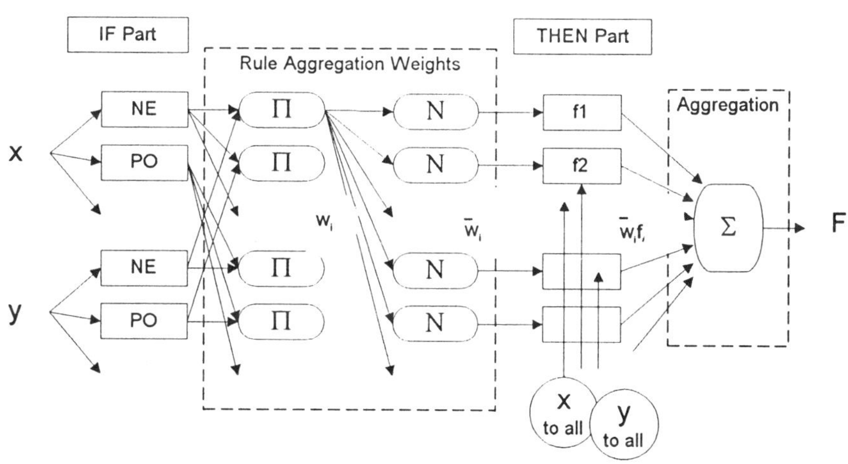

Figure 5. ANFIS network architecture; square nodes are adaptive and circular nodes are not (from Faravelli and Yao [1996]).

on a computer study of a two-degree-of-freedom hysteretic system, Faravelli and Yao found that the ANFIS controller performance is comparable with that of the conventional fuzzy controller in terms of response reduction. The former is computation efficient, requiring only ¼ of the effort for the normal fuzzy controller.

6.4.4.3 Fuzzy expert system with genetic algorithm and neural network elements (Furuta et. al. [1996]). An efficient knowledge-acquisition system is developed for improving and maintaining the knowledge base in evaluation of reinforced-concrete bridge decks. The system can acquire fuzzy production rules automatically based on joint use of genetic algorithms and neural networks methodologies: New production rules are first generated using all possible combinations of the antecedent attributes; they are then screened, and only the meaningful rules are extracted and cultivated using genetic algorithms. The evaluation (objective) function that governs meaningfulness is, in turn, determined by a neural network that has been trained by previous data and existing valid rules.

Table 4 is a simple example taken from said reference and will be used as illustration. In inferring the cause of damage from "extreme-wheel -load", many factors are considered including: structural form, crack configuration, crack location, wheel-load location and design specification, which are listed as the antecedent-part of rule in the table. Since there are four possible structural forms, eight crack configurations, etc., there are 4*8*...=4096 possible rule antecedents. The truth condition, the consequent-part of the rule, can be one of seven levels. Hence, the total number of rules is 4096*7=28,672; three such rules are listed in the first part of the table for illustration.

The possible combinations of the antecedent-part are represented by 2+3+...=12 bit chromosomes. For example, when the structural form is plate girder (value 1), the crack is in the width direction (value 1), the crack is located at the side-span (value 3), the wheel-load is located at the side-span (value 3), and the design specification is unknown (value 0), the antecedent is represented in vector form as (1, 1, 3, 3, 0), and in chromosome form as: (01 001 011 011 0). The consequent-part is coded in similar fashion.

Knowledge obtained from experts such as the two rules given in the second part of the table is similarly coded and used to train a neural network. This establishes a template of meaningfulness for evaluating other rules; candidate rules that generate large prediction errors in the trained neural network are considered unfit and a large percentage of which is discarded. The deleted rules are replaced by replicating rules selected randomly from the fit pool. A certain portion of the replicated rules are then cross-overed and mutated, and the process of evaluation is repeated with the new generation of rules, until their evaluation value is comparable to that of the expert knowledge embodied in the neural network, or when they are judged acceptable by the experts.

Table 4. Rule base for damage caused by "extreme-wheel-load" (Furuta et. al. [1996]).

	Antecedent (Environmental conditions)					Consequent (Damage Cause)
	Structural Form	**Crack Config.**	**Crack Location**	**Wheel-load Location**	**Design Specification**	**Extreme-wheel Load**
Rules Generated Randomly	Plate-girder	Width-direction	Side-span	Side-span	Unknown	Absolute true
	Plate-box	Axis-direction	Middle-span	Side-span	Before 1965	True
	Plate-girder	Penetration-form	Other	Support-beam	After 1965	Very true
	......	......	......	......	......	......
Rules Acquired from Experts	Plate-girder	Width-direction	Side-span	Side-span	Before 1965	Absolute true
	Plate-girder	Axis-direction	Middle-span		Before 1965	True
	......	......	......	......	......	......
Rules Obtained by Neural Networks and Genetic Algorithms	Plate-girder	Width-direction	Middle-span	Side-span	Unknown	Very true
	Plate-box		Side-span	Middle-span	Before 1965	True
	Plate-girder	Width-direction	Side-span	Side-span	Before 1965	Absolute true
	Plate-girder	Axis-direction	Middle-span	Side-span	Before 1965	True
	......	......	......	......	......	......

Furuta et. al. found that for bridge-deck assessment, good rules that did not exist initially could be generated. However, depending on the initial set of rules, rules that optimized the evaluation function might or might not be generated. They concluded that the proposed method could be useful when it was difficult to enunciate the complete rules used by experts, when all the optimal rules based on the experience of experts could not be guaranteed, and when it was difficult to obtain optimal rules due to large freedom in using fuzzy expressions.

6.5 OTHER RESOURCES

Other reviews and articles of interest to the reader include: Furuta [1995], which describes applications in seepage, tunneling and dams using the fuzzy sets as well as artificial intelligence and expert systems technologies; Yao and Natke [1995], which describes applications and potentials in civil infrastructures; Adeli and Hung [1995], which discusses the role of fuzzy sets in conjunction with neural networks and genetic algorithms, Ayyub [1991] which gives a comprehensive literature review of civil engineering applications; and Chameau [1990], which emphasizes geotechnical engineering. Earlier reviews such as Brown and Yao [1983], and Brown et. al. [1987], should also be of interest to civil engineers.

6.6 FUTURE OUTLOOK

Although civil engineers have been involved in applying fuzzy sets since early years of the development of the technology (e.g., Brown [1979]; Blockley [1979]), the application has not been as prevalent as in other fields. Nevertheless, a solid foundation has been laid. Since many practical problems in civil engineering involve fuzziness and ambiguity, it is inevitable that the application of fuzzy sets will become practical in solving real-world problems. It is encouraging to note that the concepts of fuzzy sets can be communicated with relative ease to practicing engineers who must deal with real-world problems.

Emphasis in civil engineering has shifted in recent years from new major constructions to efficient maintenance, management, and repair of existing structures. For the few major projects involving new construction, emphasis is on better (intelligent) functional and durability performance. Future applications of fuzzy sets will reflect these trends. In particular, one promising application of fuzzy sets is in making condition assessment of existing structures and facilities. Another promising applications is fuzzy control as applied to reducing vibration of buildings and bridges.

The developments will be evolutionary, not revolutionary. Existing technologies will be used ingenuously, and new technologies from other fields of science and engineering will be gleaned, to better process uncertain information, manage risk, and optimize the return on infrastructure investment. We see:

6.6.1 More application of fuzzy inference to condition assessment.

There will be more transfer of fuzzy inference (fuzzy relation) technology to structural condition assessment (following Yao and Yao [1997], see **Figure** 6). Actions required are functions of the current condition and expected loads in the future. Both factors can be expressed in fuzzy terms. Therefore, fuzzy relations are applicable as shown. More diagnostic models of mechanical systems will also incorporate fuzzy elements to be realistic (Natke and Cempel [1997]).

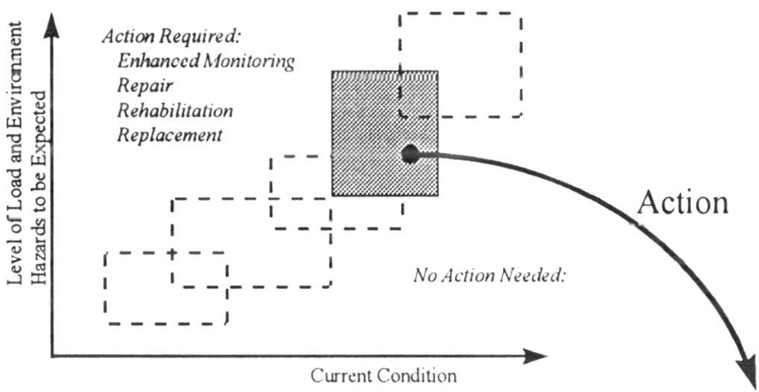

Load & Environment					
MMS	Repair	Rehab	Replace	Replace	Replace
MS	-	Repair	Rehab	Rehab	Replace
NC	-	-	-	Rehab	Replace
LS	-	-	-	Repair	Rehab
MLS	-	-	-	Repair	Repair
	VG	G	AV	P	VP

Current Condition

(MLS=much less severe; LS=less severe; NC=no change; MS=more severe; MMS=much more severe; VG=very good; G=good; AV=average; P=poor; VP=very poor)

Figure 6. Remedial action as a result of current condition of structure and loading and environmental hazards anticipated over the next inspection interval.

6.6.2 More comparative and modeling studies.

There will be more comparative studies, such as in calibrating structural model uncertainties and comparing them with results of modal identification based on seismic response (Wadia-Frascetti and Smith [1996]) and the comparison of fuzzy sets and probabilistic modeling of structural vibration (Chiang et. al. [1987]).

Integrated approaches that encompass fuzzy and probabilistic uncertainties as pioneered by Brown [1979] will be important. Two recent developments are worth mentioning. In *IRAS* (Dong and Wong [1996]), technological advances in computer science and engineering, as well as a systematic approach to uncertainty management are used to implement a comprehensive system that addresses all aspects of catastrophe risk management, from the underlying physical phenomenology to its impact on insurance and financial decisions (**Figure** 7).

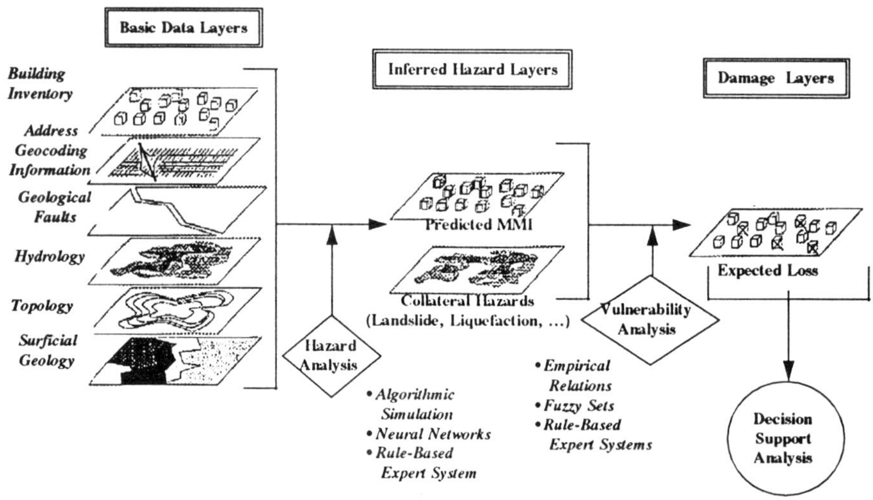

Figure 7. The *IRAS* framework (from Dong and Wong [1996]).

IRAS adopts the probabilistic approach for hazard analysis to handle uncertainties in occurrence and hazard prediction. An uncertainty model based on fuzzy set theory and expert systems is used to model judgmental knowledge and incomplete information. In addition, artificial intelligence techniques such as artificial neural networks are used to extract hidden relations in the raw database that cannot be detected readily otherwise. Fuzzy sets are also used in many other aspects of the assessment chain, notably importance factors when the effects due to various factors are to be aggregated.

Under the context of re-qualification of important structures such as power plants, bridges and offshore structures, a general decision support system was built and implemented into an intelligent software called RESTRUCT (BRITE EURAM Project BE5935 [1996]). Three other aspects of the work are most unique: (1) It represents a cooperative effort involving eleven European partners and the framework is design to address several classes of existing structures; (2) It combines established deterministic, fuzzy and Bayesian decision techniques into a multi-criteria decision methodology; this is done through combining and merging the individual techniques with joint utility functions (cf. Brown's entropy function) or Saaty's analytical hierarchy process (AHP); and (3) It interfaces with a variety of structural databases and analysis tools in order to be useful to the participating institutions.

6.6.3 More comparison of models with experiments.

There will be more comparison and verification of model studies with laboratory and field experiments, especially in the area of control. For example, in

Subramaniam [1996], fuzzy control is applied to a ¼-scale structural model with (1) a sliding base-mat, (2) a sliding base-mat supporting a single-degree-of-freedom flexible structure, and (3) a three-degree-of-freedom structure on a sliding base-mat (**Figure** 8). The base-mat is mounted on four Teflon-stainless steel isolation pads with shear springs placed on either side of the base-mat. A servo-hydraulic actuator is placed between the base-mat and the ground to provide active control forces. The servo-valves of the actuator are controlled with an analog controller based on the fuzzy control paradigm. Nine linguistic rules are used based on a single input, the measured acceleration, and the control force as output. The specimen is placed on the shake table and tested with the simulated 1940 El Centro, 1952 Taft, 1968 Hachiohe, and 1971 Pacoima Dam earthquake ground motions.

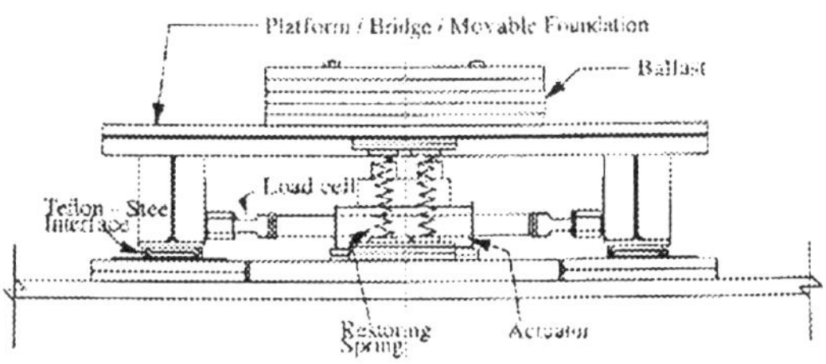

(a) Sliding rigid basemat

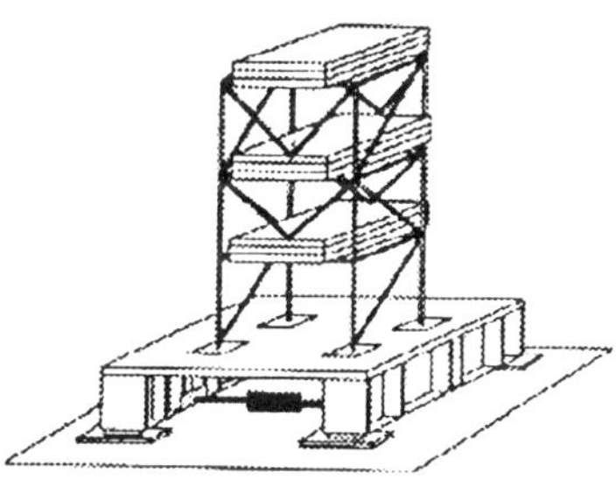

(b) Flexible structure with top two stories braced - SDOF

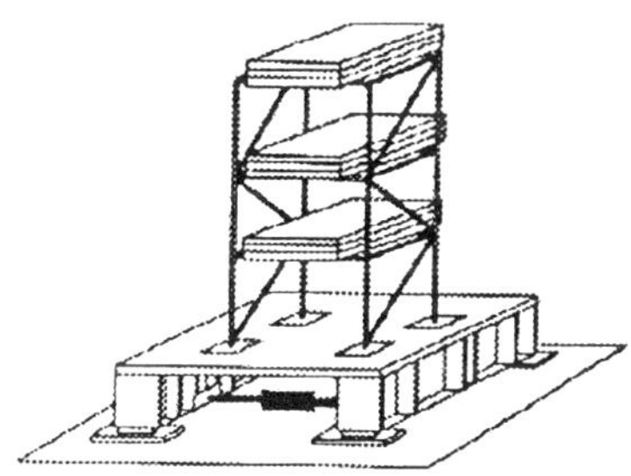

(c) Flexible structure with all stories unbraced - 3DOF

Figure 8. One-quarter scale models of building with sliding basemat and controlled by fuzzy logic (from Subramaniam et. al. [1996]).

6.6.4 More applications of genetic and intelligent algorithms.

New approach to optimization and design using genetic algorithms such as the works of Adeli and Cheng [1993, 1994a, 1994b] will continue to be explored.

Acknowledgment

Many colleagues have volunteered their help in preparing this chapter, and offered their data lists that are compiled for other purposes. In particular, we wish to acknowledge: Dr. Weimin Dong of Risk Management Solutions, Menlo Park, CA; Dr. Timothy Yao of Lucent Technologies, Naperville, IL; Prof. Wei Ling Chiang of Central University, Chung-li, Taiwan; Prof. Hiroshi Kawamura of Kobe University, Kobe, Japan; Prof. Hitoshi Furuta of Kansai University, Osaka, Japan; Dr. H. J. Zimmermann and Dipl.-Math. Karl Lieven of MIT GmbH, Aachen, Germany; and Dr. Kuei-hsiang Cheng of Stanford University, Stanford, CA. Profs. Kawamura and Furuta, and Prof. Chiang have prepared detailed reviews of civil engineering applications in Japan and Taiwan, respectively, on our behalf, and their contribution has been incorporated into the body of this chapter. Dr. (Timothy) Yao has shared with us his own review of applications in earthquake and structural engineering. Herr Lieven has kept us up to date on the multi-national project RESTRUCT in Europe. To the above and many more who are not named due to space limitation, we express our thanks and gratitude.

References

Abdel-Rohman, M., Quintana, V. H., and Leipholz, H. H., (1980), Optimal Control of Civil Engineering Structures, *Engineering Mechanics Division*, ASCE, 106, No.EM1.

Abe, M., (1996), Rule-Based Control Algorithm for Active Tuned Mass Dampers, *Journal of Engineering Mechanics*, 122, No. 8, pp. 705-713.

Adeli, H., and Cheng, N. T., (1993), Integrated Genetic Algorithm for Optimization of Space Structures, *J. Aerospace Engineering*, ASCE, 6, No.4, pp.315-328.

Adeli, H., and Cheng, N. T., (1994a), Augmented Lagrangian Genetic Algorithm for Structural Optimization, *J. Aerospace Engineering*, ASCE, 7, No.1, pp.104-118.

Adeli, H., and Cheng, N. T., (1994b), Concurrent Genetic Algorithms for Optimization of Large Structures, *J. Aerospace Engineering*, ASCE, 7, No.3, pp.276-296.

Adeli, H., and Hung, S. L., (1995), *Machine Learning: Neural Networks, Genetic Algorithms, and Fuzzy Systems*, John Wiley and Sons, New York.

ASCE, *Guidelines for Structural Condition Assessment of Existing Buildings*, (1991), American Society of Civil Engineers Publications, New York, NY.

Ayyub, B. M., and Halder, A., (1984), Project Scheduling Using Fuzzy Set Concepts, *J. Construction Engineering and Management*, 110, No. 2, pp. 189-204.

Ayyub, B. M., and Eldukair, Z. A., (1990), Safety Assessment Methodology for Construction Operations, Conference Proceedings, *Structural Safety and Reliability*, ASCE Publication, New York, pp.771-777.

Ayyub, B., (1991), Systems Framework for Fuzzy Sets in Civil Engineering, *Fuzzy Sets and Systems*, 40, pp.491-508.

Bellman, R. E., and Zadeh, L. A., (1970), Decision Making in a Fuzzy Environment, *Management Science*, 17, No.4, pp.141-164.

Blockley, D. I., (1977), Analysis of Structural Failures, *Proc. Inst. Civil Engineers*, Part 1, 62, pp.51-74.

Blockley, D. I., (1979), The Role of Fuzzy Sets in Civil Engineering, *Fuzzy Sets and Systems*, 2, pp.267-278.

Blockley, D. I., (1980), *The Nature of Structural Design and Safety*, Ellis Horwood, Chichester, England.

Blockley, D. I., and Baldwin, J. F., (1987), Uncertain Interference in Knowledge Based Systems, *Journal of Engineering Mechanics*, 113, No. 4, pp. 467-481.

Bobrow, D. G., (1985), *Qualitative Reasoning About Physical Systems*, MIT Press, Cambridge, MA.

Boinissone, P., (1987), T-Norms and T-Conorms, *Int. J. Approximate Reasoning*, 1, No.1, pp.71-101.

Boissonnade, A., Dong, W., and Shah, H. C., (1983), Fuzzy Identification of Earthquake Intensity, *Proc. 2nd Annual Workshop of the North American Fuzzy Information Processing Society (NAFIP)*, New York.

Boissonnade, A., Dong, W., Wong, S., and Shah, H., (1985), Identification of Fuzzy Systems in Civil Engineering, Conference Proceedings, *Int. Symposium on Fuzzy Mathematics in Earthquake Research*, Beijing.

Boroschek, R., and Mahin, S., (1991), Response of a Lightly-Damped Torsionally-Coupled Building, *Report UBC/EERC-91-18*, University of California, Berkeley and Earthquake Engineering Research Center, Berkeley, CA.

Bowman, M. D., Nordmark, G. E., and Yao, J. T. P., (1987), Fuzzy Logic in Metals Fatigue, *Int. J. of Approximate Reasoning*, 1, pp.197-219.

Brown, C. B., (1979), A Fuzzy Safety Measure, *ASCE J. Eng. Mech.*, 105, No.EM5, pp.855-872.

Brown, C. B., and Yao, J. T. P., (1983), Fuzzy Sets and Structural Engineering, *J. Structural Division*, ASCE, 109, No.5, pp.1211-1225.

Brown, C. B., Johnson, J. L., and Loftus, J. J., (1984), Subjective Seismic Safety Assessments, *J. Structural Engineering*, ASCE, 110, No.9, pp.2212-2233.

Brown, C. B., and Louie, D. H., (1984), Uncertainty in Civil Engineering Systems: Probability and Fuzzy Sets, Civil Engineering Systems, 1, pp.282-287.

Brown, C. B., (1987), Jaynes, Bayes and Fuzzy Information, *Civil Engineering Systems*, 4, No.3, pp.142-146.

Brown, C. B., Furuta, H., Shiraishi, N., and Yao, J. T. P., (1987), Civil Engineering Applications of Fuzzy Sets, in *Analysis of Fuzzy Information*, J. Bezdeck (editor), v.III, pp.3-15.

Casciati, F., Faravelli, L., and Yao, T., (1995), Active Structural Control toward Intelligent Infrastructures, Conference Proceedings, *Lifeline Earthquake Engineering*, ASCE Publications, New York.

Casciati, F., and Faravelli, L., (1996), Managing Multi-Degree-of-Freedom Systems in Structural Fuzzy Control, Conference Proceedings, *Probabilistic Mechanics & Structural Reliability*, ASCE Publication, New York, pp. 306-309.

Castaneda, D., and Brown, Colin, (1994), Methodology for Forensic Investigations of Seismic Damage, *J. of Structural Engineering*, 120, No. 12, pp. 3506-3524.

Chameau, J. L., and Gunaratne, M., (1984), Performance Evaluation in Geotechnical Engineering Using Fuzzy Sets, Conference Proceedings, *Probabilistic Mechanics and Structural Reliability*, Y.K. Wen, ed., ASCE Publication, New York, pp. 264-267.

Chameau, J. L., (1990), Fuzzy Sets in Civil Engineering--A Tutorial, Conference Proceedings, *Optimizing the Resources for Water Management*, ASCE Publication, New York, pp.430-433.

Chang, T. C., and Crandall, K. C., (1990), Network Resource Allocation with Support of a Fuzzy Expert System, *Journal of Construction Engineering and Management*, 116, No.2, pp. 239-260.

Chang, K. S., and Choi, J. S., (1996), Autonomous Intelligent Cruise Control Using the Fuzzy Logic, Conference Proceedings, *Applications of Advanced Technologies in Transportation Engineering*, ASCE Publication, New York, pp. 644-655.

Chao, R. J., and Ayyub, B. M., (1996), Finite Element Analysis with Fuzzy Variables, Conference Proceedings, *Building an International Community of Structural Engineers*, ASCE Publication, New York, pp. 643-650.

Chen, J., and Gupta, J., (1995), A Fuzzy Logic Model for Automated Decision Making of Pavement Maintenance Management, Conference Proceedings, *Computing in Civil Engineering*, ASCE Publication, New York.

Chiang, W. L., Dong, W., and Wong, F. S., (1987), Dynamic Response of Structures with Uncertain Parameters: A comparative study of probabilistic and fuzzy sets models, *Probabilistic Engineering Mechanics*, 2, No.2, pp.82-91.

Chou, K. C., and Yuan, J., (1993), Fuzzy-Bayesian Approach to Reliability of Existing Structures, *Journal of Structural Engineering*, 119, No. 11, pp. 3276-3290.

Diekmann, J. E., and Kraiem, Z. M., (1990), Uncertain Reasoning in Construction Legal Expert System, *Journal of Computing in Civil Engineering*, 4, No. 1, pp. 55-76.

Dong, W., (1986), *Applications of Fuzzy Sets Theory in Structural and Earthquake Engineering*, Ph.D. thesis, Department of Civil Engineering, Stanford University, Stanford, CA.

Dong, W., Lamarre, M., and Boissonnade, A., (1986), Expert System for Seismic Risk Evaluation, *Proc. 8th European Conference on Earthquake Engineering*, Laboratorio Nacional De Engenharia Civil, Lisbon.

Dong, W., and Shah, H. C., (1987), Vertex Method for Computing Functions of Fuzzy Variables, *Int. J. Fuzzy Sets and Systems*, 24, No.1, pp.65-78.

Dong, W., and Wong, F. S., (1987a), Abstraction of Knowledge Base for Expert Systems in Seismic Risk Evaluation, in *Expert Systems in Construction and Structural Engineering*, H. Adeli (editor), Chapman and Hall, Andover.

Dong, W., and Wong, F. S., (1987b), From Uncertainty to Approximate Reasoning: Part 2: Reasoning with Algorithm Rules, *Civil Engineering Systems*, 3, No.4, pp.192-202.

Dong, W., and Wong, F. S., (1987c), From Uncertainty to Approximate Reasoning: Part 3: Reasoning with Conditional Rules, *Civil Engineering Systems*, 4, No.1, pp.45-53.

Dong, W., and Wong, F. S., (1987d), Fuzzy Weighted Averages and Implementation of the Extension Principle, *Fuzzy Sets and Systems*, 21, pp.183-199.

Dong, W., Shah, H. C., and Wong, F. S., (1988), Condensation of the Knowledge Base in Expert Systems with Applications to Seismic Risk Evaluation, in *Expert Systems in Construction and Structural Engineering*, Adeli, H., (editor), Chapman and Hall, London, pp.193-223.

Dong, W., Chiang, W. L., and Wong, F. S., (1990), Non-Probabilistic Models of Uncertainty in Automatic Control of Structures, in *Intelligent Structures*, Chong, K. P., Liu, S. C., and Li, J. C., (editors), Elsevier, London, England.

Dong, W., and Wong, F. S., (1996), Catastrophe Risk Management, Conference Proceedings, *11th World Conf. on Earthquake Engineering*, Acapulco, Mexico.

Dubois, D., and Prade, H., (1980), *Fuzzy Sets and Systems: Theory and Applications*, Academic Press, New York.

Faravelli, L., and Yao, T., (1996), Use of Adaptive Networks in Fuzzy Control of Civil Structures, *Microcomputers in Civil Engineering*, 11, pp.67-76.

Feng, D., et. al., (1982a), Application of Fuzzy Mathematics in Evaluating Earthquake Intensity, *J. Earthquake Engineering and Engineering Vibration*, 2, No.3.

Feng, D., et. al., (1982b), The Application of Retrieval Method of Fuzzy Information in Earthquake Prediction Based on Seismic Activity, *Journal of Earthquake*, 5, pp.11-14.

Feng, D., et. al., (1982c), An Application of Fuzzy Sets Method to Earthquake Prediction Based on Seismic Activity, *J. of Earthquake Engineering*, 4, pp.31-34.

Fisher, D. J., and Rajan, N., (1996), Automated Constructibility Analysis of Work-Zone Traffic-Control Planning, *Journal of Construction Engineering and Management*, 122, No 1, pp.36-43.

Frank, P. M., (1994), Application of Fuzzy Logic to Process Supervision and Fault Diagnosis, Conference Proceedings, *SAFEPROCESS '94*, 2, pp.591-612.

Furuta, H., Fu, K. S., and Yao, J. T. P., (1985), Structural Engineering Applications of Expert Systems, *J. Computer-Aided Design, 17*, No.9, pp.410-419.

Furuta, H., Furukawa, K., and Shiraishi, N., (1989), Applications of Fuzzy Mathematical Programming in Structural Optimization, Conference Proceedings, *Computer Utilization*; James K. Nelson, ed., ASCE Publication, New York, pp. 201-210.

Furuta, H., Umano, M., Kawakami, K., Ohtani, H., and Shiraishi, N., (1990), A Fuzzy Expert System for Durability Assessment of Bridge Decks, Conference Proceedings, *ISUMA '90*, Maryland, pp.522-527.

Furuta, H., (1995), Application of Fuzzy Logic in Civil Engineering in Japan, Conference Proceedings, *Computing in Civil Engineering*, ASCE Publication, New York, pp. 1900-1907.

Furata, H., Kawakami, S., Kaneyoshi, M., and Tanaka, H., (1995), Practical Application of Fuzzy Tension Adjustment to Bridge Construction, Conference Proceedings, *Restructuring: America and Beyond*, ASCE Publication, New York, pp. 1651-1654.

Furuta, H., He, J.-H., and Watanabe, Eiichi, (1996), A Fuzzy Expert System for Damage Assessment Using Genetic Algorithms and Neural Networks, *Microcomputers in Civil Engineering*, 11, pp.37-45.

Fujitani, H., Miyoshi, T., Kawamura, H., Tani, A., and Mochio, T., (1996), Seismic Response Control of Building Structure by Fuzzy Optimal Logic, Conference Proceedings, *11th World Conference on Earthquake Engineering*, Acapulco, Mexico.

Hadipriono, F. C., (1988), Fuzzy Set Concepts for Evaluating Performance of Constructed Facilities, *Journal of Performance of Constructed Facilities*, 2, No. 4, pp. 209-225.

Hadipriono, F. C., and Ross, T. J., (1988), Fuzzy Set Evaluation of Lunar Structures, Conference Proceedings, *Engineering, Construction, and Operations in Space*, Stewart W. Johnson and John P. Wetzel., eds., ASCE Publication, New York, pp. 1250-1259.

Hinkle, A. J., and Yao, J. T. P., (1988), Assessment of Welded Structures with Fatigue Damage, *Int. J. of Approximate Reasoning*, 2, pp.47-63.

Ishizuka, M., Fu, K. S., and Yao, J. T. P., (1980), *Inference Method for Damage Assessment Systems of Existing Structures*, Report CE-STR-80-17, School of Civil Engineering, Purdue University, W. Lafayette, IN.

Ishizuka, M., Fu, K. S., and Yao, J. T. P., (1983), Rule-based Damage Assessment System for Existing Structures, *Archives*, 8, No.2, pp.99-118.

Issermann, R., (1992), Integration of Fault Detection and Diagnosis Methods, Conference Proceedings, *SAFEPROCESS '94*, 2, pp.591-612.

Itoh, S., and Itagaki, H., (1990), Application of Fuzzy-Bayesian Analysis to Structural Reliability, Conference Proceedings, *Structural Safety and Reliability*, ASCE Publication, New York, pp.1771-1774.

Jang, J.-S. R., (1992), ANFIS: Adaptive Network-Based Fuzzy Inference System, *IEEE Transactions on Systems, Man and Cybernetics*, 20, No.2, pp.404-435.

Jowitt, P., and Lumbers, J., (1982), Water Quality Objectives, Discharge Standards and Fuzzy Logic, in *Optimal Allocation of Water Resources*, Proc. of the Exeter Symposium, IAHS Publication No.135.

Juang, C. H., and Elton, D. G., (1986), Fuzzy Logic for Estimation of Earthquake Intensity Based on Building Damage Records, *Civil Engineering Systems*, 3, pp.187-191.

Juang, C. H., and Wey, J. L., (1991), Model for Capacity of Single Piles in Sand Using Fuzzy Sets, *J. Geotechnical Engineering*, 117, No. 12, pp. 1920-1931.

Juang, C. H., Lee, D. H., and Sheu, C., (1992), Mapping Slope Failure Potential Using Fuzzy Sets, *Journal of Geotechnical Engineering*, 118, No. 3, pp. 475-494.

Juang, C. H., and Amirkhanian, S. N., (1992), Unified Pavement Distress Index for Managing Flexible Pavements, *J. Transportation Engineering*, 118, No. 5, pp. 686-699.

Juang, C. H., Huang, X. H., Holtz, R. D., and Chen, J. W., (1996), Determining Relative Density of Sands from CPT Using Fuzzy Sets, *J. Geotechnical Engineering*, 122, No. 1, pp. 1-6.

Juang, C. H., and Elton, D. J., (1996), A Practical Approach to Uncertainty Modeling in Geotechnical Engineering, Conference Proceedings, *Uncertainty in the Geologic Environment: from Theory to Practice*, ASCE Publication, New York, pp. 1269-1283.

Kam, T.-Y., and Brown, C. B., (1983), Updating Parameters with Fuzzy Entropies, *Journal of Engineering Mechanics*, 109, No. 6, pp. 1334-1434.

Kam, T.-Y., and Brown, C. B., (1984), Subjective Modification of Aging Stochastic Systems, *Journal of Engineering Mechanics*, 110, No. 5, pp. 743-751.

Kaneyoshi, M., Tanaka, H., Kamei, M., Furuta, H., (1990), New System Identification Technique Using Fuzzy Regression Analysis, Conference Proceedings, *First Int. Symp. on Uncertainty Modeling and Analysis*, College Park, MD.

Kaneyoshi, M., Tanaka, H., Kamei, M., Furuta, H., (1993), Extended Fuzzy System Identification Method for Cable Adjustment Work in Daytime, *Reliability and Optimization of Structural Systems*, V, B-12, pp.105-112.

Kangari, R., and Bakheet, M., (1994), Linguistic Analysis of Risk in Project Planning, Conference Proceedings, *Computing in Civil Engineering*, ASCE Publication, New York, pp. 216-219.

Katayama, T., Yamazaki, F., Nagata, S., and Isoyama, R., (1991), Fuzzy Reasoning for Earthquake Damage Assessment of Large-Scale City Gas Systems, Conference Proceedings, *Lifeline Earthquake Engineering*, ASCE Publication, New York, pp. 639-650.

Kauffmann, A. and, Gupta, M., (1985), *Introduction to Fuzzy Arithmetic: Theory and Applications*, Van Nostrand Reinhold, New York.

Kawamura, H., Tani, A., Matsumoto, S., and Yamada, M., (1988), Formulation of Structural Safety and Damage Assessments by Analytic Hierarchy Process and Fuzzy Confluence Rule, *Transactions of the Architectural Institute of Japan*, 398, pp.42-51 (in Japanese).

Kawamura, H., and Yao, J. T. P., (1990), Application of Fuzzy Logic to Structural Control Motion of Civil Engineering Structures, Conference Proceedings, *NAFIPS '90*, Quarter Century of Fuzziness, 1, pp.67-70.

Kawamura, H., Tani, A., and Kambara, H., (1992), Aseismic Structural Planning System by Fuzzy Network, Conference Proceedings, *10th World Conf. on Earthquake Engineering*, Balkema, Rotterdam.

Kikuchi, S., and Donnelly, R. A., (1992), Scheduling Demand-Responsive Transportation Vehicles Using Fuzzy-Set Theory, *Journal of Transportation Engineering*, 118, No. 3, pp. 391-409.

Klisinski, M., Plasticity Theory Based on Fuzzy Sets, (1988), *J. Engineering Mechanics*, 114, No. 4, pp. 563-582.

Lai, F. S., Shenoi, S., and Fan, L. T., (1986), Fuzzy Fault Tree Analysis: Theory and Application, *Engineering Risk and Hazard Assessment*, CRC Press.

Lee, D. H., and Juang, C. H., (1992), Evaluation of Failure Potential in Mudstone Slopes Using Fuzzy Sets, Conference Proceedings, *Stability and Performance of Slopes and Embankments II*, ASCE Publication, New York, pp. 1137-1151.

Liu, Z., (1982), Computational Investigation on Fuzzy Relation Between Earthquake Intensity and Peak Ground Acceleration, *J. Earthquake Engineering and Engineering Vibration*, 2, No.3.

Liu, S., and Frank, A., (1991), Simulation of a Vehicle Platoon Control System for Automatic Highway Using the Fuzzy Control Concept, Conference Proceedings, *Applications of Advanced Technologies in Transportation Engineering*, ASCE Publication, New York, pp. 433-437.

Lorterapong, P., and Moselhi, O., (1996), Project-Network Analysis Using Fuzzy Sets Theory, *Journal of Construction Engineering and Management*, 122, No. 4, pp.308-318.

Mandani, E., (1974), Application of Fuzzy Algorithm for Control of Simple Dynamic Plant, *Proc. IEEE*, 121, No.12, pp.1585-1588.

Mandani, E., (1977a), Application of Fuzzy Logic to Approximate Reasoning Using Linguistic Systems, *IEEE Trans. On Computers*, C-26, pp.1182-1191.

Mandani, E., (1977b), Applications of Fuzzy Sets Theory to Control Systems: a Survey, in *Fuzzy Automata and Decision Processes*, Gupta, M., Saridis, G., and Gaines, B. (editors), North-Holland, Amsterdam, pp.77-88.

Mar, B. W., and Brown, C. B., (1984), Subjective Probability and Membership Functions, Conference Proceedings, *Probabilistic Mechanics and Structural Reliability*, Y.K.Wen, ed., ASCE Publication, New York, pp. 65-67.

Melhem, H., and Aturaliya, S., (1994), Bridge Condition Rating Using Fuzzy Sets and an Eigenvector of Priority Settings, Conference Proceedings, *Computing in Civil Engineering*, ASCE Publication, New York, pp. 1908-1915.

MIT GmbH, (1996), *Decision-Making for Requalification of Structures (RESTRUCT)*, Synthesis report by the Project BE5935 group, The BRITE/EURAM program of the European Community.

Miyasato, G., Dong, W., Levitt, R., Boissonnade, A., and Shah, H. C., (1986), Seismic Risk Analysis Systems, in *Expert Systems in Civil Engineering*, Kostem, N. and Maher, M., (editors), ASCE Publication, New York, pp.121-132.

Mullarkey, P., Fenves, S., and Sangrey, D., (1985), *CONE: An Expert System for Interpretation of Geotechnical Characterization Data from Cone Penetrometers*, Report R-85-147, Department of Civil Engineering, Carnegie-Mellon University, Pittsburgh, PA.

Munro, J., and Chuang, P.-H., (1986), Optimal Plastic Design with Imprecise Data, *Journal of Engineering Mechanics*, 112, No. 9, pp. 888-903.

Murlidharan, T. L., (1991), Fuzzy Behavior of Beams on Winkler Foundation, *J. Engineering Mechanics*, 117, No. 9, pp. 1953-1972.

Natke, H. G., Doll, H., Hilderbrantdt, P., and Schultz, M., (1995), Bridge Condition Assessment Using an Expert System, Structural Engineering Review, 7, No.3, pp.165-180.

Natke, H. G., and Ben-Haim, Y., (1997), Uncertainty: Models and Measures, *Mathematics Research*, v.99, Akademie-Verlag, Berlin.

Natke, H. G., and Cempel, C., (1997), *Model-Aided Diagnosis of Mechanical Systems: Fundamentals, Detection, Localization and Assessment*, Springer-Verlag, Berlin.

Natke, H. G., (to appear in 1999), *Systemtechnik und Gwaesserguete: Operationalisierung (System Engineering and Water Grade: Mathematical Formalization).*

Nguyen, V. U., (1985), Tender Evaluation by Fuzzy Sets, *Journal of Construction Engineering and Management*, 111, No. 3, pp.231-243.

Ni, S. H., Lu, P. C., and Juang, C. H., (1996), A Fuzzy Neural Network Approach to Evaluation of Slope Failure Potential, *Microcomputers in Civil Engineering*, 11, pp.59-66.

OECD, *Bridge Inspection*, (1976), Organization for Economic Cooperation and Development, Paris, France.

Ogawa, H., Fu, K. S., and Yao, J. T. P., (1984), *SPERIL-II: An Expert System for Structural Damage Assessment of Existing Structures*, Report CE-STR-84-11, School of Civil Engineering, Purdue University, W. Lafayette, IN.

Ogawa, K., and Koike, H., (1994) An Application of Fuzzy Set Theory to Preference Analysis in Bus Route Choices, Conference Proceedings, *Computing in Civil and Building Engineering*, ASCE Publication, New York, pp. 1501-1508.

Paek, J. H., Lee, Y. W., and Napier, T. R., (1992), Selection of Design/Build Proposal Using Fuzzy-Logic System, *J. Construction Engineering and Management*, 118, No. 2, pp. 303-317.

Paek, J. H., Lee, Y. W., and Ock, J. H., (1993), Pricing Construction Risk: Fuzzy Set Application, *J. Construction Engineering and Management*, 119, No. 4, pp.743-756.

Pandey, P. C., and Barai, S. V., (1994), Sensitivity-Based Weighted-Average in Structural Damage Assessment, *J. Performance of Constructed Facilities*, Vol. 8, No. 4, pp. 243-263.

Pedrycz, W., (1983), Numerical and Application Aspects of Fuzzy Relational Equations, *J. Fuzzy Sets and Systems*, 11, pp.1-18.

Pedrycz, W., (1984), An Identification Algorithm in Fuzzy Relational Systems, *J. Fuzzy Sets and Systems*, 13, pp.153-167.

Prechaverakul, S., and Hadipriono, F. C., Green, R. L., Boyle, R. D., and Ray, J. A., (1994), Pavement Treatment in Ohio Transportation System, Conference Proceedings, *Computing in Civil Engineering*, ASCE Publication, New York, pp. 1892-1899.

Rao, S. S., (1994), Multi-objective Optimization of Actively Controlled Structure, Conference Proceedings, *Analysis and Computation*, ASCE Publication, New York, pp. 276-285.

Reinhorn, A. M., Subramaniam, R. S., and Riley, M. A., (1996), Fuzzy Logic Based Control for Sliding Structures, Conference Proceedings, *Analysis and Computation*, ASCE Publication, New York, pp. 298-309.

Ross, T., Wong, F., Savage, S., and Sorensen, H., (1986), DAPS: An Expert System for Damage Assessment to Protective Structures, in *Expert Systems in Civil Engineering*, Kostem, C. N., and Maher, M. L., (editors), ASCE, New York, pp.109-120.

Ross, T. J., and Sunil, D., (1994), A Fuzzy Logic Paradigm for Fault Trees and Event Trees in Risk Assessment, Conference Proceedings, *Computing in Civil Engineering*, ASCE Publication, New York, pp. 369-375.

Ross, T. J., and Hasselman, T. K., (1995), Fuzzy Classification of Modal Vibration Data, Conference Proceedings, *Restructuring: America and Beyond*, ASCE Publication, New York, pp. 1647-1650.

Ruspini, E., (1977), A Theory of Fuzzy Clustering, Conference Proceedings, *IEEE Conf. Decision and Control*, New Orleans, pp.1378-1383.

Russell, A. D., and Fayek, A., (1994), Automated Corrective Action Selection Assistant, *Journal of Construction Engineering and Management*, 120, No. 1, pp. 11-33.

Russell, A. D., (1995), Automated Interpretation of Job Site Records, Conference Proceedings, *Computing in Civil Engineering*, ASCE Publication, New York, pp. 989-996.

Saaty, T., (1978), Exploring the Interface Between Hierarchies, Multiple Objectives, and Fuzzy Sets, *J. Fuzzy Sets and Systems*, 1, pp.57-68.

Sayed, T., Abdelwahab, W., and Navin, F., (1995), Identifying Accident-Prone Locations Using Fuzzy Pattern Recognition, *Journal of Transportation Engineering*, 121, No. 4, pp. 352-358.

Schwartz, D. I., and Chen, S. S., (1993), Order of Magnitude Reasoning for Qualitative Matrix Structural Analysis, Conference Proceedings, *Computing in Civil and Building Engineering*, ASCE Publication, pp. 1267-1274.

Shingu, K., and Funamoto, D., (1992), Fuzzy Control of Multi-Degree-of-Freedom System Structures, *Japanese J. of Fuzzy Theory and Systems*, 4, No.4, pp.427-435.

Shiraishi, N., and Furuta, H., (1983), Reliability Analysis Based on Fuzzy Probability, *Journal of Engineering Mechanics*, 109, No. 6, pp. 1445-1459.

Sobanjo, J. O., (1993), Fuzzy Sets Model for Evaluation of Bridge Projects, Conference Proceedings, *Infrastructure: Planning and Management*, ASCE Publication, New York, pp. 442-451.

Sobanjo, J. O., Stukhart, G., and James, R. W., (1994), Evaluation of Projects for Rehabilitation of Highway Bridges, *Journal of Structural Engineering*, 120, No. 1, pp. 81-99.

Song, B., Hao, S., Murakami, S., and Sadohara, S., (1996), Comprehensive Evaluation Method on Earthquake Damage Using Fuzzy Theory, *J. Urban Planning and Development*, 122, No. 1, pp. 1-17 .

Subramaniam, R. S., (1994), *Control of Structures Using Algorithms Based on Fuzzy Sets Theory*, Ph.D. dissertation, Department of Civil Engineering, State University of New York at Buffalo, NY.

Subramaniam, R. S., Reinhorn, A. M., Riley, M. A., and Nagarajaiah, S., (1996), Hybrid Control of Structures Using Fuzzy Logic, *Microcomputers in Civil Engineering*, 11, pp.1-17.

Takada, S., Sun, J. S., Ogawa, Y., and Oka, S., (1991), An Expert System for Diagnosis of Earthquake Proof for Underground Lifelines, Conference Proceedings, *Lifeline Earthquake Engineering*, ASCE Publication, pp. 737-746.

Takagi, T., and Sugeno, M., (1983), Derivation of Fuzzy Control Rules from Human Operator's Control Actions, Conference Proceedings, *IFAC Symposium on Fuzzy Information, Knowledge Representation and Decision Analysis*, pp.55-60.

Tani, A., Kawamura, H., and Watari, Y., (1992), Optimal Adaptive and Predictive Control of Seismic Structures by Fuzzy Logic, Conference Proceedings, *10th World Conference on Earthquake Engineering*, Madrid, Spain, pp.2155-2160.

Tani, A., and Kawamura, H., (1995), Fuzzy Optimal Control System of Buildings by Neural Network, in *Computing in Civil and Building Engineering*, Pahl, P., and Werner, H., (editors), Balkema, Rotterdam, pp.253-260.

Tee, A. B., (1988), *The Application of Fuzzy Mathematics to Bridge Condition Assessment*, PhD Dissertation, Purdue University, West Lafayette, IN.

Tee, A. B., Bowman, M. D., and Sinha, K. C., (1988), A Fuzzy Mathematical Approach for Bridge Condition Evaluation, *Civil Engineering Systems*, 5, pp.17-24.

Thint, M. P., Wong, F. S., and Dong, W., (1996), Fuzzy-Logic Control with Applications to Intelligent Buildings, *Civil Engineering Systems*, 14, pp.19-54.

Tong, R. M., (1977), A Control Engineering Review of Fuzzy Systems, *Automatica*, 13, pp.559-569.

Tulbure, I.-C., (1997), Zustandsbeschreibung und Dynamik Umweltrelevanter Systeme (State Description and Dynamics of Environmental-Relevant Systems), Dissertation, CUTEC-Schriftenreihe Nr.25, Clausthal Technical University.

Wadia-Fascetti, S., and Smith, H. A., (1996), Calibration of Structural Models Using Fuzzy Mathematics, *Microcomputers in Civil Engineering*, 11, pp.19-35.

Wang, P., (1983), *Fuzzy Set Theory and Its Applications*, Science and Technology Press, Shanghai.

Weber, S., (1983), A General Concept of Fuzzy Connective, Negation and Implication Based on t-norms and t-conorms, *Fuzzy Sets and Systems*, 11, No.2, pp.115-134.

Wong, F. S., Dong, W., Boissonnade, A., and Ross, T. J., (1986), Expert Opinions and Expert Systems, in *Electronic Computations*, K. M. Will (editor), ASCE Publications.

Wong, F. S., and Dong, W., (1986), A Linguistic, Fuzzy Analyzer for Knowledge-based Systems, Conference Proceedings, *Int. Conf. On Information Processing and Management of Uncertainty (IPMU)*, Paris.

Wong, F. S., Ross, T. J., and Boissonnade, A., (1987), Fuzzy Sets and Survivability Analysis of Protective Structures, in *The Analysis of Fuzzy Information*, v.3, Bezdek, J., (editor), CRC Press.

Yager, R. R., (1980), On a General Class of Fuzzy Connectives, *Fuzzy Sets and Systems*, 4, No.3, pp.235-242.

Yamada, T., Suzuki, K., and Kobayashi, H., (1994), Elasto-Plastic Response Control by a Fuzzy Control System of Building Structures Under Seismic Excitation, in *Structural Safety and Reliability*, Schueller, M., Shinozuka, M., and Yao, J., (editors), Balkema, Rotterdam, pp.1599-1606.

Yang, J. N., and Akbarpour, A., (1987), New Optimal Control Algorithms for Structural Control, *Engineering Mechanics Division*, ASCE, 113, No.EM9.

Yao, J. T. P., (1980), Identification and Control of Structural Damage, *Solid Mechanics Archives*, 5, No.3, pp.325-345.

Yao, J. T. P., (1980), Damage Assessment of Existing Structures, *J. of Engineering Mechanics*, ASCE, 106, No.4, pp.785-799.

Yao, J. T. P., (1985), *Safety and Reliability of Existing Buildings*, Pitman Advanced Publishing, Boston, USA.

Yao, J. T. P., and Fu, K. S., (1985), Civil Engineering Applications of Expert Systems, Proc. *4th Int. Offshore Mechanics and Artic Engineering Symposium*, Vol.II, ASME, New York.

Yao, J. T. P., (1988), Approximate Reasoning in Engineering Decision Analysis, Conference Proceedings, *Probabilistic Methods in Civil Engineering*, Pol Spanos, ed., ASCE Publication, New York, pp. 452-455.

Yao, J. T. P., and Natke, H. G., (1993), Effect of Active Control to Structural Reliability, Conference Proceedings, *Probabilistic Mechanics and Structural and Geotechnical Reliability*, ASCE Publication, New York, pp. 373-376.

Yao, J. T. P., and Natke, G., (1995), Fuzzy Logic and Civil Infrastructure Systems Research, Conference Proceedings, *Restructuring: America and Beyond*, ASCE Publication, New York, pp. 1643-1646.

Yao, J. T. P., and Yao, T. H.-J.,(1997), Uncertainties in Safety Analysis of Existing Structures, in *Uncertainty Modeling and Analysis in Civil Engineering*, Ayyub, B. M., (editor), CRC Press, Boca Raton, FL, pp.233-241.

Yeh, Y. C., Hsu, D. S., and Kuo, Y. H., (1991), Expert System for Diagnosing Damage of Prestressed Concrete Pile, *J. Construction Engineering and Management*, 117, No. 1, pp. 13-26.

Zadeh, L, (1965), "Fuzzy Sets", *Information and Control*, 8, pp.338-353.

Zhang, X. J., and Yao, J. T. P., (1986), *The Development of SPERIL Expert Systems for Damage Assessment*, Report CE-STR-86-29, School of Civil Engineering, Purdue University, W. Lafayette, IN.

Zimmermann, H. J., and Zysno, P., (1980), Latent Connectives in Human Decision Making, *Fuzzy Sets and Systems*, 4, No.1, pp.37-51.

7 ECOLOGICAL MODELING AND DATA ANALYSIS

Arkadiusz Salski

"The strength of the fuzzy set approach is that it starts from the premise that nature may be inherently vague or imprecise, and does not try to pretend that the real world, which has been modelled by data entities by human or machine observation, is more exact or more perfect than it really is."
(Burrough, 1989)

Abstract: The problem of uncertainty often appears in ecological research, in particular it concerns uncertainty of ecological data and vaguely defined expert knowledge. The Fuzzy Set Theory as a possible way to handle uncertainty is particularly useful for processing vague expert knowledge and uncertain or imprecise data. This chapter focuses on two large application areas of the fuzzy set theory in ecological research, namely data analysis (in particular fuzzy cluster analysis) and ecological modeling. Some application examples are presented, namely a fuzzy cluster analysis of chemicals according to their ecotoxicological properties, a fuzzy model of the breeding success of Skylarks and a fuzzy model of the population dynamics of the Yellow-necked mouse.

7.1 UNCERTAINTY PROBLEMS IN ECOLOGICAL RESEARCH

Ecological processes are very complex and often insufficiently investigated. Ecologists collect and use the information from all possible data and knowledge sources (Salski, 1996):

· sources of objective (mostly quantitative) information, like measurement and calculation

· sources of subjective (often only qualitative) information, like expert knowledge and subjective evaluations instead of measurement data. Not all ecological parameters are measurable (for example the number of fish in a particular lake); the values of such parameters can be obtained by special estimation or evaluation methods which usually also have a subjective character.

The ecological knowledge is still incomplete and vague. Therefore in many fields of ecological research ecologists have to work with both quantitative and qualitative information. A large inherent uncertainty of this information results from:

- presence of random variables
- incomplete data or inaccurate data (inaccuracy of measurement)
- approximate estimations instead of measurements (due to technical or financial problems)
- incomparability of data (varying measurement or observation conditions)
- qualitative instead of quantitative information (due to technical or financial problems)
- incomplete or vague expert knowledge (difficult to structure and formalize)
- subjectivity of the information obtained from an expert.

There are a number of ways to deal with uncertainty problems, e.g. probabilistic inference networks (Pearl, 1988) or belief intervals (Shafer et al., 1990), but the most successful method of dealing with uncertainty or imprecision is the fuzzy approach. Compared with conventional methods of information processing the application of the fuzzy set theory enables us to make better use of imprecise ecological data and vague expert knowledge in two ways:

- the representation and processing of imprecise data defined as fuzzy sets
- the representation and processing of vague knowledge in the form of linguistic rules with imprecise terms defined as fuzzy sets.

Statistical or stochastic aspects of uncertainty problems are not taken into account in this chapter.

7.2 FUZZY APPROACH TO ECOLOGICAL MODELING AND DATA ANALYSIS

Fuzzy logic can be used to handle inexact reasoning and fuzzy sets to handle data uncertainty. Fuzzy logic allows working with uncertain knowledge about relations between ecosystem components and building models based on this type of information. Ecological data or classes of ecological objects can be defined as fuzzy sets with not sharply defined boundaries, that reflects better the continuous character of nature.

Data analysis and *ecological modeling / expert systems* are the main application areas of the fuzzy set theory in ecological research. The problems of environmental management and engineering (e.g. fuzzy logic control of ecological processes) are not considered in this paper.

The problem of classifying a number of objects into classes is one of the main problems of *data analysis* and arises in many areas of ecology. Conventional classification methods based on Boolean logic ignore the continuous nature of ecological parameters and the uncertainty of data, which can result in misclassification. *Fuzzy classification,* which means the division of objects into classes that do not have sharply defined boundaries, can be carried out in various ways, for example:

· application of fuzzy arithmetical and logical operations, e.g. to determine land suitability (Burrough et al., 1989, 1992)

· fuzzy clustering, e.g. to classify some crop growth parameters (Marsili Libelli, 1989), to identify fuzzy soil classes (Odeh et al., 1992) or to classify existing chemicals according to their ecotoxicological properties (Friederichs et al., 1996).

Conventional clustering methods which definitely place an object within only one cluster are not particularly useful for a classification of ecological data. Compared with conventional classification methods fuzzy clustering methods enable a better interpretation of the data structure.

Spatial data is an essential part of ecological data. The fuzzy extension of the interpolation procedure for spatial data, the so-called *fuzzy kriging*, can be mentioned here as an example of fuzzy approach to spatial data analysis (Bárdossy, 1988 and 1989; Diamond, 1989; Piotrowski et al., 1996). Fuzzy kriging is a modification of the conventional kriging procedure; it utilizes exact (crisp) measurement data as well as imprecise estimates obtained from an expert, which are defined as fuzzy numbers. Regionalisation of ecological parameters based on fuzzy kriging reflects the imprecision of input data.

Modeling is the next main application area of fuzzy sets and fuzzy logic in ecology. The integration of the fuzzy inference mechanisms and the expert system technique provides development tools for *fuzzy expert systems* and *fuzzy knowledge-based models* of ecological processes (e.g. Salski, 1992; Salski et al., 1996; Daunicht et al., 1996; Li, 1996). Fuzzy knowledge-based modeling can be particularly useful where there is no analytical model of the relations to be examined or where there is insufficient amount of data for statistical analysis (Bárdossy & Duckstein, 1995, Pedrycz 1996). In these cases the only basis for modeling is the expert knowledge, which is often uncertain and imprecise. The evolution of expert systems into fuzzy expert systems (adding imprecision or uncertainty handling to expert systems) makes the extension of their application area for complex ecological problems possible (Aminzadeh, 1994).

There are also some other *combined approaches* which result from linking the fuzzy approach with other techniques, e.g.:

· fuzzy approach with neural networks for habitat modeling in agricultural landscapes (Wieland et al., 1996)

· fuzzy modeling with conventional dynamic programming to optimal biological control of a greenhouse mite (Cheng et al.,1996)

· fuzzy approach with probabilistic uncertainty to model climate-plant-herbivore interactions in grassland ecosystems (Wu et al., 1996)

· fuzzy approach with three-dimensional modeling technique (Ameskamp, 1997)

· fuzzy clustering with GIS (Geographic Information System) technique (Edwards, 1994)

· fuzzy logic with GIS and expert system development techniques to infer soil series from environmental conditions (Zhu et al., 1996).

A fuzzy cluster analysis of chemicals according to their ecotoxicological properties and two fuzzy knowledge-based models (the reproduction of Skylarks and the population dynamics of the Yellow-nacked mouse) are presented below as examples of fuzzy applications in ecological modeling and data analysis.

7.3 FUZZY CLUSTER ANALYSIS

The problem of classifying a number of objects into different groups arises in many areas of ecology. The cluster analysis has always been an important method of data analysis. The aim of cluster analysis is to find such a partition of a number of objects that the objects of one group (cluster) are largely similar and the objects of the other groups are largely different. Besides the usual problems of sharp cluster analysis, such as the handling of hybrid objects or external values, there is the additional difficulty with the cluster analysis of ecological data caused by the uncertainty of the data. Ecological data are often presented with a semblance of accuracy when exact values cannot be ascertained. They can be estimated subjectively (e.g. the plants cover about 20% of the surface area) or inter- or extrapolated. Using the usual sharp cluster analysis these difficulties often cannot be taken into account properly.

Conventional clustering methods definitely place an object within only one cluster. With fuzzy clustering it is no more essential to definitely place an object within one cluster since the membership value of this object can be split up between different clusters. The most common clustering method, the so-called fuzzy c-mean or fuzzy ISODATA-method (Bezdek, 1980), is based on the minimization of the following functional (the least-squared errors-functional):

$$J_d(c) = \sum_{i=1}^{c} \sum_{k=1}^{s} F(x_k, v_i)^m \sum_{j=1}^{s} \left\| x_k(j) - v_i(j) \right\|^2 \tag{1}$$

where:

x_k is the object in R^s, s is the number of object parameters, n is the number of objects

v_i is the cluster center of the cluster X_i,

$c \in N$ is a desired number of clusters $(2 \le c \le n)$,

m is a weighting exponent (the so-called fuzzifier), $m \ge 1$,

$F(x_k, v_i)$ represents the membership of the object x_k to the cluster X_i

which satisfy the following conditions:

$$F(x_k, v_i) \in [0,1] \qquad 1 \le i \le c, \quad 1 \le k \le n$$

$$\sum_{i=1}^{c} F(x_k, v_i) = 1 \qquad 1 \le k \le n$$

$$\sum_{k=1}^{n} F(x_k, v_i) > 0 \qquad 1 \leq i \leq c$$

Using the weighting exponent m (fuzzifier) one can determine the degree of partition fuzziness. In comparison to conventional clustering methods the distribution of the membership value provides additional information from which the membership value can be interpreted as the degree of similarity between properties of a particular object and properties characterizing particular clusters. If the number of clusters is not known a priori then the evaluation of the quality of the partition by means of the partition efficiency indicators is of special importance (Roubens, 1982).

7.3.1 A Fuzzy Cluster Analysis of Chemicals According to their Ecotoxicological Properties

Both the costs of ecotoxicological testing procedures and the large number of existing chemicals make it necessary to select representative chemicals which faithfully reflect the relevant properties of possibly a major group of compounds. Compared with conventional clustering methods the fuzzy clustering technique is more appropriate to handle the uncertainty of a part of ecotoxicological data. Information about the degree of similarity between properties of a particular chemical and properties characterizing particular clusters (deduced from the distribution of the membership function) is particularly important since there are quite a few chemicals with more or less overlapping properties which would not be registered by conventional clustering methods. The main tasks of a fuzzy cluster analysis of chemicals according to their ecotoxicological properties formulated by Friederichs et al. (1996) are:

- to find distinguishable clusters with characteristic properties
- to find chemicals representative for each cluster
- to examine the role of different parameters for clustering.

The main data basis for this analysis is the reports of the Advisory Committee on Existing Chemicals of Environmental Relevance (BUA) which select and evaluate existing chemicals according to their environmental relevance (BUA reports, 1985-1993). The data material shows a lot of gaps; therefore the choice of parameters for clustering purposes can sometimes be more a question of data availability than of data suitability. Where degradability and toxicity data were not available the gaps have been filled with alleged worst-case values (i.e. low degradability and high toxicity). Since the fuzzy c-mean procedure works with numerical data, some data, e.g. biodegradability data, have to be transformed from symbolic into numerical values (Figure 1).

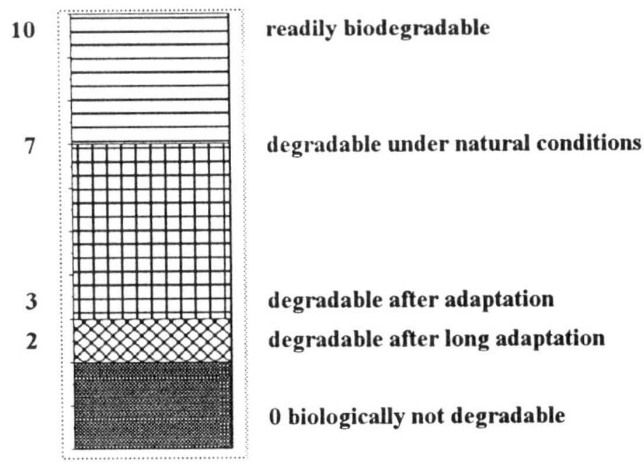

Figure 1: Transformation of biodegradability data into scores used for clustering (FRIEDERICHS et al., 1996).

The parameters selected for clustering have been divided into three groups:

- parameters relevant for the distribution of a compound between different compartments (log P_{OW} and Henry constant)
- parameters representing the potentials for hydrolysis, photolysis and biodegradability
- toxicity indicators: toxicity for a) microbes, b) aquatic invertebrates, c) aquatic vertebrates, d) mammals and a combined indicator term for cancerogeneity, mutageneity and teratogeneity.

Table 1 provides a part of the data set used for the cluster analysis (FRIEDERICHS et al., 1996). The numbers in brackets are not continuous numbers but refer to the number of the BUA-report. For the analysis of the data material described above the Fuzzy Clustering System ECO-FUCS, which was developed in the Ecology Center of the University of Kiel (Salski, 1993), has been used. The

Table 1: Data set for 30 chemicals (a part of the data set used in FRIEDERICHS et al., 1996)

Parameter: 1: log P_{OW}
2: Henry constant [Pa m³ mol^{-1}]
3: biodegradability (score)
4: degradation in water [1/ t½] in days
5: degradation in the air [1/ t½] in days
<01> number of BUA report
6: toxicity for microbes (score)
7: toxicity for aquatic invertebrates (score)
8: toxicity for aquatic vertebrates (score)
9: mammalian toxicity (score)
10: cancerogeneity/mutageneity/teratogeneity (score)

	1	*2*	*3*	*4*	*5*	*6*	*7*	*8*	*9*	*10*
<01>Chloroform---------------	1.9	310	3	0	0.0083	10	3	10	3	5
<03>Pentachlorophenol---------	3	0.04	2	1	1	0	0	0	0	10
<08>Dichlorobenzene----------	3.60	3.1	2	0	0.04	3	3	3	10	10
<10>p-Nitromethoxybenzene--	2.03	0.2	2	0	0.196	10	3	3	0	5
<13>Nonylphenol--------------	3.27	136	2	0	0	3	0	0	10	5

<15>Diphenylamine------------	3.62	0.09	2	0	24	3	3	10	10	5
<17>1,2,4-Trichlorobenzene---	4.21	1.08	2	0	0.033	0	3	0	10	5
<18>Ditolyl ether---------------	5.54	15.7	2	0	0	10	3	3	20	15
<20>Tris-(2-chloroethyl)phosph.-	0.3	0.00	2	0.0003	0.1	20	-	3	10	5
<22>Dibutylphthalat-----------	4.7	0.27	10	0.0003	0.56	10	3	0	20	5
<23>Tributylamine-------------	4.6	20.3	2	0	3.85	10	3	10	10	0
<24>Benzene------------------	2.1	440	10	0	0.077	10	10	3	0	0
<25>Hexachloropentadiene----	5.04	2736	2	144	6.85	3	0	0	0	15
<26>o-Tolidine----------------	2.4	0.00	3	0.33	12	3	-	-	10	0
<27>o-Dianisidine--------------	2.0	0.00	2	0	12	3	-	-	10	0
<30>3,3Dichlorobenzidine-----	3.55	0.00	0	0	12	10	20	0	20	0
<38>o-Chlorotoluol-------------	3.45	970	0	0	0.18	10	3	3	0	5
<51>N-Ethylaniline----------------	3	5	3	-	0.003	10	3	3	10	10
<55>Chlorotoluidine----------------	2.58	0.1	7	0	8	10	0	3	10	5
<59>Nitrobenzene------------------	1.6	2.4	3	0.04	0.014	3	3	3	0	5
<67>Diethylenglykoldimethyleth.	-0.36	0.04	2	0	1	20	20	20	20	5
<68>Hexanedioic acid-----------	0.08	0.10	10	-	0.227	20	-	10	10	20
<70>Acetic acid anhydride--------	-0.2	-	10	232	0.045	20	3	10	10	15
<72>Trichloromethylbenzene--	2.92	39.2	10	655	-	10	3	20	3	0
<73>Benzoyl chloride----------	0	-	10	400	-	10	-	3	10	5
<75>4-Nitrophenol------------	1.9	0.00	7	0	0.143	3	3	3	10	15
<84>N,N-Dimethylformamide-----	-1.01	0.00	10	0	0.25	10	20	20	10	5
<86>1,2,4,5-Tetrachlorobenzene-	4.6	15290	2	0.02	0.0065	3	3	3	3	5
<89>Triethylentetramine----------	-1.44	0	0	0	14.04	20	3	10	10	5

analysis of the partition efficiency indicators (Figure 2) shows that 5 clusters can be taken as the "optimal" value of the cluster number (minimal value of entropy, maximal value of the partition coefficient and non-fuzziness index). Table 2 shows the results of clustering with chosen chemicals in 5 clusters for the diagonal norm with a fuzzifier value of 1.3 and an ISODATA initial partition. The numbers in bold-face show the highest membership values whereas the underlined values indicate that these chemicals tend towards a membership of more than one cluster with values of 0.10. The highest membership value relates a chemical to a particular cluster while the distribution of a membership between other clusters can be interpreted as a measure of similarity with other chemical groups. The analysis of these results (Table 2 and Figure 3) permits to recognize the following properties of chemicals within their particular clusters:

- Cluster 1: This cluster is characterized by low biological degradability, high photolysis and low volatility. All biphenyls with 2-Mercaptobenzothiazole, Chlorotoluidine and Diphenylamine may serve as representatives since their membership values are closest to 1.
- Cluster 2 (two chemicals only): Extremely high hydrolysis rates and high biodegradability group Trichlormethylbenzene and Benzoylchloride in this cluster, while all other parameters are of distinctly lesser importance.
- Cluster 3: The cluster contains chemicals with low log P_{OW}, low overall toxicity and a slightly increased biodegradability. This cluster contains the smallest number of aromatic compounds, none of which is chlorinated. Representatives are N,N-Dimethylformamide and Diethylenglykoldimethyl ether.

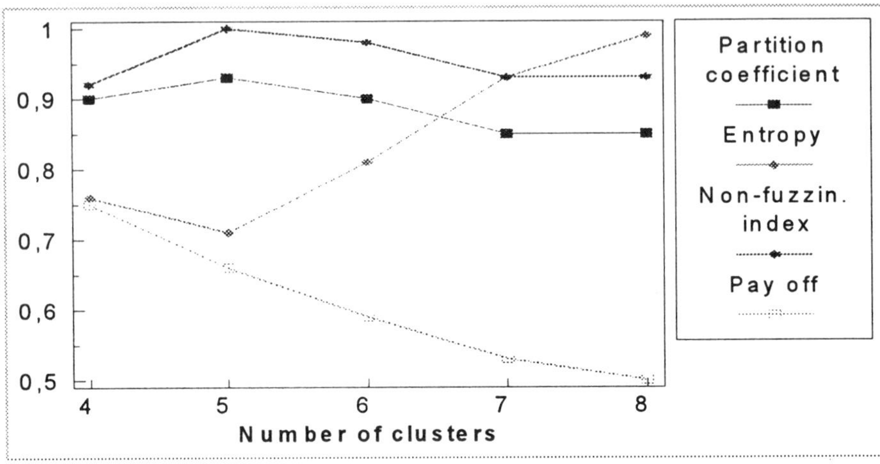

Figure 2: Partition efficiency indicators for a fuzzifier value of 1.3 (FRIEDERICHS et al. 1996).

· Cluster 4: A comparison of this cluster center with those of the other clusters shows for each parameter a neighborhood to a center of another cluster and in particular to cluster 5. So there are no significant parameters besides a slightly increased aquatic toxicity. Representatives are p-Nitromethoxybenzene, Nitrobenzene and Epichlorohydrin. The low aquatic toxicity probably lets Diethyl ether tend toward partial membership in cluster 3.

· Cluster 5: The highest log P_{OW}, highest volatility and a combination of high aquatic and low mammalian toxicity are the characteristics of this cluster. Nonylphenol, Dichlorobenzene and Biphenyl may serve as representatives. A high potential for accumulation in soil organic matter or in biota is indicated by the log P_{OW} value.

Table 2: Final clustering partition of the 30 chemicals for the data set shown in Table 1 (FRIEDERICHS et al., 1996)

The numbers in bold-face show the highest membership values (membership values with a membership ≥ 0.10 to different clusters are underlined).

Cluster 1		**1**	**2**	**3**	**4**	**5**
<15>Diphenylamine----------------		**0.98**	0.00	0.00	0.01	0.01
<26>o-Tolidine----------------------		**0.96**	0.00	0.00	0.03	0.01
<27>o-Dianisidine------------------		**0.96**	0.00	0.00	0.03	0.01
<30>3,3Dichlorobenzidine---------		**0.69**	0.00	0.08	0.12	0.11
<55>Chlorotoluidine----------------		**0.98**	0.00	0.01	0.01	0.00
<74>2-Mercaptobenzothiazole------	**0.99**	0.00	0.00	0.01	0.00	
<89>Triethylentetramine-----------		**0.93**	0.00	0.00	0.00	0.00
Cluster 2		**1**	**2**	**3**	**4**	**5**
<72>Trichlormethylbenzene--------		0.00	**1.00**	0.00	0.00	0.00

<73>Benzoylchloride---------------		0.00	**0.99**	0.00	0.01	0.00
Cluster 3		**1**	**2**	**3**	**4**	**5**
<67>Diethylenglykoldimethylether		0.01	0.00	**0.98**	0.01	0.00
<68>Haxanedioic acid--------------		0.04	0.00	**0.60**	0.16	0.19
<70>Acetic acid anhydride---------		0.05	0.21	**0.52**	0.11	0.11
<84>N,N-Dimethylformamide------	0.00	0.00	**0.99**	0.01	0.00	
Cluster 4		**1**	**2**	**3**	**4**	**5**
<01>Chloroform--------------------		0.00	0.00	0.00	**0.96**	0.04
<03>Pentachlorophenol------------		0.04	0.00	0.00	**0.68**	0.28
<10>p-Nitromethoxybenzene------		0.00	0.00	0.00	**0.99**	0.01
<20>Tris-(2-chloroethyl)phosphate		0.11	0.00	0.04	**0.73**	0.12
<24>Benzene-----------------------		0.03	0.00	0.04	**0.72**	0.21
<38>o-Chlorotoluol---------------		0.01	0.00	0.00	**0.78**	0.21
<59>Nitrobenzene-----------------		0.00	0.00	0.00	**0.99**	0.01
Cluster 5		**1**	**2**	**3**	**4**	**5**
<08>Dichlorobenzene-------------		0.00	0.00	0.00	0.02	**0.98**
<13>Nonylphenol-----------------		0.00	0.00	0.00	0.01	**0.99**
<17>1,2,4-Trichlorobenzene------		0.00	0.00	0.00	0.07	**0.93**
<18>Ditolyl ether----------------		0.01	0.00	0.01	0.04	**0.94**
<22>Dibutylphthalat--------------		0.02	0.00	0.02	0.07	**0.89**
<23>Tributylamine----------------		0.07	0.00	0.01	0.36	**0.56**
<25>Hexachloropentadiene--------		0.09	0.01	0.01	0.25	**0.64**
<51>N-Ethylaniline---------------		0.00	0.00	0.00	0.18	**0.82**
<75>2-Nitrophenol----------------		0.02	0.00	0.03	0.17	**0.78**
<86>1,2,4,5-Tetrachlorobenzene-		0.01	0.00	0.00	0.31	**0.68**

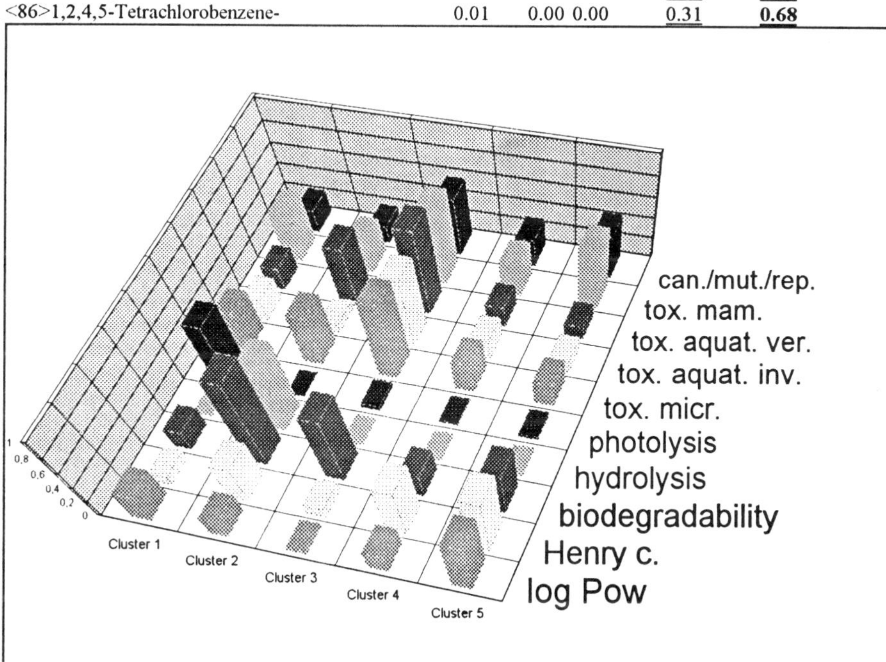

Figure 3: Cluster centers (normalized scales) of the final partition with worst-case assumption for gaps of toxicity data (FRIEDERICHS et al., 1996).

7.4 FUZZY KNOWLEDGE-BASED MODELS

In cases where the relations between the components of an ecosystem are not exactly known or where we do not have any analytical models for these relations, or we have an insufficient amount of data for statistical analysis, the idea is to build a fuzzy model based on expert knowledge. Ecologists often use vague and ill-defined natural language to describe their knowledge (Salski, 1992; Salski & Kandzia, 1991). Therefore this knowledge can be represented by a set of linguistic "IF - THEN" rules which can be interpreted as a linguistic description of the relation between the input and output of a model. There are quite a few good books containing details about fuzzy rule-based modeling such as Bárdossy & Duckstein (1995) and Pedrycz (1996).

In the process of creation of a fuzzy model the following steps must be carried out:

·determination of model structure
·creation of fuzzy knowledge base
·choice of fuzzy inference method
·calibration and validation.

First one has to determine the model structure, that is input and output variables, the number of sub-models, connections between sub-models etc. The next step is the creation of a fuzzy knowledge base which is the main part of a fuzzy model. It contains a set of linguistic rules and definitions of fuzzy sets which correspond to the linguistic terms (like "low", high", etc.) in these rules.

The major problem of the creation of the fuzzy knowledge base is to find an appropriate set of linguistic rules which describes the ecosystem or ecological process to be modelled. That can be difficult because the expert's knowledge might be too complex to be written as a limited set of rules. The set of linguistic rules should be complete and it should provide a correct answer to every possible input value. To satisfy this condition the sum of all input values in the premise of rules (union of fuzzy sets) should "cover" the value space of the input variable. If this condition is not satisfied the number of linguistic rules should be increased or one can try to increase the so-called degree of fuzziness of the fuzzy sets. It should be mentioned that the formulation of these linguistic rules and the definition of fuzzy sets have a subjective character.

The set of linguistic rules is a basis for the calculation of the relation between the output and input of a model. Then the model output can be calculated from the following fuzzy relational equations (Pedrycz, 1981):

$$C = B \circ \mathrm{R} \tag{2}$$

or

$$A_{k+1} = A_k \circ D_k \circ \mathrm{R} \qquad (\mathrm{k} = 0,1,2,....) \tag{3}$$

in the case of dynamic models (with feedback),

where: B and C are fuzzy sets of the input variable x and output variable y, respectively (defined in the spaces X and Y, respectively),

A_{k+1}, A_k are fuzzy sets of state variable z defined in the space Z ,

D_k is a fuzzy set of the control (forcing) variable *u* defined in the space U,
R is fuzzy relation defined in the spaces $X \times Y$ or $Z \times Z \times U$, described as
a set of linguistic rules ,
$\circ$ is a composition operator (e.g. the max-min composition).

Using one of the fuzzy inference methods one can compute output values for certain input values. The input values can take a crisp or fuzzy set form. Linguistic terms (represented by fuzzy sets) are also allowed for the input. The output values have the form of a fuzzy set. Then this fuzzy set can be transformed into a numerical value (defuzzification process) or approximated to one of the linguistic terms we have defined for the output variable. This so-called linguistic approximation can be accomplished by means of the calculation of the distance between fuzzy sets.

The calibration process means here any change within the database (e.g. the change of the fuzzy set definitions) in order to fit observations and expert knowledge better. The validation process (further simulation tests for a wide range of independent data) should give the answer to the question how well the model fits the real system.

Two fuzzy knowledge-based models are presented in the following sub-chapters as examples of static and dynamic (see equations 2 and 3) ecological models, namely the model of the relation between the breeding success of Skylarks and the evaluation of the vegetation structure at the beginning of the breeding period and the model of the population dynamics of the Yellow-necked mouse.

7.4.1 A Fuzzy Knowledge-Based Model of the Annual Production of Skylarks in Pure Crop Areas

The breeding success of Skylarks is determined by several important factors, namely (Daunicht et al., 1996):

- weather (especially temperature and rain)
- agricultural management like the choice of the vegetation cultivated, the spraying etc.
- mortality during migration and in the wintering areas
- the individual reaction of the birds themselves, etc.

The relationships between these factors and the breeding success of Skylarks are not exactly know; an analytical model of these relations does not exist, there are also insufficient data for statistical analysis. But there are still a lot of important data which show the bounds and limits for the modeling of these relations: data about the influence of the seasonal change in the ground covering to the usage of farmland as a foraging area, data about the different degrees of acceptance by the female Skylark of the special types of territories for breeding, data about the preferred vegetation height for nesting places, data about the different breeding success in different structure-elements of the farmland and data about the breeding success of a few years for different types of territories. These data and also the expert knowledge can be used as a basis for modeling although the knowledge related to this problem is incomplete and vague. Therefore a fuzzy logic approach has been chosen by Daunicht et al. (1996) to solve the problems of uncertainty of expert knowledge and uncertainty of data.

The vegetation structure (diversity, vegetation height, covering degree) is the main factor which affects the number of breeding pairs, the breeding success and in effect the reproduction of this bird. In pure crop areas the influence of the vegetation structure dominates all the other factors extremely. The structure of the fuzzy knowledge-based model of the annual production of Skylarks in pure crop areas (described in Daunicht et al., 1996) is presented in Figure 4. Three variables which describe a vegetation structure are defined as input variables of the model, namely:

- the diversity of the vegetation types within a study area
- the vegetation height of the earliest growing type of vegetation within a study area at the beginning of the breeding period
- the covering degree of the ground for the earliest growing type of vegetation within a study area at the beginning of the breeding period.

The values of the input variables "vegetation height" and "covering degree" should be evaluated at the beginning of the breeding period on May 20th for spring crops and on April 20th for winter crops and for a mixture of both of them. The next input variable of the model is the number of territories, i.e. the number of subareas accepted by the males of the male stock as a territory. The submodels describe the influence of the vegetation structure on the number of breeding pairs in a special type of territory (submodel 1) and on the reproduction rate (submodel 2). The output variable of the first submodel, "the breeding pairs" is multiplied by the other input variable, "the number of territories". The output variables of both submodels are combined by scalar multiplication to obtain the production value of the whole area. This value is calculated for the whole breeding period (one prediction step, no feedback in the model structure).

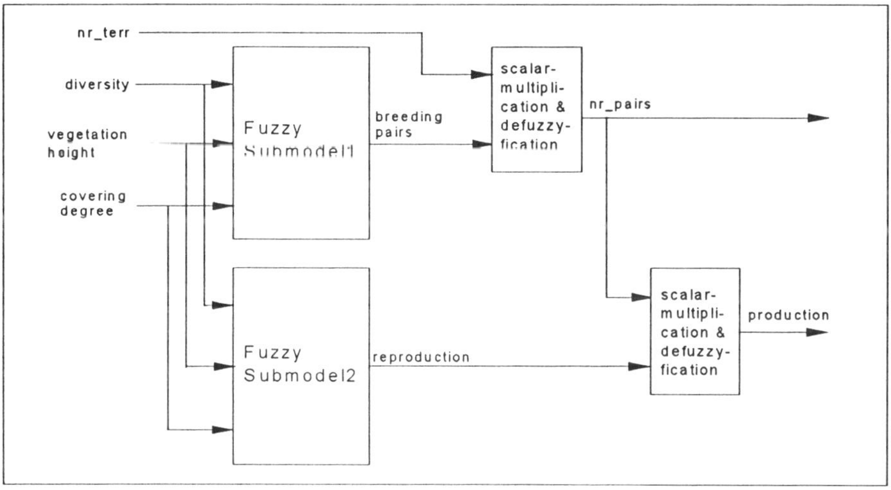

Figure 4. The structure of the fuzzy knowledge-based model of the annual production of Skylarks in pure crop areas (Daunicht et al., 1996).

The knowledge base of the model presented here has been formulated for single vegetation types like "early fast", "early slow", "middle" and "late" and for combinations of two of them with each other or with the type "grassland-line". The input variable "diversity" has been defined as a symbolic variable, which means it can take only symbolic values (e.g. "early fast" or "early slow - grassland-line"). The input variable "number of territories" is a crisp variable and can take any natural value in the defined interval. The input variables "vegetation height" and "covering degree" are defined as linguistic variables. For each of them three fuzzy sets have been formulated. The fuzzy sets "low", "average" and "high" for the variable "vegetation height" are presented in Figure 5. The knowledge base contains more than 100 rules based on five years of observations in a study area in Northern Germany. The values of the output variable "reproduction" in the rule conclusions can take the form of fuzzy sets like "low" or "high" or can be defined as fuzzy numbers (e.g. "about 1"), for example:

"IF ("diversity" is "early slow - grassland-line")
AND ("vegetation height" is "average")
AND ("covering degree" is "high")
THEN ("reproduction" is "about 1").

The values of the input variables "vegetation height" and "covering degree" in the premise should be defined as fuzzy sets formulated for these variables. The knowledge base of the model has been created using the Modeling Support System FLECO (developed in the Ecosystem Research Center at the University of Kiel and implemented on a CRVAX) which has been developed in order to facilitate the

construction of fuzzy knowledge-based models of complex ecological processes (Salski and Kandzia, 1993).

The values of the output variables can be calculated using one of the fuzzy inference methods offered by FLECO (e.g. Assilian/Mamdani, which was used for the simulation presented below). The input values of the variables "vegetation height" and "covering degree" can take a crisp or fuzzy set form. Linguistic terms (represented by fuzzy sets that we have defined for these variables) or fuzzy numbers (e.g. "about 30 cm") are also allowed as the input. The output values of "breeding pairs" and "reproduction" have the form of fuzzy sets.

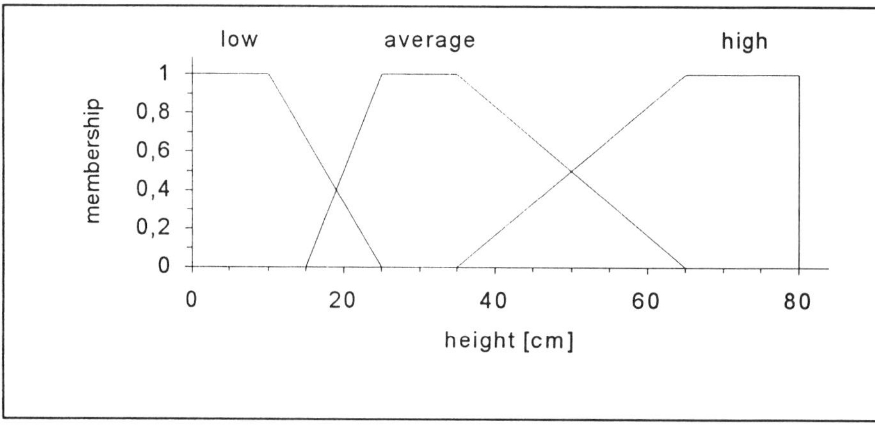

Figure 5. Fuzzy sets for the variable "vegetation height" (Daunicht et al., 1996).

Figure 6 presents the simulation results for the combination of two vegetation types, namely for the combination of the "early slow" growing winter-wheat and the "grassland-line" alongside of the field-roads. This figure shows the relation between "number of pairs" / "production" and "covering degree" for the "vegetation height" of about 30 cm and the relation between "number of pairs" / "production" and "vegetation height" for the "covering degree" of about 65%. The values of "number of pairs" and "production" have been transformed into crisp values (center-of-gravity method) in order to make them comparable with some data of the field research (for above-mentioned combination of vegetation types) which are also shown in Figure 6. The simulation results for this combination show the faster decrease of the values of "number of pairs" and "production" for the bigger values of the covering degree. The relation between "number of pairs" / "production" and "vegetation height" is almost linear in the interval 10 -70 cm.

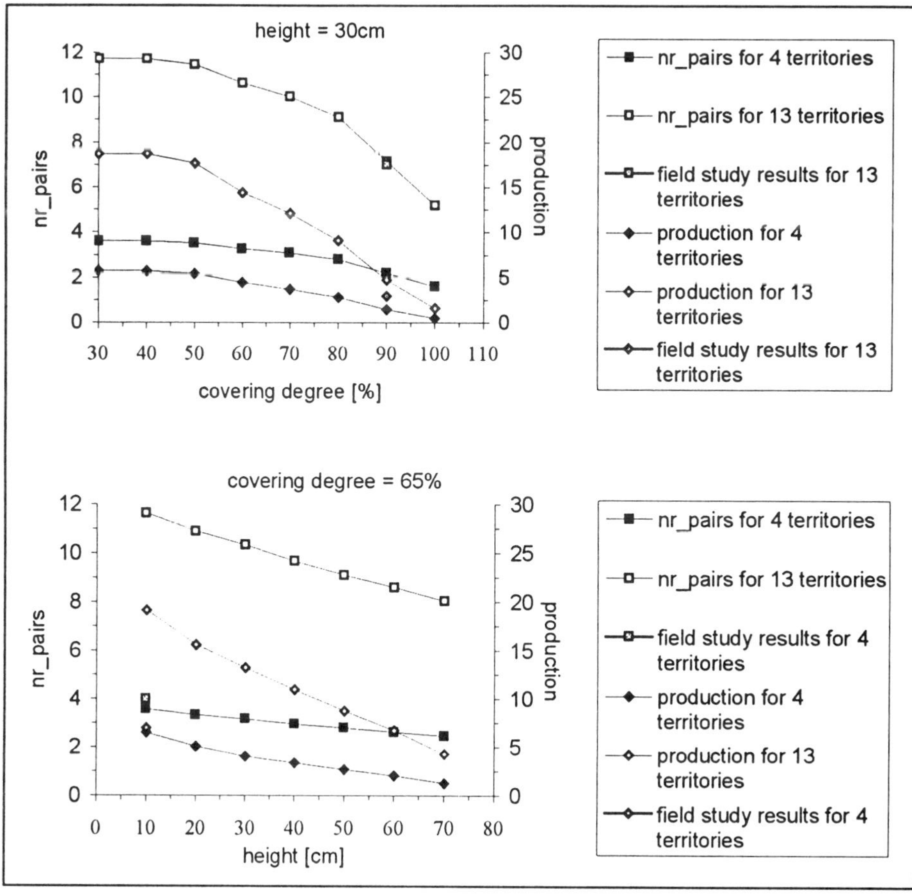

Figure 6. Simulation results for the combination of two vegetation types "early slow" growing winter-wheat and the "grassland-line" aside the ways (Daunicht et al., 1996).

7.4.2 A Fuzzy Knowledge-Based Model of the Population Dynamics of The Yellow-Necked Mouse (*Apodemus Flavicollis*) in a Beech Forest

Animal weights, food availability and soil surface moisture are the most important factors affecting the population dynamics of the Yellow-necked mouse in a beech forest. The relationships between these factors and the population dynamics of these small mammals are not exactly known. Due to some technical problems associated with collecting data for a free ranging animal population there was a high degree of uncertainty of part of the available data. That was the reason for employing unconventional modeling methods based on the linguistic description of the process dynamics given in the form of linguistic rules which are formulated by an expert

(Bock & Salski, 1996).

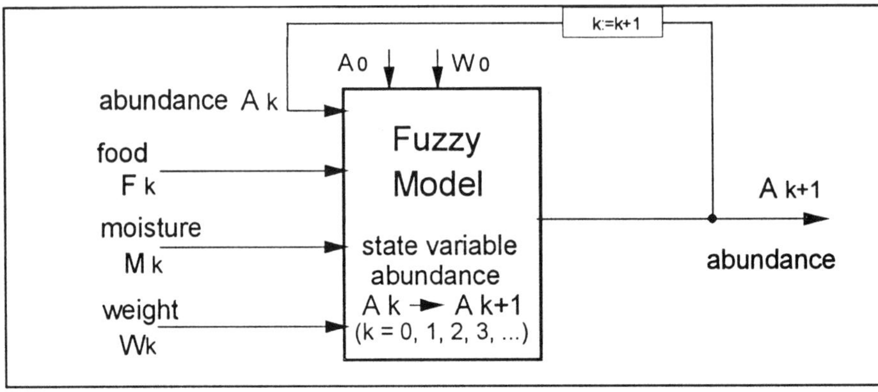

Figure 7. The structure of a fuzzy knowledge-based model of the population dynamics of the Yellow-necked mouse (*Apodemus flavicollis*) in a beech forest (Bock & Salski, 1996).

Figure 7 shows the structure of the fuzzy dynamic model (developed by Bock & Salski, 1996) with the state variable "abundance". The prediction of abundance at time k+1 is based on the values of abundance, food availability, soil surface moisture and animal weights at time k. The initial value A_0 of the state variable and the initial values of the input variables "food", "moisture" and "weight" (F_0, M_0 and W_0, respectively) have to be provided. Then we can calculate the values of "abundance" in successive moments in time (k = 1,2 ,3,....) for given values of the input variables for each prediction step. Each prediction step (difference between moments in time) represents a period of two months.

The state variable "abundance" and the variables "weight" and "food" are defined as linguistic variables. Seven fuzzy sets were determined for the variable "abundance" and three fuzzy sets for the input variables "food" and "weight" (Figure 8). The input variable "moisture" is defined as a symbolic variable, that means its values can only be symbolic statements (like dry, average and wet).

The knowledge base of the model contains about 100 linguistic rules in the "IF-THEN" form, for example:

IF the current value for "abundance" is "low"
AND "weight" is "average"
AND "food" is "high"
AND "moisture" is "average"
THEN "abundance" in the next prediction step is "high".

The linguistic terms "low", "high", etc. in the premise and conclusion parts of the rules are determined as fuzzy sets, as shown in the Figures 8. As mentioned earlier, the definition of fuzzy sets and the formulation of linguistic rules are subjective in character. The knowledge base of the model has been created using the same modeling support system FLECO which has been used for the creation of the model presented in the previous sub-chapter.

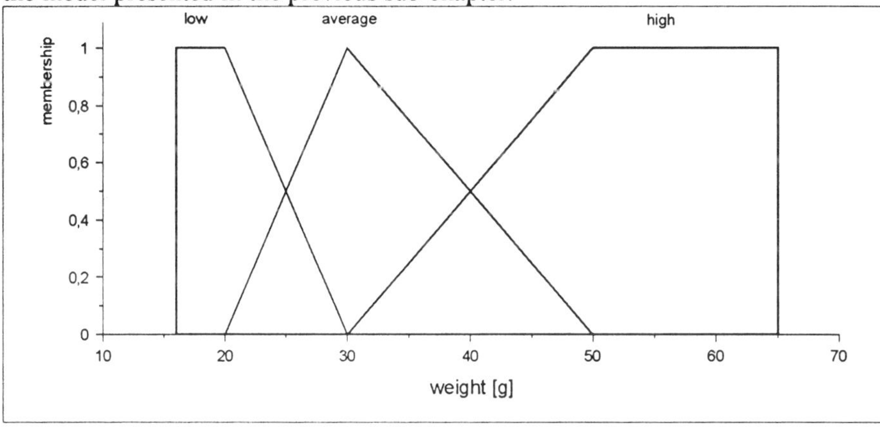

Figure 8. Fuzzy sets for the variable "weight" (Bock & Salski, 1996).

The prediction of abundance, based on an actual state, can be made in one or many steps. As mentioned above, in the case of a multistep prediction we need not only the initial values (k=0) of the state and the input variables but also the values of the input variables expected for the next calculation steps (k = 1, 2, 3, ...). These values can be more or less precise, e.g. we can expect "about 1300 kg/ha" food or we can estimate this value less precisely as "high". This can be particularly useful for further prediction steps. The simulation results (in the fuzzy set form) can be approximated by linguistic expressions (like "high", "low", etc.) or transformed into crisp values.

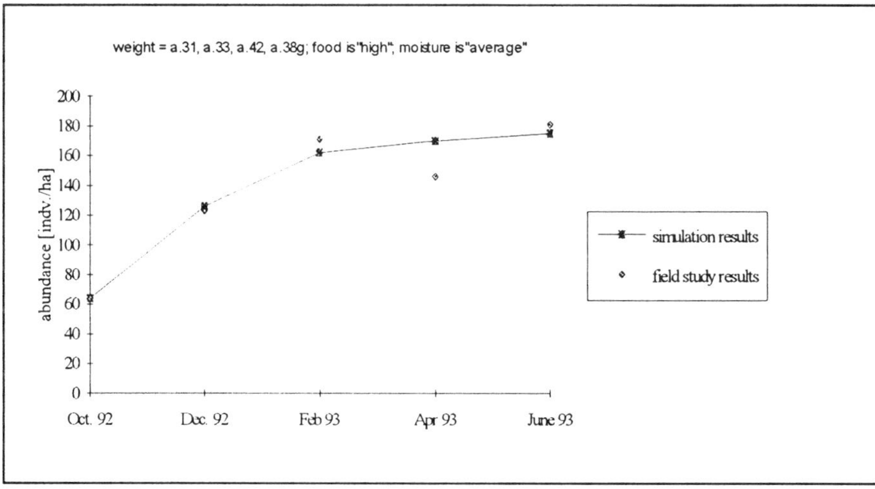

Figure 9. Simulation and field study results for the period from October 1992 - June

1993 (Bock & Salski, 1996).

Figure 9 shows the simulation results for the time period from December 1992 - June 1993 and the initial values for abundance (A_0 = 64 Ind./ha) and weight (W_0 is "about 31g"). For a better comparison with the field study results the same values of the input variables were taken for the next simulation steps as the observed values of these variables. The abundance of mice was very high in spring 1993 because of a high beech-nut crop in autumn 1992. These high values for abundance were also predicted by the model (due to high values of the variable "food", see Figure 9). The difference between simulation and field study results during this time period is no bigger than 10-15%. It should be mentioned here that the simulation results were calculated for an imprecise input value ("high") for the variable "food". It was difficult to estimate this value more precisely, however, a fuzzy logic approach enables us to make use of such imprecise information.

7.5 FINAL REMARKS

The problem of processing uncertain or imprecise information in ecological research appears to be essential. Fuzzy interpretations of data structure and a fuzzy representation of expert knowledge are a very natural and intuitively plausible way to formulate and to solve some uncertainty problems in ecological data analysis and in ecological modeling. Fuzzy logic provides the means to combine linguistic statements with numerical data and to process both of them in one simulation step. Heterogenial and often imprecise ecological data and vague expert knowledge can be integrated more effectively using fuzzy approach.

The development of easy-to-use tools for research or for practical tasks is very important for the promotion of fuzzy logic applications in ecology. The number of these applications is constantly growing. An increasing interest on applications of the fuzzy expert systems in environmental management and engineering can be expected in the near future.

There is an increasing number of applications of hybrid systems which combine the fuzzy approach with other techniques, e.g. neural networks, probabilistic approach, dynamic and linear programming, cellular automata or GIS technique. The integration of the fuzzy concept into other approaches seems to be very promising for the use in ecological applications.

References

Ameskamp, M. (1997). „Three-dimensional rule-based continuous soil modelling“. Ph.D. Thesis, Institut für Informatik und Praktische Mathematik, Christian-Albrechts-Universität, Kiel, p. 206.

Aminzadeh, F. (1994). „A applications of fuzzy expert systems in integrated oil exploration.“ In A Minzadeh, F. & J Amshidi, M. (Editors): Soft Computing.

PTR Prentice Hall, New Jersey, 29-43.

BUA - Beratergremium für umweltrelevante A ltstoffe der GDC h (Editor) (1985-1993): BUA-Stoffberichte 1 - 90. VCH Verlagsgesellschaft Weinheim, New York, Basel, Cambridge

Bárdossy, A., Bogardi, I., & Kelly, W. E. (1989). „Geostatistics utilizing imprecise (fuzzy) information." *Fuzzy Sets and Systems* v. 31, no. 3, 311-327.

Bárdossy, A., Bogardi, I., & Kelly, W. E. (1990). „Kriging with imprecise (fuzzy) variograms, I: theory." *Math. Geology* v. 22, no. 3, 63-79.

Bárdossy, A. & Duckstein L. (1995). *Fuzzy rule-based modeling with applications to geophysical, biological and engineering systems.* Boca Raton: CRC Press.

Bezdek, J.C. (1980). „A convergence theorem for the fuzzy c-means clustering algorithms." *IEEE Trans. PAMI*, PAMI-2(1): 1-8.

Bock, W. & Salski, A. (1998). „A fuzzy knowledge-based model of population dynamics of the Yellow-necked mouse (*Apodemus flavicollis*) in a beech forest." *Ecological Modelling* v. 108, 155-161.

Burrough, P.A. (1989). „Fuzzy mathematical methods for soil survey and land evaluation." *Journal of Soil Science* v. 40, 477-492.

Burrough; P.A.; Macmillan, R.A. & Van Deursen (1992) „Fuzzy classification methods for determination land suitability from soil profile observations and topography." *Journal of Soil Science,* 43, 193-210

Cheng, Z.; Horn, D.J.; Lindquist, R.K. & Taylor, R.A.J. (1996). „Automated soil inference under fuzzy logic." *Ecological Modelling* v. 90, No. 2: 111-122.

Daunicht, W.; Salski, A.; Nöhr, P. & Neubert, C. (1996). „A fuzzy knowledge-based model of the annual production of Skylarks." *Ecological Modelling* v. 85, No. 1, 67-74.

Diamond, P. (1989). „Fuzzy kriging." *Fuzzy Sets and Systems* v. 33, no. 3, 315-332

Edwards, G. (1994). „Characterizing and maintaining polygons with fuzzy boundaries in geographic information systems." In Waugh, T. and Healey, R. (editors): Advances in GIS research, Proc. of the 6th International Symposium on Spatial Data Handling, Edinburgh, Taylor & Francis, London, 223-239.

Friderichs, M.; Fränzle, O. & Salski, A. (1996). „Fuzzy clustering of existing chemicals according to their ecotoxicological properties." *Ecological Modelling* v. 85, No. 1, 27-40.

Li, B.-L. (Editor) (1996). „Fuzzy Modelling in Ecology." *Ecological Modelling* special issue, v. 90, No. 2.

Marsili-Libelli, S. (1989). „Fuzzy clustering of ecological data." *Coenoses,* 4 (2), 95-106.

Odeh, I.O.A.; McBratney A.B. & Chittleborough D.J. (1992). „Soil pattern recognition with fuzzy-c-means: application to classification and soil-landform interrelationship. " *Soil Sci. Soc. Am. J.*, 56, 505-516.

Piotrowski, J.A.; Bartels, F.; Salski, A. & Schmidt, G. (1996). „Geostatistical regionalization of glacial aquitard thicknessin northwestern Germany, based on fuzzy kriging. " *Mathematical Geology* v. 28, No. 4, 437-452.

Pearl, J. (1988). *Probabilistic reasoning in intelligent systems: networks of plausible inference.* San Mateo, CA: Morgan Kaufmann.

Pedrycz, W. (1981). „An approach to the analysis of fuzzy systems." *Int. Jour. of Control*, 34, 403-421.

Pedrycz, W. (1996). *Fuzzy modelling, Paradigms and practice.* Boston: Kluwer Academic Publishers.

Roubens, M. (1982). „Fuzzy clustering algorithms and their cluster validity." *European Journal of Operational Research* , 10, 294-301.

Salski, A. (1992). „Fuzzy knowledge-based models in ecological research." *Ecological Modelling* 63, 103-112.

Salski, A. (1993) „Fuzzy clusters and fuzzy models in ecological research." Proc. of First European Congress on Fuzzy and Intelligent Technologies, Aachen, 485-490.

Salski, A.; Kandzia, P. (1993). „Modelling Support System for Ecological Application Based on Fuzzy Logic." In: Bouchon-Meunier, Valverde & Yager (Eds.): Uncertainty in Intelligent Systems, North-Holland, 269-275.

Salski, A.; Fränzle, O. & Kandzia, P. (Eds) (1996). „Fuzzy Logic in Ecological Modelling." *Ecological Modelling* special issue, v. 85, No. 1

Salski, A. (1996). „Fuzzy approach to ecological modelling and data analysis." Proc. of FUZZY'96, Zittau, 316-325.

Shafer, G. & Pearl, J. (1990). „Readings in uncertain reasoning." , San Mateo, CA: Morgan Kaufmann.

Wieland, R.; Schultz, A. & Hoffmann, J. (1996). „The use of neural networks and fuzzy-methods in landscape modelling." Proc. of FUZZY'96, Zittau, 370-379.

Wu, H.-I.; Li, B.-L.; Stoker, R. & Li, Y. (1996). „Modelling forest dynamics with vital attributes and fuzzy systems theory." *Ecological Modelling* v. 90, No. 2,147-160.

Zhu, A.-X.; Band, L.E.; Dutton, B. & Nimlos T.J. (1996). „A semi-arid grazing ecosystem simulation model with probabilistic and fuzzy parameters." *Ecological Modelling* v. 90, No. 2, 123-146.

8 FUZZY SETS APPROACH TO SPATIAL ANALYSIS*

Yee Leung

Abstract: Since its inception, fuzzy sets theory has been extensively developed and applied to the analysis and solution of complex problems in a large variety of disciplines. Human behavior in space is highly complex and is intrinsically imprecise. Modeling human systems with classical mathematics generally lead to a mechanical analysis and prediction of behavior in space and time. To capture more appropriately imprecise human behavior as well as spatial structures and processes, we need a theoretical paradigm which deals directly with complexity and imprecision in a formal manner. Over the years, fuzzy sets theory in general and fuzzy logic in particular have been successfully developed and applied to solve fundamental problems in spatial analysis. In this paper, I give a state-of-the-art examination of the fuzzy sets approach to spatial analysis. It covers the approximate characterization of imprecise spatial concepts, regional conceptualization and regionalization, spatial preference and equilibrium analysis, spatial optimization, spatial information systems and spatial decision support systems. I also outline in the end plausible directions for further research, particularly the integrative development with other new paradigms for spatial analysis.

8.1 INTRODUCTION

Human systems in general and spatial systems in particular are complex systems with omnipresence of incompleteness, imprecision, and subjectivity. While fuzziness is common in these systems, it has been avoided for the sake of precision and mathematical simplicity and we have assumed that our systems are well-defined. In the pursuit of precision, spatial scientists customarily fit spatial behavior to rigid mathematical models which make no provision for the imprecision and subjectivity in our cognition, perception, and decision making processes.

---

*For convenience purpose, mathematical notations employed in this paper may not follow the conventions adopted by the Handbook.

In the description of spatial behavior, we commonly use terms with artificial precision which fail to account for the intrinsic fuzziness of the phenomena under study. In the treatment of space, oblivious to the fuzziness of boundaries, precision in the subdivision and grouping of regions is usually enforced. In spatial economic analysis, it is generally assumed that there is no fuzziness in our preference-utility-choice structure so that different goods or places can be precisely discriminated and selected. When we need to optimize, information on the availability of resources is usually assumed to be exact and objectives and constraints can be precisely specified. In the construction of and operation on spatial databases such as geographical information systems and remote sensing, two-valued logic is a common place though imprecision is rampant in spatial information processing. To incorporate human intelligence in spatial decision support systems, binary logic again is employed as the backbone of inference albeit our reasoning is mostly based on commonsense embedded with imprecise terms or expressions.

With such relatively unrealistic assumptions, conventional spatial models often tend to be over-simplified, too mechnical, and too inflexible to explain and plan for the real world. To obtain a better approximation to and control of our spatial systems, it is essential to build theories or models which mimic human values, reasoning, and behavior. Instead of deemphasizing or avoiding fuzziness, spatial scientists should directly deal with it in a rigorous manner.

In recent years, some researchers have argued for the significance of formulating spatial models through fuzzy sets theory so that human spatial behavior can be better approximated (see for example Gale, 1972; Pipkin, 1978; Macmillan, 1978; Leung, 1979, 1983b, 1985b, 1988a; Ponsard, 1980; and Altman, 1994). It appears that fuzzy sets models are more general, realistic, and flexible in analyzing spatial structures and processes. The purpose of this paper is to provide a state-of-the-art scrutiny of the fuzzy sets approach to spatial analysis over the years. My discussion ranges from the analysis of fundamental spatial concepts, the very basic aspect of spatial analysis, to the construction of intelligent spatial decision support systems, the most recent advance in geoinformatics and spatial decision analysis.

In section 8.2, approximate characterization of some fundamental spatial concepts via the concept of a linguistic variable is discussed. Imprecision in regional concepts and regionalization, the basic building blocks of spatial analysis, is then examined in section 8.3. Fuzzy spatial preference structure and spatial equilibrium problems are scrutinized in section 8.4. In section 8.5, basic problems of spatial optimization under imprecision are investigated. Issues of spatial information processing and intelligent spatial decision support systems are discussed in section 8.6. I then conclude the paper with plausible directions for further reserach in section 8.7.

8.2 APPROXIMATE CHARACTERIZATION OF FUZZY SPATIAL CONCEPTS

Every scientific discipline has a set of fundamental concepts on which analyses are based; the field of spatial analysis is no exception. Concepts such as *distance*, *direction*, and *connection* have long been identified as basic terminologies for the analysis of spatial phenomena (Nystuen, 1968; Papageorgiou, 1972). Traditionally, the formalization of these terms are based on the definitions given in the theories of mathematical analysis such as those in the Euclidean or Minkowskian space. Under this system, the concepts are exact and their operational definitions have precise connotations. However, with the advancement of analytical and experimental results in spatial cognition and spatial choice, human behavior has been demonstrated to be a function of our cognitive-valuative structure in which imprecision prevails. Consequently, traditional mathematical definitions of the aforementioned fundamental concepts are inadequate in grasping the imprecision of human cognition and valuation. For example, distance cognition is context-dependent. A given length of physical distance may seem to be long in one spatial setting but becomes shorter in another. Therefore, in place of the physical length of a given distance, the meaning of that length to the decision maker becomes more important in determining his behavior in space. Statements such as "It is *too long* a distance to walk to that restaurant" and "the distance between my home and place of work is *rather short*" are common expressions in our cognitive and decision-making processes. The terms *too long* and *rather short*, for example, are meanings attached to a given physical distance in a person's cognitive environment. However, just how long a distance in physical term is *too long* and how short is *rather short* are fuzzy. Likewise, the cognition of direction and connection is also imprecise. Thus, a given cognitive term, such as *long*, places a restriction on the range of the possible values a base variable, such as physical distance, can take. Consequently, the study of spatial connotations of our common language is necessary in the understanding and prediction of human behavior in space. The concept of a linguistic variable based on fuzzy sets theory (Zadeh, 1975a, b, c) is thus appropriate for such a task (Leung, 1982a). In what follows, linguistic characterizations of distance, connection, and direction are selected as examples for discussion.

8.2.1 Linguistic Characterization of Distance

Since fuzzy terms such as *short* or *long* can be employed to describe cognitive distance, the fuzzy concept *distance* can then be treated as a linguistic variable characterized by the quintuple (w, $T(w)$, $\underline{U}$, G, M) detailed in the following:

$$w \triangleq \textit{distance} \tag{1}$$

$$\begin{aligned} T(w) = {} & \textit{short} + \textit{not short} + \textit{very short} + \textit{somewhat short} \\ & + \textit{long} + \textit{not long} + \textit{very long} + \textit{somewhat long} \\ & + \textit{somewhat short or long} + \textit{short and long} \\ & + \textit{not very short or not very long} \\ & + \textit{long but not very long} + \textit{extremely long} + \ldots, \end{aligned} \tag{2}$$

$T(w)$ is the term set in which *short* and *long* are primary terms defined on the universe of discourse $\underline{U} = [0, \infty)$, with $u \in \underline{U}$ stands for distance in miles. To differentiate fuzzy and ordinary sets, all ordinary sets are underlined throughout.

Hedges like *not*, *very*, *somewhat*, and *extremely* are employed as operators in modifying the primary terms to create new linguistic values. Operations such as union and intersection are also applied in a like manner to combine terms together. All of these constitute the syntactic rule G.

To associate meaning to each linguistic value, a semantic rule M which serves as a fuzzy restriction on the base variable u imposed by the linguistic value may take the following form:

If the primary terms *short* and *long* are defined by

$$M(short) = \int_0^\alpha 1/u + \int_\alpha^\infty e^{-k(u-\alpha/\alpha)^2}/u, \quad k > 0, \tag{3}$$

$$M(long) = \int_0^\beta 0/u + \int_\beta^\infty [1 - e^{-k(u-\beta/\beta)^2}]/u, \quad k > 0, \tag{4}$$

then, other terms may be defined as follows:

$$M(very\ short) = \text{CON}(short) = M^2(short) \tag{5}$$

$$M(somewhat\ short) = \text{DIL}(short) = M^{1/2}(short) \tag{6}$$

$$M(somewhat\ short\ or\ long) = (somewhat\ short) \bigcup long \tag{7}$$

$$M(short\ and\ long) = short \bigcap long. \tag{8}$$

In the case of an infinite term set, systematic algorithms for generating linguistic values and assigning their associated meanings may be necessary. Let

$$w \triangleq distance \tag{9}$$

$$\begin{aligned} \text{If } T^\infty(w) &= W_1 + W_2 + W_3 + \cdots \\ &= short + very\ short + very\ very\ short + \cdots, \end{aligned} \tag{10}$$

is the term set, then its elements may be generated by the production system of the syntactic rule G as follows:

$$T \to short \tag{11}$$

$$T \to T + very\ T. \tag{12}$$

If the primary term *short* is defined by (3), then the following semantic rule may be constructed as

$$M(very \ldots very\ short) = \int_0^\alpha 1/u + \int_\alpha^\beta [e^{-k(u-\alpha/\alpha)^2}]^{2n}/u, \quad k > 0, n = 1, 2, \ldots \tag{13}$$

8.2.2 Linguistic Characterization of Connection

In spatial analysis, the study of connectedness of objects often involves terms like contiguity and adjacency. The concept of contiguity can be exact or fuzzy. In the spatial context, when we say two objects are contiguous we usually mean that they are in actual contact along a boundary or at a point. Thus, either two objects touch or they do not. There is no fuzziness involved. However, in the temporal context, contiguous in time ordinarily means next or near in time. This definition then becomes the definition of adjacency. In describing adjacency, words such as *close* or *remote* often have fuzzy spatial connotations. Therefore, regardless of whether the context is spatial or temporal, linguistic characterization of the concepts of *connection* appears to be a more appropriate method in defining the degree of connectedness between objects.

Specifically, *connection* can be treated as a linguistic variable characterized by the quintuple ($\mathbf{R}$, T($\mathbf{R}$), $\underline{U} \times V$, G, M) detailed as follows:

$$\mathbf{R} \triangleq \textit{connection} \tag{14}$$

$$\begin{aligned} T(\mathbf{R}) = \textit{close} &+ \textit{not close} + \textit{very close} + \textit{more or less close} + \textit{remote} \\ &+ \textit{not remote} + \textit{very remote} + \textit{more or less remote} \\ &+ \textit{pretty remote} + \textit{extremely remote} \\ &+ \textit{not close and not remote} + \cdots, \end{aligned} \tag{15}$$

is the term set with *close* and *remote* serving as primary terms.

The element (u, v) in the universe of discourse $\underline{U} \times \underline{V}$ may be selected as the distance between u and v.

A syntactic rule G similar to that in the approximate characterization of distance can be employed in generating the terms in the term set.

If the primary terms are defined by

$$\textit{close} = \int_{U \times V} [1 - k(u - v)^2]^{-1} / (u, v), \quad k > 0 \tag{16}$$

$$\textit{remote} = \int_{U \times V} [1 - e^{-k|u - v|}] / (u, v), \quad k > 0, \tag{17}$$

then, a semantic rule like that in the conceptualization of *distance* can be applied to derive meanings to the other terms in the term set.

8.2.3 Linguistic Characterization of Direction

Like *distance* and *connection*, the cognition of *direction* also possesses fuzziness. Thus, aproximate characterization of this concept is appropriate. Specifically, *direction* can be regarded as a linguistic variable symbolized by a quintuple (y, $T(y)$, $\underline{U}$, G, M) in which

$$y \triangleq direction \tag{18}$$

$$\begin{aligned} T(y) = {} & north + a\ bit\ north + somewhat\ north + extremely\ north \\ & + east + a\ bit\ east + somewhat\ east + extremely\ east \\ & + south + a\ bit\ south + somewhat\ south + extremely\ south \\ & + west + a\ bit\ west + somewhat\ west + extremely\ west \\ & + somewhat\ east\ and\ somewhat\ north \\ & + a\ bit\ south\ or\ west + \ldots, \end{aligned} \tag{19}$$

is the term set spanned by the primary terms *north*, *east*, *south*, and *west* through a syntactic rule *G*.

The meanings of the linguistic values on the universe of discourse $U = [0°, 360°]$ is defined by a semantic rule *M* similar to that in the characterization of *distance*. The primary terms again need to be defined first. For example, *north* can be defined by

$$north = \int_{0°}^{180°} \mathrm{Sin}\, u/u + \int_{180°}^{360°} 0/u. \tag{20}$$

By the same token, the other primary terms can be defined in a simlar way.

Therefore, approximate characterization of our fuzzy cognitive process can be achieved through the concept of a linguistic variable discussed above. Spatial connotations of our common language can also be analyzed systematically. The definitions depicted in this section may serve as fundamental concepts for the analysis of more complicated spatial patterns, processes, and decision making. For example, the direction related phrases such as *up north*, *down south*, *back east*, and *out west* have specific spatial implications on the length of the distance that needs to be traveled and the angular distance of movement. Since *distance* and *direction* are linguistic variables as defined, a composite linguistic variable may be specified as: *directed vector* = (*distance*)·(*direction*). Specifically, the meaning of the term in the term set may be defined as *M*(*up north*) = *M*(*long*) · *M*(*north*). This new concept again is a fuzzy subset which imposes a restriction on the base variables, physical distance and angles. Applying different hedges, connectives, and operations on *up north*, a new set of vocabulary can be obtained.

Since cognition is context-dependent, these definitions are by no means universal. For example, instead of the primary terms *long* and *short*, words such as *far* and *near* may serve a better purpose in *distance* characterization in some situations. However, the basic theoretical construct of a linguistic variable is general enough to analyze most of the spatial phenomena.

Though *distance*, *connection*, and *direction* are the only concepts discussed here, other spatial terminologies, such as Lynch's (1960) cognitive nodes, points, lines, paths, and edges, can also be represented by a similar procedure. In addition to defining concepts, linguistic variables may also be employed to unravel spatial images on the location of points and lines, as well as shapes of different geometric figures. For example, Lee (1976) has proposed a fuzzy sets procedure in conceptualizing ambiguous images of some geometric configurations, and Leung and Yan (1997) have formulated a framework for representing fuzzy points, lines, and polygons in geographic information systems.

8.3 REGIONAL CONCEPTS AND REGIONALIZATION

Regions are fundamental analytical units on which most spatial analyses are based. Conventionally, a region is treated as a spatial construct which can be precisely identified and delimited. Therefore, boundaries separating regions are assumed to be clear-cut. Contrary to this common conceptualization or practice, regions however may not be precisely identified and boundaries generally exist as gradients and should appear as zones rather than lines. The implication then is that a region is generally a fuzzy concept and a boundary is generally a fuzzy line.

The justification for such an argument is mainly threefold (Leung, 1985a). First, it is difficult to delimit precise boundaries separating regions and nonregions over a space on which phenomena distribute or vary in a more or less continuous manner. Regional boundaries merge into one another to a certain extent because regions are part of a continuum in space. Thus, a precise boundary does not reflect appropriately such a phenomenon. Second, terms such as *hot* and *dry* are adjectives with imprecise physical denotations. When they are used to classify regions, the corresponding boundaries become fuzzy. Third, regional differences are differences in degree rather than kind. Intermediate areas usually exist between regions. Therefore, regional boundaries concern the degree of belongingness to regions and are thus fuzzy.

To have a formal representation of imprecise regional concepts, Gale (1976) and Ponsard (1980) proposed to conceptualize a region as a fuzzy subset. A formalization, however, had not been attempted until Leung (1985a, 1987a) constructed a regional classification system through a concept of a linguistic variable and a theory of possibility. It has been applied to perform tasks in regional characterization and regional assignment.

8.3.1 Regional Characterization

Within the framework, regions are characterized by linguistic propositions with mathematical translations as follows:

Let a region Z be characterized by a set of characteristics $\{Y_k\}$, $k = 1, \ldots, \ell$. Based on Leung (1985a), a region Z can then be characterized by a linguistic proposition

$$P\text{: } Y(Z) \text{ is } F, \tag{21}$$

where $Y = (Y_1, \ldots, Y_k, \ldots, Y_\ell)$ is a ℓ-ary linguistic variable, and Y_k's are possibilistic independent, $F = F_1 \times \cdots \times F_k \times \cdots \times F_\ell$ is a fuzzy n-ary relation in the universe of discourse $\underline{U} = \underline{U}_1 \times \cdots \times \underline{U}_k \times \cdots \times \underline{U}_\ell$, which induces a possibility distribution

$$\Pi_Y = F, \tag{22}$$

such that

$$\begin{aligned} \text{Poss}(Y = u) &= \Pi_Y(u) = \mu_F(u) \\ &= \min[\mu_{F_1}(u), \ldots, \mu_{F_k}(u), \ldots, \mu_{F_\ell}(u)], \\ u &= (u_1, \ldots, u_k, \ldots, u_\ell) \in \underline{U}. \end{aligned} \tag{23}$$

Since the characterization in (21)-(23) have fuzzy spatial connotations, the core and boundary of region Z may not be precise.

Under this situation, the core of a region Z, denoted as *CORE(Z)*, is defined by

$$\begin{aligned} CORE(Z) = \Big\{ u^* \mid \mu_F(u^*) &= \text{height}(F) = \sup_{u} \mu_F(u) \\ &= \min\Big[\sup_{u_1} \mu_{F_1}(u_1), \ldots, \sup_{u_k} \mu_{F_k}(u_k), \ldots, \sup_{u_\ell} \mu_{F_\ell}(u_\ell)\Big]\Big\}. \end{aligned} \tag{24}$$

If F_k, for all k, is a normalized fuzzy subset, i.e., $\sup_{u_k} \mu_{F_k}(u_k) = 1$, then the core of Z includes all $u^* \in \underline{U}$ such that $\mu_F(u^*) = 1$. The physical interpretation of the core of a region Z is the point or area in space whose characteristics are most compatible to the linguistic proposition in (21) characterizing Z.

The boundary of a region Z, denoted as *BOUNDARY(Z)*, is defined by

$$\begin{aligned} BOUNDARY(Z) = \Big\{ u \mid 0 < \mu_F(u) < 1 \Leftrightarrow \\ 0 < \min\Big[\mu_{F_1}(u), \ldots, \mu_{F_k}(u), \ldots, \mu_{F_\ell}(u)\Big] < 1 \Big\}. \end{aligned} \tag{25}$$

This definition states the imprecision of boundaries and reflects the property of gradation. Since the regional characterization in (21)-(23) has imprecise spatial connotations, the corresponding boundary is naturally fuzzy. Thus, a boundary is in fact a zone including all points in space whose characteristics are more or less compatible to the regional characterization in (21)-(23). That is, we are indeterminate on the precise value at which an individual characteristic can be considered insignificant with respect to a designated region. Therefore, boundaries exist as gradients rather than lines. Nevertheless, the boundary in (25) becomes a crisp line if imprecision vanishes. Hence, we can treat it as a general definition of a boundary.

Sometimes, having a zone as a boundary may be impractical. For example, in the preparation of a map, a line reflecting the approximate separation of regions might be more desirable. Under this situation, a precise boundary can always be delimited by restricting $\mu_F(u)$ in (25) to a specific value α, $0 < \alpha < 1$.

The α-boundary of a region Z, denoted as α-*BOUNDARY(Z)*, is then defined by

$$\alpha\text{-}BOUNDARY(Z) = \{u \mid \mu_F(u) = \alpha\}, \quad 0 < \alpha < 1. \tag{26}$$

Based on this definition, the boundary is a line connecting all points in space whose compatibility to the characterization of region Z is α. Region Z then becomes a α-level set:

$$Z = \{u \mid \mu_F(u) \geq \alpha\}, \quad 0 < \alpha < 1. \tag{27}$$

The physical implication of restricting the boundary to a precise value is that we are certain in deciding how much of an individual characteristic a spatial unit must possess to belong to the designated region. That is, the α-boundary is a line delimiting a region whose elements are at least α-compatible to the description of region Z. Thus, elements whose compatibilities are less than α are considered as nonregion. In other words, the concept of α-boundary is a simplification of the imprecise regional patterns and can be treated as a defuzzification mechanism.

For the two-region case, the boundary between regions Z and *W*, denoted as *BOUNDARY(Z, W)*, is defined by

$$BOUNDARY(Z, W) = \{u \mid 0 < \min[\mu_F(u), \mu_G(u)] < M\}, \tag{28}$$

where F and G are respectively composite linguistic characterizations of regions Z and W.

At *BOUNDARY(Z, W)* = u^*, we have $\mu_{F\cap G}(u^*) = \sup_u \mu_{F\cap G}(u) = M$. By the separation theorem of bounded and convex fuzzy subsets (Zadeh, 1965), the highest degree of separation S = 1 − M is realized at this point and the boundary delimits a precise line that separates Z and W into two disjoint regions:

$$Z = \{u \mid \mu_F(u) > M\}, \quad \text{and} \quad W = \{u \mid \mu_G(u) > M\}. \tag{29}$$

Moving away from u^*, the boundary exists as a zone rather than a line and Z and W overlap to a certain degree. Thus, the boundary includes all points in space whose characteristics are more or less compatible to the characterizations of both regions Z and W. Its spread is restricted by the membership function $\mu_{F\cap G}(u) = \min[\mu_F(u), \mu_G(u)]$. The more similar regions Z and W are, the fuzzier the boundary becomes.

To have a controlled overlapping of regions Z and W, a separation threshold

$$\ell < \sup_u \min[\mu_F(u), \mu_G(u)] = M, \tag{30}$$

can be employed to delimit the boundary betwen Z and W. The spread of the boundary is then controlled by the separated threshold ℓ and the membership function $\mu_{F \cap G}$.

The α-boundary between regions Z and W, denoted as *α-BOUNDARY(Z, W)*, is defined by

$$\alpha\text{-}BOUNDARY(Z, W) = \{u \mid \min[\mu_F(u), \mu_G(u)] \geq \alpha\}, \quad 0 < \alpha \leq \ell. \tag{31}$$

Thus, the α-boundary between two regions is again a zone in general. Based on (31), regions Z and W become

$$Z = \{u \mid \mu_F(u) \geq \alpha\}, \quad \text{and} \quad W = \{u \mid \mu_G(u) \geq \alpha\}. \tag{32}$$

Depending on the value of α, regions Z and W overlap to a certain degree.

To have a higher level of generalization, let there be m, $m > 2$, regions $\{Z_i\}$, $i = 1, ..., m$. For each region Z_i, let the following be its characterization:

$$\begin{aligned} &P_i\colon Y(Z_i) \text{ is } F_i, \\ &\Pi_Y = F_i, \\ &\text{Poss}(Y = u) = \pi_Y(u) = \mu_{F_i}(u). \end{aligned} \tag{33}$$

Let F_i, $i = 1, ..., m$, be bounded and convex. Then $F_i \cap F_j$, $\forall i, j, i \neq j$, are bounded and convex with maximal grades M_{ij}'s obtained by

$$M_{ij} = \sup_u \min[\mu_{F_i}(u), \mu_{F_j}(u)], \quad \forall\, i, j, i \neq j. \tag{34}$$

Then, the boundary between any two regions Z_i and Z_j follows the argument of the two-region case and becomes

$$BOUNDARY(Z_i, Z_j) = \left\{u \mid 0 < \min\left[\mu_{F_i}(u), \mu_{F_j}(u)\right] \leq M_{ij}\right\}. \tag{35}$$

However, when all regions are considered, a separation threshold for all regions is

$$\ell < \min_{\substack{i,j \\ i \neq j}} \sup_u \min\left[\mu_{F_i}(u), \mu_{F_j}(u)\right] = \min_{\substack{i,j \\ i \neq j}} M_{ij}. \tag{36}$$

It again can be employed to control the overlapping of all regions. [See Leung (1982b) for a parallel argument in the separation of market areas among firms].

Under this circumstance, the α-boundary between regions Z_i and Z_j, denoted as *α-BOUNDARY(Z_i, Z_j)*, is defined by

$$\alpha\text{-}BOUNDARY(Z_i, Z_j) = \left\{u \mid \min\left[\mu_{F_i}(u), \mu_{F_j}(u)\right] \geq \alpha\right\}, \quad 0 < \alpha \leq \ell. \tag{37}$$

Consequently, regions Z_i and Z_j become

$$Z_i = \left\{u \mid \mu_{F_i}(u) \geq \alpha\right\}, \quad \text{and} \quad Z_j = \left\{u \mid \mu_{F_j}(u) \geq \alpha\right\}. \tag{38}$$

Following the idea of fuzziness in regions discussed above, Bjørke (1998) has characterized fuzzy region and boundary by point-set topology. However, the generalization through the support of a fuzzy set only reflects the membership function in the extreme cases. It does not capture the characteristics contained in the whole membership function that is pertinent to the meaning of a fuzzy set.

Once regions have been characterized, we can assign spatial units to appropriate regions. A regional assignment procedure consists of a rule, or a set of rules, through which a discriminatory assignment of elementary spatial units to a set of predetermined regions can be performed. When both spatial entities are fuzzily characterized, it is natural to expect that a spatial unit can be assigned to several regions with varying degrees of possibility.

8.3.2 Regional Assignment

Parallel to the characterization of a region, a spatial unit X can be designated by a linguistic propostion

$$Q: Y(X) \text{ is } G, \tag{39}$$

which induces a possibility distribution

$$\Pi_{Y(X)} = G, \tag{40}$$

such that

$$\text{Poss}(Y(X) = y) = \pi_{Y(X)}(y) = \mu_G(y) = \min_k \mu_{G_k}(y_k), \tag{41}$$

where $G = G_1 \times \ldots \times G_p$ is a p-ary fuzzy relation in $\underline{Y} = \underline{Y}_1 \times \ldots \times \underline{Y}_p$, and $y \in \underline{Y}$.

The likelihood of X being assigned to Z is obtained as the conditional possibility defined by

$$\begin{aligned}\mathrm{Poss}(Q \mid P) &= \mathrm{Poss}(Y(X) \text{ is } G \mid Y(Z) \text{ is } F) \\ &= \sup_y \min[\mu_G(y), \mu_F(y)] \\ &= \min\left\{\sup_{y_1} \min\left[\mu_{G_1}(y_1), \mu_{F_1}(y_1)\right], \ldots, \sup_{y_p} \min\left[\mu_{G_p}(y_p), \mu_{F_p}(y_p)\right]\right\},\end{aligned} \tag{42}$$

which states that the possibility of assigning spatial unit X to region Z equals the smallest of the minimum threshold heights among the intersections of fuzzy subsets G_k's and F_k's. That is, it depends on how well the patterns, fuzzy subsets, characterizing X match those of Z for all characteristics.

If the linguistic proposition Q in (39) consists of exact information about X, such as

$$Q\colon Y(X) \text{ is } \alpha, \quad \alpha \in R^p, \tag{43}$$

then, the assignment rule in (42) becomes

$$\mathrm{Poss}(Q \mid P) = \mathrm{Poss}(Y(X) \text{ is } \alpha \mid Y(Z) \text{ is } F) = \min_k \mu_{F_k}(\alpha_k). \tag{44}$$

Thus, the fuzzy sets assignment rule generalizes the conventional zero-one procedure. Regional assignment with precise information becomes a special case of the rule in (42).

Since what is possible about an event does not guarantee what is certain about an event, a measure of certainty is then necessary in regional assignment. With respect to the linguistic propositions P in (21) and Q in (39), we can obtain the certainty of assigning spatial unit X_j to region Z_i as follows:

$$\mathrm{Cert}(Q_j \mid P_i) = \inf_u \max\left(\mu_{G_j}(u),\ 1 - \mu_{F_i}(u)\right). \tag{45}$$

Based on Cayrol, Farrenu, and Prade (1980), a possibility/certainty duality of regional assignment is obtained as (Leung, 1985a)

$$\mathrm{Cert}(Q_j \mid P_i) = 1 - \mathrm{Poss}(\neg Q_j \mid P_i). \tag{46}$$

It means that the degree of certainty of assigning X_j to Z_i is equal to the degree of impossibility of not assigning X_j to Z_i.

In additional to regional characterization and regional assignment discussed above, fuzziness also exists in the grouping for regions. Differing from fuzzy isodata analysis (Bezdek, 1981; Key et al., 1989; Foody, 1997), partitioning of fuzzy relations has been employed to construct a flexible regionalization framework that groups spatial units into regions (Leung, 1984b).

8.3.3 Grouping for Regions

The fuzzy sets methods in Leung (1984b) are capable of deriving regions which are homogeneous or functional, hierarchic or nonhierarchic, disjoint or nondisjoint with either qualitative or quantitative information.

The foundation of the regionalization procedure is a similitude relation which is a fuzzy binary relation possessing the following properties:

$$\begin{aligned}
&\text{(a)}\quad \mu_R(x_i, x_i) = 1 \qquad\quad , \forall (x_i, x_i) \in \underline{X} \times \underline{X} \quad \text{(reflexivity)}\\
&\text{(b)}\quad \mu_R(x_i, x_j) = \mu_R(x_j, x_i), \ \forall (x_i, x_j) \in \underline{X} \times \underline{X} \quad \text{(symmetry)}\\
&\text{(c)}\quad \mu_R(x_i, x_k) \geq \max_j \ \min[\mu_R(x_i, x_j), \mu_R(x_j, x_k)],\\
&\qquad\qquad \forall (x_i, x_j), (x_j, x_k), (x_i, x_k) \in \underline{X} \times \underline{X}\\
&\qquad\qquad \text{(max-min transitivity)},
\end{aligned} \tag{47}$$

where, $\underline{X} = \{x_1, ..., x_n\}$ is the set of elementary spatial units. Conditions (a) and (b) in (47) are self-explanatory. Condition (c) is a transitive property which states that the similarity of x_i and x_k, $\mu_R(x_i, x_k)$, should be at least as high as the maximum of the lowest degree of similarity of x_i and x_j and that of x_j and x_k for all intermediate x_j's.

Since R can be decomposed as

$$R = \max_{\alpha \in [0,1]} \alpha \cdot R_\alpha, \tag{48}$$

where R_α is the α-level fuzzy binary relation defined by

$$\mu_{R_\alpha}(x_i, x_j) = \begin{cases} 1, & \text{if } \mu_R(x_i, x_j) \geq \alpha, \\ 0, & \text{if } \mu_R(x_i, x_j) < \alpha, \end{cases} \tag{49}$$

then for each degree of similarity α, R_α can be partitioned into a set of equivalence classes which are disjoint. Each equivalence class can be interpreted as a region of α-similarity. Specifically,

$$\alpha\text{-region} = \{x_k \mid \mu_R(x_i, x_k) \geq \alpha, \quad \forall x_i, x_k \in R_\alpha\}, \quad \alpha \in [0, 1]. \tag{50}$$

Within each α-region, spatial units are at least α-degree similar to each other. The similarity of spatial units of two α-regions is however less than α. Thus, for each degree of similarity α there is a set of disjoint regions. Taking all α-values together gives a regionalization pattern for all disjoint regions and all degrees of similarity.

Taking the complement of R in (49), we obtain a dissimilitude relation $\mu_{\neg R}$ satisfying

$$\begin{aligned}
&\text{(a)} \quad \mu_{\neg R}(x_i, x_i) = 1 \qquad\qquad , \forall (x_i, x_i) \in \underline{X} \times \underline{X} \quad \text{(antireflexivity)} \\
&\text{(b)} \quad \mu_{\neg R}(x_i, x_j) = \mu_{\neg R}(x_j, x_i), \ \forall (x_i, x_j) \in \underline{X} \times \underline{X} \quad \text{(symmetry)} \\
&\text{(c)} \quad \mu_{\neg R}(x_i, x_k) \leq \min_j \max [\mu_R(x_i, x_j), \mu_R(x_j, x_k)], \qquad (51) \\
&\qquad\qquad \forall (x_i, x_j), (x_j, x_k), (x_i, x_k) \in \underline{X} \times \underline{X} \\
&\qquad\qquad \text{(min-max transitivity)}.
\end{aligned}$$

The value $\mu_{\neg R}(x_i, x_j)$ can be interpreted as the distance (dissimilarity) between spatial units x_i and x_j. Condition (c) in (51) states that the dissimilarity between x_i and x_k should be at most the minimum of the highest degree of dissimilarity of x_i and x_j and that of x_j and x_k for all intermediate x_j's.

Since $\neg R = 1 - R$, the regionalization pattern derived is identical to that obtained from the similarity concept. Of course, distance can always be defined differently and the regionalization pattern changes accordingly. For example, when condition (c) in (51) is replaced by the min-addition transitive property

$$\begin{aligned}
&\mu_{\neg R}(x_i, x_k) \leq \min_y [\mu_{\neg R}(x_i, x_j) + \mu_{\neg R}(x_j, x_k)], \qquad (52) \\
&\forall (x_i, x_j), (x_j, x_k), (x_i, x_k) \in \underline{X} \times \underline{X},
\end{aligned}$$

we no longer have disjoint regions in the decomposition of $\mu_{\neg R}$. They become nondisjoint maximal subrelations instead. By allowing distance or interaction to be asymmetric and to decompose, a hierarchical system of regions can be derived. The regionalization scheme based on the partitioning of fuzzy relations is summarized in Fig. 1.

It should be noted that when μ_R: $\underline{X} \times \underline{X} \rightarrow \{0, 1\}$, and the max-min operation is replaced by the Boolean-sum-Boolean-product operation, the fuzzy set regionalization procedure becomes the conventional zero-one grouping procedure. Such a generalization has also been demonstrated in the analysis of sociometric structures (Leung, 1980).

8.4 PREFERENCE STRUCTURE AND SPATIAL EQULIBRIUM ANALYSIS UNDER FUZZINESS

Conventionally, spatial economic theories and models are constructed on the assumption that individuals are capable of making clear-cut discrimination with regard to different goods or places. Consumers and producers are assumed to possess perfect information and precise cognitive and decision making capabilities

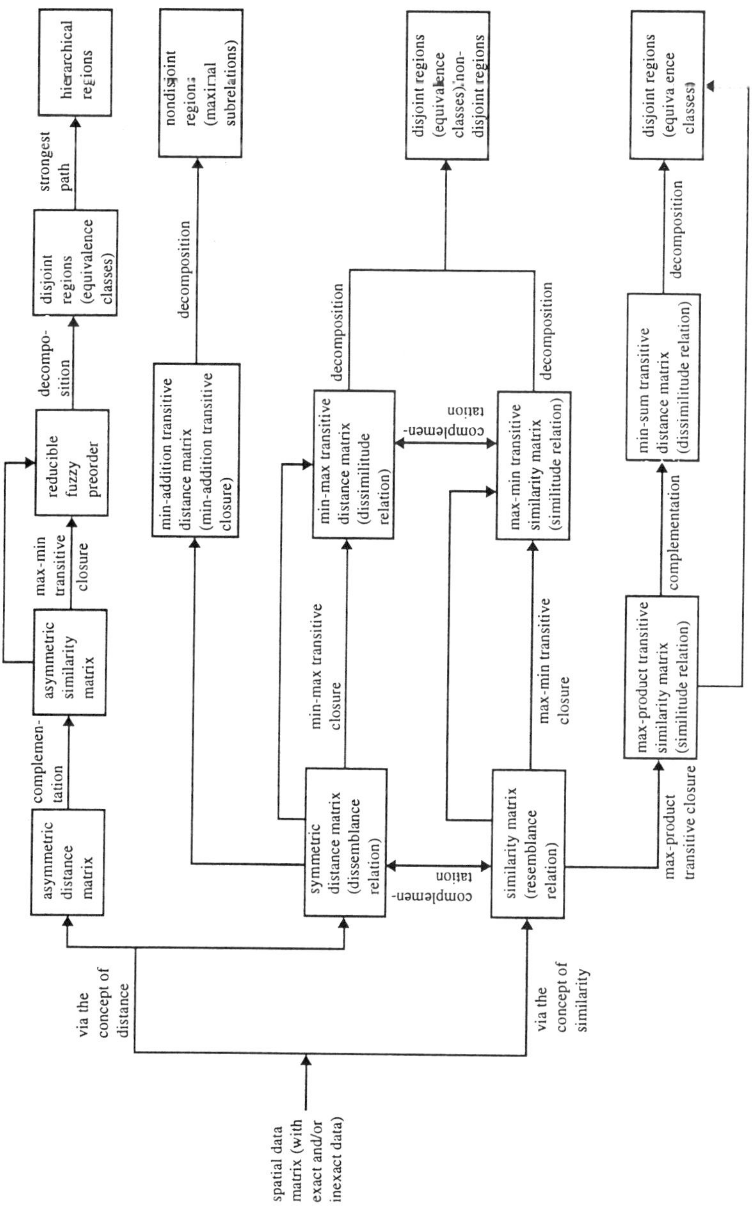

Figure 1 Schematic representation of various regionalization procedures
(adopted from Leung, 1984b)

so that individual or group preference is exact and the associated utility and choice structures are precise. In general, such preference-utility-choice structures presume too much. The information available is usually imperfect and individuals are often vague about making decisions because they are unable to discriminate perfectly between goods or places. It has been argued recently that preferences and perceived utilities are often vague and the all-or-nothing preference and utility structures are just a special case of the more overwhelming fuzzy structure found in reality (see Pipkin, 1978; Leung, 1979, 1982b; Nijkamp, 1979; Ponsard, 1979, 1980, 1982a, b).

8.4.1 Fuzzy Preference Structure

Employing fuzzy relations concepts, Ponsard (1979, 1981) defines a consumer's strict preference structure (X, $\succ$) as a binary fuzzy relation which is antisymmetric and max-min transitive, that is a fuzzy nonstrict order. The indifference structure (X, $\sim$) is expressed as a similitude relation which is reflexive, symmetric, and max-min transitive (see (47)). Combining the strict preference and indifference relations, a consumer's preference-indifference structure (X, $\geq$) can be defined as a fuzzy preorder R, such that

$$\begin{aligned}
&\text{(a)} \quad \mu_R(x_i, x_i) = 1 \qquad , \ \forall (x_i, x_i) \in \underline{X} \times \underline{X} \quad \text{(reflexivity)} \\
&\text{(b)} \quad \mu_R(x_i, x_k) \geq \max \min(\mu_R(x_i, x_j), \mu_R(x_j, x_k)), \\
&\qquad\qquad \forall (x_i, x_j), (x_j, x_k), (x_i, x_k) \in \underline{X} \times \underline{X} \\
&\qquad\qquad \text{(max-min transitivity)},
\end{aligned} \tag{53}$$

where $\mu_R(x_i, x_j)$ indicates the degree of preference or indifference between two complexes of goods in the set of all possible consumptions X. The transitivity property (b) in (53) states that the degree of preference for complex x_i relative to x_k is at least the maximum of the lowest degree of preference for complex x_i relative to x_j and that for all the intermediate complexes of x_j's relative to x_k.

Thus, the fuzzy preference structure allows for the gradation of preference intensity. If the max-min operation in (b) of (53) is substituted by the Boolean-sum-Boolean-product operation, the all-or-nothing preference structure in the classical theory of consumption is obtained, with the exception that reflexivity and max-min transitivity do not guarantee the congruence of indifference with strict and strong preference. Thus the fuzzy set version is a generalization of its counterpart in classical theory.

To obtain a reducible preference-indifference relation, the following two conditions have to hold (Ponsard, 1979):

(a) Strong preferences for the indifferent complexes x_i and x_j relative to the indifferent complexes x_ℓ and x_m at least equal the degree of indifference between complexes x_i and x_j. Specifically, for $\mu_R(x_i, x_j) = \mu_R(x_j, x_i) = \beta$,

$$\begin{aligned} \mu_R(x_i, x_\ell) \geq \beta \quad &\text{and} \quad \mu_R(x_j, x_\ell) \geq \beta, \\ \mu_R(x_i, x_m) \geq \beta \quad &\text{and} \quad \mu_R(x_j, x_m) \geq \beta. \end{aligned} \tag{54}$$

(b) The weak preferences for the indifferent complexes x_ℓ and x_m relative to the indifferent complexes x_i and x_j strictly equal the degree of indifference between the complexes x_ℓ and x_m. Specifically, for $\mu_R(x_\ell, x_m) = \mu_R(x_m, x_\ell) = \beta'$,

$$\begin{aligned} \mu_R(x_\ell, x_i) \geq \beta' \quad &\text{and} \quad \mu_R(x_m, x_i) \geq \beta', \\ \mu_R(x_\ell, x_j) \geq \beta' \quad &\text{and} \quad \mu_R(x_m, x_j) \geq \beta'. \end{aligned} \tag{55}$$

Similar to the regionalization procedure discussed in section 3, the congruent preference-indifference relation can be decomposed into disjoint preference-indifference subrelations (similitude relations). Parallel to the α-region argument, the degree of preference for one complex relative to another complex is α within the same equivalence class. Since preference between complexes of goods are asymmetric, a hierarchical preference structure can be established.

8.4.2 Fuzzy Utility and Optimization

Specification of a preference-indifference structure allows a fuzzy utility function to be defined as any homomorphism from the preference-indifference structure (X, $> \sim$) to ($[0, 1]$, $> \sim$), (Ponsard, 1979). Specifically, a fuzzy utility function f may be characterized by the following conditions:

$$\begin{aligned} &\mu_R(x_i, x_j) = \mu_R(x_j, x_i) \Rightarrow f(x_i) = f(x_j), \quad \forall x_i \in \underline{X}_R, \quad \forall x_j \in \underline{X}_R, \\ &\mu_R(x_i, x_j) > \mu_R(x_j, x_i) \Rightarrow f(x_i) > f(x_j), \quad \forall x_i \in \underline{X}_R, \quad \forall x_j \in \underline{X}_R. \end{aligned} \tag{56}$$

That is, complexes of goods belonging to the same preference-indifference subrelation generate the same degree of utility. If one preference-indifference subrelation contains a complex which is preferred to a complex belonging to another subrelation, then the utility of the former is higher than that of the latter. Again, such a formulation is a generalization of the classical utility function. Continuity of the utility function can also be generalized in the fuzzy set framework. However, the theory of fuzzy sets only has a weak theorem of separation and we cannot make the assumption in classical theory that the possible set of consumption is countable.

In classical spatial economic analysis, a consumer maximizes his utility with respect to a budget constraint, while a producer maximizes his profit or utility of profit subject to a technological constraint. Due to imperfect information, inexact cognitive and decision making processes, and incomplete control of inputs and outputs, the budget and technological constraints often cannot be precisely specified. Thus, a consumer may have to maximize his fuzzy utility function subject to a fuzzy budget constraint, and a producer may have to maximize his fuzzy utility of profit subject to a fuzzy technological constraint. In general, this specification involves the maximization of a fuzzy objective function subject to a fuzzy constraint.

By application of the framework of fuzzy optimization proposed by Bellman and Zadeh (1970) together with a result obtained by Tanaka, Okuda, and Asai (1974), Ponsard (1980) shows that the spatial equilibrium problem in a fuzzy environment leads to the main result of fuzzy optimization which is an extremum problem of a scalar function:

$$\begin{aligned} \sup_{x_i \in X} \mu_D(x_i) &= \sup_{x_i \in X} \min[\mu_O(x_i), \mu_C(x_i)] \\ &= \sup_{\alpha \in [0,1]} \min[\alpha, \sup_{x_i \in C_\alpha} \mu_O(x_i)]. \end{aligned} \tag{57}$$

Here, $\underline{X}$ is the set of all possible actions, μ_O: $\underline{X} \to [0, 1]$ and μ_C: $\underline{X} \to [0, 1]$ are respectively the fuzzy objective function (utility function) and fuzzy constraint (budget or technological constraint) defined on $\underline{X}$, $\underline{C}_\alpha$ is the α-level set with respect to the fuzzy subset C, and D is the decision space obtained as the intersection of fuzzy subsets O and C.

When a consumer's utility function is exact but the budget constraint is fuzzy, or a producer's utility of profit is exact but the technological constraint is fuzzy, the spatial equilibrium problem becomes the solution of a fuzzy integral (Ponsard, 1982b). In the same way general spatial equilibrium and dynamic spatial equilibrium in a fuzzy environment can also be analyzed, albeit mathematically more complicated.

Following the fuzzy-set line of argument, Billot (1995) has developed an axiomatic analysis of fuzzy equilibria. A fuzzy-set version of value has been formulated to study aggregated models in microeconomics with fuzzy economic behavior.

In addition to spatial equilibrium analyses, a variety of fuzzy spatial economic problems has also been analyzed (Ponsard and Fustier, 1986). For example, the concepts of fuzzy preference have been employed in market area analysis. Carlucci and Donati (1977) have demonstrated that when a consumer's preference for a firm is imprecise, market areas become fuzzy clusters with a certain degree of overlapping. By defining preferences for a firm as a convex fuzzy subset and then applying the separation theorem and the concept of a separation threshold, Leung (1982b) has shown that market areas in general overlap and can only be separated to a certain extent. The mutually exclusive property in classical theory turns out to be a special case of the fuzzy sets theoretic formulation.

In relation to fuzzy preference and choice, Leung (1979) proposes a fuzzy-set locational choice rule with hierarchical objectives. It is demonstrated that site preference is by no means clear-cut, but is a matter of degree. Imprecision in locational choice is also reflected in the residential choice model formulated by Bona, Inaudi and Mauro (1980). Residential choice is derived through the maximization of a fuzzy satisfaction function.

Therefore, by modelling fuzziness in a rigorous manner, flexibility and realism of spatial economic models can be substantially enhanced. Analytically, its higher level of generalization should make the fuzzy sets approach both philosophically and mathematically appealing.

8.5 SPATIAL PLANNING THROUGH FUZZY OPTIMIZATION

Urban and regional planning is similar to the regionalization and spatial equilibrium problems in that it often involves quantitative and/or qualitative information, subjective valuation, and imprecise decision making processes. Conventional methods are not flexible enough to effectively analyze problems with inexact objectives and constraints. For example, instead of imposing a maximum permissible resource usage α, a real number, decision makers may only be able to restrict it to a tolerance region *approximately* α, a fuzzy subset. Instead of maximizing an objective function, planners may prefer to specify a fuzzy satisfaction region within which the optimal value of the objective function may be achieved. The maximization of a fuzzy objective function subject to fuzzy constraints is naturally a fuzzy optimization problem. Such an approach is especially important when the original nonfuzzy problem fails to yield an optimal solution due to the incompatibility of objectives and constraints, or when the time-varying parameters are to be incorporated in the optimization framework.

Since its inception, fuzzy mathematical programming (Bellman and Zadeh, 1970) has become one of the research focuses in the optimization of fuzzy systems (see for example Dubois, 1987; Hannan, 1981; Inuigachi and Ichihashi, 1990; Inuiguchi et al, 1990, 1992; Leung, 1983a, 1984a; Negoita, 1981; Orlovski, 1984; Stancu-Minasian, 1984; Tanaka and Asai, 1984; Tanaka et al, 1974; Zimmermann, 1976, 1978). In spatial analysis, four basic classes of fuzzy spatial optimization problems can be identified with respect to the fuzziness or non-fuzziness of objectives and constraints (Leung, 1988a). (See Fig. 2).

In class (1), both the objective(s) and constraints are precise. This class belongs to the conventional optimization problems with precise objective(s) and constraints which take on the general format:

$$\begin{aligned} &\max\ c^T x \quad (\text{or } \max\ \{c_i^T x,\ i = 1, 2, \ldots, m\}, \\ &\qquad\qquad \text{for multiple-objective problems}) \\ &s.t.\ \ x \in \underline{X} \end{aligned} \tag{58}$$

where $c, x \in R^N$ and $\underline{X} \subset R^N$. This class is a special case of fuzzy optimization in which no fuzziness is involved. Its solution can be obtained by conventional optimization methods.

Problems in class (2) involve precise objective(s) and fuzzy constraints having the format:

Objective(s) \ Constraints	Precise	Fuzzy
Precise	(1)	(2)
Fuzzy	(3)	(4)

Figure 2 Classification of fuzzy optimization problems
(adopted from Leung, 1988a)

$$\begin{aligned} &\max\ c^T x \quad (\text{or } \max\ \{c_i^T x,\ i = 1, 2, \ldots, m\}, \\ &\qquad\qquad \text{for multiple-objective problems}) \\ &s.t.\quad x \in \underset{\sim}{\underline{X}} \end{aligned} \tag{59}$$

Fuzziness in the constraint set may be due to imprecise coefficients and/or imprecise RHS. In general, a fuzzy constraint in $x \in \underset{\sim}{\underline{X}}$ (here ~ is used to indicate the existance of fuzziness) can be in any of the three forms: $a^T x \underset{\sim}{=} b$, $a^T x \underset{\sim}{\geq} b$, or $a^T x \underset{\sim}{=} \leq b$, where a^T and b are vectors with fuzzy and/or non-fuzzy coefficients.

Objective(s) in class 3 are fuzzy while constraints are precise. The general format of the problems is:

$$\begin{aligned} &\max\ \underset{\sim}{c^T x} \quad (\text{or } \max\ \{c_i^T x,\ i = 1, 2, \ldots, m, \}, \\ &\qquad\qquad \text{for multiple-objective problems}) \\ &s.t.\quad x \in \underline{X} \end{aligned} \tag{60}$$

where $c^T x$ can be in any of the three forms: $c^T x \underset{\sim}{=} Z$, $c^T x \underset{\sim}{\geq} Z$, or $c^T x \underset{\sim}{\leq} Z$, with fuzzy coefficients in $c^T x$ and/or fuzzy goal(s), Z.

Class (4) is a pure fuzzy optimization problem in which the objective(s) and constraints are fuzzy. They can be represented as:

$$\begin{aligned} &\max\ \underset{\sim}{c^T x} \quad (\text{or } \max\ \{c_i^T x,\ i = 1, 2, \ldots, m\}, \\ &\qquad\qquad \text{for multiple-objective problems}) \\ &s.t.\quad x \in \underset{\sim}{\underline{X}} \end{aligned} \tag{61}$$

Over the years, a variety of solution methods have been proposed to solve the above classes of problems and their variants. A common solution approach is to reformulate the following basic fuzzy optimization problem

$$\sup_{x} \left[\mu(c^T x) \wedge \mu(x \in \underset{\sim}{X}) \right] \tag{62}$$

into a conventional optimization problem and then solve it by well-established algorithms (Zimmermann, 1976, 1978; Leung, 1987b, 1988b, c; Inuiguchi et al., 1990). For resolving spatial conflicts in multiobjective optimization, a concept of a fuzzy ideal has been proposed as a basis for deriving compromise solutions (Leung, 1982c, 1983a, 1984a). For situations involving hierarchical fuzzy objectives, a hierarchical programming method has also been formulated (Leung, 1987b). For the analysis and control of dynamic spatial systems, fuzzy dynamic programming methods (Bellman and Zadeh, 1970; Kacprzyk, 1983) have also been developed under the spatial context (Leung, 1988a).

Though fuzzy optimization brings flexibility to static and dynamic spatial planning by mathematical programming, the formulations are still confined by the conventional optimization framework which is sometimes considered to be too rigid and structured for real-life decision making. Flexibility gained by introducing elastic goals and constraints still could not free us from the rigidity of the conventional mathematical programming framework. In place of a mathematical function, real-life optimization process may just involve a set of fuzzy IF-THEN rules as constraints and the end result may require an optimal set of fuzzy IF-THEN rules to guide the search for an optimal solution. Derivation of a set of control rules which satisfies a set of initial-condition rules may be a plausible step to free fuzzy optimization problems from the restriction of the conventional optimization framework. However, more theoretical and applied research need to be done to demonstrate under what circumstances such an approach is valid and useful.

Moreover, fuzzy optimization methods should be able to entertain any form of data, fuzzy or non-fuzzy, coming from the database. We should also be able to integrate them with rules (Leung and Leung, 1995).

8.6 SPATIAL INFORMATION AND INTELLIGENT SPATIAL DECISION SUPPORT SYSTEMS

In spatial decision making, information comes in volumes. It can be obtained by various means such as maps, aerial photographs, remotely sensed images, and digitized signals. Advancement of spatial information technology, in particular geographic information system (GIS), has made spatial and aspatial data largely accessible to decision makers. The effective integration of knowledge with information, in turn, enables us to develop powerful spatial decision support system (SDSS) for real-life problem solving.

Management of uncertainty is a key issue in the development and applications of GIS and SDSS. Fuzzy set and fuzzy logic have thus become a major paradigm for the handling of uncertainty, especially due to imprecision, in these systems (Zhang and Leung, 1996; Leung, 1997a).

8.6.1 Geographic Information Systems

In recent years, GISs have become a subject of intensive investigations in the field of spatial analysis. Relatively successful software and hardware developments have enabled the public and private organizations to use GIS as a means for information management and decision analysis. Present day GISs are a georeference system for the storage, retrieval, manipulation, analysis, and display of data. A standard GIS consists of a data input and storage subsystem, a data retrieval and analysis subsystem, and a data output subsystem (see for example Burrough, 1986; Macguire et al., 1991; Laurini and Thompson, 1992).

Though existing GISs appear to be quite satisfactory in performing most of the above functions, they are still just a relatively sophisticated data processing system with conceptual shortcomings and very limited human intelligence and experience. Automation of data input, retrieval, analysis, and output is still primitive in terms of capturing the complicated and imprecise real world and replicating human wisdom and expertise.

With respect to logical foundations, present day GISs are predominantly based on Boolean logic. It is especially so in data retrieval and overlay operations. The all-or-nothing mathematical system of Boolean logic has however been demonstrated to be inadequate for spatial analysis and planning in general (Leung, 1988a) as well as spatial classification and boundary delimitations in particular (Leung, 1984b, 1985a, 1987a; Burrough, 1996; Lagacherie et al., 1996). It is well-known in the GIS community that overlay by Boolean logic generally ends up with loss and/or misrepresentation of information. It forces artificial precision on our otherwise imprecise thought processes. It arbitrarily screens out intrinsic imprecision in our database which might be essential in our analysis. It fails to communicate the extent of imprecision or error to users. To the extremes, it might mesmerize users with speedily generated erroneous pictures. Therefore, results of overlay unravel more questions about data quality and boundary mismatch than they solve. Similar problems also exist in other data analysis and cartographic modeling exercises. It is thus imperative to have a more appropriate logical system for the design of GIS. Such a system should be able to handle and to communicate precise and imprecise information. It should also be instrumental in the design of intelligent spatial decision support system (Leung, 1997a). Fuzzy sets in general and fuzzy logic in particular can play an important role in this area of research.

The use of fuzzy sets in cartography (Bouille, 1982; Muller, 1987), database design (Robinson and Strahler, 1984), representation of fuzzy spatial concepts, data, and relationships (Robinson, 1984; Robinove, 1986, Altman, 1994), query of spatial information (Robinson et al, 1985; Leung, 1989f, 1997a), classification of spatial objects in GIS or remotely sensed data (Kent and Mardia, 1988; Burrough, 1989; Wang, 1990; Burrough et al, 1992; Hall et al., 1992; Maselli et al., 1995; Molenaar, 1996; Fisher, 1996; Usery, 1996), and visualization of fuzzy spatial images (Leung et al, 1992) has been investigated in recent years. All these works, in one way or another, can be viewed as the materialization of fuzzy spatial concepts (discussed in Leung, 1982a, 1984b, 1985a, 1987a) in the context of spatial information systems.

To better handle uncertainty in GIS, it is often necessary to have an integrative utilization of fuzzy sets theory and probability theory. Taking the point-in-polygon query as an example, since a point or a polygon can take on any of the three forms: precise, fuzzy, and random, then point-in-polygon queries can basically be entertained in nine situations (Fig. 3). It is essential to understand the theoretical and practical issues involved in each situation. For example, we need to know in Case 2 how to entertain the point-in-polygon query: "Is a point in a polygon?" in the situation (P, A), i.e. "Is P in A?", where a point is known with certainty but a polygon is fuzzy.

		Polygon P	Polygon F	Polygon R
Point	P	Case 1 $(\underline{P}, \underline{A})$	Case 2 $(\underline{P}, A)$	Case 3 (P, A^*)
Point	F	Case 4 $(P, \underline{A})$	Case 6 (P, A)	Case 7 (P, A^*)
Point	R	Case 5 $(P^*, \underline{A})$	Case 8 (P^*, A)	Case 9 (P^*, A^*)

P : precise; F : fuzzy (imprecise), R : random (with error)

Figure 3 Basic classes of point-in-polygon queries
(adopted from Leung and Yan, 1997)

To sum up, we have the following nine cases:

Case 1. Query involving precise point and precise polygon, $(\underline{P}, \underline{A})$.
Case 2. Query involving precise point and fuzzy polygon, $(\underline{P}, A)$.
Case 3. Query involving precise point and random polygon, $(\underline{P}, A^*)$.
Case 4. Query involving fuzzy point and precise polygon, $(P, \underline{A})$.
Case 5. Query involving random point and precise polygon, $(P^*, \underline{A})$.
Case 6. Query involving fuzzy point and fuzzy polygon, (P, A).
Case 7. Query involving fuzzy point and random polygon, (P, A^*).
Case 8. Query involving random point and fuzzy polygon, (P^*, A).
Case 9. Query involving random point and random polygon, (P^*, A^*).

Without going through the derivations (see Leung and Yan, 1997, 1998 for details), I summarize the answers to the above queries in Fig. 4. Except for the deterministic answer in Case 1, answers to the rest of the cases are attached with some kinds of an uncertainty argument. It has also been shown that Cases 3, 4, and 5 are generalizations of Case 1. Case 6 is a generalization of Cases 2 and 4. Case 7 is a generalization of Cases 3 and 4. Case 8 is a generalization of Cases 2 and 5. And, Case 9 is a generalization of Cases 3 and 5.

Case 1 ($\underline{P}$, $\underline{A}$) all-or-nothing: 1 (yes), 0 (no)	Case 2 ($\underline{P}$, A) degree of membership: $\mu_A(a,b)$	Case 3 (P, A*) upper-bound probability : $1 - \text{prob}(A^* \subset R((a,b)))$ at the most
Case 4 (P, $\underline{A}$) degree (possibility): $\bigvee_{(x,y) \in A} \mu_P(x,y)$	Case 6 (P, A) degree (fuzzy integral): $\bigvee_{(x,y)} [\mu_P(x,y) \wedge \mu_A(x,y)]$	Case 7 (P, A*) upper-bound plausibility: $\bigvee_{(x,y)} \mu_P(x,y) \wedge [1 - \text{prob}(A^* \subset R(x,y))]$ at the most
Case 5 (P*, $\underline{A}$) probability : $\iint_A f_{P*}(x,y)\, dxdy$	Case 8 (P*, A) plausibility: $\bigvee_{\alpha \in [0,1]} [\alpha \wedge \iint_{A_\alpha} f_{P*}(x,y)\, dxdy]$	Case 9 (P*, A*) upper-bound expectation: $\iint_{R^2} f_{P*}(x,y)\ [1-\text{prob}(A^* \subset R((x,y))]dxdy$ at the most

Figure 4 The answers to point-in-polygon query in nine cases
(adopted from Leung and Yan, 1997)

8.6.2 Intelligent Spatial Decision Support Systems

To make GIS a more effective tool for spatial analysis and decision making, human intelligence has been regarded as a must and knowledge-based GIS has thus emerged in the research agenda (Marble, 1984; Tomlinson, 1984; Green et al., 1985; Burrough, 1986; Onsrud, 1986; Abler, 1987; Ripple and Ulshoefer, 1987; Robinson and Frank, 1987; Smith et al., 1987; Fisher et al., 1988; NCGIA, 1989).

Domain specific expert systems for geo-information processing and inference have been discussed or constructed in recent years (Robinson and Jackson, 1985; Wang and Newkirk, 1988; Qian et al., 1990; Ton et al., 1991; Kartikeyan et al., 1995). The possibilities of employing fuzzy sets and fuzzy logic to build knowledge-based GIS have been extensively investigated (Leung, 1989a, b, c, d, e, f). To facilitate the construction of expert systems for a variety of domain specific spatial problems, a fuzzy-logic-based expert system shell (FLESS) has been developed (Leung and Leung, 1993a, b). Any mix of fuzzy and non-fuzzy terms as well as uncertainties in rules (major premises) and facts (minor premises) is allowed in the shell. It employs fuzzy logic to handle approximate reasoning and fuzzy numbers to handle imprecision and uncertainty factors. To simplify knowledge engineering, FLESS is menu-driven with restricted natural language interfaces.

FLESS consists of four subsystems. They are the knowledge acquisition subsystem, the consultation driver, the fuzzy knowledge base, and the maps display module (see Fig. 5). The knowledge acquisition subsystem consists of management modules for objects, facts, fuzzy terms, rules and system properties. These modules are responsible for acquiring and managing knowledge in rules and facts. The fuzzy knowledge base stores all knowledge entities such as objects, facts, fuzzy terms, rules and system properties. The consultation driver contains three modules referred to as the inference engine, the linguistic approximation routine and the review management module. The inference engine extracts knowledge from the fuzzy knowledge base and makes inferences via rules and facts. The linguistic approximation routine maps fuzzy subsets (mathematical expressions) onto linguistic expressions (natural-language expressions) for any conclusions drawn by FLESS. The review management module handles various reviews such as "what if" and "how" requested by end-users. Lastly, the maps display module can display inference or analysis results in map form. A typical example is regions with fuzzy boundaries.

To enable an integrative use of structured knowledge (e.g. mathematical models, statistical routines, algorithms), loosely-structured knowledge (e.g. values, experience and expertise captured in rules), and spatial information systems (e.g. GIS and remote sensing systems), an intelligent SDSS shell has further been developed for effective spatial decisionmaking in a complex environment (Leung, 1993; Leung et al., 1997a, b). The basic features of the shell is depicted in Fig. 6. In addition to rule-based inference under certainty and uncertainty, the shell is capable of handling spatial data and models in an integrative and interactive manner. It runs on the windows environment with extensive use of the DLL and DDE technologies for dynamic model and data communication. Employing the tool, a variety of domain specific SDSS can be efficiently constructed.

8.7 CONCLUSION

I have examined in this paper the development and applications of fuzzy sets theory in spatial analysis and planning. It has been demonstrated that the fuzzy sets approach provides an effective means to model human behavior in space. It is especially versatile when our systems are embedded with intrinsic imprecision. We have seen major theoretical and practical advancements in the last two decades or so, particularly in recent years, with impressive results.

However, the complexity of our spatial systems cannot be handled by fuzzy sets alone. It is especially the case when dynamics and learning become the focal point of research. The complementary use of fuzzy sets in general and fuzzy logic in particular with other paradigms such as neural networks, genetic algorithms, evolutionary computation, and probability theory will be a profitable direction for further research (Fischer, 1994; Leung and Leung, 1995; Leung, 1997a). For

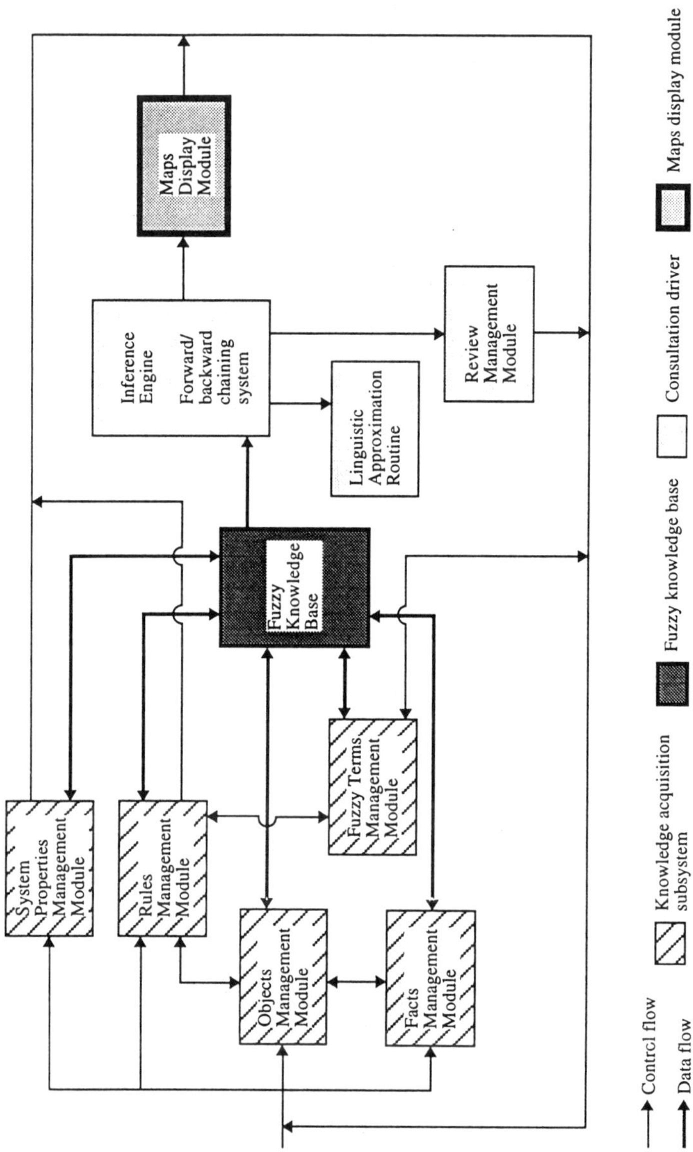

Figure 5 Overall structure of the expert system shell, FLESS.

(adopted from Leung and Leung, 1993a)

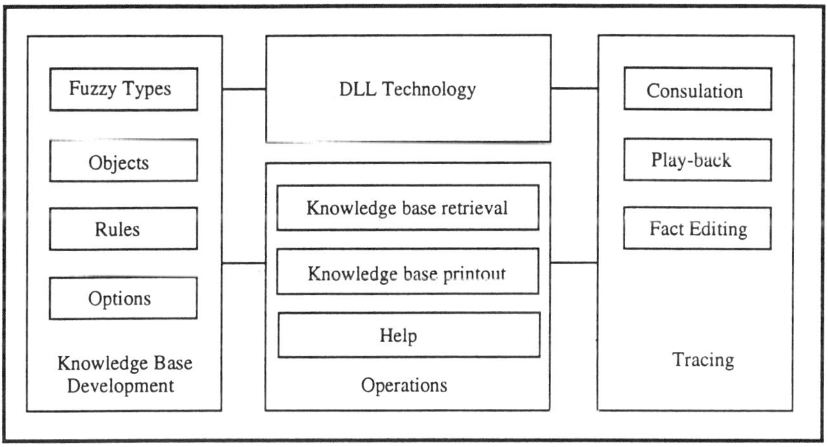

Figure 6 Features of the SDSS shelll

(adopted from Leung et al., 1997a)

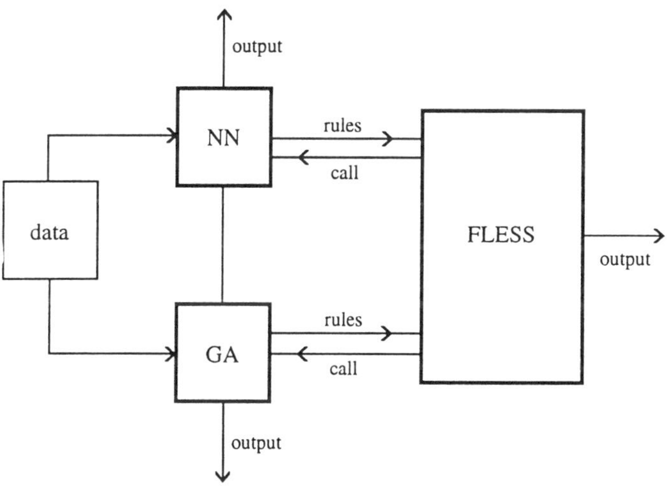

FLESS (rule-based expert system shell)
NN (neural-networks subsystem)
GA (genetic-algorithms subsystem)

Figure 7 General architecture of the hybrid connectionist expert system

(adopted from Leung, 1998)

example, the integration of fuzzy logic and neural network for spatial inference (Leung, 1994; Huang and Leung, 1998), spatial data classification and rule learning (Leung and Lin, 1996; Leung, 1997b) have been attempted in recent years. The development of genetic algorithms for fuzzy rule extraction has also been investigated (Leung et al., 1996). To make spatial decision making efficient and effective, Leung (1998) has constructed a general framework for the development of hybrid connectionist expert system which integrates rule-based inference, neural-networks and genetic-algorithms into a seamless whole (Fig. 7). Based on the trend of development, more significant results in the areas of spatial knowledge acquisition, spatial learning, spatio-temporal change detection, data compression, and spatial reasoning are expected in the years to come.

REFERENCES

Abler, R.F. "The National Science Foundation National Center for Geographic Information and Analysis", International Journal of Geographical Information Systems, 1(1987), 303-326.

Altman, D. "Fuzzy Set Theoretic Approaches for Handling Imprecision in Spatial Analysis", International Journal of Geographical Information Systems, 8(1994), 271-289.

Bellman, R.E. and L.A. Zadeh. "Decision-making in a Fuzzy Environment", Management Sciences, 17(1970), B-141-164.

Bezdek, J.C. Pattern Recognition with Fuzzy Objective Function Algorithms. New York: Academic Press, 1981.

Billot, A. Economic Theory of Fuzzy Equilibria, an Axiomatic Analysis. Berlin: Springer-Verlay, 1995.

Bjørke, J.T. "Fuzzy Regions: Topological Properties and Binary Topological Relations", 1998. (unpublished paper)

Bona, B., D. Inaudi, and V. Mauro. "A Fuzzy Approach to Residential Location Theory", in R. Trappl, G.J. Klin, and L. Ricciardi (eds.), Progress in Cybernetics and Systems Research, Vol. III, New York: John Wiley, 1980, pp. 314-320.

Bouille, F. "Actual Tools for Cartography Today", Cartographica, 19(1982), 27-32.

Burrough, P.A. Principles of Geographical Information Systems for Land Resources Assessment. Oxford: Clarendon Press, 1986.

Burrough, P.A. "Fuzzy Mathematical Methods for Soil Survey and Land Evaluation", Journal of Soil Science, 40(1989), 477-492.

Burrough, P.A. "Natural Objects with Indeterminate Boundaries", in P.A. Burrough and A.U. Frank (eds.), Geographic Objects with Indeterminate Boundaries, London: Taylor & Francis, 1996, pp. 3-28.

Burrough, P.A., R.A. MacMillan, and W. Van Deursen. "Fuzzy Classification Methods for Determining Land Suitability from Soil Profile Observations and Topography", Journal of Soil Science, 43(1992), 193-210.

Carlucci, D. and F. Donati. "Fuzzy Cluster of Demand within a Regional Service System", in M.M. Gupta, G.N. Saridis, and B.R. Gaines (eds.), Fuzzy

Automata and Decision Processes, Amsterdam: North-Holland, 1977, pp. 379-385.

Cayrol, M., H. Farreny, and H. Prade. "Possibility and Necessity in a Pattern-Matching Process", in Proceedings of the 9th International Congress on Cybernetics, Namur: International Association of Cybernetics, 1980, pp. 53-65.

Dubois, D. "Linear Programming with Fuzzy Data", in J.C. Bezdek (ed.), Analysis of Fuzzy Information, Vol.III: Application in Engineering and Science, Boca Raton: CRC Press, 1987, pp. 241-263.

Fischer, M.M. "Expert Systems and Artificial Neural Networks for Spatial Analysis and Modeling: Essential Components for Knowledge-Based Geographical Information Systems", Geographical Systems, 1(1994): 221-235.

Fisher, P. "Boolean and Fuzzy Regions", in P.A. Burrough and A.U. Frank (eds.), Geographic Objects with Indeterminate Boundaries, London: Taylor & Francis, 1996, pp. 87-94.

Fisher, P.F., W.A. MacKaness, G. Peacegood, and G.G. Wilkinson. "Artificial Intelligence and Expert Systems in Geodata Processing", Progress in Physical Geography, 12(1988), 371-388.

Foody, G.M. "A Fuzzy Set Approach to the Representation of Vegetation Continua from Remotely Sensed Data: An Example from Lowland Heath", Photogrammetric Engineering and Remote Sensing, 58(1992), 221-225.

Gale, S. "Inexactness, Fuzzy Sets and the Foundation of Behavioral Geography", Geographical Analysis, 4(1972), 337-349.

Gale, S. "A Resolution of the Regionalization Problem and its Implications for Political Geography and Social Justice", Geografiska Annaler, 58-B(1976), 1-16.

Green, N.P., S. Finch, and J. Wiggins. "The 'State of the Art' in Geographical Information Systems", Area, 17(1985), 295-301.

Hall, G.B., F. Wang and Subaryono. "Comparison of Boolean and Fuzzy Classification Methods in Land Suitability Analysis by Using Geographical Information Systems, Environment and Planning A, 24(1992), 497-516.

Hannan, E.L. "Linear Programming with Multiple Fuzzy Goals", Fuzzy Sets and Systems, 6(1981), 235-248.

Huang, C.F. and Y. Leung. "Estimating the Relationship Between Isoseismal Area and Earthquake Magnitude by a Hybrid Fuzzy-Neural-Network Method", Fuzzy Sets and Systems, 1998. (to appear)

Inuiguchi, M. and H. Ichihashi. "Relative Modalities and Their Use in Possibilistic Linear Programming", Fuzzy Sets and Systems, 35(1990), 303-323.

Inuiguchi, M., H. Ichihashi, and Y. Kume. "Relationships between Modality Constrained Programming Problems and Various Fuzzy Mathematical Programming Problems", Fuzzy Sets and Systems, 49(1992), 243-259.

Inuiguchi, M., H. Ichihashi, and H. Tanaka. "Fuzzy Programming: A Survey of Recent Developments", in R. Slowinski and J. Teghem (eds.), Stochastic Versus Fuzzy Approaches to Multiobjective Mathematical Programming Under Uncertainty, Dordrecht: Kluwer, 1990, pp. 45-68.

Kacprzyk, J. Multistage Decision-making under Fuzziness. Koln: Verlag TUV Rheinland, 1983.

Kartikeyan, B., K.L. Majumder, and A.R. Dasgupta. "An Expert System for Land Cover Classification", IEEE Transactions on Geoscience and Remote Sensing, 33(1995), 58-66.

Kent, J.T. and K.V. Mardia. "Spatial Classification Using Fuzzy Membership Models", IEEE Transactions on Pattern Analysis and Machine Intelligence, 10(1988), 659-671.

Key, J.R., J.A. Maslanik, and R.G. Barry. "Cloud Classification from Satellite Data Using a Fuzzy Sets Algorithm: A Polar Example", International Journal of Remote Sensing, 10(1989), 1823-1842.

Lagacherie, P., P. Andrieux, and R. Bouzigues. "Fuzziness and Uncertainty of Soil Boundaries: From Reality to Coding in GIS", in P.A. Burrough and A.U. Frank (eds.), Geographic Objects with Indeterminate Boundaries, London: Taylor & Francis, 1996, pp. 275-286.

Laurini, R. and D. Thompson. Fundamentals of Spatial Information Systems. N.Y.: Academic Press, 1992.

Lee, E.T. "An Application of Fuzzy Sets to the Classification of Geometric Figures and Chromosome Images", Information Sciences, 10(1976), 95-114.

Leung, K.S. and Y. Leung. "Fuzzy Knowledge-based Systems: Reviews and Perspectives", in Y. Yam and K.S. Leung (eds.), Future Directions of Fuzzy Theory and Systems, Singapore: World Scientific, 1995, pp. 19-39.

Leung, Y. "Locational Choice, a Fuzzy Set Approach", Geographical Bulletin, 15(1979), 28-34.

Leung, Y. "A Fuzzy Set Analysis of Sociometric Structure", Journal of Mathematical Sociology, 7(1980), 159-180.

Leung, Y. "Approximate Characterization of Some Fundamental Concepts of Spatial Analysis", Geographical Analysis, 14(1982a), 19-40.

Leung, Y. "Market Area Separation in a Fuzzy Environment", in R.R. Yager (ed.), Fuzzy Set and Possibility Theory: Recent Developments, New York: Pergamon, 1982b, pp. 551-561.

Leung, Y. "Dynamic Conflict Resolution through a Theory of a Displaced Fuzzy Ideal", in M.M. Gupta and E. Sanchez (eds.), Approximate Reasoning in Decision Analysis, Amsterdam: North-Holland, 1982c, pp. 381-390.

Leung, Y. "A Concept of Fuzzy Ideal for Multicriteria Conflict Resolution", in P.P. Wang (ed.), Advances in Fuzzy Sets, New York: Plenum Press, 1983a, pp. 387-403.

Leung, Y. "Fuzzy Sets Approach to Spatial Analysis and Planning — A Nontechnical Evaluation", Geografiska Annaler, 65B (1983b), 65-75.

Leung, Y. "Compromise Programming under Fuzziness", Control and Cybernetics, 13(1984a), 203-214.

Leung, Y. "Towards a Flexible Framework for Regionalization", Environment and Planning A, 16(1984b), 203-215.

Leung, Y. "A Linguistically-based Regional Classification System", in P. Nijkamp, H. Leitner, and N. Wrigley (eds.), Measuring the Unmeasurable, Dordrecht: Martinus Nijhoff, 1985a, pp. 451-486.

Leung, Y. "Basic Issues of Fuzzy Set Theoretic Spatial Analysis", Papers of the

Regional Science Association, 58(1985b), 35-46.
Leung, Y. "On the Imprecision of Boundaries", Geographical Analysis, 19(1987a), 125-151.
Leung, Y. "Hierarchical Programming with Fuzzy Objectives and Constraints", in J. Kacprzyk and S.A. Orlovski (eds.), Optimization Models using Fuzzy Sets and Possibility Theory, Dordrecht: D. Reidel, 1987b, pp. 245-257.
Leung, Y. Spatial Analysis and Planning under Imprecision. Amsterdam: North-Holland, 1988a.
Leung, Y. "Interregional Equilibrium and Fuzzy Linear Programming: 1", Environment and Planning A, 20(1988b), 25-40.
Leung, Y. "Interregional Equilibrium and Fuzzy Linear Programming: 2", Environment and Planning A, 20(1988c), 219-230.
Leung, Y. Fuzzy Logic and Knowledge-based Geographic Information Systems: 1, A Critique and a Proposal for Intelligent Geographic Information Systems. Occasional Paper No. 99, Hong Kong: Department of Geography, The Chinese University of Hong Kong, 1989a.
Leung, Y. Fuzzy Logic and Knowledge-based Geographic Information Systems: 2, Knowledge Representation and Measure of Confidence. Occasional Paper No. 100, Hong Kong: Department of Geography, The Chinese University of Hong Kong, 1989b.
Leung, Y. Fuzzy Logic and Knowledge-based Geographic Information Systems: 3, Translation Rules and Truth-values of Fuzzy Proposition. Occasional Paper No. 101, Hong Kong: Department of Geography, The Chinese University of Hong Kong, 1989c.
Leung, Y. Fuzzy Logic and Knowledge-based Geographic Information Systems: 4, Rules of Inference. Occasional Paper No. 102, Hong Kong: Department of Geography, The Chinese University of Hong Kong, 1989d.
Leung, Y. Fuzzy Logic and Knowledge-based Geographic Information Systems: 5, Acquisition of Knowledge. Occasional Paper No. 103, Hong Kong: Department of Geography, The Chinese University of Hong Kong, 1989e.
Leung, Y. "Fuzzy logic and Knowledge-based GIS: A Prospectus", Proceedings of IGARSS'89 12th Canadian Symposium on Remote Sensing, Canada: IGARSS, 1989f, pp. 47-50.
Leung, Y. "Towards the Development of an Intelligent Spatial Decision Support System", in M.M. Fischer and P. Nijkamp (eds.), Geographic Information Systems, Spatial Modeling, and Policy Evaluation, Berlin: Springer-Verlag, 1993, pp. 131-145.
Leung, Y. "Inference with Spatial Knowledge: An Artificial Neural Network Approach", Geographical Systems, 1(1994): 103-121.
Leung, Y. Intelligent Spatial Decision Support Systems. Berlin: Springer-Verlag, 1997a.
Leung, Y. "Feedforward Neural Network Models for Spatial Data Classification and Rule Learning", in M.M. Fischer and A. Getis (eds.), Recent Developments in Spatial Analysis, Spatial Statistics, Behavioural Modelling and Computational Intelligence, Berlin: Springer-Verlag, 1997b, pp. 337-359.

Leung, Y. Hybrid Connectionist Expert System for Spatial Inference, 1998. (unpublished paper)

Leung, Y., Y. Gao and Z.B. Xu. "A Novel Genetic-Based Method for Training Fuzzy Systems", 1996. (unpublished paper)

Leung, Y., M.F. Goodchild and C.C. Lin. "Visualization of Fuzzy Scenes and Probability Fields", Proceedings of the Fifth International Symposium on Spatial Data Handling, Vol. 2, 1992, pp. 480-490.

Leung, Y. and K.S. Leung. "An Intelligent Expert System Shell for Knowledge-based Geographic Information Systems: 1, The Tools", International Journal of Geographic Information Systems, 7(1993a), 189-199.

Leung, Y. and K.S. Leung. "An Intelligent Expert System Shell for Knowledge-based Geographic Information Systems: 2, Some Applications", International Journal of Geographic Information Systems, 7(1993b), 201-213.

Leung, Y., K.S. Leung and C.K. Lau. "A Development Shell for Intelligent Spatial Decision Support Systems: 1, Concepts and Tools", Geographical Systems, 4(1997a): 19-37.

Leung, Y., K.S. Leung, C.K. Lau and Z.P. Zhao. "A Development Shell for Intelligent Spatial Decision Support Systems: 2, An Application in Flood Simulation and Damage Assessment", Geographical Systems, 4(1997b), 39-57.

Leung, Y. and X. Lin. "Fast Extraction of Fuzzy and Non-fuzzy IF-THEN Rules by a Radial Basis Function Network with Unsupervised Competitive Learning", 1996. (unpublished paper)

Leung, Y. and J.P. Yan. "Point-in-Polygon Analysis Under Certainty and Uncertainty", GeoInformatica, 1(1997), 93-114.

Leung, Y. and J.P. Yan. "A Locational Error Model for Spatial Features", International Journal of Geographical Information Science, 12(1998), 607-620.

Lynch, K.A. The Image of the City. Cambridge: M.I.T. Press, 1960.

Macmillan, W.D. "An Introduction to the Theory of Fuzzy Sets in the Context of the Construction of Representational, Spatial Economic Theory", in R.L. Martin, N.J. Thrift and R.J. Bennett (eds.), Towards the Dynamic Analysis of Spatial Systems, London: Pion, 1978, pp. 38-52.

Maguire, D.J., M.F. Goodchild, and D. Rhind. Geographical Information Systems: Principles and Applications, Vol. 1: Principles; Vol. 2: Applications. Harlow: Longman, 1991.

Marble, D.F. "Geographic Information Systems: an Overview", in Proceedings, Spatial Information Technologies for Remote Sensing Today and Tomorrow, N.Y.: IEEE Computer Society, 1984, pp. 18-24.

Maselli, F., C. Conese, T. De Filippis, and S. Norcini. "Estimation of Forest Parameters through Fuzzy Classification of TM Data", IEEE Transactions on Geoscience and Remote Sensing, 33(1995), 77-84.

Molenaar, M. "A Syntactic Approach for Handling the Semantics of Fuzzy Spatial Objects", in P.A. Burrough and A.U. Frank (eds.), Geographic Objects with Indeterminate Boundaries, London: Taylor & Francis, 1996, pp. 207-224.

Muller, J-C. "The Concept of Error in Cartography", Cartographica, 24(1987), 1-15.

NCGIA. — National Center for Geographic Information and Analysis. "The Research Plan of the National Center for Geographic Information and Analysis", International Journal of Geographical Information Systems, 3(1989), 117-136.

Negoita, C.V. "The Current Interest in Fuzzy Optimization", Fuzzy Sets and Systems 6(1981), 261-269.

Nijkamp, P. "Conflict Patterns and Compromise Solutions in Fuzzy Choice Theory", Journal of Peace Science, 4(1979), 67-90.

Nystuen, J.D. "Identification of Some Fundamental Spatial Concepts", in B.J.L. Berry and D. Marble (eds.), Spatial Analysis: A Reader in Statistical Geography, Englewood Cliffs: Prentice-Hall, 1968, pp. 35-41.

Onsrud, H.J. "Land Information System Research Needs", Proceedings of the Urban and Regional Information Systems Association, 1986, pp. 247-253.

Orlovski, S.A. "Multiobjective Programming Problems with Fuzzy Parameters", Control and Cybernetics, 13(1984), 175-183.

Papageorgiou, G.J. "Description of a Basis Necessary to the Analysis of Spatial Systems", Geographical Analysis, 1(1972), 213-215.

Pipkin, J.S. "Fuzzy Sets and Spatial Choice", Annals, Association of American Geographers, 68(1978), 196-204.

Ponsard, C. "On the Imprecision of Consumer's Spatial Preferences", Papers, Regional Science Association, 42(1979), 59-71.

Ponsard, C. Fuzzy Economic Spaces. Working Paper No. 43, Dijon: Institute of Economics Mathematics, University of Dijon, 1980.

Ponsard, C. "An Application of Fuzzy Subsets Theory to the Analysis of the Consumer's Spatial Preferences", Fuzzy Sets and Systems, 5(1981), 235-244.

Ponsard, C. "Producer's Spatial Equilibrium with a Fuzzy Constraint", European Journal of Operational Research, 10(1982a), 302-313.

Ponsard, C. "Partial Spatial Equilibria with Fuzzy Constraints", Journal of Regional Science, 22(1982b), 159-175.

Ponsard, C. and B. Fustier (eds.). Fuzzy Economics and Spatial Analysis. Dijon: Librairie de l'Universite, 1986.

Qian, J.Z., R.W. Ehrich, and J.B. Campbell. "DNESYS — An Expert System for Automatic Extraction of Drainage Networks from Digital Elevation Data", IEEE Transactions on Geoscience and Remote Sensing, 28(1990), 29-45.

Ripple, W.J. and V.S. Ulshoefer. "Expert Systems and Spatial Data Models for Efficient Geographic Data Handling", Photogrammatric Engineering and Remote Sensing, 53(1987), 1431-1433.

Robinove, C.J. Principles of Logic and the Use of Digital Geographic Information Systems. U.S. Geological Survey Circular 977, Denver: U.S. Geological Survey, 1986.

Robinson, G. and M. Jackson. "Expert Systems in Map Design", Proceedings, Auto-Carto-7, 1985, pp. 430-439.

Robinson, V.B. "Modeling Inexactness in Spatial Information Systems", Proceedings, Modeling and Simmulation, Vol. 15, 1984, pp. 157-161.

Robinson, V.B. and A.U. Frank. "Expert Systems for Geographic Information Systems", Photogrammetric Engineering and Remote Sensing, 53(1987), 1435-

1441.

Robinson, V.B. and A.H. Strahler. "Issues in Designing Geographic Information Systems under Conditions of Inexactness", Proceedings, Machine Processing of Remotely Sensed Data, West Lafayette: Purdue University, 1984, pp. 198-204.

Robinson, V.B., D. Thongs, and M. Blaze. "Machine Acquisition and Representation of Natural Language Concepts for Geographic Information Retrieval", Proceedings, Pittsburg Modeling and Simulation Conference, Vol. 16, 1985, pp. 161-166.

Smith, T., D. Peuquet, S. Menon, and P. Agarwal. "KBGIS-II: A Knowledge-based Geographical Information System", International Journal of Geographical Information Systems, 1(1987), 149-172.

Stancu-Minasian, I.M. Stochastic Programming with Multiple Objective Functions. Dordrecht: Reidel, 1984.

Tanaka, H. and K. Asai. "Fuzzy Linear Programming Problems with Fuzzy Numbers", Fuzzy Sets and Systems, 13(1984), 1-10.

Tanaka, H., T. Okuda, and K. Asai. "On Fuzzy Mathematical Programming", Journal of Cybernetics, 3(1974), 37-46.

Tomlinson, R.F. "Geographical Information Systems — a New Frontier", Proceedings of the International Symposium on Spatial Data Handling 1, Zurich: Geographisches Institut, Abteilung Kartographie/EDV, 1984, pp. 2-3.

Ton, J.C., J. Sticklen, and A.K. Jain. "Knowledge-based Segmentation of Landsat Images", IEEE Transactions on Geoscience and Remote Sensing, 29(1991), 222-232.

Usery, E.L. "A Conceptual Framework and Fuzzy Set Implementation for Geographic Features", in P.A. Burrough and A.U. Frank (eds.), Geographic Objects with Indeterminate Boundaries, London: Taylor & Francis, 1996, pp. 71-85.

Wang, F. and R. Newkirk. "A Knowledge-based System for Highway Network Extraction", IEEE Transactions on Geoscience and Remote Sensing, 26(1988), 525-531.

Wang, F.J. "Fuzzy Supervised Classification of Remote Sensing Images", IEEE Transactions on Geoscience and Remote Sensing, 28(1990), 194-201.

Zadeh, L.A. "Fuzzy Sets", Information and Control, 8(1965), 338-353.

Zadeh, L.A. "The Concept of a Linguistic Variable and its Application to Approximate Reasoning, I", Information Sciences, 8(1975a), 199-249.

Zadeh, L.A. "The Concept of a Linguistic Variable and its Application to Approximate Reasoning, II", Information Sciences, 8(1975b), 301-357.

Zadeh, L.A. "The Concept of a Linguistic Variable and its Application to Approximate Reasoning, III", Information Sciences, 9(1975c), 43-80.

Zhang, W.X. and Y. Leung. Principles of Inference Under Uncertainty. Xi'an: Xi'an Jiaotong University Press, 1996. (in Chinese)

Zimmermann, H.-J. "Description and Optimization of Fuzzy Systems", International Journal General Systems, 2(1976), 209-215.

Zimmermann, H.-J. "Fuzzy Programming and Linear Programming with Several Objective Functions", Fuzzy Sets and Systems, 1(1978), 45-55.

9 CHEMISTRY AND CHEMICAL ENGINEERING

Willi Meier

Abstract: In this chapter an overview is given on recent applications of Fuzzy Logic in chemistry and Chemical Engineering. Applications are presented in the field of Analytical Chemistry, Theoretical Chemistry, Medical Chemistry and Chemical Engineering. In these aeras Fuzzy Logic is already widely used. Due to the fact that Fuzzy Control aspects are covered in various other chapters of this volume (see chap. 1, 2, 3, 4) only applications of Chemical Engineering are presented here, where the chemical aspect is in the foregound.

9.1 INTRODUCTION

Fuzzy Logic has already become a standard tool for control applications in Chemical Engineering. Sophisticated control strategies which are able to use linguistic variables like „high“ or „low“ make it possible to run complex chemical plants such as distillation columns very efficiently.

Also in the basic sciences linguistic variables play an important rule. In chemistry terms like reactivness, chirality or symmetry are used, which can be considered as linguistic variables.

Common mathematical methods treat these variables as ordinary numbers. Therefor many theories and concepts are only valid in an abstract laboratory environment and not in practice.

This is summed up in the words of Einstein, who wrote that *„as far as the propositions of mathematics refer to reality, they are not certain; and as far as they are certain, they do not refer to reality.“* [Einstein, 1928]

Fuzzy Logic tries to fill the gap between theory and practice by using membership functions to model linguistic variables, such that uncertainty can be implemented in theoretical concepts.

Some areas of chemistry where Fuzzy Logic methods are already extensively applied are shown in figure 1. Chemical Engineering is catagorized as a subset of chemistry. This is due to the fact that the pure engineering aspects of Chemical Engineering like Control or Process Design are covered in other chapters of this volume (chap.1, 2, 3, 4). In this chapter only the chemical aspects of Chemical Engineering are presented.

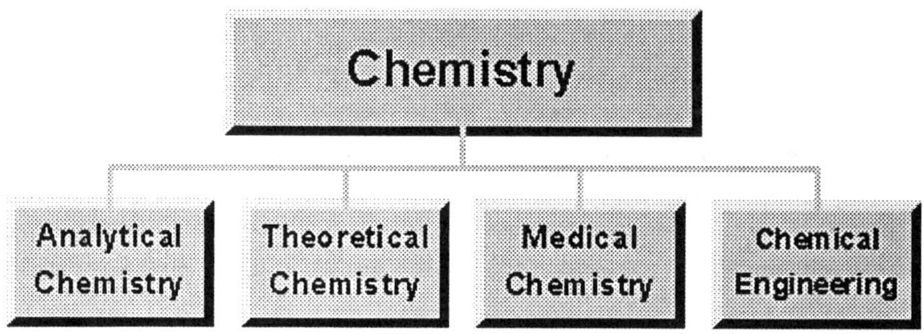

Figure 1: Some Application Areas of Fuzzy Logic in chemistry

In the following sections an overview is given about the state of the art of Fuzzy Logic applications in chemistry and Chemical Engineering.

9.2 APPLICATION AREAS

9.2.1 Analytical Chemistry

The field of mathematical treatment of analytical data is called chemometrics. Common methods are least squares or clustering algorithms. Fuzzy Logic and Neural Networks are becoming more and more integrated in the repertoire of chemometric methods [Otto, 1997]. Some applications are described below.

9.2.1.1 Quality control of diesel fuel

Identification and interpretation of spectral data plays an important role in Analytical Chemistry. Cheap, but powerful, computers produce large amounts of data, which must be analysed and stored. The data sets are examined mostly with well-known statistical methods. These treat the data as exactly defined mathematical properties. Uncertainty or incompleteness of the set often remains unconsidered. Fuzzy Logic is suitable to describe vague data in an adequate manner. Spectral data of samples, where composition differs because of normal variability (eg. effluent samples), can be compared using these methods.

One example is the interpretation of gas chromatograms [Zimmermann, 1994]. The data of such chromatograms consists of signals, described by location and intensity. Figure 2 shows two gas chromatograms of diesel fuel samples. The reference sample is that of a specimen, which fulfills certain quality standards. The task of an analytical chemist is to decide which specimens have a similar gas chromatogram and fulfill the quality standards and which do not.

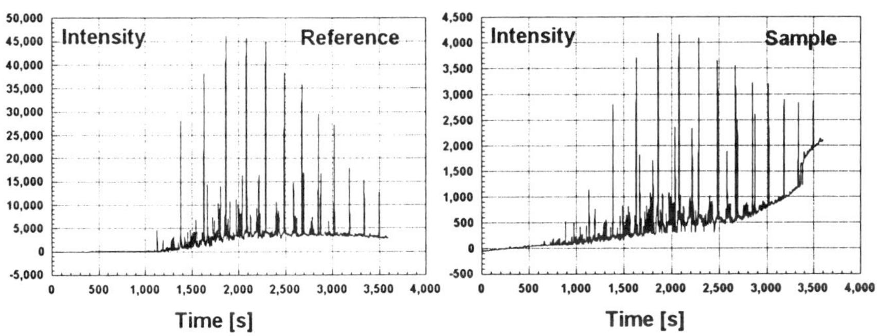

Figure 2: Gas Chromatograms of two fuel samples

A treatment of these signals as a mathematical vector will not succeed because form and intensity of the peaks can vary with the normal variability of the samples. So simply taking the difference between reference and specimen will fail.

The application of Fuzzy Logic allows the comparison of functions instead of vectors. Membership values z are calculated, which are a measure of the similarity of two signals (see eq. 1).

$$z(x,y)=MAX[0,(1-(x-x_r)^2/u_r^2-(y-y_r)^2/v_r^2)] \quad (1)$$

The maximum of this function is one. The minimum is zero. A value of one for the membership means, that the signals are completely similar, a value of zero means the opposite. x_r and y_r are location and intensity of signals in the reference, x and y are the parameters for the specimen. u_r and v_r are values for the uncertainty of the considered peak. Figure 3 shows the membership functions for different values of u_r and v_r. The membership functions have the shape of ellipses. With u_r and v_r one can adjust adequate uncertainties for location and intensity of peaks. For every peak of the chromatogram a membership value is obtained which describes the similarity of this peak to the related peak in the reference chromatogram. Finally the minimum of these values describes the similarity of the specimen to the reference.

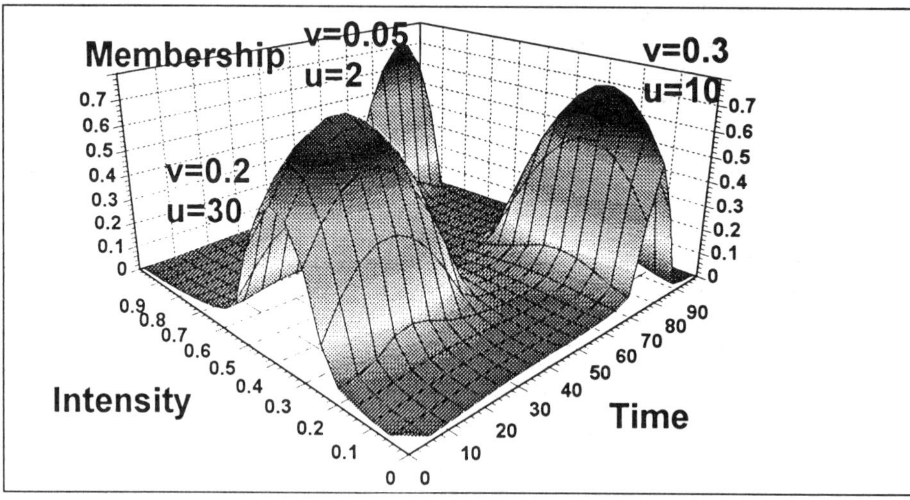

Figure 3: Fuzzy Sets of retention time and signal intensity

With a training set of specimens with known quality the values of u_r and v_r are adjusted and afterwards this classificator can be used for checking the specimens of unknown quality.

9.2.1.2 Analysis of tropical timbers with fuzzy clustering methods

Cluster algorithms try to find structure in data sets. They are widely used in Analytical Chemistry. Objects, which are described by features are compared by calculating the distance between them. Objects can be things like chromatograms, molecules or atoms. Features are descriptive parameters of the objects. Features of molecules can be molecular weight or number of atoms of the molecule.

The cluster algorithms divide the data set into classes. Similar objects should belong to the same class. If substances are clustered, which are for example candidates for herbicides, substances with similar biological activity should be together in the same class.

Crisp methods assign objects completely to a class. Fuzzy methods regard the classes as fuzzy sets to which the objects have a degree of membership. The membership degree is quantified by values between zero and one. The value one means that the object belongs completely to the class and zero means that it does not. Values between zero and one show that the object lies between two classes and cannot be classified in a crisp way. This is the advantage of fuzzy clustering algorithms compared to crisp ones. Objects can be classified in a more sensitive way and the classification results reflect more precisely the structure of the data set.

Some representatives of fuzzy cluster algorithms are:

Fuzzy C-Means [Bezdek, 1981]
Fuzzy Divisive Hierarchical Clustering [Dumitrescu, 1997]
Generalized Fuzzy C-Means [Dumitrescu, 1997]
Possibilistic Clustering Algorithms [Krishnapuram, 1993]

In this application tropical timbers were investigated by pyrolyses connected with a gas chromatographic procedure [Meier, 1997]. Tropical timbers, such as teak or mahogany, grow mainly in South America and South East Asia. But the clear cutting of topical woods especially in Brazil may lead to serious changes of the climate. On the other hand industrial countries in North America and Europe import a large amount of these timbers.

In South East Asia these timbers are cultivated in plantations. Plantation timber, stemming form these countries, may be importet, while import of timbers from native forests especially from South America should be forbidden, for the protection of the tropical forest there.

Therefore it is important to have an analytical method, which can determine the origin of the timbers. The chemical composition of the timbers differs slightly due to their origin. Pyrolysis of timber samples combined with gas chromatography is one method to detect these differences. In a case study 15 gas chromatograms of timber samples were analysed, stemming from five different countries. Three features were taken from the chromatograms by calculating the sum of peak intensities over three intervals.

The Fuzzy C-Means Algorithm could clearly distinguish between samples stemming from different countries. So this technique could be employed to classify and interpret gas chromatograms. This is an alternative fuzzy method for analysing gas chromatograms compared to the procedure described in 9.2.1.1.

9.2.1.3 Classification of mineral water

The Fuzzy Divisive Hierarchical Clustering Algorithm (FDHC) was used for the classification of mineral waters [Dumitrescu, 1997]. The amounts of dissolved ions like CO_2, HCO_3^-, Cl^-, Ca^{2+}, Mg^{2+} Fe^{2+} and Na^+ were taken as features. The FDHC found three classes. The detected classes refer to the total hardness of the mineral water.

Class one contains only waters with a hardness less than 50 German hardness degrees. The mineralisation is 1000-2000 mg/liter. CO_2 and HCO_3^--ion concentrations are small. The degree of membership to this cluster lies between 0.965 and 0.995.

Class two waters show a degree of hardness between 50 and 100. The mineralisation lies between 3000 and 6000 mg/liter. The membership ranges between 0.544 - 0.992. So this cluster is less compact.

Class three waters have a degree of hardness above 100. Mineralisation lies over 6000 mg/liter. The waters have also a very high degree in CO_2 and HCO_3^- concentration.

9.2.1.4 Rise of product yield of an electrochemical synthesis via fuzzy clustering methods

Nitriles are important substances for the production of certain polymers and rubbers.

Experimental data of a new propionitrile synthesis was investigated [Dumitrescu, 1997]. 2000 electroreduction processes were performed and analysed. The data to be analysed consists of gas chromatographic data, represented as six component vectors. The components represent the concentrations of the electroreduction process. These are propionitrile, adiponitrile, succinonitrile, methylglutaronitrile, bis(cyanoethyl)-ether and tricyanohexanes. The authors used the FDHC to catagorize 112 concentration vectors. Eight final clusters were received. It turned out that the most important component for classification was the concentration of propionitrile. A significance analysis show the importance of the other products by establishing the clusters. The goal here is to develop a valuable technique to monitor the progress of the electrochemical reaction occurring during the nondimerization electroreduction of acrylonitrile and to ascertain the success of the selective orientation of the process towards propionitrile formation.

9.2.1.5 Application of Fuzzy Logic in Thin Layer Chromatography

Thin Layer Chromatography (TLC) is a standard separation technique in chemistry. For a good separation of two components it is very important to choose the right solvent or combination of solvents, because this separation technique is based on the differing solubilities of the two components in a solvent system measured by R_f values. The FHDC was used to investigate the TLC of carotenoids [Dumitrescu, 1997]. Seven solvent systems were investigated. Each system was characterized by their R_f values of 11 carotenoids. Three clusters were found in a fuzzy hierarchy with two decomposition levels. Effective systems can now be combined, which are sufficient dissimilar. If one system has a large membership to one class, it will be far removed from the other classes. It may be efficient to combine systems, which are as far apart as possible and for which of each the system has the greatest membership to its own class. The hierarchical structure of the FHDC allows to detect relationship between detected classses.

So the FHDC is a tool to find the best system and the optimal combination of two or more TLC-systems according to their positions in the obtained fuzzy hierarchy.

9.2.2 Application of Fuzzy Logic in Theoretical Chemistry

9.2.2.1 Application of Fuzzy Logic in Chiral Studies

The limonen molecule can have two different configurations. One configuration is the reflection of the other. Though this difference seems to be small the molecules may have for example completely different smells (see figure 4.). Such molecules are called enantiomorphous. Enantiomers have identical scalar properties such as melting or boiling points and pseudoscalar properties that are identical in magnitude but opposite in sign, such as optical rotation. Such systems are called chiral. Pseudoscalar properties of achiral systems are zero.

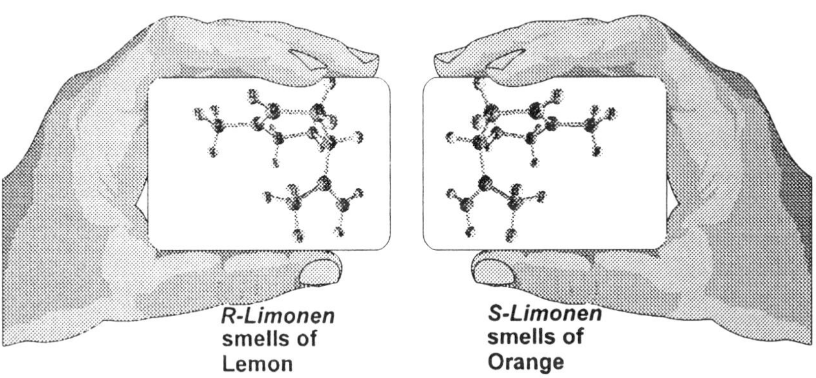

Figure 4: Enantiomers of Limonen

Chiral molecules can also have completely different biological activities. The contergane catastrophe was caused by neglecting this (see figure 5).

R-configuration, not teratogene **S-configuration, extreme teratogene**

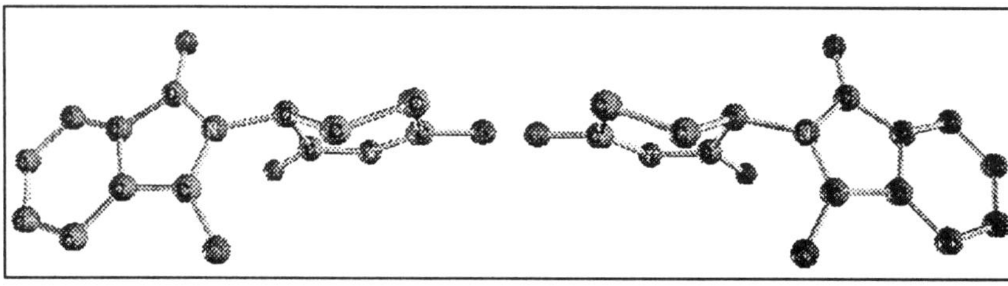

Figure 5: Enantiomers of Contergan

Chirality can be considered as a primitive fuzzy concept [Mislow, 1997]. A mixture of 2-butanol (see figure 6) is called racemic, if it consists of equal numbers of enantiomers. If we imagine that 2-butanol is produced by hydrogenation of butanone both enantiomers will be procuced in equal amounts. To have a strict racemic mixture the number of molecules has to be even, but the probability is exactly one half that any macroscopic sample has an odd number of molecules. So there is an even chance that at least one molecule in the sample remains uncompensated. Even if the system has an even number of molecules the probability is small that it is strictly racemic. Chiral may be expressed on the scalar of a numerically valued quantity that expresses some manifestation of chirality, e.g. optical rotation or chemical shift difference. So therefore one may talk of a more or less achiral system. Chiral and achiral are linguistic variables which can be modelled by fuzzy sets.

The application of Fuzzy Logic may lead to a better description of chirality than the traditional crisp concepts.

R-Butanol(2) ← **Butanon(2)** → **S-Butanol(2)**

Figure 6: Formation of both 2-butanoles via hydrogenation of butanone

9.2.2.2 Application of Fuzzy Logic in Molecular Recognition

Molecular Modeling plays a more and more important role in chemistry. With modern techniques it is possible to visualize chemical reactions on a molecular level. In biochemistry one has to deal with large molecules like proteins. They have a complicated structure and an uneven surface. One way to describe how a substrate can react with such a biomolecule is the key-in-lock principle. Size and shape of the substrate has to be of the magnitude that it fits into the so called binding niche or active site of the receptor (see figure 7). Substrates are substances like herbicides or insecticides, receptors could be enzymes that are responsible for a special biological function in the plant. When the substrate docks on the receptor the biological functions of the receptor get blocked, and for example the plant stops growing and dies.

The chemist has the possibility to see the docking process of a substrate to a receptor on his computer screen. This is a good way to see which substrate could fit into the binding niche. But this method is limited to the three dimensional space. There are certainly more than three variables to consider for a successful docking process. Not only the size of the molecule plays an important role, but also the charges of the atoms.

Figure 7: Key-in-lock principle

Humans are excellent in pattern recognition. Fuzzy Logic tries to model these human ability by sophisticated pattern recognition techniques like fuzzy clustering. So it may get possible to apply this human ability modeled by Fuzzy Logic in high dimensional space.

One step in this direction is the segmentation of protein surfaces using Fuzzy Logic [Brickmann, 1994]. The surface is automatically divided into triangular molecular segments, which represent distinct domains on the basis of physical and topographical molecular properties. Properties like the molecular electrostatic potential are described as linguistic variables with the linguistic terms „highly negative, negative, neutral, positive, highly positive“. Other used linguistic variables are the molecular lipophilicity potential and a new defined surface topographic index.

The new method was applied for the investigation of Trypsin and Trypsinogen. The analysis of the local lipophilicity results in several distinct lipophilic surface domains.

Three of them are located at the active site, indicating the contribution of hydrophobic interactions to the substrate binding of Trypsin.

9.2.3 Application of Fuzzy Logic in Medical Chemistry

The search for new drugs and pharmaceuticals is a long term task. The cost of developping a new drug will be 150.000.000 US$. A new herbicide or insectizide will cost about 100.000.000 US$. The methods and tools to find an efficient substance remain nearly unchanged over the last several decades. Random search is still mainly applied such that 6.000 to 10.000 substances have to be synthesized and tested to find one new therapeuticum. Thousands of chemists in chemical companies are working on this problem without finding one good candidate for a pharmaceutical product.

For pharmaceuticals different phases of development have to be considered. First there are experiments in the chemical and biological labs which give a first impression about the new synthesized substance. Afterwards preclinical experiments are performed mainly on cell culturs and animals. Finally when the substance has passed all these examinations the clinical phases start. The risk of failure of a substance reduces from phase to phase.

The high risks in the preclinical phase show clearly the high potential for using efficient search strategies and tools.

Already 100 years ago the key-lock prinicple was developed by Emil Fischer (s. 9.2.2.2). This principle describes very illustratively the biological effects of molecules. Enzymes are very big molecules which initiate biological processes in organisms. They have a 3D-structure with holes and also smooth sufaces. When a potential drug, normally a smaller molecule fits in one of the holes, the so called binding niche, the biological function of this enzyme is blocked. In cases of plants growing may be inhibited or in cases of viruses, their whole metabolism is blocked such that the virus dies.

Structural parameters play an important role in this docking process, but also charging parameters are very important. It is not sufficient that a molecule fits geometrically in such a binding hole. If the molecule does not reach it due to repulsive charging forces the docking process will fail. Also the metabolism of the organism makes things very complicated. If the docking process in the lab was very successful, it could be that the molecule fails completely in an organism.

The efficacy of a herbicide depends upon several parameters. Some of them lie in the molecular structure of the substance. Quantitative Structure Activity Relationships (QSAR) investigations try to find mathematical relationships between the molecular parameters and the efficacy. Unfortunately biological experiments are very uncertain. One cannot exactly define how strong a substance interacts with the plant. A lot of experience is necessary to look at the plant and to decide and quantify the damage caused by the treatment with herbicides. So the results of such experiments are fuzzy and can be expressed by fuzzy sets.

Moriguchi et al. worked out a fuzzy algorithm, which is based on a least squares technique [Moriguchi, 1992].

The so called „Fuzzy adaptive least squares" (FALS) is a pattern recognition method for analysing structure activity rating data to generate QSAR models. The method was used to investigate the structure activity correlation of α-methylene-γ-butyrolactone dervatives with allergenic activity. These substances occur naturally in many plants and can cause severe allergic contact dermatitis in man. The allergenic activity on the skin of guinea pigs sensitized to alantolactone was tested. The structures of the substances investigated are shown in figure 8. Best results were obtained with a three descriptor function.

The difference of the number of oxgen atoms from that in alantoacetone, the difference in the first-order molecular connectivity index [Kier, 1986] and the difference in the first-order atomic connectivity of the β'-carbon atom in the γ-substituent of lactone rings (see figure 8) were taken as descriptors. The average skin reaction of the substances were classified into three classes, which refer to different activation. Only two substances out of 29 were misclassified. So this structure-activity relationship is suitable to see the dependencies of biological activity on molecular descriptors.

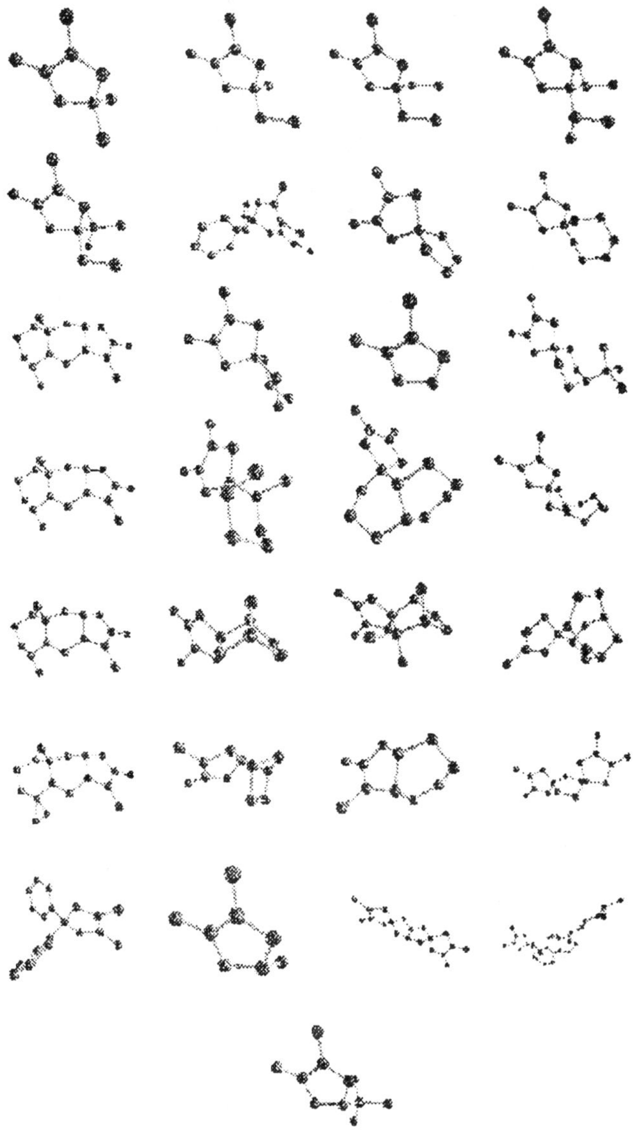

Figure 8: Structures of 29 Methylene-butyrolactones

9.2.4 Chemical Engineering

9.2.4.1 Introduction

Chemical Engineering was first evolved in Europe in the second half of the last century to meet the specialized engineering requirements of the chemical and allied industries. It now occupies a clearly defined place among the various disciplines of the engineering profession. Significant contributions have been made by chemical engineers in, for example, nuclear energy, automatic control, energy conversion, space travel, solid state physics, synthetic materials, food processing, biological control, and so on.

Fuzzy Logic has been applied in Chemical Engineering for nearly 15 years. First PID-controllers were substituted by fuzzy controllers and later sophisticated control strategies such as model predictive control were developed.

9.2.4.2 Application of Fuzzy Clustering Methods for Investigation of a Steamcracker Plant

Over 97% of the worldwide annual commercial production of ethylene is based on thermal cracking. Naphtha, which is obtained by distillation of crude oil, is the principal raw material. It is heated in cracking furnaces up to 830°C. During the cracking process also acetylenic, diolefenic and aromatic compounds are produced, which are known to deposit coke on the inside surfaces of the furnace tubes. This coke layer inhibits heat transfer from the tube to the process gas, so that at some time the furnace must be shut down to remove the coke. Several furnaces are parallel integrated into the production process to guarantee a continuous run of the whole plant. The crude online measured process data is not suitable for determining the degree of coking.

The Fuzzy-C Means clustering algorithm was applied, to cluster the process data measured at the furnace. The data consits of 20 quanitities like temperatures, pressures or data from analysis of the exhaust fumes. Two classes were obtained by clustering, which symbolise the coked and decoked state of the furnace. When the current state of the furnace is classified one get a membership to the coked and decoked class. Is the membership value close to one regarding the coked class, the operator knows, that he soon has to shut down the furnace for decoking.

The classifier makes the information visible by reducing the dimension of the problem. So this system supports the operator to make the right decisions in due time.

9.2.4.3 Overview on recent applications

The chapter provide an overview on recent applications of Fuzzy Logic in Chemical Engineering. The applications are briefly described. For a detailed description the reader is refered to the original paper.

Separation technologies like distillation and extraction are difficult to control by classical strategies.

Fuzzy Logic is used here, for example, to synthesize sharp multicomponent separation sequences, such as those encountered in distillation columns. [Floquet, 1994]. This approach is based on the fuzzification of two heuristic rules and combines the values of the estimate separation mass load coefficients and the difference in normal boiling points of the components in a fuzzy rule based procedure. A control program called DSEQSYS, distillation sequencing system has already been developed [Harrison, 1994].

Refinery reaction processes are studied by neural nets and fuzzy sets [Wang, 1996].

A fuzzy model-based system controls the moisture content of spent grain leaving an industrial steam dryer [Bremner, 1995].

Fuzzy Logic is also used for saftey assessment in chemical plants [Huagong Jixie, 1995], [Cogsy Club, 1995].

Fermentation processes are characterized by a great uncertainty concerning the influence of reaction parameters on the reaction path. In the optimisation of fermentation processes Fuzzy Logic can replace process models, which are very cost intensive and time consuming to develop [Froese, 1995].

Recent applications in the biotechnological area are the control of growth of immobilized *Catharathus roseus* cells in polyurethane foams [Hu, 1995] or the control of the production of baker's yeast [Simes, 1995].

Further recent applications are listed below:

Batch process control [Reeve, 1994]
Process Design [Sawyer, 1994]
Pulping operations [Broderick, 1995],
Sludge processes [Fu, 1995],
Qualitative Process Modelling [Wang, 1995],
Dynamic Simulation [Parkinson, 1995],
PVC polymerisation reactor [Goulao, 1996],
Combustion control [Gierend, 1996],
Long-range predictive control [Linkens, 1996],
Fuzzy relational model-based control [Sing, 1996],
Nonlinear pH process [Kavsek-Biasizzo, 1997],
Fault diagnosis [Tarifa, 1997],
Process Operational Decision Support [Wang, 1997],
Simultaneous reaction and distillation [Wilson, 1997].

The integration of economic aspects into the production process is one of the trends in modern process control. Here Fuzzy Logic can also help to fill the gap between process and business control [Krings, 1995, Tarifa, 1995, Tessier, 1995, Petley, 1995].

9.3 SUMMARY AND OUTLOOK

Fuzzy Logic is already widely used in the chemical industry as well as in basic chemical science. In the Chemical Engineering area it has become a standard tool for control and modeling purposes and it is part of a sustainable development as pointed out by the AllChemE-Group (Alliance for Chemical Sciences and Technologies in Europe) [AllChemE, 1996].

Chemometric is also one field where Fuzzy Logic already has a considerable number of applications. Commercial fuzzy tools for investigating analytical data are already available.

In basic science there are only few applications. Concepts in chemistry are often empirical and Fuzzy Logic can help to put linguistic variables like chirality or acidity into an appropriate form for computers and scientific software programs.

Applications in basic science theory are still at the beginning. In the future Fuzzy Logic can help to turn ideas of chemists into mathematical concepts and theories and improve empirical concepts in chemistry in such a way that they can be quantified and used for computation.

References

AllChemE-Group [1996]:"Chemistry Europe and the Future", p. 27, Royal Society of Chemistry, London 1996.

Bezdek, J. [1981]:"Pattern Recognition with Fuzzy Objective Function Algorithms", Plenum, New York 1981.

Bremner, H., Postlethwaite, B. [1995]:"Developing a relational fuzzy model-based controller for an industrial process", in: IChemE Res. Event, 1, 40-42, 1995.

Brickmann J., Heiden W. [1994]:"Protein Surface Segmentation Using Fuzzy Logic", in: J. Mol. Graphics, 12, 106-115, 1994.

Broderick, G., Paris, J., Valade, J.L. [1995]:"Factors affecting the optimal performance of a high-yield pulping operation", in: Can. J. Chem. Eng., 73(3), 391-399, 1995.

Cogsys Club [1995]:"Fuzzy tools help interpret alarms", in: Process Eng., 76(1), 23-27, 1995.

Dumitrescu, D. [1997]:"Fuzzy Classification in Analytical Chemistry" in: Rouvray H. „Fuzzy Logic in Chemistry", Academic Press, New York 1997.

Edgar, C. [1997]:"Fuzzy models come into focus", in: Chem. Engr., 6, 11-14, 1997.

Einstein, A. [1928]:"Geometrie und Erfahrung, p. 3, Springer, Berlin 1928.

Floquet, P., Pibouleau, L., Aly, S., Domenech, S. [1994]:"Fuzzy heuristic approach for sharp separation sequence synthesis", in: Comput. Chem. Eng., 18(9), 785-795, 1994.

Froese, T., Heimbucher, K. [1995]:"Optimisation of fermentation processes - fuzzy logic replaces process models", in: Chem.-Tech., 24(8), 59-60, 1995.

Fu, Chungsheng, Poch, M. [1995]:"Fuzzy modelling and pattern recognition for dynamic processes and its application for an activated sludge process", in: Chem. Eng. Sci., 50(23), 3715-3725, 1995.

Gierend, C. [1996]:"Combustion Control by Fuzzy Control", in: VGB Kraftswerktech., 76(6), 487-497, 1996.

Goulao, M.L., Saraiva, P., [1996]:"Fuzzy temperature control of a simulated PVC polymerization reactor", in: IChemE res. Event, 2, 766-768, 1996.

Harrison, B. K., Flowers, T. L., Niccolai, M.J. [1994]:"Automated synthesis of distillation sequences using fuzzy logic and simulation": in: AICHE j., 40(8), 1341-1348, 1994.

Hu, Zong Ding, Yuan Ying Jin [1995]:"Fuzzy growth kinetics of immobilized *Catharanthus roseus* cells in polyurethane foams", in: Chem. Eng. Sci., 50(20), 3297-3301, 1995.

Kavsek-Biasizzo, K., Skrjanc, I., Matko D. [1997]:"Fuzzy Predictive Control of Highly Nonlinear pH process", in: Comput. Chem. Eng., 21, 613-618, 1997.

Kier, L., Hall L., [1986]"Molecular Connectivity in Structure-Activity Analysis", Research Studies Press, New York 1986.

Krings, L. [1995]:"Advance Cement System - closing the gap between process and business control", in: Abb Re., 6, 22-31, 1995.

Krishnapuram R. Keller J. [1993]:"A Possibilistic Approach to Clustering", in: IEEE Transactions ob Fuzzy Systems, 1, 98-110, 1993.

Linkens, D. A., Kandiah, S. [1996]:"Long-range predictive control using fuzzy process models" in: Chem. Eng. Res. Des., 74(A1), 77-88, 1996.

Meier W., Weber R., Zimmermann H.-J. [1994]:"Fuzzy data analysis - Methods and industrial applications", in: Fuzzy Sets and Systems, 61, 19-28, 1994.

Meier W. [1997], unpublished results.

Mislow, K. [1997]:"Fuzzy Restrictions and Inherent Uncertainties in Chirality Studies" in: Rouvray H. „Fuzzy Logic in Chemistry", Academic Press, New York 1997.

Moriguchi, I:, Hirono, S:, Liu Q., Nakagome, I. [1992]:"Fuzzy Adaptive Least Squares and its Application to Structure-Activity Studies": in: Quant. Struct.-Act. Relat., 11, 325-331, 1992.

Otto, M. [1997]:"Chemometrie", p. 304, VCH-Verlag, Weinheim 1997.

Parkinson, G., Fouhy, K. [1995]:"Advanced process control: ingenious, yet accessible", in: Chem. Eng. 102(5), 28-31, 1995.

Petley, G.J., Edwards, D.W. [1995]:"Further developments in chemical plant cost estimating using fuzzy matching", in: Comput. Chem. Eng., 19, 675-680, 1995.

Reeve, A. [1994]:"Control. Past and future", in: Control Instrum., S21-S23, 1994.

Sawyer P. [1994]:"Fuzzy logic rules OK?", in: Chem. Engr., 577, 21-22, 1994.

Simes, T., Linko, P., Numers, C. von, Nakajima, M., Endo, I. [1995]:"Real-time fuzzy-knowledge-based control of baker's yeast production", in: Biotechnol. Bioeng., 45(2), 135-143, 1995.

Sing, C.H., Postlethwaite, B.:"Fuzzy relational model-based control applying stochastic and iterative methods for model identification", in: Chem. Eng. Res. Des., 74(A1), 70-76, 1996.

Tarifa, E., Chiotti, O. [1995]:"Flexibility vs. costs in multiproduct batch plant design: a fuzzy calculation algorithm", in: Chem. Eng. Res. Des., 73(A8), 931-940, 1995.

Tarifa E., Scenna, N., J. [1997]:"Fault diagnosis, direct graphs, and fuzzy logic", in: Comput. Chem. Eng., 21, 649-654, 1997.

Temeng, K.O., Schnelle, P.D., McAvoy, T.J. [1995]:"Model predictive control of an industrial packed bed reactor using neural networks", in: J. Process Control, 5(1), 19-27, 1995.

Tessier, P.J.C., Qian Yu [1995]:"Application of fuzzy relational modelling to industrial quality control", in: Chem. Eng. Technol., 18(5), 330-336, 1995.

Wang, X.Z., Yang, S.A., Veloso, E., Lu, M.L., McGreavy, C. [1995]:"Qualitative process modelling - a fuzzy signed directed graph method", in: Comput. Chem. Eng., 19, 735-740, 1995.

Wang, X. Z., Chen, B.H., Yang, S.H., McGreavy, C. [1996]:"Neural nets, fuzzy sets and digraphs in safety and operability of refinery reaction processes", in: Chem. Eng. Sci., 51(10), 2169-2178, 1996.

Wang X.Z., Chen B.H., Yang, S.H., McGreavy, C. [1997]:"Fuzzy Rule Generation from Data for Process Operational Decision Support", in: Comput. Chem. Eng., 21, 661-666, 1997.

Wilson, J.A., Martinez E.C. [1997]:"Neuro-fuzzy modeling and control of a batch process involving simultaneous reaction and distillation", in: Comput. Chem. Eng., 21, 1233-1238, 1997.

Zimmermann H.-J., Meier, W. [1994]:"Einsatz von Fuzzy Logic in der Analytischen Chemie", in: GIT, 38, 502-505, 1994.

II Medicine

10 FUZZY LOGIC AND POSSIBILITY THEORY IN BIOMEDICAL ENGINEERING

Kurt Becker

Abstract

Living biological systems are consisting of countless self organised structures and the underlying characteristics of their interaction is often not completely understood. Biomedical engineering can employ different modelling techniques to describe these complex systems in a generalised way. Most of the physiologic models today are defined quantitatively with techniques which were developed for linear systems and control theory.

Because of the complexity in biological systems, accurate mathematical models fail and the fuzzy approach offers a fully deterministic solution on a higher level of abstraction. The expert's knowledge of both the experienced physicians and the biomedical engineers is an important source of information for the design of intelligent machines.

Two applications where fuzzy sets are employed successfully are described. The first example comes from the field of intelligent real time monitoring in anaesthesia and supports the anaesthesiologist in his decision making process on the patient's haemodynamic state. The second example describes the implementation of a fuzzy controller for a total artificial heart (TAH). After these introductory examples some other applications from different medical fields which employ the fuzzy set theory are briefly discussed.

10.1 Introduction

Biomedical Engineering is an interdisciplinary discipline, in which engineering principles are applied to understand, modify, replace or control biologic systems, as well as design and manufacture products that can monitor physiologic functions and assist in the diagnosis and treatment of patients (Bronzino 1995).

Most of the physiologic models today are defined quantitatively with techniques which were developed for linear systems and control theory. These quantitative, analytical models can be used to improve the general understanding of functionality of the system and to simulate general dynamic system behaviour, but for the diagnosis of an individual patient's disease these classic quantitative methods are not sufficient. An adaptation of these models for diagnostic purposes in an individual patient is very difficult to perform, because there are inter- and intra- individual differences between patients and the most important state variables cannot be measured directly.

Applying their expert knowledge, experienced physicians diagnose the patient state inferring information and data from different sources. With the advent of modern general purpose computer systems this complex diagnostic task can be supported employing soft computing techniques like fuzzy logic and neural networks or a combination of both. The introduction of the fuzzy set theory provided a better framework to handle biological process states which up to then could not be described in a precision that is needed for conventional models (Pis and Mesiar, 1989).

The main objective to introduce computer based medical decision systems was to relieve the physician from "encyclopaedic" aspects of medicine and to enable him to concentrate on patient care. In the early seventies it was proposed, that automatic decision making would help to formalise the as yet informal rules of medical practice and also help disseminate rare and costly speciality expertise far beyond the narrow confines of academic medical centres to the general practice settings. Kulikowski presents a comprehensive overview on the evolution of artificial intelligence in medicine (Kulikowski 1995).

Once medical knowledge is programmed into a computer system the knowledge model can be used for different purposes. Fields in biomedical engineering where fuzzy logic is applied are data pre-processing, process diagnosis, process control and expert systems.

In the following, two example applications are described. The first example comes from the field of intelligent real time monitoring in anaesthesia and supports the anaesthesiologist in his decision making process on the patient's haemodynamic state. The second example describes the implementation of a fuzzy controller for a total artificial heart (TAH). After these introductory examples some other applications from different medical fields which employ the fuzzy set theory are briefly discussed.

10.2 Example I: An intelligent decision support- and alarm system for cardiac anaesthesia

The process of patient care during high invasive surgery requires fundamental knowledge of the physiologic processes and a long lasting experience in patient management to cope with the inter- individual variability of the patients. In order to reduce cognitive strain and to support the anaesthesiologist in intra- operative monitoring an online decision support system has been proposed (Schecke et al. 1991) and implemented (Becker et al. 1994) which presents an automatic evaluation of the patients haemodynamic state during cardiac surgery.

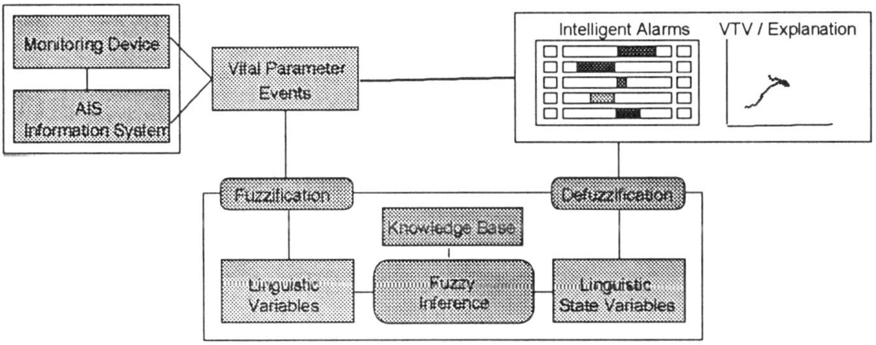

Figure 1 Structure of the intelligent decision support- and alarm system (Becker et al. 1994).

To derive an anaesthesiologic, mental model for the evaluation of the haemodynamic state of a patient undergoing cardiac surgery, the decision making process of the anaesthesiologist was analysed and the most important parameters and relations were identified.

The anaesthesiologist makes an estimation about the patient state recognising various invasively measured blood pressures, blood gases and visual impressions of the patient. A model of five so called haemodynamic state variables was defined:

- "blood volume"
- "myocardial contractility"
- "heart rate"
- "afterload" and
- "depth of anaesthesia"

The model serves as a basis for the anaesthesiologic treatment of the patient. For the estimation of these haemodynamic state variables the values of systolic arterial pressure "APsys", left atrial pressure "LAP", heart rate "HR", but also the knowledge about administered anaesthetic drugs are considered in the diagnostic process.

To automate the state variable evaluation a fuzzy logic approach was chosen. Figure 1 shows the structure of the intelligent decision support- and alarm system.

The mental state variable model can be expressed by qualitative evaluation rules based on the anaesthesiologist's experience. The rules do not consider exact values of some input parameters, like a systemic arterial pressure of 103 mmHg, but use more qualitative expressions like the „blood pressure of the patient is a little to high" to imitate the evaluation process of the anaesthesiologist.

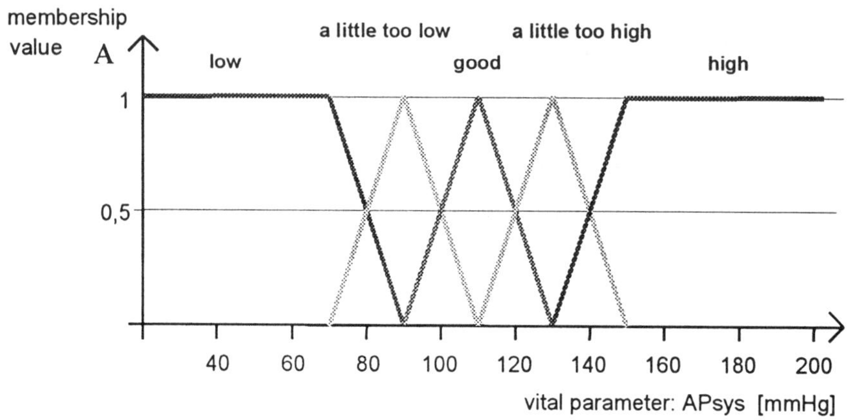

Figure 2 Definition of the membership functions for the linguistic variable "systolic arterial pressure" (Becker et al. 1994)

The input parameters are gathered online from the patient monitoring system and from the anaesthesia information system AIS (Klocke et al. 1986). These input parameters are transformed to the qualitative, linguistic level in a fuzzification procedure. As an example, figure 2 shows the definition of the membership functions for the linguistic variable systolic arterial pressure. This way of defining membership functions for the linguistic variable systolic arterial pressure was also used by Asbury and Tzabar (1995). Then, the input statements are aggregated in the fuzzy inference engine. The fuzzy inference rules were acquired from thirteen different anaesthesiologists in a standardised knowledge acquisition procedure (Becker et al., 1993). The knowledge base consists of 188 fuzzy rules. Two example rules are printed below:

- If LAP is too high and APsys is good then blood volume is too high (0,2)
- If LAP is too high and APsys is a little too high then blood volume is extremely high (0,8)

For inferencing the input parameters the fuzzy associative memory inference approach (von Altrock 1995) was used, where a degree of support is attached to each rule which stands for the plausibility of this rule in the knowledge base. The value of the degree of support is between 0 and 1, the 1 stands for full plausibility and the 0 indicates that the rule is not plausible. For representation of the linguistic "and" in the prerequisite section of the rules, the compensatory γ-operator and for aggregation of the rules the product operator is used (Zimmermann 1996). Defuzzification of the inference output is performed with the centre of area method.

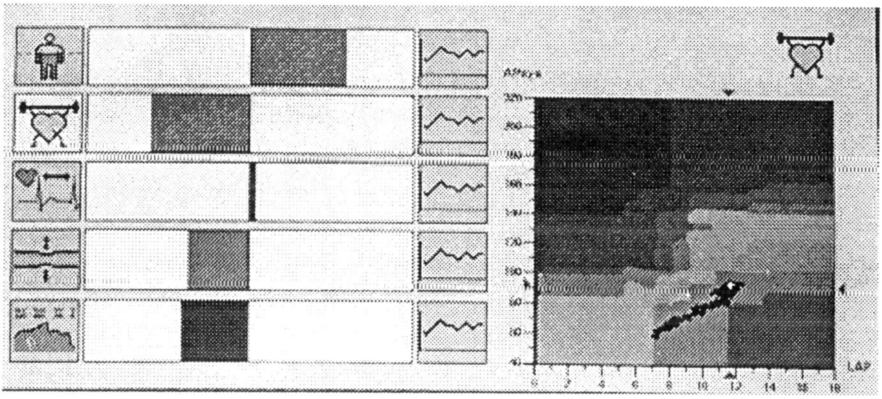

Figure 3 Screenshot of the user interface of the intelligent decision support- and alarm system. Left side: Profilogram; Right side: State plane plot and inference explanation component

A real time decision support system must not increase cognitive strain of the user, so the presentation of the inference resulted in a user interface which was adapted to the user and integrated into the system in an system ergonomic way. The user interface of the intelligent alarm system is divided into two sections (Figure 3).

In the left section the haemodynamic state variable evaluation is presented in a so called "profilogram" visualisation. The icons on the left side represent the state variables "blood volume", "myocardial contractility", "afterload", "heart rate" and "depth of anaesthesia". In a rectangular field besides each icon a colour coded bar graph shows the actual evaluation of the state variable. Bars to the left side of the centre line indicate a state variable becoming "too low"; in the same way a bar to the right indicates a state variable becoming "too high". When the length of the bar increases, the colour of the bar changes from green to signal red. The arrangement of these coloured bars in a profilogram form yields a structured presentation of the complex and dynamically changing information on the patients haemodynamic state variables. The colour coded, vertical profile enables the anaesthetist to catch the situation at a glance and to react instantaneous to dangerous trends in the patient's hemodynamics.

An explanation of the system inference can be accessed by touching and activating one of the state variable icons. Then a complete state plane visualisation of the activated state variable inference by means of the two most important input parameters LAP and APsys is presented. The black line represents the vital parameter trend of the last ten minutes. The underlying contour plot of the inference state plane is coloured similar to the profilogram. This visualisation allows the anaesthetist to comprehend the system evaluation and to predict the behaviour of the state variable after blood volume or drug application by a visual extrapolation of the trend curve.

The system must not make therapy suggestions. It is very important that the anaesthesiologist is kept in the loop and makes his therapy decisions by himself. Normally, the anaesthesiologist will have the same notion of the patient haemodynamic state as the intelligent alarm system. If the anaesthetist has another impression than the system the anaesthesiologist will check the patient state immediately and this will lead to a faster recognition of dangerous situations of the patient's haemodynamic state.

An online evaluation of the knowledge base of the system with real patient data from the operating room showed promising results (Becker 1996). The sensitivity of alarm recognition is 99,3% and the specificity is 66%, which is a very good performance compared to commercial available, conventional alarm systems for intensive care devices.

A functional prototype of the intelligent decision support- and alarm system was installed in the operation theatre. The system showed a very good performance during seven cardiac-surgical, seven neuro-surgical and three urological operations. 684 alarms were generated by the system and compared to the notion of two female and 10 male anaesthetists. Mean sensitivity and specificity of the alarm recognition were about 99% (Becker et al. 1997).

10.3 Example II: A fuzzy controller for a total artificial heart (TAH)

Modern artificial organs need to be controlled with sophisticated control strategies. The fuzzy control technique seems suitable especially for this application. Artificial heart assist devices and total artificial hearts are needed for patients, whose own poor heart performance cannot guarantee sufficient organ perfusion and where a biological transplant replacement is missing (bridge to transplant).

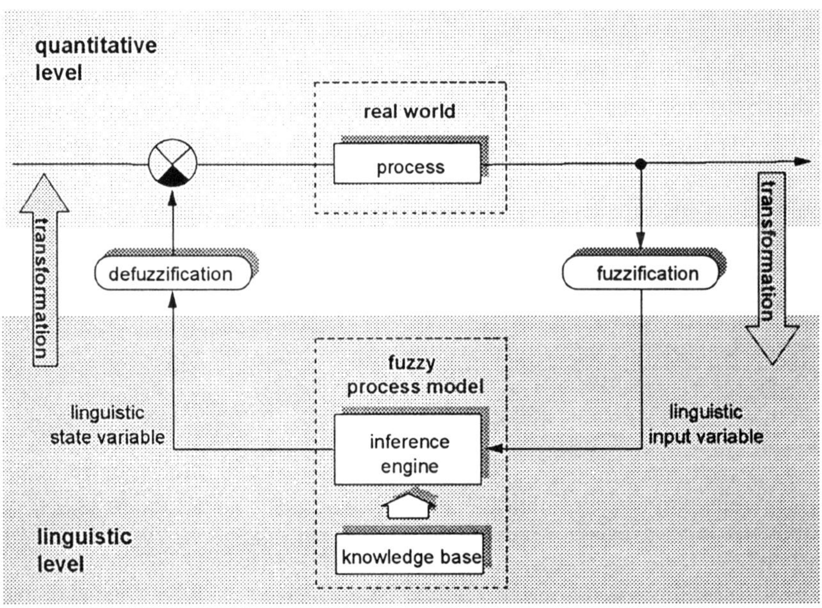

Figure 4 Control loop of the fuzzy total artificial heart (TAH) controller. Input variables are filling state of the left pump chamber and the current pump rate, output variable is the change of the pump rate. Input parameter aggregation: minimum operator; rule aggregation: product operator; defuzzification: centre of area method (Kaufmann et al. 1995)

A fuzzy controller for a total artificial heart (TAH) was developed at the Helmholtz-Institute for Biomedical Engineering (Helmholtz TAH) (Kaufmann et al. 1994). The most important requirement to any artificial heart control system is the automatic adaptation of the artificial heart's output to the perfusion demand, that means a timely adaptation of pump output to ensure sufficient organ perfusion. Further requirements are the protection of the organs from high peak pressures (Rau et al. 1995).

The control system uses an indirect method to detect left atrial pressure which controls pumping speed and output. Depending on the preload and the actual pump rate, the filling state of the heart is determined. The pump has three main filling states (Kaufmann et al. 1995). The "fast filling" state means, that the filling procedure of the chamber is completed long time before the blood output starts. Fast filling is not dangerous but cardiac output is not optimised (more blood is offered to the ventricle than can be ejected with this pump rate). "Slow filling" means, that the blood output starts before the chamber is filled up completely. This is a potentially dangerous state because the pusher-plate of the pump collides mechanically with the diaphragm. Collision can lead to a severe damage of the diaphragm and therefore this situation has to be detected early and prevented. In between these states is the "optimal filling" state where the output cycle starts immediately when the pump chamber is filled up. Discussing this condition, the heart rate is optimally adapted to the preload of the heart. The fuzzy controller has to ensure, that pump rate adaptation of the heart is maintained in the optimal filling state.

The shapes of the membership functions are adapted to the requirements of the control process. "To keep it simple", the functions are linearized in the transition sections between the membership degrees B = 0 and B = 1.

The fuzzy rulebase consists of 25 rules covering the whole state space of the fuzzy-controller. Three example rules are printed below:

- If filling is fast and pump-rate is good then turn pump-rate a little faster.
- If filling is optimal and pump-rate is good then keep pump-rate.
- If filling is slow and pump-rate is good then turn pump-rate slower.

The controller's dynamic behaviour can be influenced directly by the rule formulation. For example, the reaction on a slow filling event is always a large decrease in pump-rate, by contrast the reaction on a fast filling event is only a small increase in pump-rate. The current value for the change in the pump rate is determined by the MIN/PROD inference and the centre of area method for defuzzification.

In-vitro investigations in the laboratory using a TAH-labtype and an appropriate mock up circulation system showed the high efficiency of the fuzzy controller. During all experiments the adaptation of the controller to a preload change was stable and deterministic. An adaptation to preload change is achieved in 4 to 8 beats, which is sufficient. Increasing preload corresponds to the state of "fast filling" whereas a preload decrease corresponds to the dangerous "slow filling" state where the pusher plate impacts on the diaphragm. It is important to ensure, that the adaptation to a decreasing preload is very fast. The fuzzy controller fulfils this task in only 2 to 3 beats

10.4 Other applications of fuzzy logic and possibility theory in biomedical engineering

This section discusses further examples of applications from different medical fields, sorted by the topics: data pre-processing and interpretation, fuzzy control applications and expert systems. It is not intended to be a comprehensive and complete review of all biomedical applications employing fuzzy set theory.

10.4.1 Data pre-processing and classification systems

An important field for the application of fuzzy models in medicine is data pre-processing in order to filter and classify biological data or to enhance medical images. Modern patient monitoring systems like electrocardiographs (ECG), electroencephalographs (EEG) or similar devices produce a high amount of time series data of the patient. The data has to be interpreted by the physician and has to be checked for pathologic signs. This is a very time consuming task, because a lot of normal data has to be checked to find signs for abnormalities and evidence for illness of the patient. This task can be supported by modern pattern recognition systems which filter the data stream and present only those curves which contain hints for abnormalities. Since these systems must not miss any abnormal signal, these pattern recognition system should have a very high sensitivity. The data reduction and the grade of support for the physician is correlated with the specificity of the classification.

Very early applications using the fuzzy set theory for the task mentioned above were proposed by Kalmanson (1973), Kalmanson and Stegall (1975) and Hiramatsu et al. (1974). They tried fuzzy modelling techniques for the classification of aortic valve insufficiency and cardiovascular disease diagnosis. These early applications started a controversial discussion on the applicability of the fuzzy set theory in medical diagnosis (Fox, 1977) and led to a more fundamental research of human cognition and medical decision making processes (Wagner, 1980), (Cerutti and Timo, 1981), (Bartolin et al., 1982), (Zadeh, 1982).

Today, fuzzy models for data pre-processing and interpretation are widely accepted and many applications have been presented by different research groups (Bezdek and Pal 1992).

One of the main fields for pattern recognition system is ECG screening, fuzzy pattern recognition is used to reduce the amount of data to be reviewed and finally diagnosed by a physician as already discussed above (Degani and Bartolan, 1987), (Holzmann et al., 1990), (Cabello et al., 1991), (Degani, 1992), (Kweon et al., 1995), (Cohen et al. 1995). Similar approaches are presented to interpret and diagnose brainstem auditory evoked potentials (Brai et al., 1994) or to classify oesophageal motility records (Abou-Chadi et al., 1994). The performance of these systems is not significantly different regarding sensitivity and specificity compared to a human investigator.

Ruttkay and Riecansky (1994) present a pattern recognition system based on fuzzy set theory, that allows the recognition of five classes of ventricular activation out of instantaneously measured dipolar electrocardiotopographic images. This data preprocessing supports the physician in classifying various degrees of coronary artery stenosis. Another approach for the non-invasive detection of coronary artery disease (CAD) is described by Akay et al. (1994). They record the acoustic signals of the turbulent blood flow in partially occluded arteries and isolate the diastolic portions of the signal by synchronisation with the ECG-Signal. Diastolic heart sound recordings from 30 patients (15 normal, 15 abnormal) were selected as training cases for a fuzzy min-max neural network realised with the winner-take-all decision rule. This fuzzy neural net found 47 of 55 abnormal cases and 24 of 27 normal cases in a test series which is a good result for a non-invasive method.

Jaulent and Degoulet (1994) present a fuzzy classification procedure to classify visual types of renal artery lesions from radiological signs. This classification takes into account inter- and intra-observer variability in the perception of different categories. Weighted aggregation operations from the framework of possibility theory allow to take into account the different semantic levels in the defined classes. The authors plan to evaluate the objectiveness of the classification with a clinical case database.

Dove et al. (1994) describe a method for the online detection of endocardial and epicardial borders of the left ventricular walls that is used for ultra fast computed tomographic images of the heart. The method uses a fuzzy hough-transformation of the image data. A comparison of the automatically traced images with manually traced images showed a good correlation. With this method the physician is aided in the calculation of chamber volume and ejection fraction to evaluate the patient's heart performance.

Clark et al. (1994) employed a fuzzy c-means algorithm to segment and to label magnetic resonance image (MRI) volumes of the brain. This hybrid approach combines knowledge based techniques with unsupervised fuzzy clustering to detect tumour abnormalities. The segmentation system was tested with 17 (10 normal and 7 pathologic) volumes from different volunteers and patients. Normal and abnormal volumes were correctly identified but no quantitative comparison was performed.

Kuncheva (1990) presents a fuzzy multi level classifier to classify data with different classification methods and to combine the results to get an optimal result. In the special case of a fuzzy two level classifier (Kuncheva, 1994), different classification methods were used on the same data set in a first level and in the second level the classification results were aggregated by a fuzzy approach. Kuncheva compared the classification performance of this hybrid approach with different single level classification methods, i.e., linear discriminant analysis (LDA) and fuzzy k-nearest-neighbour (Fk-NN) in an experimental setting to predict cardiac disorders in pilots during centrifuge training. The results show that the fuzzy two level classifier has a higher classification accuracy compared with single level LDA and Fk-NN. Sittig et al. (1992) describe a clinically robust method for the detection and rejection of artefacts and trend detection in heart rate signals data based on fuzzy logic.

Arzi and Magnin (1989) presented a fuzzy set theoretic approach to the automatic analysis of nystagmic eye movements. In the conventional analysis of nystagmic eye movements, an operator has to distinguish between slow- and fast phases of eye movements before the data can be evaluated. To automate this process of data pre-processing a fuzzy clustering of slow- and fast phases is made. This approach permits a complete automation of experimental or clinical investigations with a high level of reliability. Even non trained operators are able to initialise the necessary parameters and to perform data acquisition and analysis.

Loslever et al. (1993) describe a multidimensional signal analysis procedure for human gait analysis. One step of this procedure was to represent the linear and non-linear relationships between corresponding time windows as fuzzy modalities. This data pre-processing makes it easier to perform the multiple correspondence analysis to get gait patterns which include both time and space aspects. By comparing these findings to normal gait patterns this method can be used to validate the rehabilitation of a particular patient.

10.4.2 Fuzzy control applications

In anaesthesia and intensive care medicine different research groups are working on closed loop drug delivery systems. Martin (1994) gives a short introduction to fuzzy set theory and fuzzy logic and discusses the clinical use of closed loop drug delivery controllers. He criticises the often poorly defined models of the input-output relationship between the pharmacocinetics of the drugs and the control variable. He comes to the conclusion that there are qualitatively defined systems in anaesthesiology where applications of fuzzy logic could be beneficial, such as controlling the depth of anaesthesia, safety shells and intelligent alerts and alarms.

Continuous application of a controlled amount of a vasoactive drug with an infusion pump causes a dilation or constriction of muscle cells that leads to a decrease or an increase of arterial and/or venous blood pressures. A comprehensive overview on the general application of automated drug delivery in intensive care is provided by Woodruff (1995). Fuzzy control systems for this kind of control task are described by Ying et al. (1988), Ying and Sheppard (1990), Borches et al. (1991), Ying et al. (1992), Ruiz et al. (1993), Ying and Sheppard (1994)(sodium nitroprusside), Fukui and Masuzawa (1989) (angiotensin II), Oshita et al. (1993), Oshita et al. (1994) (nicardipine).

The management of anaesthesia during non critical surgical procedures can be supported by similar approaches. Instead of vasoactive drugs, anaesthetic drugs are applied with only arterial pressure as a control variable (Tsutsui and Arita, 1994). Zbinden et al. (1995) found promising results in testing their feedback controller for anaesthetic depth by adjusting the concentration of inhalation anaesthetic gas (isoflourane) with fresh gas in 10 patients. Especially during skin incision the anaesthesiologist was outperformed by the fuzzy controller. For unstable patient states or during high invasive surgery these very simple process models have to be improved. The main problem of these approaches is, that the currently measured patient parameters are insufficient to determine the patients haemodynamic and respiratory state. The interaction between the patient's dynamical subsystems is very complex and such simple control strategies will only work in patients where most of these subsystems are stable.

Newer approaches consider not only one control variable like blood pressure but also cardiac output and adjust the infusion rate of two drugs, i.e. sodium nitroprusside as a vasodilator and dopamine to increase the contractility of the heart. Held and Roy (1995) tested the system in non-linear simulations and found a problem in controlling critical states of the model. For that reason they implemented an hierarchical control system, where the fuzzy controller is only employed in non critical control situations.

For high frequency ventilation Noshiro et al. (1994) implemented a simple fuzzy PID controller to control the rotational frequency of the ventilator depending on the end tidal pCO2 error and the change in the error. They came to the conclusion that the fuzzy PID controller is easier to implement than an conventional PID controller because no mathematical model of the dynamic characteristic of the ventilator is needed. They state that the fuzzy PID control method should be replaced by a more sophisticated fuzzy technique to produce better control characteristics.

The complex relationships during general anaesthesia are considered by Greenhow et al. (1993). They used a more complex fuzzy model to evaluate the anaesthetic state of the patient and to give dosage advice's to the anaesthesiologists for the inhaled volatile anaesthetics. During a clinical trial with seven patients most of the dosage advice's were followed by the anaesthesiologists but no quantitative evaluation of the system was performed.

The automated delivery of muscle relaxants (atracurium) in anaesthesia, controlled by a self organising fuzzy logic controller SOFLC, is described by Mason et al. (1994). Input parameter for the controller are evoked electromyography signals of the patient. The system was tested in simulation experiments and on patients which were not critically dependent on muscle relaxation. The anaesthetists were satisfied with the performance of the system.

Carollo et al. (1993) presented results of simulation experiments of a postoperative pain controller. Pain is quantified by the patient on a numerical scale with 10 pain levels and depending on the pain level an analgesic drug (alfentanil) is infused. The control task is subdivided into two phases. In the first phase a closed loop fuzzy controller brings the patient to a zero-pain state, in the second phase, an open loop continuos infusion controller maintains the analgesic concentration in plasma depending on the patient's input.

A nested loop controller with a fuzzy module to analyse the filling conditions of a centrifugal blood pump is presented by Schima et al. (1994). They compared different controller types, i.e. non-linear, fuzzy and neuronal approaches and found, that applying a fuzzy controller the pump performance can be increased by 20% compared to a conventional control theoretical approach.

Yoshizawa et al. (1994) use fuzzy clustering techniques to assess cardiovascular dynamics and to control ventricular assist devices (Yoshizawa et al., 1992). Other fields where the fuzzy control approach proved to be very sufficient are the process control of haemodialysis (Röher et al. 1995) and the rate control of cardiac pacemakers (Sugiura et al., 1991).

Another approach of intelligent real-time monitoring is introduced by Sedbrook et al. (1993). They presented a visualisation system which creates a decision support for the physician who has to diagnose acute respiratory infections. Based on the respiratory dynamics of a healthy patient, which are presented on a three dimensional animated display, the physician can interactively explore and refine a fuzzy cluster analysis on the patient data.

10.4.3 Expert systems

Using so called consultation expert systems, the physician enters data and information to be evaluated and the expert system generates diagnoses and may give therapeutic hints. Medical diagnosis is based on analyses of patient populations in terms of etiological and pathological subtypes in order to assist decision making of the individual patient and his disease. Conventional method use the all-or-nothing concept of cluster membership and are therefore insensitive to inter individual patient differences. On the basis of ageing patterns it was shown that fuzzy set-theoretic models are more sensitive to changes in subject level of functioning over time (Clive et al., 1983). Hudson and Cohen (1994) provide an overview on fuzzy logic in medical expert systems. Applications of this important application domain are described in the next chapter of this handbook.

10.5 Discussion

The discussed applications indicate that fuzzy logic and control become important tools for biomedical engineers to make modern medical equipment more intelligent and easy to use. All these systems rely on quantitative and qualitative process models of the biological subsystem in focus. Unlike in manmade systems like cars or aeroplanes, where all important state variables can be reported and controlled by sensors and actuators, biological process models can only describe certain state variables and relations that are needed for solving a particular problem. The second important point are the inter-individual differences between patients which can only be covered by a problem oriented modelling technique.

Because of the complexity in biological systems, accurate mathematical models fail and the fuzzy approach offers a fully deterministic solution on a higher level of abstraction. The expert's knowledge of both the experienced physicians and the biomedical engineers is an important source of information for the design of intelligent machines. Today sufficient, commercially available tools support the process of building and testing fuzzy models and therefore the time to come to a working solution is decreased by rapid prototyping and simulation techniques.

But even when the fuzzy approach seems so transparent and easy to use, like with all other scientific models, a fuzzy model is always a simplification of reality and it should be complex enough to imitate the demanded problem space. This has to be verified in living biological systems to finally test their performance. But on the other hand, the fuzzy model has to be as simple as possible to allow easy understanding, handling and working with it. Coming back to Martin (1994): "However, fuzzy logic, like any other scientific tool, should not be considered the holy grail. It should be applied only when, and where, it is appropriate".

10.6 References

Abou-Chadi-FE, Ezzat-FA, Sif-el-Din-AA: A fuzzy pattern recognition method to classify oesophageal motility records. Ann-Biomed-Eng, 1994 Jan-Feb, 22(1), 112-9

Akay-YM, Akay-M, Welkowitz-W, Kostis-J: Non-invasive detection of coronary artery disease. IEEE Eng Med Biol Magazine, 13(5), 1994, 761-4

Arzi-M, Magnin-M: A fuzzy set theoretical approach to automatic analysis of nystagmic eyemovements, IEEE-Trans-Biomed-Eng. 1989 Sep, 36(9), 954-63

Asbury-AJ, Tzabar-Y: Fuzzy logic: New ways of thinking for anaesthesia, Br-J-Anaesth, 1995, 75(1), 1-2

Bartolin-R, Bouvenot-G, Soula-G, Sanchez-E: The fuzzy set theory as a biomedical diagnostic aid (author's transl), Sem-Hop. 1982 Jun 3, 58(22), 1361-5

Becker-K, Kaesmacher-H, Juffernbruch-K, Rau-G, Kalff-G, ZimmermannHJ: Acquisition of a "fuzzy" knowledge base for an intelligent alarm system. Helmholtz-Institute for Biomedical Engineering, Aachen, Research Report 1991/92 8, 176-83, 1993.

Becker-K, Rau-G, Käsmacher-H, Petermeyer-M, Kalff-G, Zimmermann-HJ: Fuzzy logic approaches to intelligent alarms. IEEE Eng Med Biol Magazine, 13(5), 1994, 710-6

Becker-K: The employment of quantitative and qualitative models during implementation and validation of an intelligent decision support and alarm system for cardiac-anaesthesia. PhD-Thesis, RWTH-Aachen, Germany 1996, (in german)

Becker-K, Thull B.Rau-G, Käsmacher-Leidinger-H, Stemmer-J, Rau-G, Kalff-G, Zimmermann-H-J: Design and validation of an intelligent patient monitoring and alarm system based on a fuzzy logic process model. Artif. Intell. Med. 11, 1997, 33-53

Bezdek-JC, Pal-SK (eds.), Fuzzy models for pattern recognition - Methods that search for structures in data. IEEE Press, 1992

Borches-D, Ruiz-R, de-Miguel-E: Hypotension induced by sodium nitroprusside administered via an automatic-adaptive dose regulating system. Rev-Esp-Anestesiol-Reanim. 1991 Jan-Feb, 38(1), 3-7

Brai-A, Vibert-JF, Koutlidis-R: An expert system for the analysis and interpretation of evoked potentials based on fuzzy classification: application to brainstem auditory evoked potentials, Comput-Biomed-Res. 1994 Oct, 27(5), 351-66

Bronzino-JD (ed.), The biomedical engineering handbook. IEEE Press 1995

Cabello-D, Barro-S, Salceda-JM, Ruiz-R, Mira-J: Fuzzy K-nearest neighbor classifiers for ventricular arrhythmia detection., Int-J-Biomed-Comput. 1991 Feb, 27(2), 77-93

Carollo-A, Tobar-A, Hernandez-C: A rule-based postoperative pain controller: simulation results, Int-J-Biomed-Comput. 1993 Nov, 33(3-4), 267-76

Cerutti-S, Timo-Pieri-C: A method for the quantification of the decision-making process in a computer-oriented medical world, Int-J-Biomed-Comput. 1981 Jan, 12(1), 29-57

Clark-MC, Hall-MO, Golgof-DB, Clark-LP, Velthuizen-RP, Silbiger-MS: MRI segmentation using fuzzy clustering techniques. IEEE Eng Med Biol Magazine, 13(5), 1994, 730-42

Clive-J, Woodbury-MA, Siegler-IC: Fuzzy and crisp set-theoretic-based classification of health and disease. A qualitative and quantitative comparison, J-Med-Syst. 1983 Aug, 7(4), 317-32

Cohen-KP, Tompkins-WJ, Djohan-A, Webster-JG, Hu-YU: QRS detection using a fuzzy neural network. Proc 17th IEEE Conf Eng Med Biol, 1995

Darling-CB: Database technology for medical records, Instr-Course-Lect. 1992, 41, 521-6

Degani-R: Computerized electrocardiogram diagnosis fuzzy approach, Methods-Inf-Med. 1992 Nov, 31(4), 225-33

Degani-R, Bartolan-G: Fuzzy numbers in computerized electrocardiography. Fuzzy Sets Systems 1987, 24, 345-62

Dove-EL, Philip-K, Gotteiner-NL, Vonesh-MJ, Rumberger-JA, Reed-JE,Stanford-W, McPherson-DD, Chandran-KB: A method for automatic edge detection and volume computation of the leftventricle from ultrafast computed tomographic images, Invest-Radiol. 1994 Nov, 29(11), 945-54

Fox-J: Some observations on fuzzy diagnosis and medical computing, Int-J-Biomed-Comput. 1977 Oct, 8(4), 269-75

Fukui-Y, Masuzawa-T: Development of fuzzy blood pressure control system. Iyodenshi-To-Seitai-Kogaku. 1989 Jun, 27(2), 79-85

Greenhow-SG, Linkens-DA, Asbury-AJ: Pilot study of an expert system adviser for controlling general anaesthesia, Br-J-Anaesth. 1993 Sep, 71(3), 359-65

Held-CM, Roy-RJ: Multiple drug hemodynamic control by means of a supervisory-fuzzy rule-based adaptive control system: validation on a model, IEEE-Trans-Biomed-Eng. 1995 Apr, 42(4), 371-85

Hiramatsu-K, Kabasawa-K, Kaihara-S: Application of the fuzzy logic to medical diagnosis. Iyodenshi-To-Seitai-Kogaku. 1974 Jun, 12(3), 148-55

Holzmann-C, Hasseldieck-U, Rosselot-E, Estevez-P, Andrade-A, Acuna-G: Interpretation module for screening normal ECG. Med Prog Technol 1990, 16(3), 163-71

Hudson-DL, Cohen-ME: Fuzzy-logic in medical expert systems. IEEE Eng Med Biol Magazine, 13(5), 1994, 693-8

Jaulent-MC, Degoulet-P: Diagnosing renal atery lesions with a fuzzy logic model. IEEE Eng Med Biol Magazine, 13(5), 1994, 699-704

Kalmanson-D: Cardiovascular research and fuzzy sets theory. For an open policy inmedical research. Nouv-Presse-Med. 1973 Nov 17, 2(41), 2757-60

Kalmanson-D, Stegall-HF: Cardiovascular investigations and fuzzy sets theory, Am-J-Cardiol. 1975 Jan, 35(1), 80-4

Kaufmann-R, Reul-H, Rau-G: Electromechanical artificial heart with a new gear type and angled pump chambers. Int J Artif Organs, 1994, 8, 481-7

Kaufmann-R, Becker-K, Nix-C, Reul-H, Rau-G: Fuzzy control concept for a total artificial heart, Artif-Organs. 1995 Apr, 19(4), 355-61

Kulikowski-CA: History and development of artificial intelligence methods for medical decision making. in: Bronzino JD (ed.), The biomedical engineering handbook. IEEE Press 1995 2681-98

Kuncheva-LI: Fuzzy multi-level classifier for medical applications, Comput-Biol-Med. 1990, 20(6), 421-31

Kuncheva-LI: A fuzzy two-level classifier for high-g analysis. IEEE Eng Med Biol Magazine, 13(5), 1994, 717-23

Klocke-H, Trispel-S, Rau-G, Hatzky-U, Daub-D: An anesthesia information system for monitoring and record keeping during surgical anesthesia. J Clin Monit 1986, 2, 246-61

Kweon-HJ, Suk-JW, Song-JS, Lee-MH: Intelligent QRS typification using fuzzy clustering. Proc 17th IEEE Conf Eng Med Biol, 1995

Loslever-P: Error and data coding in the multi-dimensional analysis of human movement signals, Proc-Inst-Mech-Eng-H. 1993, 207(2), 103-10

Martin-JF: Fuzzy control in anesthesia [editorial, comment]: J-Clin-Monit. 1994 Mar, 10(2), 77-80

Mason-DC, Linkens-DA, Abbod-MF, Edwards-ND, Reilly CS: Automated delivery of muscle relaxants using fuzzy-logic control. IEEE Eng Med Biol Magazine, 13(5), 1994, 678-86

Noshiro-M, Matsunami-T, Takakuda-K, Ryumae-S, Kagawa-T, Shimizu-M, Fujino-T: Fuzzy and conventional control of high-frequency ventilation, Med-Biol-Eng-Comput. 1994 Jul, 32(4), 377-83

Oshita-S, Nakakimura-K, Kaieda-R, Murakawa-T, Tamura-H, Hiraoka-I: Application of the concept of fuzzy logistic controller for treatment of hypertension during anesthesia. Masui. 1993 Feb, 42(2), 185-9

Oshita-S, Nakakimura-K, Sakabe-T: Hypertension control during anesthesia. IEEE Eng Med Biol Magazine, 13(5), 1994, 667-70

Pis-P, Mesiar-R: Fuzzy model of inexact reasoning in medicine, Comput-Methods-Programs-Biomed. 1989 Sep, 30(1), 1-8

Rau-G, Becker-K, Kaufmann-R, Zimmermann-HJ: Fuzzy logic and control: principal approach and potential applications inmedicine, Artif Organs. 1995 Jan, 19(1), 105-12

Röher-O, Schmidt-R, Korth-S: Fuzzy-controlled drug infusion during extracorporal blood purification. Proc 3rd EUFIT, Verlag Mainz, Aachen, Germany, 1995, 1626-32

Ruiz-R, Borches-D, Gonzalez-A, Corral-J: A new sodium-nitroprusside-infusion controller for the regulation of arterial blood pressure, Biomed-Instrum-Technol. 1993 May-Jun, 27(3), 244-51

Ruttkay-Nedecky-I; Riecansky-I: Dipolar electrocardiotopographic evaluation of ventricular activation inpatients with various degrees of coronary artery disease. J-Electrocardiol 1994, 27(2), 149-55

Schecke-T, Langen-M, Popp-HJ, Rau-G, Kasmacher-H, Kalff-G: Knowledge-based decision support for patient monitoring in cardioanesthesia, Int-J-Clin-Monit-Comput. 1992, 9(1), 1-11

Schima-H, Trubel-W, Wieselthaler-G, Schmidt-C, Muller-MR, Siegl-H,Losert-U, Wolner-E: The Vienna implantable centrifugal blood pump, Artif-Organs. 1994 Jul, 18(7), 500-5

Sedbrook-TA, Wright-H, Wright-R: A visual fuzzy cluster system for patient analysis, Med-Inf-Lond. 1993 Oct-Dec, 18(4), 321-9

Sittig-DF, Cheung-KH, Berman-L: Fuzzy classification of hemodynamic trends and artifacts: experiments with the heart rate, Int-J-Clin-Monit-Comput. 1992 Dec, 9(4), 251-7

Sugiura-T, Mizushina-S, Kimura-M, Fukui-Y, Harada-Y: A fuzzy approach to the rate control in an artificial cardiac pacemaker regulated by respiratory rate and temperature: a preliminary report, J-Med-Eng-Technol. 1991 May-Jun, 15(3), 107-10

Tsutsui-T, Arita-S: Fuzzy-logic control of blood pressure through enflurane anesthesia. J-Clin-Monit. 1994 Mar, 10(2), 110-7

von Altrock 1995: Fuzzy logic & neurofuzzy applications explained. Prentice Hall, NJ, 1995

Wagner-W: "Fuzzy sets" as a formal model of cognitive structures - an overview. Arch-Psychol-Frankf. 1980, 133(2), 85-115

Woodruff-EA: Clinical care of patients with closed-loop drug delivery systems. in: Bronzino JD (ed.), The biomedical engineering handbook. IEEE Press 1995, 2447-58

Ying-H, McEachern-M, Eddleman-DW, Sheppard-LC: Fuzzy control of mean arterial pressure in postsurgical patients withsodium nitroprusside infusion, IEEE-Trans-Biomed-Eng. 1992 Oct, 39(10), 1060-70

Ying-H, Sheppard-L, Tucker-D: Expert-system-based fuzzy control of arterial pressure by drug infusion, Med-Prog-Technol. 1988, 13(4), 203-15

Ying-H, Sheppard-LC: Real-time expert-system-based fuzzy control of mean arterial pressure in pigs with sodium nitroprusside infusion, Med-Prog-Technol. 1990 May, 16(1-2), 69-76

Ying-H, Sheppard-LC: Regulating mean arterial pressure in postsurgical cardiac patients. IEEE Eng Med Biol Magazine, 13(5), 1994, 671-7

Yoshizawa-M, Takeda-H, Watanabe-T , Miura-M, Yambe-T, Kathira-Y, Nitta-S: An automatic control algorithm for the optimal driving of the ventricular assist device. IEEE Trans Biomed Eng, 39(3), 1992, 243-52

Yoshizawa-M, Takeda-H, Yambe-T, Nitta-S: Assessing cardiovascular dynamics during ventricular assistance. IEEE Eng Med Biol Magazine, 13(5), 1994, 687-92

Zadeh-LA: A note on prototype theory and fuzzy sets, Cognition. 1982 Nov, 12(3), 291-7

Zbinden-AM, Feigenwinter-P, Petersen-Felix-S, Hacisalihzade-S: Arterial pressure control with isoflurane using fuzzy logic, Br-J-Anaesth. 1995 Jan, 74(1), 66-72

Zimmermann-H-J: Fuzzy set theory and its applications, 3rd ed., Kluwer, 1996

11 APPROXIMATE REASONING IN COMPUTER-AIDED MEDICAL DECISION SYSTEMS

Jean-Christophe BUISSON

Abstract:

A survey on existing knowledge-based medical systems using approximate reasoning is presented. It is shown that the medical goals at which such systems aim at must be carrefully designed, in order to be accepted by the medical community, and that taking into account imprecision and uncertainty in facts and knowledge is a major ingredient for success. This survey shows a large diversity of knowlege representation and reasoning models. Particular medical and mathematical pitfalls are pointed out, among them the problem of representing directly symptoms and diseases with fuzzy sets, which leads to interpretation problems. It is shown that possibility distributions are a good means to express most medical notions. Following the tradition of MYCIN, most systems represent and propagate uncertainty, whereas imprecision is rarely dealt with when it should.

11.1 INTRODUCTION

The expert systems trend created by MYCIN (Shortliffe, 1976) provoqued a great interest in the medical community at the beginning of the 80's. A large number of projects were started, which were more or less aimed at modelling the diagnostic process of the different medical specialities, very often without considering the medical suitability of their approach.

At the end of the 80's, the expert system approach met with renewed criticism because of its inability to provide robust decision models which accurately represent real-world situations (Hudson, Cohen & Anderson, 1992). Most medical projects were abandoned, because of their poor intrinsic results as well as their inability to be integrated into a therapeutic process. Typically, the systems which were doing medical diagnoses from data interactively provided were doomed to a failure, when they were used in situations where real (and better) physicians were at hand (Fieschi et al., 1982). As for systems which were to be used in remote places (boats, etc.), they were faced with the difficult problem of facts elicitation from the user (which is an expert task in itself in medicine).

Most systems presented here have been around for years, and are or have been used routinely in real therapeutic situations. They bypassed the upmentionned difficulties by being a useful part of an organized medical process. Tey are taking into account the inherent imprecision and uncertainty of medical facts and laws, and therefore the robustness of their results is another important aspect of their success.

11.2 CRITICAL SURVEY OF SIGNIFICANT SYSTEMS

11.2.1 MYCIN, the pioneer

MYCIN is the first rule-based medical expert system ever created. It's been developed mainly between 1972 and 1978 at the Heuristic Programming Project of Stanford University (Buchanan & Shortliffe, 1984). It's an old system, no longer used, but its influence on contemporary programs is still visible. Its approximate reasoning model suffered many inconsistencies, but it dealt with the key problems of approximate reasoning, namely uncertainty representation and propagation, conflicts resolution, facts elicitation and knowledge acquisition.

MYCIN gives advice on blood infectious diseases. At the time, surveys showed that a large part of patients who required antimicribial therapy were given a drug that was inappropriate for their clinical condition. An improved therapy selection in patients requiring therapy is precisely the decision task with which MYCIN is designed to assist.

MYCIN works in four steps:

1. decide whether the patient has a significant infection

2. determine the likely identity of the offending organism

3. decide what drugs are apt to be effective against this organism

4. choose the drug that is most appropriate given the patient's clinical conditions

11.2.1.1 Facts representation. Through a specific vocabulary (*clinical parameters* in *contexts*), facts in MYCIN represent variable values on finite domains. A fact 'the patient is a child', as expressed verbally by the interaction modules of MYCIN, tells with certainty that the age of the patient has the value 'child', among the three of the domain {'child', 'teenager', adult'}. It must be noted these values correspond to the crisp intervals $[0,13]$, $(13,20]$ and $(20,150]$ respectively. In fact, the domain of a clinical parameter is always a finite set of values, and continuous domains such as age are cut into mutually exclusive parts.

MYCIN associates to each value *val* of a clinical parameter p a *certainty factor* $CF \in [0,1]$, with the following meaning:

- $CF = +1$: MYCIN is certain that *val* is the value of p
- $CF = -1$: MYCIN is certain that *val* is not the value of p
- $CF = c (c > 0)$: MYCIN estimates with the degree c that *val* is the value of p
- $CF = c (c < 0)$: MYCIN estimates with the degree c that *val* is not the value of p
- $CF = 0$: MYCIN does not have any idea about *val* being the value of p (initial state)

So the certainty of a fact (p, val) can be expressed, not only with true/false, but on a continuous scale. It is easy to express ignorance about a fact by using the certainty degree 0. But it is not possible to represent facts such as 'the patient is about 15 years old', since each domain is already partitionned into a finite number of predefined subdomains.

The certainty factor CF is the difference between two belief measures:

- MB $\in [0,1]$: measure of belief in the hypothesis that *val* is the value of p
- MD $\in [0,1]$: measure of disbelief in the hypothesis that *val* is the value of p
- min(MB,MD) = 0
- CF = MB - MD
- MB(p) = MD($\neg p$) in the binary case $\{p, \neg p\}$

For single-valued clinical parameters, as soon as one of the possible values is declared as certain (CF = +1), all other values are stored with CF = -1.

11.2.1.2 Rules representation and firing. Rules in MYCIN are of the form **if** *premise* **then** *action*, where an *action* may be of the form 'the value of the clinical parameter p is *val*, with the certainty c' where $c \in [-1,+1]$ represents an *attenuation coefficient* of the conclusion. On the 850 conclusion of the 440 rules of MYCIN in 1978, only 160 had an attenuation coefficient equals to +1 or -1.

Conjunctions (resp. disjunction) of conditions are evaluated as the minimum (resp. maximum) evaluation of each condition. Elementary conditions may have many forms, such as:

- KNOWN p: 1 if $CF_{max} > 0.2$, 0 otherwise

- NOTKNOWN p: 1 if $CF_{max} \leq 0.2$, 0 otherwise
- DEFINITE p: 1 if $CF_{max} = 1$, 0 otherwise
- ...

, where CF_{max} is the greatest certainty factor for each possible value of the clinical parameter p. A very important discontinuity is necessarily observed when the input data certainty changes gradually around the arbitrary threshold 0.2.

A rule is fired when the certainty CF_c of its condition part is found greater than the arbitrary threshold value 0.2. For each rule conclusion 'p is *val* with the certainty CF_r', MYCIN computes $CF = CF_c * CF_r$, which will assess the degree of certainty of the conclusion.

If two rules are candidates for firing, and lead to two certainty factors CF_1 and CF_2 for the same conclusion 'p is *val*', the aggregated certainty factor is:

- if $CF_1 * CF_2 \geq 0$ (rules not conflicting): $CF = CF_1 + CF_2 - CF_1 * CF_2$
- if $CF_1 * CF_2 < 0$ (rules conflicting): $CF = CF_1 + CF_2$

These *ad-hoc* formulas have been very much discussed afterwards. The conflicting case has been improved in 1979 by Van Melle (1979):

$$CF = \frac{CF_1 + CF_2}{1 + \min(|CF_1|, |CF_2|)} \tag{11.1}$$

It can be shown (Dubois & Prade, 1988a) that if (11.1) is used on a binary set $\{u, \overline{u}\}$, it becomes the possibilistic rule of combination with the product, which is associative:

$$\frac{\pi_1(t)\pi_2(t)}{\max_t \pi_1(t)\pi_2(t)} \tag{11.2}$$

where $\pi_i(u) = 1 - MB(\overline{u})$ and $\pi_i(\overline{u}) = 1$ if $MB(\overline{u}) > 0$

The blind combination mechanism may lead to indesirable conclusion, especially if rules having different levels of specificity are used in parallel (as in the example 'if p then q with CF=c' and 'if p and r then not q CF=c").

However, the main problems of certainty propagation as well as combination of conflicting and unconflicting information were correctly stated.

11.2.2 CADIAG-2: general internal medicine

11.2.2.1 Clinical issues. The central goal of the CADIAG-2 project (Adlassnig, 1990) is the development of a medical consultation system for general internal medicine. Its clinical issues are to assist a physician in the differential diagnostic process, by indicating all possible diseases which might be the cause of the patient's pathological symptoms, by proposing useful further examinations to confirm or exclude diagnostic hypotheses, and by indicating all the patient's symptoms not yet accounted for by the system's diagnoses. Its knowledge base is focused mainly on rheumatic and gastroenterological diseases, currently covering about 185 and 110 diseases, respectively, with a special emphasis on rare diseases (an other aspect of its usefulness). An evaluation of CADIAG-2 on 154 patients with rheumatoid arthritis gave an accuracy rate of 88.7% (Adlassnig, 1991).

The elicitation of data is greatly facilitated by the integration of CADIAG-2 into the medical information system of the Vienna General Hospital (Adlassnig, 1986). This integration allows the collection of the patient's findings for CADIAG-2 via the routine medical documentation and laboratory databases. In addition, patient data not routinely collected can be added to CADIAG-2 through a man-machine interface which processes medical terms given in natural language.

11.2.2.2 The concept of entity and its representation in CADIAG-2. CADIAG-2 deals with the following medical entities:

- the *patients*, noted P_q, whose set is: $\Pi = \{P_1, ..., P_r\}$
- the *symptoms*, noted S_i, the set of which is: $\Sigma = \{S_1, ..., S_m\}$
- the *diseases*, noted D_j, of the set: $\Delta = \{D_1, ..., D_n\}$
- the *symptoms combinations primary*, noted SCP_k, which are logical combinations (with 'and', 'or', and 'not') of S_i and D_j
- the *symptoms combinations secondary*, noted SCS_l, logical combinations of SCP_k

Relatively to a patient P_q, each entity ξ_i (symptom, disease, etc.) is considered as a *fuzzy subset.* ξ_i corresponds to the patient P_q, not in a yes or no way, but with graduality. The value μ_{ξ_i}, between 0 and 1, expresses the degree at which ξ_i is present for the patient P_q, where 0 denotes a complete absence and 1 a complete presence. CADIAG-2 considers this degree also as a particular value of a *fuzzy relation*, noted $R_{P\xi}$, between the set of patients Π and the set of the considered entities. For instance R_{PS} is the fuzzy relation on $\Pi \times \Sigma$ between the patients (P) and the possible symptoms (S), and a given symptom S_i is present for a given patient P_q with the degree $\mu_{R_{PS}}(P_q, S_i)$ (abbreviated into $\mu_{S_i} \in [0, 1]$).

Finally, $\mu_{\xi_i} =?$ denotes a complete lack of knowledge about the presence or absence of ξ_i for the patient (initial value). This value '?' is *ad-hoc*, and cannot be any value of $[0, 1]$. Actually, there is no easy way to represent ignorance with this formalism. We'll see later that representing ignorance is a special case of the more general problem of representing *disjunctions*, and that *possibility distributions* allow this in a straightforward manner.

When acquiring a new fact, either from the patient data base or interactively from the user, a *fuzzy interpreter* translates it into membership degrees for the associated entities. It should be noted that a value such as a laboratory result is taken and considered as certain and precisely known.

Most other symptoms are of a yes/no kind. An example of a patient's symptoms pattern given in Adlassnig & Kolarz (1982) is:

symptoms	**measured or present**	**fuzzy value** $\mu_p(S_i)$
fever	37.5 Celcius Degrees	1.00
elevated fever	idem	0.60
hydrops of the knee	yes	1.00
carditis	yes	1.00
joint pain	yes	1.00
erythema	yes	1.00
previous tonsillitis	yes	1.00
synovial fluid		
staphylococci	?	none
AST elevated	yes	1.00

'fever' and 'elevated fever' are connected to the same measured value of temperature (37.5). This value has been converted by the fuzzy interpreter into the two membership values 1.00 and 0.60. All other symptoms take their value in the domain yes/no, and we can see for 'synovial fluid exam. staphylococci' the *ad-hoc* value '?' to represent ignorance, with no associated membership value.

11.2.2.3 The problem of representing medical entities with fuzzy sets. 'fever' and 'elevated fever' are called 'symptoms' by Adlassnig. Actually the symptom is 'body temperature = 37.5', and fever' plays the role of a 'pattern', or a 'fuzzy condition'. With only the membership value 0.60 for 'elevated fever' for instance, we know that the information on temperature corresponds only partially to the fuzzy subset 'elevated fever', without knowing why. We cannot guess whether the input information was precise but uncertain, or imprecise/fuzzy and certain.

This problem is mainly due to the fact that the knowledge about the patient's temperature is:

1. only represented by membership degrees w.r.t. fixed fuzzy categories
2. splitted into separated entities (*fever* and *elevated fever* for the temperature)

11.2.2.4 The inference process in CADIAG-2. Inference knowledge in CADIAG-2 is expressed by means of fuzzy relation between two entities: relations S_i/D_j and SC_k/D_j for asserting or invalidating diagnoses, relations S_i/S_j and D_i/D_j for handling co-occuring diseases and controling the consistency. A relation between a symptom S_i and a diagnosis D_j for example, is represented with two fuzzy relations:

- the *occurence* of S_i in case of D_j, noted $R^o_{S_iD_j}$

- the *confirmability* of S_i for D_j, noted $R^c_{S_i D_j}$

For instance, Adlassnig & Kolarz (1982) gives the documented relations between two rheumatic diseases and the symptoms presented in 11.2.2.2:

	symptoms	**D1: rheumatic fever**		**D2: infectious arthritis**	
		$\mu_o(S_i, D_1)$	$\mu_c(S_i, D_1)$	$\mu_o(S_i, D_2)$	$\mu_c(S_i, D_2)$
S1:	fever	often		very often	
S2:	elevated fever	often		very often	
S3:	hydrops of the knee	seldom		seldom	
S4:	carditis	very often	often	almost never	
S5:	joint pain	almost always		almost always	
S6:	erythema	often		very seldom	
S7:	previous tonsillitis	very often		very seldom	
S8:	synovial fluid staphylococci	never	never	always	always
S9:	AST elevated	almost always		seldom	

To each qualifier is associated a number which represents it. For instance, to 'very often' is associated the range 64-90, the mean value of it being 77.

So, for the relation between S_4 and D_1, we have:

$\mu_{R^o_{SD}}(S_4, D_1) = 0.77$,
$\mu_{R^c_{SD}}(S_4, D_1) = 0.64$.

Then CADIAG-2 computes four quantities:

1. the *occurence indication* of D_j for the patient P_q:

$$\mu_{R^1_{PD}}(P_q, D_j) = \max_{S_i} min(\mu_{R_{PS}}(P_q, S_i), \mu_{R^o_{SD}}(S_i, D_j)) \qquad (11.3)$$

   This formula as the next three is calculated by means of fuzzy relations compositions (see, e.g., Dubois & Prade, 1980):

$$\mu_{R_{XY}}(u, w) = \sup_{v \in Y} min(\mu_{R_{XY}}(u, v), \mu_{R_{YZ}}(v, w))$$

2. the *confirmability indication* of D_j for the patient P_q:

$$\mu_{R^2_{PD}}(P_q, D_j) = \max_{S_i} min(\mu_{R_{PS}}(P_q, S_i), \mu_{R^c_{SD}}(S_i, D_j)) \qquad (11.4)$$

3. the *non-occurence indication* of D_j for the patient P_q:

$$\mu_{R^3_{PD}}(P_q, D_j) = \max_{S_i} min(\mu_{R_{PS}}(P_q, S_i), 1 - \mu_{R^o_{SD}}(S_i, D_j)) \qquad (11.5)$$

4. the *non-symptom indication* of D_j for the patient P_q:

$$\mu_{R^4_{PD}}(P_q, D_j) = \max_{S_i} min(1 - \mu_{R_{PS}}(P_q, S_i), \mu_{R^o_{SD}}(S_i, D_j)) \qquad (11.6)$$

In a similar manner, relation degrees are computed between symptom combinations and diseases, namely:

$$\mu_{R^5_{PD}}(P_q, D_j), \mu_{R^6_{PD}}(P_q, D_j), \mu_{R^7_{PD}}(P_q, D_j), \mu_{R^8_{PD}}(P_q, D_j) \qquad (11.7)$$

Symptom combinations being logical combinations of symptoms, these values are computed using the basic formulas:

$$\mu_{\xi_i \wedge \xi_j} = min(\mu_{\xi_i}, \mu_{\xi_j})$$
$$\mu_{\xi_i \vee \xi_j} = max(\mu_{\xi_i}, \mu_{\xi_j})$$
$$\mu_{not \xi_i} = 1 - \mu_{\xi_i}$$

The *confirmed diagnoses* are all D_j for which:

$\mu_{R_2}(P_q, D_j) = 1.00$ or $\mu_{R_6}(P_q, D_j) = 1.00$

All D_j are displayed as *excluded diagnosis* if either:

$\mu_{R_3}(P_q, D_j) = 1.00$ or $\mu_{R_4}(P_q, D_j) = 1.00$ or $\mu_{R_7}(P_q, D_j) = 1.00$ or $\mu_{R_8}(P_q, D_j) = 1.00$

Finally, all diagnoses D_j are *diagnostic hypotheses* if either:

$max(\mu_{R_1}(P_q, D_j), \mu_{R_2}(P_q, D_j)) > 0.50$ or
$max(\mu_{R_5}(P_q, D_j), \mu_{R_6}(P_q, D_j)) > 0.50$

For instance, in the example presented above, CADIAG-2 finds that D_1 and D_2 are not confirmed or excluded diagnoses, but rather are *diagnostic hypotheses*. It will propose to perform the synovial fluid examination, since finding S_8 would prove D_2 as the correct diagnosis.

11.2.3 NUTRI-EXPERT: diet monitoring

11.2.3.1 Introduction. Nutri-Expert is an educational software which has been designed in cooperation with the Diabetology Department of the Toulouse Hospitals. It helps patients to improve their nutritional habits, by analysing in detail their food intakes, and by suggesting changes that result in well balanced meals. Medical experiments made on more than 1500 subjects have shown that an unsupervised 6 month use of the software improved very significantly the patients' knowledge on nutrition and their cooking habits, and even improved significantly several physiological indicators such as blood glucose (Turnin et al., 1992).

Nutri-Expert can be used as a stand-alone micro-computer program, or as a Java client. It has been extensively used for six years, one reason of this success being that it is used directly by the patient, and that it performs a task (namely balancing daily meals) which no medical practitionner would want to make. Indeed, the physician prescribes the main points of the patient's nutrition plan, and the nutritionist gives advices and sometimes describes some typical meals, but the patient at home is alone to buy, cook and eat its own foods. Thus Nutri-Expert is situated at an important and previously vacant place into the therapeutic process.

Nutri-Expert is composed of several modules closely linked together. The main one is the analysis module, which allows the patient to analyse a particular meal and get it well balanced. It uses a custom made food composition database of 1550 foods, permanently updated by a pool of nutritionists. It gives the values of 32 nutriments, for all the common cooked and raw ordinary foods and the most common dishes in France. Databases adapted to other European countries as well as North America are in progress.

11.2.3.2 Dealing with imprecise data. There are two different sources of imprecision or fuzziness in Nutri-Expert. The first is the food composition database, where some values are imprecisely known, sometimes even completely unknown. For instance, there is only one entry for *apple* in the database although there are several species of apple on sale, which can all be at various stages of maturity. In this case, imprecision is a result of the averaging of several values. For other foods, the values of some nutriments have not been measured, and the nutritionists must still express what they know about it, even with much imprecision or fuzziness.

The second source of imprecision comes from the patient, when he must feed the analysis module with the weights of his meals' foods. If the weights of some packaged items are precisely known (yogurts, etc.), all other foods must be either weighted with a kitchen scale or described in terms of portions.

Contrary to most other systems, there is no uncertainty to deal with in Nutri-Expert. Values stored or provided are imprecise or fuzzy, but certain, and there is no inference process *per-se* which introduces uncertainty. Fuzzy arithmetic is used extensively, to perform computation as well as to compare values and perform pattern-matching.

11.2.3.3 Representation of imprecise or fuzzy numbers and intervals. *Fuzzy intervals* (Dubois & Prade, 1988a) are used in Nutri-Expert to represent numbers (the values describing the meal) as well as intervals (the norms the meal must satisfy).

Fuzzy intervals allow Nutri-Expert to be pessimistic and optimistic at the same time: the support of the interval is chosen large enough to be sure that no value is unduly excluded, and the core represents the most plausible values.

What is of primary importance is to determine the set of values which are completely impossible (for which the membership degree is equal to 0) and the set of values which are completely possible (for which the membership degree is equal to 1); the remaining subsets of the domain correspond to gradual transitions. Possibility theory is not very sensitive to slight variations of the possibility degrees; what really matters is that if a value is considered more possible than another one for a variable, then a greater possibility degree must be assigned to the former value.

In other words, the piecewise linear function between the core of the fuzzy interval and its support will cope with the desired robustness.

From a computational point of view, such intervals are modeled by 4-tuples (Dubois & Prade, 1988a)): (l, r, dl, dr) where $[l, r]$ is the core and $[l - dl, r + dr]$ is the support of the fuzzy interval.

11.2.3.4 Elicitation of data from the patient. Asking the patient anything about the imprecision of his foods' weights has seemed unrealistic. When a user gives a weight value, Nutri-Expert looks up in the food database if it is a packaged item or not. If so, the weight is assumed to be a precise number, represented by a fuzzy interval $(m, m, 0, 0)$. If it is not a packaged item, the program checks if the weight has been given as a number of portions or in grams. A number of portions is assumed to imply a greater imprecision. In both cases, the precise value given by the patient is transformed into a fuzzy value, using a transforming function associated to the food and the portion/gram choice, the name of which is stored in the food database. Presently there are 15 different such functions. For instance, the $f/10/20$ function is associated to the bread's weight when expressed as a number of slices, and transforms a weight x into the fuzzy interval $(x - 10\%, x + 10\%, x * 20\%, x * 20\%)$.

As for the values stored in the food composition database, they are all fuzzy numbers, and the nutritionists have been instructed how to deal with them. A value is stored as a precise number, along with a transforming function such as the above f/10/20; the set of transforming function has been elaborated by the nutritionists themselves. When a value field is left blank, it means that absolutely nothing is known on the value. It corresponds to a possibility distribution uniformly equals to 1, which can be implemented by the fuzzy interval $(-\infty, +\infty, +\infty, +\infty)$. For instance, it is often the case for calcium in commercial foods. The total amount of calcium in a meal containing such foods can still be estimated, and the resulting imprecision will affect its upper bound.

11.2.4 Fuzzy arithmetic

Nutri-Expert needs the four basic arithmetic operations to perform its computations. It has been proved that, for continuous monotonous functions, including arithmetic operations (on the subdomains where they are monotonic), we have:

$$(M * N)_\alpha = M_\alpha * N_\alpha = \{u * v | u \in M_\alpha, v \in N_\alpha\} \tag{11.8}$$

For multiplication and division, an approximation must be made, otherwise the result is no longer trapezoidal. The most important parts of the result, namely the core and the support, can be calculated exactly, and a straight line is drawn between them on both sides to get a trapezoid-shaped result again. This approximation will leave unchanged the order of the membership values in $]0, 1[$ (see Buisson (1997) for the detailed formulas).

11.2.5 *Fuzzy pattern matching*

Using fuzzy arithmetic as described above, the program is able to compute assessments on a set of nutriments, leading to a vector of data $D_1, D_2, ...D_n$. Each D_i is a fuzzy interval, which must be compared to a corresponding norm P_i, and the different pattern matching indices must be aggregated in order to assess to normality of the whole meal.

It has been established with the medical team that the different norms P_i are also fuzzy intervals. For instance, for an ordinary lunch and for all kinds of medical diseases, they told that "the fat percentage of the caloric goal must be approximately between 25% and 35%". They added that 20% and 40% where clearly out of the norm, so their statement could be translated into the following trapezoid-shaped fuzzy interval: $(5, 25, 35, 5)$.

Two scalar measures are used in Nutri-Expert in order to estimate the compatibility between a pattern P and a data D: a degree of possibility of matching $\Pi(P; D)$ and a degree of necessity of matching $N(P; D)$ which are respectively defined by (see Dubois & Prade, 1988a):

$$\Pi(P; D) = \sup_{u \in U} min(\mu_P(u), \mu_D(u)) \tag{11.9}$$

$$N(P; D) = \inf_{u \in U} max(\mu_P(u), 1 - \mu_D(u)) \tag{11.10}$$

$\Pi(P; D)$ is a *degree of overlapping* of the fuzzy set of values compatible with P, with the fuzzy set of values compatible with D. $N(P; D)$ is a *degree of inclusion* of the set of values compatible with D into the set of values compatible with P.

The computation of $\Pi(P; D)$ and $N(P; D)$ on trapezoid-shaped intervals is a matter of lines intersection (see Buisson (1997) for a complete description).

11.2.6 *Global matching evaluation*

A set of indices $\Pi(P_i; N_i)$ and $N(P_i; N_i)$ can be computed when matching separately each data of the set $D_1, D_2, ..., D_n$ with the corresponding pattern of the set $P_1, P_2, ..., P_n$. Each D_i and P_i were defined above as an assessment and its corresponding norm for each nutriment considered for the balance of the meal.

Since the involved variables are independent (i.e., the fuzzy set of values compatible with a variable does not depend on the value given to another variable), and if it is supposed that all parts of the pattern have an equal importance, an aggregation of the different measures is (see Dubois & Prade, 1988a):

$$\Pi(P_1 \times ... \times P_n; D_1 \times ... \times D_n) = \min_{i=1,n} \Pi(P_i; D_i) \tag{11.11}$$

$$N(P_1 \times ... \times P_n; D_1 \times ... \times D_n) = \min_{i=1,n} N(P_i; D_i) \tag{11.12}$$

where $\times$ denotes the Cartesian product defined for fuzzy sets by

$$\forall u_i \in U_i, \forall u_j \in U_j, \quad \mu_{P_i \times P_j}(u_i, u_j) = min(\mu_{P_i}(u_i), \mu_{P_j}(u_j)) \tag{11.13}$$

This aggregation using the 'min' operation preserves the respective semantics of possibility and necessity degrees.

Yet in the particular case of Nutri-Expert, the different variables have not an equal importance. According to the medical team, the fat, carbohydrate and protein percentage and the caloric intake are by far the most important variables. As for the others, it depends very much on the kind of meal and of the particular medical problems of the patient: some variable may be as important as the first four, and others may be completely unimportant.

Let $w_1, w_2, ..., w_n$ be the grade of importance of patterns $P_1, P_2, ..., P_n$ respectively. It is supposed that $\forall i, \ w_i \in [0, 1]$, the greater w_i the greater the importance of P_i; we also assume that $\max_{i=1,n} w_i = 1$ (normalization), i.e., the most important patterns are graded by 1. Then if s_i denotes a degree of matching of a datum (possibility or necessity) with respect to the atomic pattern P_i, the corresponding degree of matching s_i of this datum with respect to the whole pattern $(P_1, P_2, ..., P_n)$, taking into account the importance assessement, will be given by (see Dubois & Prade 1988a; Dubois, Prade & Testemale 1988b):

$$s = \min_{i=1,n} \max(1 - w_i, s_i) \tag{11.14}$$

Note that if all the w_i's are equal to 1 (equal importance), we get $s = \min_{i=1,n} s_i$; when $w_i = 0$, the matching degree with the pattern P_i is not taken into account.

The implementation in Nutri-Expert of this weighted fuzzy pattern matching technique has been quite simple, and takes place in two stages. A two dimensional array indicates for each kind of meal and for each nutriment the norm pattern to use, and its weight in the global matching. Then the program takes into account the possible medical problems of the patient in the following way. For each disease of the patient (he may have several) and for each nutriment, another two dimensional array specifies a new norm pattern and a new weight, which may override a previous constraint of the first stage.

Finally, a meal is said well balanced if the global pattern matching with the 'well balanced' norm patterns leads to $\Pi = 1$ and $N \geq 0.7$. Otherwise, Nutri-Expert makes a comment for each nutriment responsible for the unbalance. The threshold 0.7 has been empirically adjusted in relation with the medical group.

11.2.6.1 A heuristic search algorithm to balance a meal. After having pointed out the unbalanced nutriments, the main task of Nutri-Expert is to indicate to the patients how to modify the meal to get it well balanced, and adapted to their medical problems (for a diabetic for instance, the glucide intake must be in narrower ranges than for other people). But patients happen to be generally unable to get a meal well balanced by themselves. The task is indeed a difficult one, because modifying the weight of a food often leads to unbalance several nutriments at the same time.

The problem may seem linear, and solvable using classical techniques, but several points make it impossible. First the linear equations to solve have fuzzy terms. Second, the food weights to compute must be compatible with food portions. Third, some foods are forbidden for some people; others have to be replaced. The problem is much more of an operational kind, and the notion of *possible actions on a meal* has emerged as central. These operations can be parametered by tables, setting average quantities, portions and accommodations according to diseases. Then the meal entered by the patient with these rules generates a space of derived meals where the solution is to be found.

Heuristic search is a well known method to cleverly visit such a state space in order to find a solution that matches the constraints (Nilsson, 1980). The search graph is only potentially defined, in the sense that only the root is explicitly available, together with a set of rules which specify how to build the successors of a current node.

Using applicable rules on some node ("the father") creates new nodes ("the sons"), together with an arc from the father to each son. This is called *node expansion*. Each arc is valuated by a cost which is supposedly a positive number. The cost of a path is the sum of the costs of its arcs. An optimal solution corresponds to a minimal cost path from the root to a goal node. The tree search methodology consists in applying rules, where possible, to nodes until a goal node is reached. As long as a goal node is not reached, the main problem is to select the proper node to be expanded.

This choice can be driven by the knowledge of a so-called *evaluation function* f so that its value, $f(n)$, at ant node n estimates the sum of the cost from the root to node n (denoted $g(n)$) and the minimal cost from node n to a goal state (denoted $h(n)$). Let $h^*(n)$ be the cost of an optimal path from n to t. Upper and lower bounds of $f^*(n)$ are denoted $\overline{f}(n)$ and $\underline{f}(n)$ respectively. An upper bound of $f^*(n)$ is the cost of the corresponding path and can be used to reduce the width of the search graph by pruning pending (= not expanded yet) nodes n' such that $\overline{f}(n) \leq \underline{f}(n')$. An estimate $f(n)$ which is not established as an upper or lower bound is called *heuristic*.

If $h(n) \leq h^*(n)$ then such an algorithm is called A^* by Nilsson (1980). It terminates in a finite number of steps, and provides an optimal path (exit 5) or such a path does not exist (exit 3). The algorithm is then said to be *admissible*. When $h(n)$ is only heuristic, then the optimality of a solution discovered by the algorithm is no longer guaranteed. However, the use of a heuristic evaluation may enable such a solution to be discovered faster than using a bad lower bound $\underline{h}(n)$ (for instance, $\underline{h}(n) = 0$ only produces a uniform-cost algorithm).

Dubois, Farreny and Prade (1987) presented an extension of the basic A^* algorithm which tries to synthetize all the available information concerning the value of $h^*(n)$ (in particular the knowledge of a lower bound $\underline{h}(n)$), with a fuzzy interval, in order to allow for a more elaborate reordering of the potential nodes to expand.

11.2.6.2 Choice of an heuristic. The first version of the balancing algorithm used the following forward estimation function (Buisson, 1997):

$$h(meal) = \frac{\sum_{i=1,p}(1 - N_i)}{2} \tag{11.15}$$

where N_i is the matching necessity degree of the hypothetical meal n with the norm pattern of nutriment i, for the p nutriments involved for the patient. This formula is based on the experimentally observed fact that a randomely chosen action performed on a randomly chosen meal gives way to two nutriment mismatches, on average.

$f(n)$ is clearly heuristic, and the algorithm is not of the A^* kind. From a set of about 3480 real patients' meals, this first algorithm found a solution for about 2510 meals. 'finding a solution' here means finding a node which satisfy the constraints after less than 1000 trials. The main problem of this heuristic is that it characterizes how much wrong is the meal rather than it estimates the cost to the closest solution.

Recently, a new forward estimation has been used (Buisson & Garel, 1998):

$$h(meal) = \max_{n \in nutriments} \Delta_n(n, goal(n)) \tag{11.16}$$

where $\Delta_n(n, goal(n))$ estimates the minimum cost, in number of operations, it demands for improving the nutriment n, in other words, the number of operations it takes to have this nutriment normalized using the better food for it.

The 'max' in (11.16) is a 'fuzzy max' of the fuzzy interval Δ_n (Dubois & Prade, 1988a). Such an operation can be very hard to work out completly.

A simplified version of this new heuristic has recently been implemented in Nutri-Expert (Buisson & Garel, 1998). From the test database of 3480 meals, 3060 meals are now corrected. Most of the 420 remaining meals have a minimal cost to improve them so high that it is very difficult to reach the solution within the 1000 trials.

11.2.7 RENOIR/MILORD

RENOIR (Belmonte-Serrano et al., 1994) aids the nonspecialized physician in diagnosing arthritis and collagen diseases. It is a rule-based expert-system which has been programmed using the MILORD environment (Godo et al., 1989), and its knowledge base contains more than 1000 rules. Uncertainty is modeled in MILORD by fuzzy intervals representing the meaning of linguistic certainty values. These labels are a closed set of choices for the user to express his confidence in the data he is introducing in the system. They are preferred to scalar uncertainty factors, the authors arguing that most physicians are unable to make a fair estimation of the inaccuracy of their judgments, making far larger estimation errors that the boundaries accepted by themselves as feasible. All computations are based on possibility theory, and many compatibility degrees are computed off-line, since the number of combinaisons of linguistic labels is finite.

Here is an example of an inference process with MILORD:

- IF
  1. Community acquired pneumonia is *almost sure*
  2. Bacterial disease is *possible*
  3. (No aspiration) is *very possible*
- THEN Enterobacteria is *quite possible*

Observed facts:

1'. Community acquired pneumonia is *very possible*

2'. Bacterial disease is *almost sure*

3'. Aspiration is *slightly possible*

Inference steps:

a. Compatibility between (1) and (1') gives: *moderately possible*

b. Compatibility between (2) and (2') gives: *almost sure*

c. Negation of (3'): (no aspiration) is *very possible*

d. Compatibility between (3) and (c) gives: *possible*

e. [(a) and (b) and (d)] gives: *moderately possible*

f. Inference: (e) and the rule value (*possible*) gives: *moderately possible*

g. Composition between (f) and the conclusion label (*quite possible*) gives: **possible**

11.2.8 Cardioanaesthesia monitoring

An example of fuzzy rule based systems oriented towards mediacl process control rather than diagnosis is provided by an alarm system in cardioanaesthesia (Becker et al., 1993, 1994, 1997) which helps the anesthesist to manage the stabilisation of the patient's haemodynamic state during open-heart surgery. The system works on-line, gathering all required data from a general anaesthesia information system. It supports monitoring of five haemodynamic 'state variables', which are estimated by the intelligent alarm system using fuzzy rules. There are approximately 50 rules for each 'state variable', which have been acquired from 13 anaesthesists.

The input and output parameters of the alarm system are transformed into 'linguistic variables' and their terms corresponding to their linguistic use in anaesthesia. The terms are 'too low', 'a little too low', 'good', 'a little too high', 'too high'. The fuzzification transforms the vital parameters into linguistic variables using these membership functions. Rules are then used to compute a result, and a defuzzification into a crisp output value is performed by the 'center of gravity' method. The five state values are displayed as colored bars on a monitoring equipment situated in front of the anaesthesist.

An example of a rule is: "if systolic arterial pressure is a little too high, and left atrial pressure is too low, then myocardial contractility is good". To the rule is also associated a value of support between 0 and 1.

The acquisition of the membership functions used during the fuzzification step has been made by summing up for each fuzzy label the ranges given by ten experienced anesthesists, each having a weight of 1/10.

11.2.9 Other Systems

11.2.9.1 PROTIS and SPHINX. The rule-based system PROTIS (Soula, Vialettes & San Marco, 1983; Soula & Sanchez, 1982) gives therapeutic advice to nonspecialized physicians about non insulo-dependent diabetic patients.

A rule in PROTIS has the following form:

FR: $< condition > \rightarrow < decision >$ (e, r)

$< decision >$ is an element in a finite set of possible decisions; e and r are respectively called *suggestion degree* and *rejection degree*, and have the following meaning:

- the more e is near 1 and the more $< condition >$ is satisfied, the more $< decision >$ is confirmed
- the more r is near 1 and the more $< condition >$ is satisfied, the more $< decision >$ is rejected

Rule conditions are a combination of elementary conditions, using the $\wedge$ and $\vee$ connectors. Elementary conditions are represented by possibility distributions, among three classes of predefined shapes. Facts are also represented by possibility distributions of a parabolic shape, modelling imprecision and fuzziness, but not uncertainty.

A fuzzy pattern matching (Cayrol, Farreny & Prade, 1980) is made between known facts and elementary conditions, leading to possibility and necessity degrees which are combined using min and max for conjunctions and disjunctions of conditions respectively (so no compensation or differences in relative importance are allowed).

For a rule 'FR: $c \rightarrow d(e, r)$' whose condition c is satisfied with the possibility and necessity degrees P and N respectively, PROTIS computes the two following indices:

$\alpha = \max(0, P - (1 - e))$: evocation degree of d by c

$\beta = \min(1, N + (1 - r))$: rejection degree of d by c

When several rules give several α_i and β_i for the same decision d, they are combined into $\alpha = \max_i \alpha_j$ and $\beta = \min_i \beta_i$. Like MYCIN, this kind of blind combination may lead to confusing results for conflicting or non independent rules.

There is only one level of rules; after the inference process, PROTIS displays for the physician the possible therapeutic decisions with their associated evocation and rejection degrees.

An interesting feature in PROTIS is the presence of meta-rules which allow the translation of condition possibility distributions in particular situations, or the modify the weights e and r. It permits for instance to take into account the exceptions of a rule by forcing to (0,1) the weights (e, r).

The rule-based system SPHINX (Fieschi et al., 1982; Soula & Sanchez, 1982) has been applied to diabetes and icterus. Rules in SPHINX have the same form and meaning as in PROTIS. The inference process is almost exactly the same, leading to evocation and rejection degrees computed as above, with the difference that P and N are replaced by a single number t, called *semantic conformity degree*, computed in an *ad-hoc* manner.

11.2.9.2 AUDIGON. AUDIGON (Weierich, Wetze & Niemann, 1993) makes a diagnosis of osteaoarthrosis, based on MR-images of the knee. The knee cartilage is segmented in small regions, and 80 membership values CF from 15 different features are automatically extracted from the image, for each region. The diagnostic process can then be started from this symbolic description.

It is implemented using fuzzy rules with a min-max inference. Here is an example of a rule and its use:

certainty factor("segment-grade-1-damaged") =
min (CF(slightly contrasted),
1 - CF(little thickness),
CF(surface very smooth),
CF(surrounding intensity values slightly higher),
CF(small contrast in the surrounding))

The certainty factors are computed for each segment and inserted into three-dimensional maps.

11.2.9.3 Ultasonic prenatal examination and digital mammography. These are applications where programs work from medical images, and try to detect or reject abnormal measurements.

For prenatal examination, a simple fuzzy inference (Smith & Arabshahi, 1996) has been implemented for classifying a foetus as normal, slightly abnormal or abnormal, based on measurements of head circumference, abdominal circumference, and femur length.

In mammography analysis (Kovalerchuk et al., 1997; Bothorel, Bouchon Meunier & Muller, 1997), the meaning of terms such as 'size of undulations', 'deepness of lobularity' and others are redefined as fuzzy subsets. The related primitives are automatically extracted from a mammogram, and a mass found is classified as lobulated or microlobulated, by combining the various membership degrees using the 'min' operator.

11.2.9.4 Cancer radiotherapy. Another type of applications in cancer radiotherapy aims at defining the target-volume to be irridiated (Waschek, 1996). This volume must be chosen large enough to enclose the whole tumor including all the microscopic tumor cell spread. On the other hand, the surrounding healthy tissues have to be spared as much as possible.

Diagnostic imaging techniques are not sensitive and specific enough to visualize all individual tumor cells infiltrating the normal tissues beyond the boundaries of the visible gross tumor mass. Due to this lack of information, different interpretations among radiotherapists are justifiable concerning the extent of the planning target volume (PTV) for certain tumor types. This can be interpreted as fuzziness of the target volume.

Medical images are segmented into *voxels* (smallest volume elements), and an automatic estimatation for each voxel of the value of TCP (Tumor Control Probability) and NTCP (Normal Tissue Complication Probability) is made. A NTCP of 5% is considered as the maximum acceptable complication risk, whereas TCP must be as near 1 as possible. These two values are transformed into linguistic fuzzy labels, and fuzzy rules using the max-min inference are applied in order to estimate the membership degree of the considered voxel to PTV.

The center of gravity defuzzification with symmetrically extended outer sets is then performed, and the results are displayed using 3D-plots.

11.2.9.5 Diagnostic of diabetes using glucose tolerance test. Some systems have an output which is a matter of degree of intensity (rather than an output with a certainty level). This is the case of the system (Arita, Yoneda & Hori, 1993) which considers diabetes as a fuzzy concept, and proposes a diagnostic process using a fuzzy inference from data gathered during the glucose tolerance test. This test consists of measuring blood glucose 6 times, every 30 minutes after a 75g of dextrol intake. For a given time t_k (k: 0, 30, 60, 90, 120, 180), a set of rules of the kind "if the level of BG is comparatively large and the IRI is comparatively small, then the patient is diagnosed as moderately diabetic".

11.2.9.6 EMERGE. EMERGE (Hudson & Cohen, 1986) is an expert system for the analysis of patient complaints in the emergency room environment. The system has been successfully evaluated on 179 retrospective cases and 100 prospective cases. Each symptom, sign or test result may be present to a certain degree, modeled as a membership value into a fuzzy subset. Each elementary condition may have a particular weight in the antecedent of a rule. A conjunction is evaluated as:

$$\sum_{i=1}^{n} w_i * c_i \tag{11.17}$$

where w_i is the weight of a condition and c_i is its degree of presence. So a relative compensation is made between conditions, contrary to the standard use of 'min'. The firing of a rule is determined by the degree of presence s of its antecedent exceeding a threshold. The value s becomes the certainty factor associated with that rule.

11.2.9.7 Diagnostic interpretation of electrocardiograms. A electrocardiogram interpretation has been performed, first based on medical rules and on automatically extracted signal features (Bortolan & Degani, 1988). The imprecision in data and rule antecedents and the uncertainty of rule conclusions are managed in the framework of the fuzzy set theory, and a 'best fit' method is used to linguistically express fuzzy and uncertain results with words used in the cardiology community.

Recently, the same diagnostic interpretation has been made using a neural network with a fuzzy preprocessing stage (Silipo, Bortolan & Marchesi, 1995). A study based on a large validation database of ECG signals has been performed, and has shown a very good diagnostic accuracy.

11.2.9.8 Diagnosis of Nephropathia Epidemica. An architecture for a generic tool which supports data analysis and development of diagnostic modules in clinical medicine has been developed (Eklund, Fogström & Forsström, 1992). It uses a fuzzy neural net architecture with only one layer, which is fed with data from a preprocessing fuzzifying module. This approach has been applied to the diagnosis of Nephropathia Epidemica, and proved the importance of the preprocessing stage.

11.2.9.9 Closed-loop delivery of muscle relaxant. A model-based controller for the closed-loop delivery of a muscle relaxant has been realised, relying on a model of the process under control (Kern, Johnson & Westenskow, 1997). The resulting system has been shown to perform well under simulated and real conditions.

11.3 CONCLUDING DISCUSSION

Most medical facts, from symptoms to conclusions, are related to variables whose domains are continous. Often when the domain is reduced to a set of finite values, an oversimplification is made. In a previous work (Buisson et al., 1987), we had been working on the diagnosis of diabetes and its classification in two classical categories, namely *insulino-dependent diabetes* and *non insulino-dependent diabetes.* It appeared later than this dichotomy was more blurred than we had thought, and that each type of diabetes was indeed a fuzzy category.

MYCIN dealt only with finite sets of values for its clinical parameters, and it leads unavoidably to sudden changes of conclusions when input data are modified continuously. However, it allowed a simple and efficient representation of uncertainty, and even ignorance.

CADIAG-2 deals only with continuous variables, for symptoms as well as diagnoses. But they are represented and stored as a single membership value for a set of predefined fuzzy entities, and a lot of information is lost in this process. Moreover, the model does not permit to represent ignorance, and *ad-hoc* means have to be used.

Thorough comparisons have been made in (Daniel, Hajek & Nguyen, 1997; Dubois, Lang & Prade, 1991) between CADIAG-2 and MYCIN. Daniel, Hajek & Nguyen (1997) have shown how CADIAG-2 can be embedded into MYCIN-like systems, and proposed an approach of inclusion of negative knowledge into CADIAG-2.

As advocated in Buisson, Farreny & Prade (1985), possibility distributions can store all kinds of information available on a variable, in a simple and computationaly efficient way. Imprecision, fuzziness, uncertainty and even ignorance can be represented in a straightforward way, as well as pieces of information which are both imprecise and uncertain: 'X is A is certain to the degree $1 - \epsilon$'

$$\pi_X(u) = max(\mu_A(u), \epsilon) \tag{11.18}$$

where ϵ stands for the level of possibility of being outside of A

When dealing with a conjunction of independent conditions, formula (11.11) should be used, or (11.14) if the conditions have not an equal importance. However, we have seen with the EMERGE system that sometimes, physicians allow a compensation between conditions, and then an additive combinaison such as (11.17) must be used.

The possibilistic framework also offers representation capabilities for expressing different kinds of rules including purely gradual rules (of the form 'the more X is A, the more Y is B'), and rules whose condition parts may also be a matter of degree, but whose conclusions are pervaded with uncertainty (Dubois & Prade, 1996), as well as nonmonotonic reasoning capabilities for managing rules with exceptions having different levels of specificity (Ben Ferhat, Dubois & Prade, 1998).

In all the reviewed systems, fuzzy set theory, even when used in a rather elementary way, improved significantly the quality of the programs, by taking into account the inherent uncertainty and imprecision of both data and medical knowledge. Thus the results can change gradually with data, and there are no more discontinuity near thresholds or when reaching the border of a class of values (Nutri-Expert). Uncertainty can be elicited in a qualitative way, and qualitative results can also be obtained (MILORD, CADIAG-2, etc.). Finally, many medical norms which are crisply defined can be flexibly expressed in a more natural way (e.g. in cardioanesthesia alarms, prenatal examinations, mammography analysis).

The interest of researchers in knowledge-based systems in medicine has apparently somewhat declined in the last ten years, whereas mathematical frameworks such as fuzzy sets and possibility theory are now ready to handle problems in a sound and efficient way. In a near future, computerized patients' records will be available at the physicians's desks, and this will create a need for new developments.

References

Adlassnig K.-P., Kolarz G. (1982). CADIAG-2: Computer-assisted medical diagnosis using fuzzy subsets. In: M.M. Gupta and E. Sanchez (Eds.) *Approximate Reasoning in Decision Analysis*, North-Holland Publishing Company, 219-247.

Adlassnig K.-P. (1982). A survey on medical diagnosis and fuzzy subsets. In: M.M. Gupta and E. Sanchez (Eds.) *Approximate Reasoning in Decision Analysis*, North-Holland Publishing Company, 203-217.

Adlassnig K.-P. (1986). Fuzzy set theory in medical diagnosis. *IEEE Transactions on Systems, Man and Cybernetics* SMC-16, pp. 260-265.

Adlassnig K.-P. (1990). Update on CADIAG-2: a fuzzy medical expert system for general internal medicine. In: W. H. Janko et al. (Eds.) *Progress in Fuzzy Sets ans Systems*, Kluwer Academic Publishers, Netherlands, pp. 1-6.

Adlassnig K.-P. (1991). Knowledge acquisition study and accuracy rate evaluation for CADIAG-2/RHEUMA with 308 clinical cases, *Proc. Med. Inform. Eur. 1991*, Springer-Verlag, Vienna, pp. 332-336.

Arita S., Yoneda M., Hori Y. (1993). Supporting system for the diagnosis of diabetes mellitus based on glucose tolerance test responses using a fuzzy inference. In: R. Lowen and M. Roubens (Eds.) *Fuzzy Logic*, Kluwer Academic Publishers, Netherlands, pp. 301-310.

Becker K., Käsmacher H., Juffernbruch K., Rau G., Kalff G., Zimmermann H.-J. (1993). An intelligent alarm system for decision-support in cardioanesthesia: knowledge base and user interface. *Proc. First European Congress on Fuzzy and Intelligent Technologies (EUFIT'93)*, Aachen, Sept. 7-10, pp. 1023-1026.

Becker K., Käsmacher H., Rau G., Kalff G., Zimmermann H.-J. (1994). A fuzzy logic approach to intelligent alarms in cardioanesthesia. *Proc. Int. Conf. on Fuzzy Systems IEEE 1994*, pp. 2072-2076.

Becker K., Thull B., Käsmacher-Leidinger H., Stemmer J., Rau G., Kalff G., Zimmermann H.-J. (1997). Design and validation of an intelligent patient monitoring and alarm system based on a fuzzy logic process model. *Artificial Intelligence in Medicine*, 11, pp. 33-53.

Belmonte-Serrano M., Sierra C., López de Mántaras R. (1994). RENOIR: An expert system using fuzzy logic for rheumatology diagnosis. *International Journal of Intelligent Systems*, 9, pp. 985-1000.

Ben Ferhat S., Dubois, D., Prade H. (1998). Practical handling of exception-tainted rules and independence information in possibilistic logic. *Applied Intelligence*, 9, pp. 101-127.

Bortolan G., Degani R. (1988). Linguistic approximation of fuzzy certainty factors in computerized electrocardiography. In: M.M. Gupta, T. Yamakawa (Eds.) *Fuzzy Computing*, North-Holland, pp. 243-261.

Bothorel S., Bouchon Meunier B., Muller S. (1997). A fuzzy logic based approach for semiological analysis of microcalcifications in mammographic images. *International Journal of Intelligent Systems*, 12, pp. 819-848.

Buchanan G. G., Shortliffe E. H. (1984). *Rule-based expert systems. The MYCIN experiments of the Stanford Heuristic Programming Project*, Addison-Wesley, Reading.

Buisson J.-C., Farreny H., Prade H. (1985). The development of a medical expert system and the treatment of imprecision in the framework of possibility theory. *Information Sciences*, 37, pp. 211-226.

Buisson J.-C., Farreny H., Prade H., Turnin M.C., Tauber J.P., Bayard F. (1987). Toulmed, an inference engine which deals with imprecise and uncertain aspects of medical knowledge. *Proc. European Conference on Artificial Intelligence in Medicine (AIME 87)*, In: J. Fox, M. Fieschi, R. Engelbrecht (Eds.) *Lecture Notes in Medical Informatics*, 33, Springer-Verlag, pp. 123-140.

Buisson J.-C. (1997). Nutri-Expert, an educational software in nutrition. *International Journal of Intelligent Systems*, 12, 915-933.

Buisson J.-C., Garel A. (1998). Balancing meals using fuzzy arithmetic and heuristic search algorithms, Int. Rep. IRIT.

Cayrol M., Farreny H., Prade H. (1980). Fuzzy pattern matching. *Kybernetes*, 11, pp. 103-116.

Daniel M., Hajek P., Nguyen P.H. (1997). CADIAG-2 and MYCIN-like systems. *Artificial Intelligence in Medicine*, 9, pp. 241-259.

Dubois D., Prade H. (1980). *Fuzzy sets and systems: theory and applications*, Academic Press, New-York.

Dubois D., Farreny H., Prade H. (1987). Combinatorial search with fuzzy estimates. In J. Kacprzyk, S.A. Orlovski (eds.), *Optimization Models Using Fuzzy Sets and Possibility Theory*, pp. 171-185.

Dubois D., Prade H. (1988a). (with the collaboration of H. Farreny, R. Martin-Clouaire, C. Testemale). *Possibility theory - An approach to the computerized processing of information.* Plenum Press, New-York.

Dubois D., Prade H. (1988b). Representation and combinaison of uncertainty with belief functions and possibility measures. *Comput. Intell.*, 4, pp. 244-264.

Dubois D., Prade H. Testemale C. (1988c). Weighted fuzzy pattern matching, *Fuzzy Sets and Systems*, 28, pp. 313-331.

Dubois D., Lang J., Prade H. (1991). Fuzzy sets in approximate reasoning, Part 2: Logical approaches. *Fuzzy Sets and Systems*, 40, 203-244.

Dubois D., Prade H. (1996). What are fuzzy rules and how to use them. *Fuzzy Sets and Systems*, 84, pp. 169-185.

Eklund P., Fogström M., Forsström J. (1992). A generic neuro-fuzzy tool for developing medical decision support. Proceeding of the MEPP'92, Abo Akademi Press, Abo, pp. 11-27.

Fieschi M., Joubert M., Fieschi D., Soula G., Roux M. (1982). Sphinx: an interactive system for medical diagnosis aids. In: M.M. Gupta and E. Sanchez (Eds.) *Approximate Reasoning in Decision Analysis*, North-Holland.

Godo L., López de Mántaras R., Sierra C, Verdaguer A. (1989). MILORD: The architecture and the management of linguistically expressed uncertainty. *International Journal of Intelligent Systems.*, vol. 4, pp. 471-501.

Hudson D. L., Cohen M. E. (1986). Management of uncertainty in a medical expert system. In: B. Bouchon and R.R. Yager (Eds.) *Uncertainty in Knowledge-Based Systems.*, Springer-Verlag, pp. 283-293.

Hudson D. L., Cohen M. E., Anderson M. F. (1992). Approximate reasoning with if-then-else rules in a medical expert system, *International Journal of Intelligent Systems*, 7, pp. 71-79.

Kern S.E., Johnson J.O., Westenskow D.R. (1997). Fuzzy logic for model adaptation of a pharmacokinetic-based closed loop delivery system for pancuronium. *Artificial Intelligence in Medicine*, 11, pp. 9-31.

Kovalerchuk B., Triantaphyllou E., Ruiz J. F., Clayton J. (1997). Fuzzy logic in computer-aided breast cancer diagnosis: analysis of lobulation. *Artificial Intelligence in Medicine*, 11, pp. 75-85.

Nilsson, N. (1980). *Principles of artificial intelligence.* Tioga Pub. Co., Palo Alto, Ca.

Shortliffe E. H. (1976). *Computer-based medical consultation, MYCIN*, Elsevier/ North-Holland, New-York, 1976.

Sanchez E. (1998). Fuzzy logic and inflammatory protein variations. *Clinica Chimica Acta*, 270, pp. 31-42.

Silipo R., Bortolan G., Marchesi C. (1995). Fuzzy proprocessing and artificial neural networks classification for the diagnostic interpretation of resting

ECG. In: A. Murray, R. Arzbaecher (Eds.) *Computers in Cardiology 95*, IEEE Computer Society, Los Alamitos, pp. 365-368.

Soula G., Sanchez E. (1982). Soft deduction rules in medical diagnosis processes. In: M.M. Gupta and E. Sanchez (Eds.) *Approximate Reasoning in Decision Analysis*, North-Holland, pp. 77-88.

Soula G., Vialettes B., San Marco J.L. (1983). PROTIS, a fuzzy deduction-rule system: application to the treatment of diabetes. *Proc. MEDINFO 83*, Amsterdam, pp. 533-255.

Smith B. J., Arabshahi P. (1996). A fuzzy decision system for ultrasonic prenatal examination enhancement. *Proc. Int. Conf. on Fuzzy Systems IEEE 1996*, pp. 1712-1717.

Steinmann F. (1997). Fuzzy set theory in medicine. *Artificial Intelligence in Medicine*, 11, pp. 1-7.

Turnin M.C., Beddok R., Clottes J, Abadie R., Martini P., Buisson J.C., Soule-Dupuy C., Bayard F., Tauber J.P. (1992). Telematic expert system Diabeto : a new tool for diet self monitoring for diabetic patients. *Diabetes Care*, 15 (2).

Van Melle W. (1979). A domain-independant production-rule system for consultation programs. Proc. IVth IJCAI Tokyo, Aug. 1979, 923-925.

Waschek T., Levegrun S., Schlegel W., Van Kampen M., Engenhart-Cabillic R. (1996). Target volume definition for three-dimensional radiotherapy of cancer patients with a fuzzy rule based system. *Proc. Int. Conf. on Fuzzy Systems IEEE 1996*, pp. 1719-1725.

Weierich P., Wetzel D., Niemann H. (1993). AUDIGON - A medical expert system with new judgement functions for the fuzzy rulebase. *Proc. First European Congress on Fuzzy and Intelligent Technologies (EUFIT'93)*, Aachen, Sept. 7-10, pp. 472-477.

Zadeh L.A. (1978). Fuzzy sets as a basis for a theory of possibility. *Fuzzy Sets and Systems*, 1 (1), pp. 3-28.

Zimmermann H.-J., Zysno P. (1980). Latent connectives in human decision making. *Fuzzy Sets and Systems*, 4, pp. 37-51.

12 IMAGE PROCESSING IN MEDICINE

James C. Bezdek
Melanie A. Sutton

Abstract. Image processing covers a lot of territory, including its use as an enabling technology for the more ambitious undertaking which is nowadays called *computer vision* (Jain et al., 1995). A comprehensive survey of fuzzy models for image processing and computer vision would require its own volume (cf. Chapter 5 of Bezdek et al., 1999). This chapter is confined to applications of image processing in medical domains that are *not based on neural network models or mathematical morphology.* Others chapters of this handbook address these topics. Even these strict constraints leave us in the unenviable position of having to choose from more excellent work than we have space to review, and the perhaps inexcusable position of being ignorant about some work of which we are simply unaware. With apologies to the many authors of papers in either category, we offer a snapshot of the use of fuzzy models for image processing in computational medicine.

12.1 MEDICAL IMAGE PROCESSING

Image processing in medicine is very different, more specialized, and in some sense less utilized than it is in most application domains that involve imaging sensors. To appreciate this, we quote from the very readable paper by Silvana Dellipiane (1997), who offers a concise and realistic assessment of the current relationship between image processing and medicine:

"Apart from basic image processing tools (e.g., histogram modification (windowing), reslicing and cutting or 3D volumes), and some graphical tools (e.g. polygon and ellipse drawing), very few results obtained by image processing research have so far been exploited in medical practice."

This quote emphasizes the wide gap between the use of a processed image to find, for example, a wrench in a bin picking environment, and a brain tumor for preoperative surgical planning. Interpretation and efficient utilization of medical imagery demands specialized skills, including knowledge of sensor peculiarities, anatomy, pathology, and even potential treatments. Consequently, information possessed by medical images is almost always extracted by a human expert who is a practicing clinician in some specialized area of medicine.

Figure 12.1 illustrates the areas most frequently discussed in connection with fuzzy image processing in medicine. The modules (with their acronyms) in Figure 12.1 are the areas which we will focus on in this chapter.

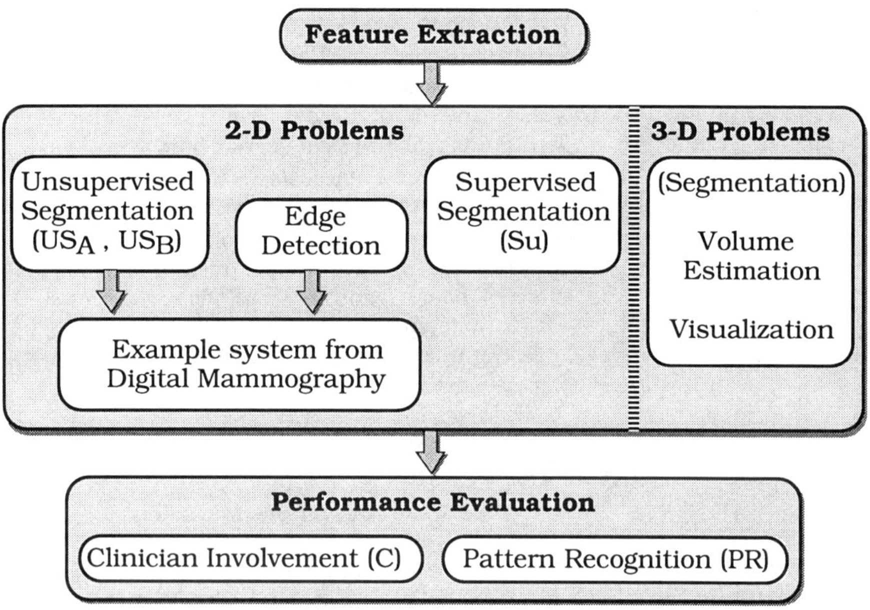

Figure 12.1 Typical modules in medical image analysis

We begin in the next section with a brief review of numerical pattern recognition. Section 12.3 discusses pattern recognition methods and clinician involvement as used to evaluate automated medical analysis systems. Sections 12.4-12.5 discuss specifics of the feature extraction and segmentation processes when applied to the medical domain. We do not provide a separate section for edge detection, which is sometimes presented as a segmentation technique, but more often appears in conjunction with region based segmentation. In Sections 12.6-12.8, we review current research efforts involving unsupervised and supervised segmentation of medical images with two spatial dimensions (*2D images*) in applications as diverse as cardiology (e.g., to detect left and right ventricles), electron microscopy (e.g., to analyze E coli), brain tissue analysis (e.g., to detect pathology), and AIDS research (e.g., to detect HIV-positive lesions).

In Sections 12.9-12.11 we focus on the increasingly important domain of digital mammography. We address the complexity of design decisions faced by researchers in this field and the diversity of electronically available mammographic databases. Section 12.11 concludes our discussion of mammography with an overview of a typical mammographic image processing system which uses fuzzy modules for edge detection and segmentation.

Section 12.12 discusses fuzzy models for problems that - at least in principle - involve images with three spatial dimensions (*3D images*). Most of these models are aimed towards just two 3D applications: visualization and (segmentation for) volume estimation, both of which can be used for surgical planning and therapy. We lump together two distinct groups of 3D models in this section: "real" 3D models that process voxel data directly, and what Dellipiane (1997) calls "pseudo 3D" models, i.e., those by which a 3D volume is (re) constructed by processing computationally independent 2D slices and binding their segmentations by some system of interslice constraints. Finally, we offer some conclusions and possible topics for future research in Section 12.13.

12.2 NUMERICAL PATTERN RECOGNITION

This section quickly reviews some basic ideas and notations that are needed for our chapter. The meaning of various parameters for models, algorithms and functions is discussed at length in Bezdek et al., (1999), and will not be repeated here. Notations used here are exactly the same as notations for the corresponding items in that volume.

There are four types of *class labels* - crisp, fuzzy, probabilistic and possibilistic - that can be associated with pixels or windows in an image. Let integer c denote the number of classes, $1 < c < n$. Vector $\mathbf{e}_i$, with elements $e_{ij} = \{1, i = j; 0, i \neq j\}$, is the *crisp label vector* for class i, $1 \le i \le c$. The vector $\mathbf{y} = (0.1, 0.6, 0.3)^T$ is a constrained label vector for c = 3 classes; its entries lie between 0 and 1, and sum to 1. If $\mathbf{y}$ is a label vector for some $\mathbf{x} \in \Re^p$ generated by, say, the fuzzy c-means model, we call $\mathbf{y}$ a *fuzzy label* for $\mathbf{x}$. If $\mathbf{y}$ came from a method such as maximum likelihood estimation in mixture decomposition, $\mathbf{y}$ would be a *probabilistic label*. Vectors such as $\mathbf{z} = (0.7, 0.2, 0.7)^T \in [0, 1]^3 - \{\mathbf{0}\}$ are called *possibilistic label vectors* for c = 3 classes. Possibilistic labels are produced, e.g., by possibilistic clustering algorithms (Krishnapuram and Keller, 1993). Extensions of these ideas to any number of classes is straightforward.

Image data are represented here as sets of *feature vectors* $X = \{\mathbf{x}_1, \mathbf{x}_2, ..., \mathbf{x}_n\}$ in *feature space* $\Re^p$, where p is the number of measurements associated with each spatial location (pixel, voxel or window) in the image. Don't confuse p, the dimension of a feature space associated to X with the spatial dimension of X itself. Thus, we may have 7D feature vectors for 2D images, 2D feature vectors for 3D images, etc. If the spatial location of a feature vector is important, we write $\mathbf{x}_{ij}$ for the feature vector associated with pixel address (i,j) in a 2D image. Feature vector $\mathbf{x}_{ij}$ may or may not include the coordinate values (i, j) as elements. We discuss the derivation of features more thoroughly in Section 12.4.

When X is unlabeled, the assignment of label vectors to its elements is called *clustering* (or unsupervised learning). Sets of (cn) values $\{u_{ik}\}$ that are arrayed as a (cxn) matrix $U = [\mathbf{U}_1 \ldots \mathbf{U}_k \ldots \mathbf{U}_n] = [u_{ik}]$, where $\mathbf{U}_k$ denotes the k-th column of U, are called *c-partitions* of X. U is a *possibilistic, fuzzy, probabilistic* or *crisp c-partition* of X when each of its columns is a label vector of the corresponding type. If U is crisp or fuzzy, u_{ik} is the *membership* of $\mathbf{x}_k$ in the i-th fuzzy cluster of X.

Clustering algorithms are functions $\mathbf{C}: X \mapsto \mathcal{R}_\mathbf{C}$, where $\mathcal{R}_\mathbf{C}$ is the range of $\mathbf{C}$. The output of $\mathbf{C}$ can be *just* a partition, but many clustering algorithms produce outputs besides partitions. The most common example is a second set of parameters called *point prototypes* (or cluster centers) $\mathbf{V} = \{\mathbf{v}_1, \ldots, \mathbf{v}_c\}, \mathbf{v}_i \in \Re^p \; \forall i$.

Since definite class assignments (tissue types) for each pixel or window are the usual goal in image processing, non-crisp labels $\mathbf{y}$ are often transformed into crisp labels $\mathbf{H}(\mathbf{y})$ using the conversion or *hardening* function

$$\mathbf{H}(\mathbf{y}) = \mathbf{e}_i \Leftrightarrow \left\|\mathbf{y} - \mathbf{e}_i\right\| < \left\|\mathbf{y} - \mathbf{e}_j\right\| \Leftrightarrow y_i > y_j \quad ; \; j \neq i. \tag{1}$$

In equation (1), $\|*\|$ is the Euclidean norm, $\|\mathbf{y} - \mathbf{e}_i\| = \sqrt{(\mathbf{y} - \mathbf{e}_i)^T (\mathbf{y} - \mathbf{e}_i)}$, on $\Re^c$, and ties are broken arbitrarily. $\mathbf{H}$ finds the crisp label vector $\mathbf{e}_i$ closest to $\mathbf{y}$ by finding the *maximum coordinate* of $\mathbf{y}$, and assigning the corresponding crisp label to the object $\mathbf{z}$ that $\mathbf{y}$ labels. When $\mathbf{y}$ is a fuzzy label, $\mathbf{H}(\mathbf{y})$ is called *defuzzification* of $\mathbf{y}$ by the maximum membership principle.

Let $\boldsymbol{\mathcal{P}} = \{U_i : 1 \leq i \leq N\}$ denote N different partitions (with or without extra parameters such as $\mathbf{V}$) of a fixed unlabeled data set X that may arise as a result of one of the following:

(i) clustering X with one algorithm **C** at various values of c ; or
(ii) clustering X over other algorithmic parameters of **C** ; or
(iii) applying $\{\mathbf{C}_i\}$ to X, each with various parameters ; or
(iv) all of the above .

Cluster validity is the study (selection or rejection) of which $U_i \in \boldsymbol{P}$ best represents the data in some well defined sense. Visual examination (e.g., shown as track US_A in Figure 12.5) of an algorithmically suggested, artificially colored structure in an image is termed *human validation*. In this case, a clinically knowledgeable operator inspects, say, a segmentation of a medical image, and either rejects it or accepts it and assigns a physical label to each region (cluster). This approach can be successful in terms of labeling the segmented image correctly only if the operator can imagine tissue structure as it must be in the data. Since X is not labeled, the "true" substructure in the data is unknown. Thus, human validation is subjective and to some extent non-repeatable. Nonetheless, this method is extremely important, as clinicians historically trust human judgment more than computational evidence.

Another possibility is to instead use *rule-based validation* (e.g., as shown in track US_B in Figure 12.5), in which clinical knowledge is encapsulated in the form of a rule-based system. This circumvents the need for a human on-line, and probably represents the best hope at present to realize a truly unsupervised medical assistant.

A third strategy - *mathematical validation* - uses a *validity functional* $\mathcal{V}: D_{\mathcal{V}} \mapsto \Re$ to rank $U_i \in \boldsymbol{P}$ (e.g., as shown in the top box of the left hand side of Figure 12.5). The domain $D_{\mathcal{V}}$ of $\mathcal{V}$ is usually (but not necessarily) chosen to match the range of **C**, $D_{\mathcal{V}} = R_{\mathbf{C}}$. The dashed horizontal line in the top clustering box in Figure 12.5 (track US_A) indicates that some researchers involved in image segmentation don't bother with $\mathcal{V}$, preferring instead to rely on either a human or a rule base to decide how good the segmentation is. At the other extreme, some investigators not only use $\mathcal{V}$, but try to incorporate it into the control of **C**, in which case the resultant algorithm is said to possess *dynamic* (direct or indirect) cluster validity - the automatic adjustment of c and therefore U during the operation of the clustering algorithm.

The "true" parameters of a model that represents substructure in any data set are unknown. Consequently, validity functionals have little chance of being generally useful for identifying a "best" solution. More typically, $\mathcal{V}$ is used instead to eliminate *badly wrong* solutions. Thus, validity functionals are usually used <u>prior</u> to validation by humans or rule bases.

A *classifier* is any function **D** that carries real p-vectors to one of the three label sets (cf. Bezdek et al., (1999)). The value of **D** at $\mathbf{z} \in \Re^p$, $\mathbf{y} = \mathbf{D}(\mathbf{z})$, is the *label vector* for **z**. **D** is a *crisp classifier* if and only for any $\mathbf{z} \in \Re^p$ there is an i so that $\mathbf{D}(\mathbf{z}) = \mathbf{e}_i$. Training **D** means finding its parameters and is called *supervised learning.*

Supervised methods need (usually crisp) labels for at least some of the feature vectors in each subclass of X. X is often crisply partitioned into a *training set* X_L, $n_L = \sum_{i=1}^{c} n_i = |X_L|$ and a *test set* X_T, $n_T = |X_T|$, $X = X_L \cup X_T$, , $X_L \cap X_T = \emptyset$ and $n = |X| = n_L + n_T$. When the data in X have crisp labels, X has the following form:

$$X = \left\{ \overbrace{\underbrace{\mathbf{x}_1^1, \ldots, \mathbf{x}_{n_1}^1}_{\text{labeled } 1}, \ldots, \underbrace{\mathbf{x}_1^c, \ldots, \mathbf{x}_{n_c}^c}_{\text{labeled } c}}^{X_L} \;\middle|\; \overbrace{\underbrace{\mathbf{x}_1^{(i)} \ldots, \mathbf{x}_{n_T}^{(k)}}_{\text{labeled}}}^{X_T} \right\} = X_L \cup X_T, \tag{2}$$

where superscript i shows the class label, superscript (i) means one of the c labels, and n_i is the number of training data identified by label vector $\mathbf{e}_i$ for i = 1,2, ...c.

In the conventional approach to classifier design, X_L is used to train **D** and then **D** is used to label X_T. In this approach, which is depicted as track Su in Figure 12.5 for the special case of image segmentation, X_L is constructed by judicious selection of training data by an operator who is familiar with clinically meaningful substructure in the data. For each observed tissue class the operator assigns every pixel (voxel, window) in the chosen subregion a (crisp) physical label (color). Since each patient has different anatomical structure, it is necessary in this approach to obtain labeled data for algorithmic training for each subject. This is a specialized, expensive and time-consuming procedure, and is thus not the preferred technique for automated analysis of medical imagery.

12.3 PERFORMANCE EVALUATION

Many researchers in computer vision and image processing develop algorithms with scant regard to expert knowledge about objects and object classes. This is natural when images are processed with the aim, say, of finding chairs, tables, cars or coffee cups. After all, most of us are experts in domains such as these. Medical image analysis, however, presents a vastly more complicated problem when visual evaluation is relied upon to assess the quality and utility of algorithmic outputs.

Usually, clinicians are provided with a set of images and asked to evaluate one or more "properties" of the enhancement process, such as judging the faithfulness of the replication of the film or sensor information, the utility of the enhancement (edges, regions, etc.), or the implication of *region of interest* (ROI) prompts (Hume et al., 1996). We illustrate these points with an example from digital mammography. As shown in Figure 12.2, radiologists (more generally, clinicians with expertise in the appropriate medical domain) can be involved in performance evaluation in three ways. For example, radiologists can compare:

C1. original sensor outputs (e.g., films) to unenhanced digitizations of the outputs (digital images) ;
C2. unenhanced digital images to enhanced digital images ;
C3. original sensor outputs to enhanced digital images .

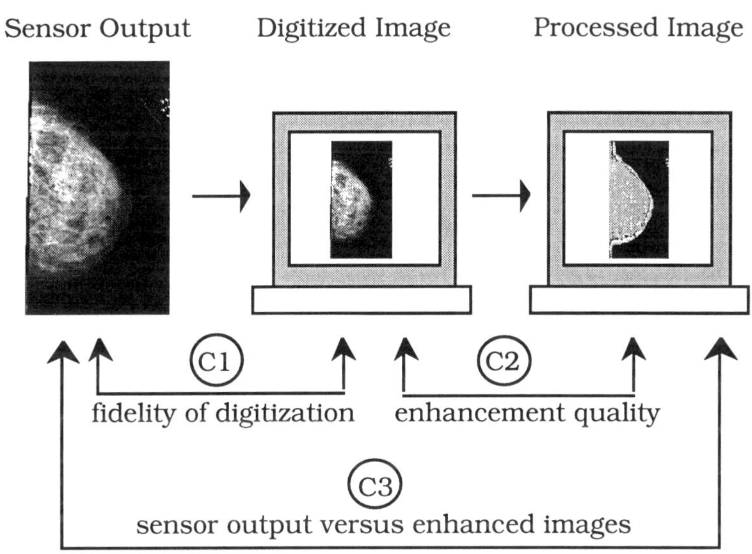

Figure 12.2 Clinician involvement in algorithm evaluation

As an example, when radiologists examine mammogram films, a light box display unit is used, and the entire mammogram suite can be viewed. However, when a digital version of the film is viewed on a computer monitor, unless the image is significantly subsampled, only a fraction of the digitized film can be viewed at one time (on commonly used monitors), often with 256 shades of gray (assuming an 8 bit display). When printed, these digitized images must be further cropped or compressed to satisfy printer constraints. For example, an average 600 dpi (dots per inch) laser printer can resolve only 122 gray levels (Apple, 1997). Research indicates that humans themselves can only resolve 32 gray levels (Green,

1989). This is an especially important point for fuzzy image analysis, because fuzzy labels for pixel vectors offer the possibility of an essentially infinite number of color shades via continuous membership values.

While radiologists feel most comfortable with films or other forms of direct sensor outputs, non-clinicians such as computer scientists are most comfortable with displays on computer monitors and printouts of digitized images. Consequently, some compromise is involved when evaluating the performance of a medical imaging system. Having access to all three of the data types displayed in Figure 12.2 (original films, unenhanced digital images and processed digital images) maximizes flexibility when determining the nature of this compromise, while still ensuring the development of a sound and repeatable evaluation methodology.

In addition to clinician involvement in performance evaluation, mammographic image databases often contain ground truth information which may include, e.g., *American College of Radiology* (ACR, 1995) ratings for characterizations of breast tissues, the size and locations of lesions, etc. *Ground truth information* is derived by one or more domain experts in one of two ways:

GT1 (primary). visually guided hand labeling of clinical features on sensor outputs; or

GT2 (secondary). algorithmically determined locations of ROIs that are visually assessed and accepted by clinical experts.

When ground truth of either type is available, performance analysis can be based on various pattern recognition methods such as (Huo et al., 1996):

PR1. using labeled test data to estimate error rates ;

PR2. correlation between computed and ground truth labels ;

PR3. analyzing receiver operating characteristic curves .

There are several ways to use these measures. For example, algorithm **A** can be used with training data X_L to parametrize a classifier D, and then the labels the classifier assigns to test data X_T can be used to estimate the apparent error rate (generalization potential) of D. If ground truth information is available, estimated regions of interest can be compared to the ground truth to assess the quality of the algorithm. Standard measures of agreement or goodness of fit such as correlation coefficients can be used to compare the relative quality of several algorithms **A, B ... Z** on common data sets that possess ground truth.

The scoring method used to determine a "hit" versus a "miss" when using ground truth information deserves careful consideration (Keller et al., 1998). For example, comparisons can be based on matching centroids of observed versus expected regions, intersections of bounding boxes for detected and benchmark regions, etc. Standardization of the technique employed for a given data set is critical to making unbiased assessments of an algorithm's performance. The general situation for assessment by pattern recognition techniques is summarized in Figure 12.3 (for the mammographic image domain; "micro" stands for microcalcifications).

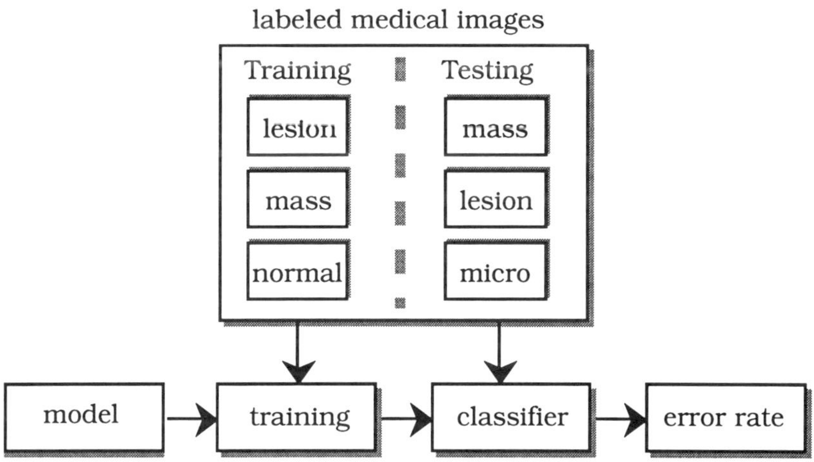

Figure 12.3 Evaluation of supervised approaches

Evaluation without clinician involvement such as illustrated in Figure 12.3 can provide insight into the success or utility of a proposed technique. However, clinician involvement is vital to developing a generalizable, non-database specific, repeatable methodology that will be accepted by health care personnel.

12.4 FEATURE EXTRACTION

This section considers the "front end" of medical image processing systems, where features are chosen. Figure 12.4 depicts several methods for extracting features from an input image. Medical imaging systems use various sensors to collect spatial distributions of measurements which represent the underlying tissue characteristics. As an example, 2D *magnetic resonance images* (MRIs) are composed of several relaxation times as well as proton density at each spatial location. These raw data support pixel, edge and region-based segmentation (the distinction being what numerical features extracted from the image are used as the basis for processing).

Figure 12.4 shows the extraction of a vector $\mathbf{x}_{ij}$ from just the ij-th pixel on the left hand side; and from a region (window) surrounding the ij-th pixel on the right hand side. The left side thus illustrates pixel-based features, while the right side illustrates either edge or region-based features.

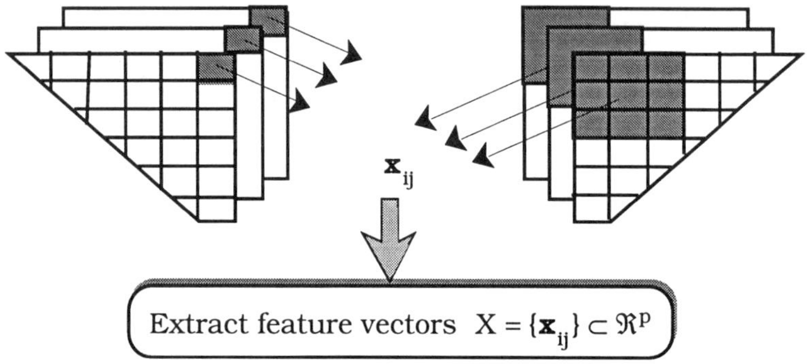

Figure 12.4 Extraction of features for image processing

By way of illustration, let $T1_{ij}$, $T2_{ij}$ and ρ_{ij} denote the spin lattice relaxation, transverse relaxation, and proton density of pixel (i,j) in an MR slice of overall dimensions $(m \times n)$. We can aggregate these three measurements into *pixel vector* $\mathbf{x}_{ij} = (T1_{ij}, T2_{ij}, \rho_{ij})$ in $\Re^3$; and the pixel vectors so constructed comprise a data set X that supports *pixel-based* methods.

On the other hand, if we estimated the horizontal and vertical gradients, say $g_{v,ij}$ and $g_{h,ij}$ of the intensity function at pixel (i,j) from intensities in some neighborhood of (i,j) in each of the three MR slices, there would be either three sets of features in $\Re^2$ or one set of features in $\Re^6$ to support *edge-based* segmentation of X. (Some writers call gradients *texture* features.) Finally, we might instead extract and order the nine intensities from a 3×3 window centered at pixel (i,j) in each of the three slices. This would result in either three sets of features in $\Re^9$ or one set of features in $\Re^{27}$ to support *region-based* segmentation of X.

It is both possible and sometimes profitable to combine pixel and window-based features to make up each vector in X. In this chapter we highlight several studies that also use the *spatial coordinates* of the pixels either as part of the feature vector, or in the extraction of the features selected for use. In the final analysis the choice of features is very important, but the "goodness" of features is extremely domain dependent. In this sense, it is difficult to quantitatively describe a "generally useful" feature space. This is especially true in the medical domain, where clinicians must be relied upon to supply insight and recommendations about the features they use when visually processing a medical image.

12.5 IMAGE SEGMENTATION

Once features have been extracted, the most frequently used low level image processing operation is *image segmentation*. More generally, some authors also refer to edge images as segmented images. We prefer to classify edge detection in a separate image processing category. Most texts define an image segmentation as a *crisp c-partition of the image*. Thus, for a 2D m×n image, a segmentation is a c×mn matrix U whose columns are crisp (pixel) label vectors for the mn pixels.

An additional constraint usually placed on segmentation is that each crisp cluster in U identify a region in the image that is "homogenous" with respect to some property or predicate (Jain et al., 1995). For example, each image cluster might be required to have a specified intensity range, or area, or be fully connected, etc. Most of the segmentation algorithms reviewed here do *not* satisfy this additional requirement because clustering algorithms are usually developed for non-domain specific applications, i.e., without this special constraint in mind. The lack of universally accepted validation methods for comparing various segmentations of an image lessens the severity of this omission.

Many studies of non-fuzzy segmentation methods have been published. For example, Morrison and Attikiouzel (1994) describe segmentation by statistical and neural network models; other multispectral methods are discussed by Just and Thelen (1988), Hyman et al. (1989), Vannier et al. (1991), Raman et al. (1991), Dellipiane (1991) and Liang (1993). Jain and Flynn (1996) provide a wonderful survey of image segmentation by non-fuzzy cluster analysis. Fuzzy clustering and early work in fuzzy image processing is discussed in many papers reprinted in Bezdek and Pal (1992).

There are many ways to classify segmentation methods, none of which leads to a crisp partition of them. For example, Figure 92 in Dellipiane (1997) shows a classification based on a tree rooted at image segmentation that subdivides algorithms based on the parameters that guide them to their goal. She identifies three main groups based on *density*, *topology* and *geometry*.

Almost all of the methods reviewed in our chapter fall into the first and perhaps oldest category (density), where leaves at a depth of 5 levels in the tree include segmentation approaches for regions, 2D region or 3D volume, and for boundaries, 2D edge or 3D surface.

Our discussion concerning the importance of human experts in medical image processing in Section 12.1 leads us to a somewhat different means of classifying segmentation algorithms. Our focus is primarily on how and where knowledge beyond that possessed by the sensor data is injected into the system. Based on this criterion, traditional pattern recognition methods for segmentation can be (roughly) classified into two groups, *supervised* (Su) and *unsupervised* (US) methods, depending on whether the vectors in X are labeled or unlabeled.

Figure 12.5 illustrates the distinction we make here between the two groups of segmentation methods that seem popular in the fuzzy modeling domain, and further subdivides unsupervised methods into two subgroups based on whether a human (US_A) or a rule base (US_B) is used to assign labels (tissue classes) to the unlabeled clusters (or regions) in the image. In the sequel we refer to the three vertical paths (left to right) in Figure 12.5 as *tracks* US_A, US_B and Su, respectively.

The bottom portion of Figure 12.5 reemphasizes the need for final evaluation of processed images by medically trained experts. Only a very few of the papers we review actually carry the system to this stage routinely, and as part of the algorithm itself (cf. Delliapiane, 1997).

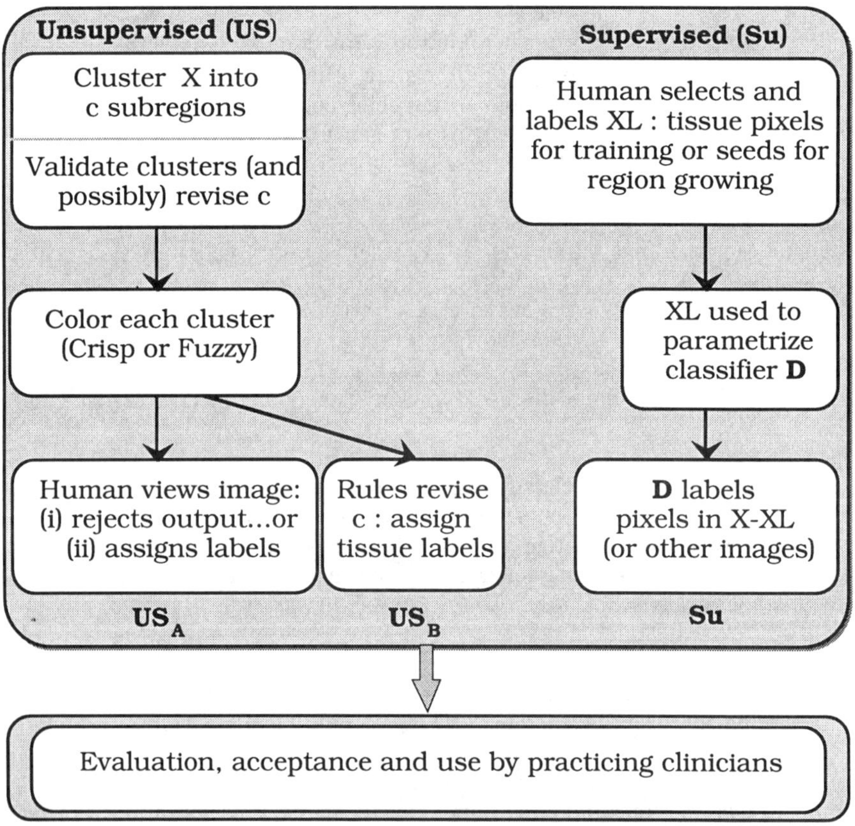

Figure 12.5 Segmentation with classical pattern recognition

Note carefully that it is the labels of the data (and their use during training) which divides the methods illustrated in Figure 12.5 into supervised and unsupervised approaches. The fact that humans or rules are used to assign physical labels to tissue clusters in the two US tracks is of course supervision in some broader (and in medicine, crucially more important) sense, but here we use the term supervised in the usual context of classifier design - that is, when training data are used prior to segmentation of test images.

We will also encounter segmentation algorithms that use human knowledge about the image domain as a basis for non-traditional classifier designs (Herndon et al., 1996; Hata et al., 1997, 1998). These models, and several others that appear in the section on 3D problems, don't fit in the framework of Figure 12.5 very well. For convenience, we will discuss these models in Section 12.8 along with more traditional track Su models.

After clustering to find a partition U of unlabeled X in either US track, a crisp color is usually assigned to each tissue class. If U is already crisp, this is straightforward. Otherwise, the simplest way to do this is to harden each column of U with **H** as in equation (1). Another possibility is to assign "fuzzy" colors to each pixel by mixing c basic colors in proportion to their memberships. Lighter shades are usually assigned to the pixels with strong membership values and darker shades are used for lower membership values. This has the effect of outlining borders where classes are intermixed and has been preferred by physicians (Bezdek et al., 1993).

The choice of which color to use for which region, and how to shade regions in images is a seemingly trivial part of segmentation. However, visual displays often have pre-established expectations in the medical community. The coloring scheme chosen significantly affects the utility of computed outputs, so this issue deserves careful attention.

Several fuzzy approaches to the tracks US_A and Su in Figure 12.5 are reviewed by Bezdek et al. (1993). Several fuzzy models for track US_B were recently surveyed by Clark et al. (1994). For readers interested in image processing (not specifically for applications in the medical domain) based on the fuzzy reasoning paradigm that has emerged from fuzzy control, we highly recommend the survey by Keller et al. (1996) as a companion to this chapter.

12.6 UNSUPERVISED SEGMENTATION: TRACK US_A

This section is about track US_A of Figure 12.5. The basic ingredients of most of the papers we review in this section are: the *hard, fuzzy and possibilistic c-means* (HCM, FCM, PCM) clustering algorithms and their variants; the hard and fuzzy *k-nearest neighbor* (k-nn) rules; and the inclusion of various cluster validity functionals for these. Mathematical details for these methods appear in Bezdek et al. (1999).

The earliest fuzzy segmentation of *medical* images along track US_A seems to be the work of a group at Stanford and IBM/Palo Alto, De La Paz et al. (1986). Their work concentrated exclusively on pixel-based MR features such as those described previously. This group published perhaps a dozen papers that compared MR segmentation by FCM to other approaches, most notably unsupervised decomposition of normal mixtures with the *expectation-maximization* (EM) algorithm. Di Gesu et al. (1991) summarize and update much of this work. These studies were based on images from about 300 patients, and the basic conclusion was that, although time consuming, FCM provided the best segmentations (as validated by humans, track US_A).

Di Gesu et al. eventually recommended an approximate version of FCM known as AFCM, the first technique studied to speed up FCM (Cannon et al., 1986). Improved implementation schemes for FCM and much better computational platforms have relegated AFCM to the (scientific) graveyard. On the other hand, several aspects of automated analysis of medical imagery were pioneered by this group, and their papers deserve close attention.

Carazo et al. (1990) at the New York State Department of Health spearheaded studies of electron microscopy images with a variety of fuzzy clustering methods. FCM, the Gustafson-Kessel (1978) modification of FCM and a clustering algorithm due to Roubens (1982) were the fuzzy clustering methods used. This study investigated the use of a large number of well known validity indices $\mathcal{V}$ for mathematical validation of fuzzy clusters in the images. Carazo et al. (1990) also introduced several new validity measures.

In the Carazo et al. study 100 (64×64) images were created by slicing a 3D reconstruction of the 50S ribosomal subunit of E coli 25 times. Post-processing included projection compression, filtering, scaling and the addition of white noise. Correspondence analysis was then used to extract one 8D vector from each image. Thus X was apparently composed of 100 region-based feature vectors in $\Re^8$, one feature vector per image. This is a good, multi-disciplinary article, and this group was one of the first to discuss all the important aspects of clustering for image segmentation (feature extraction, clustering and validation) for track US_A in an integrated fashion.

de Oliveira and Kitney (1992) used 5D features extracted from 64 MR images (single feature slices) as the basis for their track US_A segmentation studies. We mention this work to emphasize the wide variety of features that can be and are extracted from medical images to support segmentation. The 5 features in this work were the pixel intensity and 4 region-based measures of neighborhood texture (including the mean and two gradients). The texture features were smoothed twice with measures of neighborhood energy before submission to the FCM clustering algorithm with c fixed at 8. This investigation used hardening of fuzzy label vectors as shown in equation (1) to produce black and white images. These authors concluded that their method could support 3D volume estimation, but we are unaware of subsequent work along this line.

A team at the biophysics laboratory at UFR Alexis CARREL performed a very thorough study of segmentation of digital cardiographic scintigraphic images for quantitative analysis of cardiac function. Boudraa et al. (1992, 1993) combined the elements of track US_A as follows. The basic image data were 64×64 single intensity (Gamma camera) images collected over 16 equi-spaced times for 105 patients. The raw data from each image was linearly filtered and normalized. Then, each temporal sequence was compacted into a single *phase image* Φ using Fourier analysis and the 16 intensities at each spatial location. Three *weighted* features were then extracted, resulting in the feature vector $\mathbf{x}_{ij} = (\alpha x_{ij}, \beta y_{ij}, \gamma \varphi_{ij})$ in $\Re^3$, where (x_{ij}, y_{ij}) are the spatial coordinates of pixel (i,j), and (α, β, γ) are coefficients that determine the relative importance of each feature for the segmentation.

FCM was applied to the 3D data, and the Xie-Beni (1991) validity index was used - not to choose c, the proper number of clusters, which was known and fixed at c = 3 - but to find optimal values for (α, β, γ). This interesting use of a validity index falls within the clustering module shown in the left hand side of Figure 12.5, but in a very different way. After choosing (α, β, γ), the ventricle cluster was identified and isolated, a new reduced image with 2D features was generated, and FCM was applied to recluster it into left and right ventricles. (The idea of successively finer reclustering on reduced images takes other forms that we will meet later in this chapter.) Finally, the left ventricle boundary was isolated, and the ejection fraction was estimated.

Boudraa et al. evaluated their results in two ways. First, a team of trained clinicians performed human validation (method C3 in Section 12.3), and deemed the results on all 105 images quite satisfactory. Second, results were compared with two commercially available semi-automatic Gamma camera systems (Philips and SOPHA). Correlation coefficients (method PR2 in Section 12.3) between estimates of the ejection fraction from the fuzzy scheme and the two commercial methods were 0.91 and 0.94.

Another French group based at INRIA uses fuzzy models for segmentation of X-ray and MR images for ventricular endocardiogram detection. Boujeema et al. (1992a, b, 1994) use FCM as the first processing step in a two stage algorithm for edge detection which helps to sharpen ambiguous pixels using minimization of an energy functional. The second step is a fuzzy k-nn rule (Keller et al., 1985, Bezdek et al., 1986) type algorithm that is used to refine region boundaries at pixels that have weak FCM memberships from the first stage processing. The features used in these studies were pixel-based. This interesting use of a fuzzy k-nn rule for edge detection is very different from traditional approaches.

Papers by Di Gesu and Romeo (1994) and Di Gesu (1994) discuss MR segmentation using four *crisp* clustering algorithms operating on the data either in series or in parallel - an architectural style that is currently called information fusion or multilevel classification. The four algorithms were single linkage, histogram partitioning, sequential c-means and a two phase clumping procedure. In the parallel case, each clustering method produces a crisp c-partition U_i of X, which leads to a fuzzy c-partition of X, $U = \sum_{i=1}^{4} \alpha_i U_i$, $\alpha_i \in (0,1)$; $\sum_{i=1}^{4} \alpha_i = 1$. The convex weights $\{\alpha_i\}$ are estimated algorithmically, and the segmented image is colored by applying equation (1) to each column of U. In this study c = 8 was fixed, and the data were slices of an MR image from a patient with a hemorrhage. There were four measured MR parameters (T1, T2, TI, ρ), which were augmented with the spatial coordinates (as integers) of each pixel, so the data were a mixture of pixel-based and spatially relevant feature vectors in $\Re^6$.

Two measures of crisp validity were discussed by Di Gesu and Romeo, but were not used to alter c or U. Crisp ISODATA (Ball and Hall, 1967) was used as a surrogate "expert" (ground truth method GT2 in Section 12.3), so this hybrid model is in the style of track US_B, but we discuss it here because no rules from human experts were used.

A somewhat different version of the same platform appears in Di Gesu (1994). Conclusions drawn by these authors were mixed. However, the general idea of using mixed or hybrid models for image segmentation has become pervasive, and this paper is a nice example of multiple classification schemes.

A group at the University of Houston (UH) medical center led by Brandt and Kharas (1993) has also studied the use of FCM for completely unsupervised segmentation of MR images. The paper referenced here is interesting because it is, to our knowledge, the first attempt to apply PCM - possibilistic c-means - to medical imagery. Brandt and Kharas compared the effectiveness of HCM, FCM and PCM for separation of three simulated clusters as the amount of boundary overlap increased, and reported that FCM enjoyed a slight advantage in this study.

Much recent work in track US_A has been done at the University of South Florida (USF). Bensaid et al. (1996a) introduced a technique called *validity-guided (re)clustering* (VGC) and illustrated its use on MR images of patients with and without pathology. The VGC index is a generalization of the Xie-Beni (1991) index. The aim of this method is to enhance the quality of unsupervised fuzzy partitions of X by examining each FCM-derived cluster individually. When FCM terminates at U for a fixed c, the general procedure begins by selecting a particular cluster in X (say $\mathbf{U}_{(i)}$ = row i of U) for splitting. After hardening this row with **H**, FCM with c = 2 is applied to the points corresponding to $\mathbf{H}(\mathbf{U}_{(i)})$. To preserve the chosen value of c, two other clusters in U are merged at the same step.

The overall effect of VGC is to join tightly coupled clusters, split loosely coupled ones and preserve the chosen value for c. Hence, VGC is NOT a cluster validation scheme in the same sense as defined in Section 12.2, since it is not applied to different partitions of X, and is not used to choose a best value for c. However, it *is* driven by a validity functional which assesses the i-th cluster, and then sums this measure over i, providing an overall validity function for U. VGC continues reclustering until improvement becomes small.

Bensaid et al. (1996a) used 30 MR images to illustrate that VGC really improves outputs of FCM segmentation. For each image, an *optimized supervised* segmentation was constructed through an iterative process whereby the training set was repeatedly reselected in order to optimize the segmentation quality, as determined by two clinically knowledgeable investigators. Segmentation as in track Su of Figure 12.5 was done with the crisp 7-nn rule using the Euclidean norm. The 7-nn rule was chosen as a way to construct GT2 type ground truth (see Section 12.3) because this classifier is reportedly superior to various neural-like networks and probabilistic designs (Vaidyanathan et al. 1995).

The optimized 7-nn, FCM and VGC segmentations were subsequently evaluated by three expert radiologists in a blind study discussed in Bezdek et al. (1997a). Each radiologist was provided with three views (T1, T2 and ρ) of each unenhanced raw image and the three segmentations, and was asked to fill out a survey form (method C2 in Section 12.3). Individual panelists were asked to rate the quality of the first four performance indicators shown in Table 12.1 on a scale from 0 to 10, where 0 = very bad and 10 = excellent.

Each radiologist was also asked to rate the last two items (5 and 6 in Table 12.1) on a percentage basis. That is, to estimate the percentage of true positive tumor (correctly classified tumor pixels) and the percentage of false positive tumor (pixels incorrectly classified as tumor). Acronyms used in this table and later: *white matter* (WM), *gray matter* (GM), and *cerebro spinal fluid* (CSF).

Table 12.1 Comparison of three MR segmentation methods

Item	Description	7-nn rule	FCM	FCM +VGC
1	WM vs. GM	7.8	7.4	7.5
2	Normal vs. pathology	8.3	7.5	7.8
3	Tumor vs. edema	8.6	6.7	8.3
4	CSF quality	8.5	7.3	7.8
5	% True positive tumor	90.61	80.56	86.78
6	% False positive tumor	12.44	17.61	5.17

For each of the first four items in Table 12.1, the scores corresponding to each algorithm are averages over all 30 segmentations and three radiologists. The percentages of true and false positives reported for items 5 and 6 are simply averaged for each algorithm. Table 12.1 shows that VGC does enhance segmentations made by FCM. For example, with 99% confidence, VGC better differentiates between WM and GM (row 1); between normal and pathological tissues (row 2); between tumor and edema (row 3); and VGC is superior to FCM for identifying CSF (row 4).

VGC segmentations are not quite as good as those obtained with the optimized, supervised 7-nn rule. This is as expected, since supervision by labeled data usually improves (or at least should !) algorithmic outputs. VGC shows significant promise for differentiating between tumor and other tissues compared to FCM, and moreover, the ratio between its percentages of true positive and false positive tumor pixels is higher than the corresponding ratio for the supervised 7-nn rule. With 99% confidence, VGC produces significantly less false positive tumor than 7-nn. On the other hand, even with confidence as low as 90%, 7-nn's true positive tumor percentage is not significantly better than VGC's rate. So, this model seems to be a positive step on the way to truly unsupervised designs.

Velthuizen et al. (1997) discuss the use of a modification of the *mountain clustering method* (MCM, Yager and Filev, 1994) they call M3 for segmentation of MR images. The mountain method produces c point prototypes $\mathbf{V} = \{\mathbf{v}_1, \ldots, \mathbf{v}_c\}, \mathbf{v}_i \in \Re^p \ \forall i$ from unlabeled data. MCM begins at c = 1, and continues to add prototypes to **V** until an objective function threshold is exceeded. Thus, c is neither fixed nor explicitly chosen, but is implicitly specified by the termination threshold used to stop the algorithm.

Let $(\mathbf{V}, \mathbf{E}) = \{(\mathbf{v}_i, \mathbf{e}_i) : i = 1, \ldots, c\}$, where **V** is a set of c crisply labeled prototypes, one per class, ordered so that $\mathbf{e}_i$ is the crisp label for $\mathbf{v}_i$, $1 \le i \le c$; and let δ be any distance measure on $\Re^p$. The *crisp nearest prototype* (1-np) classifier $\mathbf{D}_{\mathbf{V},\mathbf{E},\delta}$ is defined, for $\mathbf{z} \in \Re^p$, as

$$\text{Decide } \mathbf{z} \in i \Leftrightarrow \mathbf{D}_{\mathbf{V},\mathbf{E},\delta}(\mathbf{z}) = \mathbf{e}_i \quad \Leftrightarrow \quad \delta(\mathbf{z}, \mathbf{v}_i) \le \delta(\mathbf{z}, \mathbf{v}_j) \ \ \forall \ \ j \ne i . \tag{3}$$

Equation (3) says: find the closest prototype to **z**, and assign its label to **z**. Ties are broken randomly. The most familiar choice of dissimilarity is Euclidean distance. It is a misnomer to call MCM (or M3) a *clustering* method because no partition U of X is produced by either of these algorithms, which are more properly viewed as prototype generator methods. However, after the terminal prototypes $\mathbf{V}_{MCM}$ or $\mathbf{V}_{M3}$ are found, they can be used to generate a crisp c-partition of X by applying the nearest prototype classifier (3) based on them to each point in the (unlabeled) image.

Velthuizen et al. (1997) use various sets of MR pixel intensities as features. Segmentation by M3 proceeds as follows:

[M3.1] run M3 on X to find $\mathbf{V}_{M3}$;

[M3.2] construct U, a crisp partition of X with equation (3): the label assigned to $\mathbf{x}_{ij}$ is the label (index) of the closest prototype;

[M3.3] physically label each cluster in U as a tissue class by matching the pixels in an algorithmic cluster to each of the ground truth tissue clusters. Assign the algorithmic cluster to the tissue class that enjoys maximum pixel matching;

[M3.4] artificially color the labeled image.

Segmentation by FCM was done with two initializations: a "standard" initialization $\mathbf{V}_o$ (cf. (9) in Velthuizen et al., 1997); and with $\mathbf{V}_{M3}$. We write FCM($\mathbf{V}$) to indicate FCM initialized with $\mathbf{V}$. FCM generates a terminal fuzzy c-partition U_{FCM} of X which is hardened using equation (1), and finally, steps [M3.3] and [M3.4] are performed on the resultant crisp partition.

Velthuizen et al. (1997) evaluated segmentations of 13 MR images using both types of ground truth (methods GT1 and GT2 in Section 12.3). Segmentations were produced by four methods: a supervised k-nn rule (used to construct the type GT2 ground truth), unsupervised M3, unsupervised FCM($\mathbf{V}_o$) and unsupervised FCM($\mathbf{V}_{M3}$). Comparisons were made visually (method C3 in Section 12.3) and quantitatively (method PR2 in Section 12.3).

Figure 12.6 shows the T1 (weighted) view of the input data for a patient that has a brain tumor. Figure 12.6(b) is the color key for the original image: CSF = *cerebro spinal fluid*; WM = *white matter*; GM = *gray matter*; GM-2 = *(falsely labeled) gray matter*. Notice that - inconveniently - the colors for tissue classes in all of the segmentation views are different than the color key labels for view (a).

A supervised k-nn segmentation is shown in Figure 12.6(c). This image results from an operator choosing labeled subsets of pixels from each tissue class, and then using the standard k-nn rule to label the remaining pixels. This is repeated until a panel of radiologists agree that the k-nn segmentation is good enough to be used as type GT2 ground truth. Ten of the thirteen images discussed in this study used this method (GT2) as a basis for comparing the results of the three algorithms (unsupervised M3, unsupervised FCM($\mathbf{V}_o$) and unsupervised FCM($\mathbf{V}_{M3}$). The other three images had manual ground truth (method GT1 in Section 12.3).

Figure 12.6(d) shows a segmentation achieved by FCM($\mathbf{V}_o$). The tumor is not detected. Instead, FCM($\mathbf{V}_o$) finds two gray matter regions that do not correspond to anatomical tissues. The M3 segmentation in Figure 12.6(e) is much better - it finds many of the tumor pixels and does not have a gm-2 tissue region. Finally, panel 12.6(f) exhibits the segmentation resulting from the initialization of FCM with the output of M3. This view should be compared to Figure 12.6(c). It's hard to see on a printed copy, but there is excellent correspondence between the tumor regions in these two views.

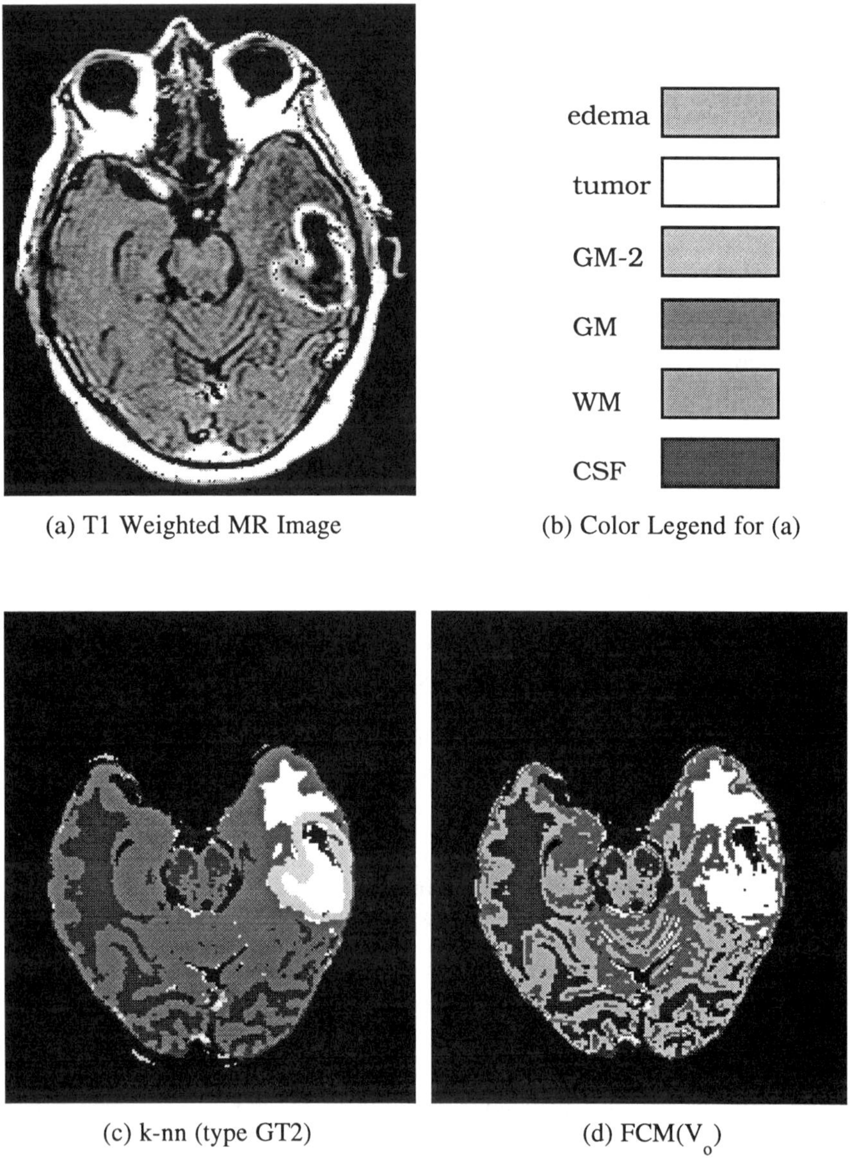

(a) T1 Weighted MR Image

(b) Color Legend for (a)

(c) k-nn (type GT2)

(d) FCM(V_o)

Figure 12.6 MR segmentations (Velthuizen et al., 1997)

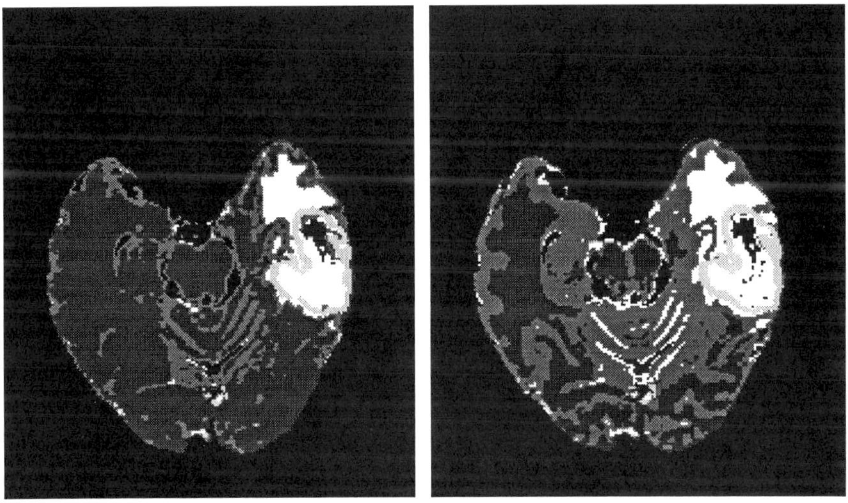

(e) M3 (f) FCM(V_{M3})

Figure 12.6 (cont.) MR segmentations (Velthuizen et al., 1997)

Table 12.2, adapted from Velthuizen et al. (1997), shows the *average* performance on pathological tissues for segmentations of thirteen images made by unsupervised M3, unsupervised FCM(V_o) and unsupervised FCM($\mathbf{V}_{M3}$).

Table 12.2 Average true and false positive pixel counts (in %) for pathological tissues (Velthuizen et al., 1997, Table 1)

	False Positives			True Positives		
	FCM($\mathbf{V}_o$)	M3	FCM($\mathbf{V}_{M3}$)	FCM($\mathbf{V}_o$)	M3	FCM($\mathbf{V}_{M3}$)
Tumor	10.3	5.6	5.2	59.4	66.1	75.5
Edema	5.9	5.9	8.7	75.9	77.9	81.2

When FCM is initialized with $\mathbf{V}_o$, segmentation is not as good as M3 - it has nearly 5% more false positives and about 7% less true positives in tumor. In edema, the recognition rates are about the same. When FCM is initialized with $\mathbf{V}_{M3}$, there is substantial improvement in the true positive rate for both tissue classes and a slight decrease in edema false positives.

Clark et al. (1998) have recently reported results from a completely automated system which extracts glioblastoma-multiforme tumors from sequences of transaxial MR images (for the purpose of tumor volume tracking during treatment). FCM is used during preprocessing to remove extracranial areas (air, bone, skin, fat, muscles, etc.), which helps to limit processing in subsequent steps. A rule-based expert system then extracts the intracranial region and initiates the process of iteratively refining the segmentation. Multispectral histogram analysis of the intracranial region followed by region analysis provides the final tumor labeling. Results are provided for thirteen unseen volume data sets, with comparisons to supervised, radiologist-labeled "ground truth" tumor volumes, and supervised *k*-nearest neighbor tumor segmentations.

The last segmentation method we discuss in this section is due to Sameti and Ward (1996). Their model begins with an initial fuzzy membership function m_{X0} whose domain is the set of pixels in image X. First, these authors normalize the gray levels in X, then find T, the value at which the histogram of the normalized intensities $\{\hat{x}_{ij}\}$ minimizes. T is used to set the crossover point where $m_{X0}(T)=0.5$. The graph of m_{X0} is displayed, but its equation is not. The function shown bears a striking resemblance to a truncated unipolar sigmoid. We think there is no harm in saying that Sameti and Ward initialize $m_{X0}(\hat{x}_{ij}) = 1/(1+e^{-(\hat{x}_{ij}-T)})$.

Following initialization, an iteration scheme that mimics gradient descent for back propagation of errors in multilayered perceptrons (Zurada, 1992) updates m_{X0}. After k iterations, the update has the form of the delta rule, $m_{X,k+1}(\hat{x}_{ij}) = m_{X,k}(\hat{x}_{ij}) + \Delta m_{X,k}(\hat{x}_{ij})$. Iteration is employed in the hope of minimizing the error function $E[m_{X,k}] = \Sigma\Sigma\left(m_{X,k}(\hat{x}_{ij})\right)\left(1-m_{X,k}(\hat{x}_{ij})\right)$. This function is zero only if $m_{X,k}(\hat{x}_{ij})=0$ or $1\,\forall i,j$ - that is, only if $m_{X,k}$ determines a crisp 2-partition of X. This segments X into a binary image. According to the authors, termination this way can occur. This procedure is subsequently applied repeatedly to each of the two crisp subsets created by the successive phase of the processing until a satisfactory segmentation of X is obtained. Consequently, X is segmented into c crisp regions where $c = 2^k\ \exists k$. Sameti and Ward segment 20 MR images into c = 4 crisp regions this way, and allude to comparing suspicious regions in them to known suspicious regions. Details of evaluation for this track US_A method are incomplete, but it is the basis for a more complete study by the same authors that we discuss in Section 12.9 (digital mammography).

Bombardier et al. (1997) have investigated automated enhancement strategies in digital subtraction angiography. In this work, two cooperating fuzzy segmentation operators based on textural and geometric properties are used to successively enhance aorta and renal artery boundaries. First, fuzzy linguistic rules are derived from the definition of an edge as "...a high transition near an homogeneous region". These rules are applied as a set of 5x11 masks to the whole image to find characteristic homogeneous and heterogeneous regions indicative of aorta outlines. Second, bifurcation points along these outlines then determine the regions of interest where subsequent analysis using an FCM-based edge operator extracts renal artery boundaries. Results are provided for a real 2D angiogram. The final edge image created would still need to be post-processed to characterize any lesion boundary abnormalities (e.g., narrowing of the artery diameter, as in stenosis).

There are many other studies that perform segmentation with fuzzy models along track US_A. Without ground truth images that can be used for quantitative assessment as in Table 12.2, none of these studies will carry much weight in terms of establishing the clinical efficacy (or lack of it) of a particular model. Human expertise is needed, and it is more efficiently used in the other two tracks shown in Figure 12.5. Nonetheless, investigations in track US_A are numerous and popular; we conjecture that this is because it is the easiest track to work in without the aid of medically knowledgeable clinicians.

12.7 UNSUPERVISED SEGMENTATION: TRACK US_B

One of the earliest (non-fuzzy) articles discussing the use of rules in the context of MR segmentation was by Menhardt and Schmidt (1988). The use of fuzzy rule-based guidance for the segmentation process, shown in track US_B of Figure 12.5, apparently began with Li et al. (1993). These authors used FCM as the first step in a *knowledge-based* (KB) system that automatically segments and labels glioblastoma-multiforme tumors in successive MR slices of the human brain and subsequently estimates total tumor volume. We call the approach discussed in Li et al. (1993) the USF-KB system.

Initial clustering of image sections in the USF-KB approach is performed by FCM with a strategy of overclustering, i.e., the image is deliberately segmented into more clusters than are known to exist. When FCM uses for c the number of clusters determined by an operator during manual labeling, FCM separates tissue types incorrectly.

Overclustering is based on the premise that multiple clusters containing the same tissue type are easier to merge than separation of tissue regions in undersegmented clusters. This is in some sense the reverse of the VGC approach: clusters are merged using mid-level rules rather than low level pixel-based processing.

After initial FCM segmentation, the overclustered partition and the cluster center for each region are provided to a (crisp) rule-based expert system which contains knowledge gathered from a variety of sources such as radiologist expertise and empirical observations. The USF-KB system includes heuristics concerning tissue characteristics in feature space (T1, T2, ρ) and anatomical information about the internal structure of the brain.

Using knowledge-based and model-based recognition techniques, the system iteratively locates tissues of interest. These focus-of-attention tissues are analyzed by matching measured to expected characteristics. For tumor segmentation, focus-of-attention is used with additional stages of fuzzy clustering to separate normal brain tissue from pathology, then to separate tumor from non-tumor pathology. This cycle is important as it allows the results of applying FCM to the raw images to guide knowledge application, which in turns guides further clustering, making it possible to break down the more complex problem of tumor segmentation into smaller, more easily attainable goals.

Table 12.3 (Table 2 in Bezdek et al., 1997a) shows results from processed slices of a single patient with diagnosed glioblastoma-multiforme scanned over five repeat sessions. Acronyms used in this table: *false positive* (FP), *true positive* (TP), *false negative* (FN), *tumor size* (TS), and *correspondence ratio* (CR). The slices were used to refine the rules in the knowledge-based system. Ground truth for these slices in the form of tumor pixel counts were made by a human operator. False positives are non-tumor pixels that are mislabeled.

Table 12.3 Pixel counts for a glioblastoma-multiforme patient using the USF-KB system (Bezdek et al., 1997a)

Patient ID	FP	TP	FN	TS	CR
p32s19	4	667	36	703	0.920
p32s20	33	1007	54	1061	0.908
p45s17	28	420	16	436	0.913
p50s21	131	1197	41	1238	0.897
p50s22	101	960	46	1006	0.881
p52s18	17	491	37	528	0.878
p52s19	82	1010	26	1036	0.922
p56s19	112	984	47	1031	0.877
p56s20	82	892	47	939	0.881

The correspondence ratio shown in Table 12.3 is computed with the formula CR = (TP - (1/2 * (FP + FN)))/TS. The true positives and false negatives sum to the tumor size, TP+FN=TS. This is one way to assess the overall quality of the segmentation. The value of CR maximizes at 1.00 when there are no FPs or FNs. On the other hand, FPs not in the tumor can cause this number to be negative. Thus, values close to 1 indicate segmentations that closely match human estimates. Table 12.3 shows that the processed slices for this patient were relatively accurate.

Hillman et al. (1995) and Chang et al. (1994, 1995) report success at Texas A&M with a track USB scheme that is in some sense diametrically opposite to the USF-KB model. This group introduces fuzzy rules as a pre-processing stage followed by FCM clustering to segment the remaining pixels in rat and human brain MR images.

These authors assert that fuzzy rules endow the system with somewhat greater reliability than crisp rules at a significant savings in CPU time and storage. Approximately 80% of the pixels are labeled by intensity thresholding with the fuzzy rule-base before clustering. The mean vector of each set of labeled pixels (after a hardening alpha-cut) provides FCM with initial prototypes. After FCM segmentation of the unlabeled pixels, the final rat brain segmentation obtained was reportedly satisfactory.

This technique has also been used to study HIV-positive lesions in human MR images. In agreement with the USF-KB system, this group reports that using FCM alone is not sufficient to produce satisfactory segmentations of MR images. They noted again that the computational complexity of FCM was a problem and that FCM had difficulty in directly utilizing spatial (non-feature) information derived from the MR image. Furthermore, they reported that FCM sometimes mislabeled "easy" pixels of classes that did not have concave hulls in feature space.

12.8 SUPERVISED SEGMENTATION: TRACK Su

Our discussion of fuzzy methods for supervised medical image segmentation begins with the *semi-supervised FCM* (ssFCM) method. Techniques of this kind in the context of c-means clustering were first discussed by Pedrycz (1985). Algorithms in this category are (i) clustering algorithms that (ii) use a finite design set $X_L \subset \Re^p$ of *labeled* data to (iii) help clustering algorithms partition a finite *unlabeled* data set $X_T \subset \Re^p$, and then (iv) terminate without the capability to label other points in $\Re^p$. The word "semi" is used because these schemes are not supervised in the sense that labeled training data are used to find the parameters of a classifier **D** which is subsequently used to complete segmentation of X_T as shown in the Su track of Figure 12.5.

The semisupervised approach is applicable in domains such as image segmentation, where users may have a small set of manually derived labeled data, and can use it to supervise classification of the remaining pixels in a single image. Initially, partitions of X for semisupervised clustering algorithms have the form

$$\underbrace{U}_{c\times n} = [\underbrace{U_L}_{c\times n_L} \mid \underbrace{U_T}_{c\times n_T}] = [\underbrace{\mathbf{U}_1^1 \ldots \mathbf{U}_{n_1}^1}_{\mathbf{e}_1} \quad \cdots \quad \underbrace{\mathbf{U}_1^1 \ldots \mathbf{U}_{n_c}^1}_{\mathbf{e}_c} \mid \underbrace{\mathbf{U}_1^u \ldots \mathbf{U}_{1,n_T}^u}_{\text{unlabeled}}], \qquad (4)$$

where superscript u indicates "unlabeled". In general, the n_i's need not be equal, nor is it necessary that the column vectors shown in equation (4) be crisp. The basic idea is to use (X_L, U_L) and X_T to find U_T. Roughly speaking, semisupervision exchanges the generalization capability of a classifier trained with (X_L, U_L) for the right to use the structural information possessed by the points in both X_L and X_T while searching for clusters U_T of X_T.

The development and use of ssFCM for MRI segmentation is discussed by Bensaid et al. (1996b). In this model a modified version of FCM (i.e., ssFCM) is applied to X_T. The training data set X_L guides ssFCM towards improved clustering of the unlabeled pixels by splitting the FCM update conditions for U and **V** into labeled and unlabeled components. The labeled components of U are fixed as in (4), and can be weighted by class so that ssFCM effectively uses many copies of the relatively few training data in X_L. We exemplify this model in Section 12.12, where ssFCM is used to make volume estimates of tumors.

Segmentation in the USF-KB model has been recently augmented by the addition of fuzzy rules. Namasivayam and Hall (1995) have shown that over a large set of MR images from different patients, fuzzy rules perform most reliably when they are based on relative differences in pixel intensities for different tissue types. Relative fuzzy rules and ssFCM applied to the unlabeled pixels in X_T are claimed to yield more accurate and much faster (compared to FCM alone) segmentations of normal MR images. ssFCM has also been used in the reclustering stage of the USF-KB system. In this application, crisply labeled training pixels are chosen by a set of rules that identify tissue types with a high degree of confidence.

Herndon et al. (1996) discuss the use of crisply label training data created by pooling data from experts who label *every pixel* in training images. This training data is then used to derive a classifier that segments an input image into c "fuzzy tissue images", one for each labeled tissue class. Technically, this classifier is *possibilistic* since the pixel memberships assigned to the c tissue images are not constrained to sum to 1. This method of segmentation is very different than the other types discussed so far, and is subsequently used for tissue volume estimation in normalized T1 MR images. We return to this paper in Section 12.12 on 3D problems.

Hata et al. (1997, 1998) have another approach to the segmentation of medical images that does not fit into the general framework established in Figure 12.5. Their model is based almost entirely on reasoning with a set of fuzzy if-then rules. Both referenced papers describe the use of a fuzzy rule base that processes numerical pixel-based features. Numerical features include intensities, spatial locations, Euclidean distances and boundary proximities.

Membership functions for the rules are given, but no tuning or training is described. Instead, the shapes and parameters of these functions are evidently based on domain specific knowledge about human physiology such as intracranial (for MR brain images) and joint structure *computerized tomagraphic* (CT) images of the human foot.

Hata et al. combine these elements with approximate reasoning by a fuzzy rule base to produce, like Sameti and Ward (1996), a single fuzzy membership function $m_X : X \mapsto [0,1]$. Unlike Sameti and Ward, the construction of m_X is non-iterative, and its use is quite different. Next, a region growing technique based on thresholding the values $\{m_X(i,j)\}$ produces a segmentation of the image into a prespecified number of crisp clusters. This strategy is applied to several 3D problems which we will return to in the section on 3D applications.

Another very different approach to supervised segmentation is *region growing* from *user selected* seed pixels (or voxels). Delliapiane (1997) and Udupa and Samarasekera (1996) both discuss models of this type based on (different) topological notions of fuzzy connectedness. The basic feature in both of these papers is to use the pixel intensities as a basis for measures of fuzzy similarity between image elements in the same tissue class.

Delliapiane (1997) summarizes her *fuzzy isovolumes* approach to supervised segmentation of 2D and 3D images in Delliapiane et al. (1996). A connectivity measure based on fuzzy topology and homogeneity is constructed from image intensities and is used to drive the segmentation process. Supervision is begun by an expert user, who interactively chooses a single pixel (or voxel) from a known tissue class as the *seed* for a region (or volume) growing technique (the training data is a crisply labeled singleton, $X_L = \{\mathbf{x}_{ijk}, \mathbf{e}_{(i)}\}$).

One tissue class at a time is built by thresholding an image that possesses a property they call (fuzzy) intensity connectedness, which is an extension of Rosenfeld's (1984) idea of fuzzy connected components in a digital image. A number of potential fuzzy isovolumes are grown from the selected seed, and the operator then chooses the most appropriate one before proceeding to the next region (tissue class) in the image. This style of segmentation proceeds non-iteratively, one region at a time, and is always terminated by a human expert - not an algorithmic criterion.

Udupa and Samarasekera (1996) also advocate the use of fuzzy connectivity for image segmentation. The examples presented by these authors all use 3D data, so we discuss them in Section 12.12. The basic methodology, however, like that of Dellipiane et al. (1996), is applicable to 2D segmentation, and because this technique is initiated by a user chosen seed for each tissue class, it follows the basic requirements for belonging to our Su track. Image segmentation and object classification are achieved by thresholding a fuzzy relation in the given image, resulting in various output images.

Udupa and Samarasekera (1996) begin with a formal representation of fuzzy digital spaces composed of *space elements* (spels) in digital p-space, special cases being pixels (p=2) and voxels (p=3). Spel intensities are used to compute fuzzy spel affinities, which in turn enable the authors to define and study the concepts of fuzzy connected components, fuzzy object extraction and fuzzy spel labeling in images.

Two region growing dynamic programming algorithms for fuzzy object extraction and labeling are proposed and analyzed. Both algorithms are initiated at a user-defined (seed) spel and are in this sense quite similar to Dellipiane et al. (1996). However, Udupa and Samarasekera (1996) aim for a much more general framework and present a much more detailed mathematical analysis than Delliapiane et al. (1996).

12.9 DIGITAL MAMMOGRAPHY

In this section we provide a more complete picture of the complexity of designing and validating automated medical image analysis systems. Our application domain is digital mammography. Among all cancer related fatalities in women in the United States, breast cancer trails only lung cancer. Kopans (1996) has this to say about computational mammography:

> "There is no radiologist, regardless of skill and expertise who has not failed to see a significant abnormality on a mammogram that is visible in retrospect. There is a psycho visual threshold for all involved in observational endeavors that cannot be completely eliminated. Since, at any one time, different observers tend to overlook different findings, multiple observers are less prone to oversight than a single observer. In order to reduce the error rate for the detection of breast cancers, radiologists have developed double reading systems in which more than one trained radiologist reviews the same mammogram. This has been shown to reduce the error rate by 5-15%. One major benefit from CAD [*computer assisted diagnosis*] is the development of a second reader system that is tireless, consistent, and cannot be distracted."

Radiologists derive mammogram interpretation reports based on the standards of the *Breast Imaging Reporting and Data System* (BI-RADS) developed by the American College of Radiology (1995). This system defines the terminology for categorization of breast abnormalities (lesions) into two broad categories, *masses* and *calcifications*. In addition, standardized descriptions of associated or stand-alone findings are provided, which include characterizations of such features as skin or nipple retraction, skin thickening, and architectural distortion.

Microcalcification and mass detection schemes are the primary focus of work we are aware of in digital mammography which utilizes fuzzy models. Microcalcifications often appear as small bright areas in the image (i.e. tiny dots) and are taken as a potential early indication of the onset of a breast tumor. Experienced radiologists can miss detection of microcalcifications by visual examination of X-ray films. Given the importance of detection of this pathology as an early indicator of breast cancer, it is surprising how sparse the fuzzy literature is in this area.

Brzakovic et al. (1990) study a fuzzy pyramid linking scheme for the detection of microcalcifications and nodules. Lo et al. (1996) focus on the detection of clustered microcalcifications using fuzzy classification modeling. Wang and Karayiannis (1997) used *fuzzy learning vector quantization* (FLVQ) and other techniques to compress digital mammograms containing microcalcifications.

One of the central features in several microcalcification studies is the use of wavelet-based correlation filters for extracting features that can be used as a basis for discriminating clusters of microcalcifications. Wang and Karayiannis (1997) apply unsupervised clustering techniques to their wavelet features, while Strickland and Theodosiou (1998) employ a fuzzy reasoning system for supervised classification using similar extracted measurements.

Hall (1996) describes the use of the *crisp* C4.5 decision tree classifier for detecting microcalcifications. His work used 40 images from the Nijmegen data base (cf. Table 12.4). The images were paired views of breasts of 21 patients, each of whom had microcalcifications in at least one breast. Each image was 2048×2048 in size with 12 bit intensity values. One or more clusters of microcalcifications in each of the 40 images were marked (circled) by two radiologists (ground truth method GT1 in Section 12.3). Thus, each circled area ("cluster") contains pixels corresponding to both microcalcified and normal tissue.

Hall (1996) extracted a total of 63 features for each pixel: *6 sets of* area, shape, average edge strength, edge strength deviation, object standard deviation, and background standard deviation descriptors (1 set of 6 for each of 7 different segmentations of an image); and 21 Laws energy features computed by convolution over 15×15 windows. Hall then created pixel by pixel (method GT2 in Section 12.3) ground truth for $c = 2$ tissue classes (microcalcified or not) by submitting each GT1 cluster to 2 different CAD calcification detection codes.

For training and testing, Hall used the leave one out method, where the "one left out" was the set of all pixels in both classes for an entire image. A crisp C4.5 decision tree classifier was trained with the pooled labeled feature vectors from the remaining 39 images, followed by testing on the data for the image held out. Repeating this 40 times (holding each image out in its turn) yielded 40 error rates. At best, for 5 of the 40 images, the error rate was zero. At worst, the maximum error rate was 73.33%. And the average error rate over the set of 40 experiments was 11.21%. Perhaps more significant, all 40 images would be correctly flagged as abnormal. Hall (1996) then discusses the fuzzification of C4.5, and illustrates his idea with an artificial data set. His speculation that a fuzzy version of C4.5 would improve the reported performance has not, to our knowledge, been subsequently tested. So, this is "almost" a fuzzy model for mammography.

Masses typically encompass larger areas in the image than microcalcifications, and tend to have irregular or spiculated boundaries. Li et al. (1995) examine the use of fuzzy membership functions for Markov random fields for tumor detection in mammogram images. Sutton and Bezdek (1997) outline an approach to tumor detection based on segmentation with FCM. Kovalerchuk et al. (1997) illustrate how a fuzzy logic approach can be used to formalize the distinction between lobulated and microlobulated masses.

Sameti and Ward (1996, 1997) discuss their partitioning method (cf. Section 12.8) on two sets of mammogram images. The 1996 paper uses a set of 20 mammograms for patients with lesions. In this paper they mark "suspicious ROIs" in each of c = 4 regions. However, evaluation of these results is not made.

The database for Sameti and Ward (1997) consists of 35 100 micron, 12 bit images obtained as either CC or MLO views (see Table 12.4 for acronyms) from 18 patients (one view was defective). The data were obtained from the *Screening Mammography Program of British Columbia.* Only the most significant 8 bits were processed. Each mammogram is first divided into 256×256 regions, and then each of these regions is processed just once, resulting in a total of $2 \times 256 \times 256$ binary regions, each of which may have ROIs called "mass candidates". Then, 24 texture features are extracted from each ROI (2 of the 24 features are mentioned as especially useful, but it is unclear whether subsequent results rely on only these two). Apparently a linear classifier is trained by jackknifing through X_L and X_T using a single image. Evaluation of the model was based on ground truth data supplied by a single radiologist (method GT1 in Section 12.3). Sameti et al. (1997) report that this method achieves 94.3% true positive detection with 0.24% false positives per image.

12.10 DATABASES FOR DIGITAL MAMMOGRAPHY

In order for computer assisted diagnosis to contribute to the delivery of reliable health care for breast cancer, mammograms must be provided digitally, and appropriate algorithms must be developed to aid the diagnostic process. Recent insurance company support of breast cancer screening programs, government support for creating globally available electronic databases, and the advent of rapidly changing technologies for the acquisition of mammograms, create a unique opportunity for researchers to study and *quantitatively compare* applications of fuzzy models (and other approaches) on a large scale.

The work described thus far has been conducted primarily with very small (and often non-shareable) data sets. Within the last few years, a number of electronically available databases have been created that enable more thorough and systematic study and testing of automated diagnosis of mammograms. Six mammographic databases are listed in Table 12.4 (a reference for each is listed separately at the end of this chapter). Abbreviations used in Table 12.4: *mediolateral oblique* (MLO), *craniocaudal* (CC), *charge coupled device* (CCD), *optical density* (OD), *gray level* (GL), *number of patient cases* (n_p), *number of cases currently available* (n_i), and *number of planned cases in the finished database* (n_i^*).

Table 12.4 A summary of mammography databases

	Nijmegen	**Wash. U.**	**DDSM**
spatial resolution	100 microns	N/A	42, 43.5, and 50 microns
intensity resolution	12 bits	12 bits	12 and 16 bits
digitizer	Eikonix 1412 CCD	LORAD CCD-based	DBA, Lumisys and Howtek scanners
grayscale response	GL 4095=0.18 OD	N/A	DBA: logarithmic in 0-3.6 OD Lumisys: linear in 0-3.6 OD, minimal GL = 495 Howtek: linear in 0-3.06 OD, minimal GL = 820
n_p	21 (n_i = 40)	N/A (n_i =30)	1789 (n_i^*= 3000)
type of views	MLO, CC	core biopsy	MLO, CC
pathology types	microcalcifications	lesions	535 normals, 699 cancers (all types), 555 benign cases

	LLNL/UCSF	**UC/UNC**	**MIAS**
spatial resolution	35 microns	50 microns	50 microns
intensity resolution	12 bits	12 bits	8 bits
digitizer	Dupont NDT	Lumisys 200	Joyce-Loebl, SCANDIG-3
grayscale response	linear to transmitted intensity	N/A	linear in 0-3.2 OD
n_p	50 (n_i = 198)	N/A (n_i^*= 1000)	N/A (n_i = 322)
type of views	MLO, CC	MLO, CC	MLO only
pathology types	normals, microcalcifications	normals, abnormals (all types)	normals, abnormals (all types)

The diversity of sensors, experimental protocols and quality control mechanisms of database collection complicates the selection of an appropriate database. Researchers must first decide if the spatial and intensity resolution of the database are sufficient for locating a particular abnormality. For example, Woods (1994) states that microcalcifications can be as small as 100 microns in an image. If more than one database is required (to increase the sample size or to test the generalization capabilities of the approach), the databases must also supply digitization information so the data can be normalized prior to comparison. (Normalization is typically based on the sensor's grayscale response characteristics, such as those listed in Table 12.4.)

The newest and most comprehensive database listed in Table 12.4 is the *Digital Database for Screening Mammography* (DDSM), which is being developed at the University of South Florida with support from the *U.S. Army Medical Research and Materiel Command* (USAMRMC). DDSM offers researchers an opportunity to study and make thorough comparisons of various models on a large, comprehensive database with ground truth information (Bowyer et al., 1996). DDSM contains 535 normal cases, 699 cancer cases, and a set of 555 benign cases (as of June, 1999). The DDSM project will ultimately build a database of 3000 cases, each case comprising four images acquired from one patient. While this may seem like a large number, screening programs at typical regional health care facilities (e.g., Sacred Heart Hospital, Pensacola, FL) involve 10,000-12,000 mammograms per year with cancer rates of approximately 0.9 percent. The DDSM project, while not approaching this caseload, will have more cases than any other mammographic database currently under development.

DDSM images are very large and their size heavily impacts system resources. Therefore, hardware components and/or compression techniques must be available or developed to process the images efficiently. For example, (some) images in the DDSM database are 16 bit intensity images which can be as large as 6900x4100 pixels, requiring approximately 56.58 MB for storage. If processing requires floating point numbers, the storage required for one such image can reach 226.32 MB. Read and processed in its entirety, one image of this size presents the user with time and storage complexity that eclipses the capabilities of most current software for viewing, processing, and interpreting structural details to the degree required by practicing radiologists. Therefore, steps must be taken to limit the impact of size while preserving the diagnostic utility of processed outputs.

Various techniques can be used to crop mammogram images to reasonable size. For example, the background (air area) may be cropped from the breast area prior to pixel level analysis. The image can also be spatially compressed by methods such as local averaging over $n \times n$ windows, which compresses the image to $(1/n)^2$ of its original size. While reduction by cropping and compression by local averaging retain gray level resolution (i.e. final images, although smaller, are at the same gray level resolution as the input image, 16 bit), the latter method will reduce the spatial resolution of the image. For example, applying 2×2 local averaging to an original 42 micron image results in an 84 micron image.

Additional preprocessing may include histogram modification to *condition* the image by enhancing specific intensity value ranges. For example, this method can be used to enhance the distinction between pixel intensity values such as 0 and 1. These two intensities can account for up to 20% of all the pixels in a mammogram, so enhancement of this type is important. Histogram modification can also be applied to compress the data from 16 bits to 8 bits. For example, a 16 bit image can be compressed to an 8 bit image by bit shifting from the range [0,...,65535] to [0,...,255], or by applying some function to the input intensity (e.g., square root) which converts the input intensity value to an 8 bit value. These methods do not alter the spatial resolution of the image (i.e. the final image is still 42 micron), but may improve the contrast quality of relatively dark mammographic images by creating a courser gray level resolution. Many other operations can be used when particular abnormalities or artifacts are present in the images. In the next section, we integrate the ideas presented thus far into the design of prototypical system.

12.11 A TYPICAL MAMMOGRAPHIC ANALYSIS SYSTEM

The University of West Florida (UWF) mammographic analysis system uses the DDSM database as a basis for studies about fuzzy models for digital mammography. Figure 12.7 depicts the overall architecture of the UWF system. Each image undergoes a sequence of steps to enhance breast substructure. The fundamental processes are segmentation of the breast area into tissue regions using FCM and structural enhancement by edge detection using the *4 rule Takagi-Sugeno edge detector* (TS4) described in Bezdek et al. (1998b).

We discuss each of the main elements of the system in Figure 12.7 with reference to the image sequences shown in Figures 12.8 and 12.9, which are for a normal patient and a patient with a large tumor. The normal image is case A_0329_1.RIGHT_CC in DDSM, scanned with the DBA scanner. The image with a tumor is case B_9023_1.LEFT_MLO in DDSM (and part of an earlier special volume of abnormal images acquired using the Lumisys scanner). DDSM ground truth is seen in the upper left view of Figure 12.9 as the set of black pixels (manually enhanced for visual clarity) that delineate the very large lesion in this patient.

Segmentation. FCM provides reliable segmentations of many medical images (Bezdek et al., 1997a), but has not, to our knowledge, been used in the mammography domain. We applied FCM clustering to the normal and abnormal input images shown in the upper left views of Figures 12.8 and 12.9. For these runs the input feature was just pixel intensity, and the number of clusters was prespecified as $c = 5$ (assuming background, fatty tissue, fibroglandular tissue, pectoral muscle and abnormalities). The FCM weighting exponent was fixed as $m = 2$, and the FCM objective function used the Euclidean Norm.

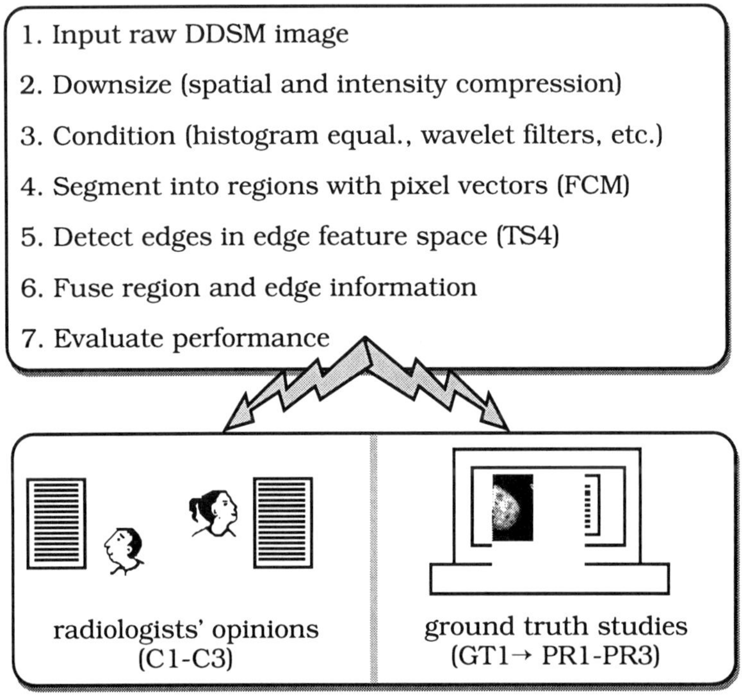

Figure 12.7 A typical mammographic imaging system

The upper right hand views in Figures 12.8 and 12.9 show artificially colored segmentations of these two input images made by hardening the terminal partitions found by FCM (by applying H as in equation (1) to each column of the partition). These results show the influence of the digitization process and the importance of normalizing the images to account for scanner differences prior to comparison of outputs. The image in Figure 12.8 was scanned on a DBA scanner (logarithmic response to OD) while the image in Figure 12.9 was scanned on a Lumisys scanner (linear response to OD). Changes in pixel intensity values do not have the same meaning for data collected by these two scanners. Our intentional failure to normalize intensities for these two images dramatizes this point: compare the processed images in either their segmented or edge image forms to see this yourself.

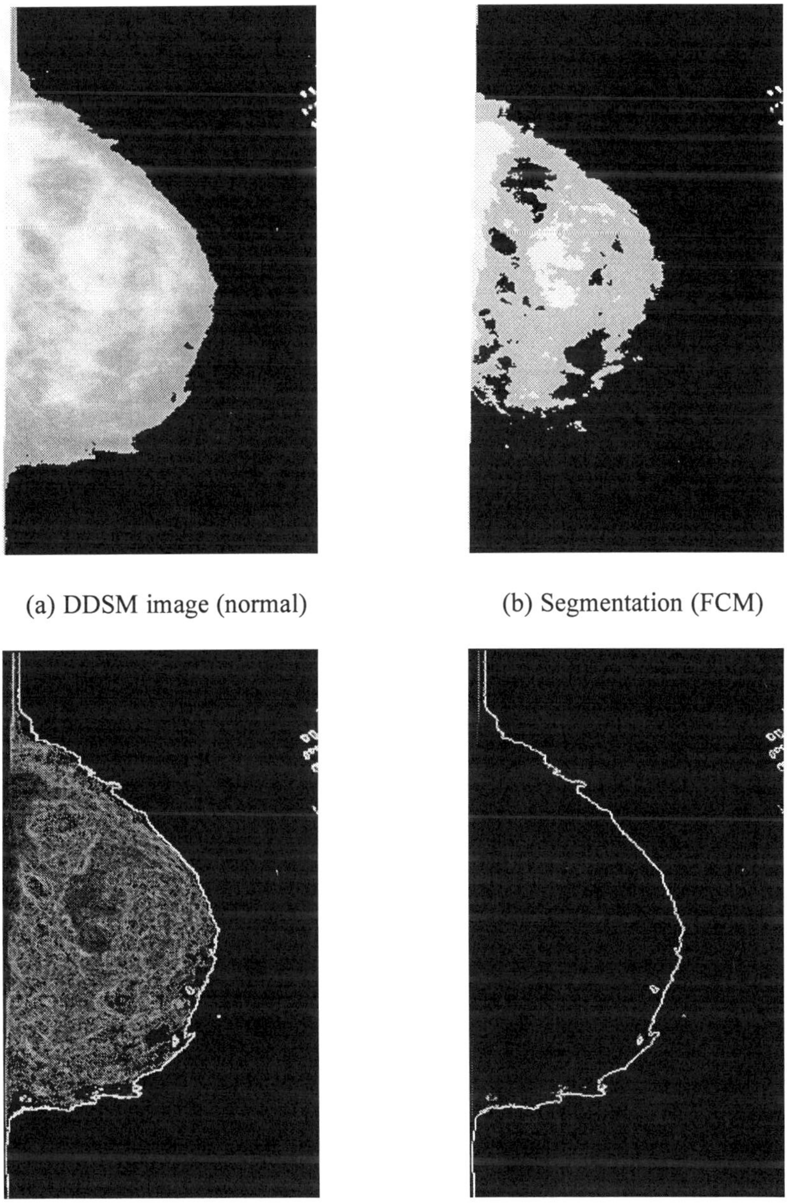

(a) DDSM image (normal)

(b) Segmentation (FCM)

(c) Edge detection (TS4, $\tau = 0.25$)

(d) edge detection (TS4, $\tau = 2.25$)

Figure 12.8 An image without abnormalities

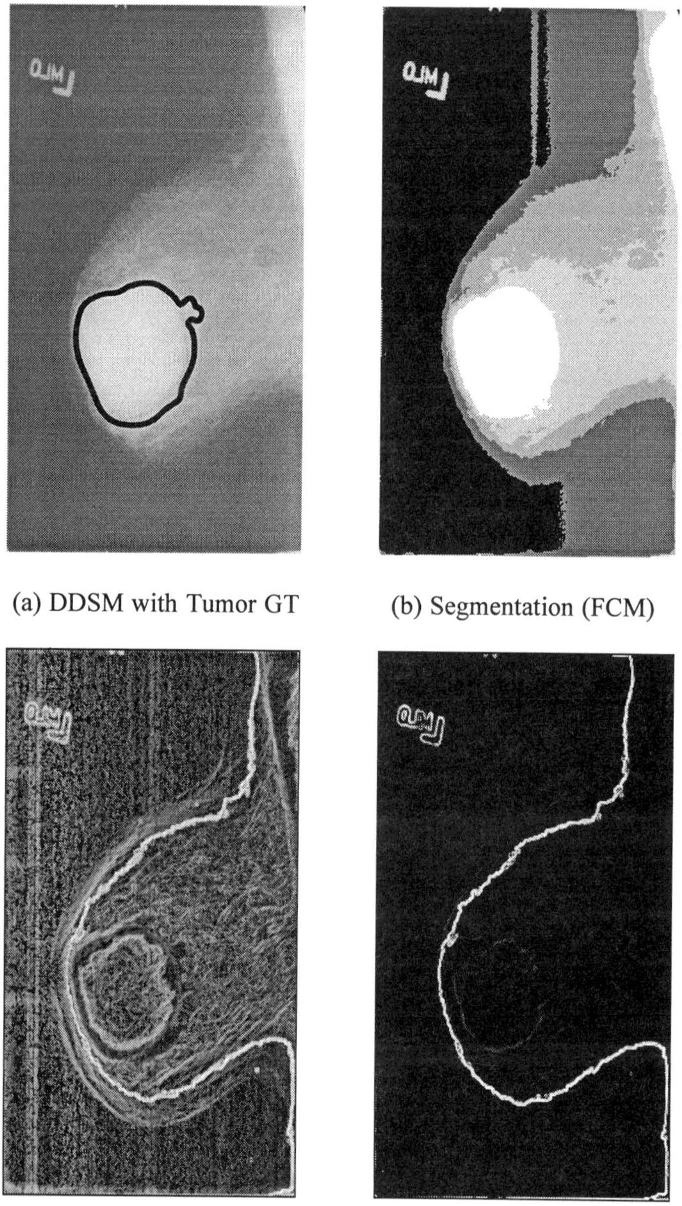

(a) DDSM with Tumor GT

(b) Segmentation (FCM)

(c) Edge detection (TS4, $\tau = 0.25$)

(d) Edge detection (TS4, $\tau = 2.25$)

Figure 12.9 An image with abnormalities

Edge detection. TS4 edge detection (Bezdek et al., 1998b) was used to enhance the substructure of the breasts shown in the lower views of Figures 12.8 and 12.9. The features used in all four edge views were the absolute values of the standard Sobel features. The pairs of views (lower left versus lower right) in Figures 12.8 and 12.9 are markedly different visually, but differ computationally by only the value of a single parameter in the TS4 edge detector (called τ in Bezdek et al., 1998b). Although the images were not normalized to account for scanner differences, the effect of variation in τ is similar in the two sets of images. The lower value of τ (0.25) enhances internal substructure, while the higher value of τ (2.25) enhances the air-skin interface or skin line.

The TS4 edge detection model seems to strengthen weak edges in the breast substructure without overlooking strong ones. Note the level of enhancement and substructure differentiation in Figures 12.8 and 12.9. Using the TS4 model this way can narrow the search for breast cancer indicators. For example, characterization of the margins of masses can provide an indication of malignancy, as can the edge distance and the intensity variation measures around the lesion. On-line viewing of images such as those in the bottom panels of Figures 12.8 and 12.9 by practicing radiologists who can vary τ in near real-time may enable them to select the best (most useful) output(s) for discriminating edge margins and other artifacts of interest within the breast substructure.

Information fusion. Output segmentations in Figures 12.8 and 12.9 are artificially colored, segmented images found by FCM at a preset value for c. The output edge images in 12.8 and 12.9 are unthresholded edge maps representing significant areas of breast substructure. The region and edge images each provide an assessment of the tissue class that should be associated with a pixel (or window of them). Our system will eventually implement a scheme somewhat akin to the fusion procedure recently described by Zugaj and Lattuati (1998).

Performance evaluation. When evaluating the results from our system, edge, region and fused outputs can be presented in sequences to several radiologists along with the original film (when available) and unenhanced digital image. Each radiologist can then provide independent ratings and comments. Visual assessment can also be augmented with quantitative measures such as centroid distance or region of interest overlap using the DDSM ground truth information.

12.12 THREE DIMENSIONAL APPLICATIONS

The last topic in our survey considers research that attempts to extend 2D techniques such as segmentation into three dimensions. Typical applications include volume estimation and visualization, both of which are needed in pre- and post-operative surgical analysis. Non-fuzzy work in this area includes Levin et al. (1989) and Xiaoping et al. (1991). Research along these lines based on either unsupervised and supervised fuzzy methods is quite sparse. We are aware of only a few groups that are currently engaged in either of these activities.

Brandt et al. (1992, 1994) studied the estimation of tissue volume changes with FCM using MR brain images of three normal children and three children of the same age who had hydrocephalus. Proton density and T2 weighted images provided the features. The method was not fully automatic, but was more in the spirit of ssFCM. An operator selected threshold (in T2) was used to set a "noise floor" to filter out low intensity noise pixels as well as pixels which belong to the area between the cerebrum and the skull. In cases where CSF and GM were very close, a user set threshold was necessary to separate them.

Images for each of the six children were segmented with FCM into c = 3 classes; WM, GM and CSF. Three slices per subject were segmented. Each slice was clustered individually and a tissue volume was created by merging the results from individual slices. How clusters were labeled was not discussed. The results showed what was expected - an increased amount of CSF in the hydrocephalic group compared to the control group. The ratio of white matter to gray matter in the hydrocephalic children was much lower than for the normal children, with the percentage of gray matter remaining stable.

These results are both predictable and encouraging. This application of FCM was compared favorably with a standard morphometric measurement approach and confirms work reported by Clark et al. (1994) that indicates FCM is effective in segmenting normal brain tissue. See Chang et al. (1995) for a similar study that investigates the use of fuzzy logic for 3D volume estimation.

The USF-KB system has also been used to estimate tumor volume (Bezdek et al., 1997a). Interslice variability makes direct volumetric clustering of pooled sets of MR slices problematic. From a knowledge perspective, developing a model of the entire structure of the brain, even qualitatively, is extremely complex (if possible at all, but see Hata et al., 1997, 1998). Consequently, the USF-KB system exploits the fact that clustering performs well on individual slices. A 3D segmentation can be constructed by combining labeled tumor areas from each processed slice (as long as there are no gaps) into a single volume.

Qualitative models called templates were added to the USF-KB system to model specific regions (in the axial plane) of the brain. These models capture changes in the brain's anatomical structure at different points in the volume. Also, slices were processed in contiguous order to allow knowledge gleaned from processing one slice to be propagated axially to assist in classification and labeling decisions in spatially adjacent slices. This is roughly equivalent to imposing an extra layer of supervision onto multiple copies of track US_B of Figure 12.5.

Figure 12.10 (a reproduction of Figure 3, Bezdek et al., 1997a) compares the total tumor volume estimated by the USF-KB system with the tumor volume based on hand labeling by radiological experts at 5 times (weeks 0, 7, 13, 16 and 20) during the course of treatment of a brain tumor patient. The patient had diagnosed Glioblastoma multiforme, and was undergoing both chemo and radiation therapy during the 20 week period. There were approximately 9 slices per volume in the axial plane (each 5 mm thick with no gaps between slices) at each of the five sampling periods. The graph shows that the unsupervised knowledge-based system very closely models physician generated ground truth.

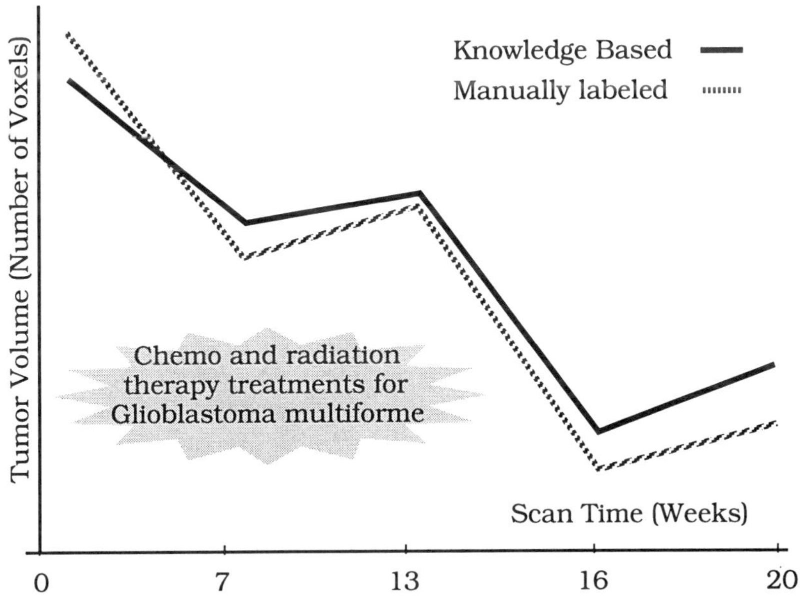

Figure 12.10 Knowledge-based vs. manual volume estimates (Bezdek et al., 1997a)

Results from a series of experiments at USF to measure tumor volume from MR brain images using ssFCM, the k-nn rule and a seed growing approach called ISG are reported in Vaidyanathan et al. (1995). The objective was to determine how sensitive these three methods were to training data chosen by observers (technicians with some medical physics training). It is difficult to choose training data from the raw images such that the measured tumor volume is consistent over multiple trials. Four patient cases were used in the study with repeat scans available for 2 patients (3 for patient 1, and 5 for patient 2). Results are given in terms of averages over the 10 tumor volumes obtained from the 4 patient cases. The tumors involved were either meningioma or glioblastoma multiforme. All patients were undergoing therapy during the course of repeat scans. Experiments reported by Vaidyanathan et al. (1995) are summarized in Table 12.5, where the columns report tumor volume variability (in percent) resulting from the choice of the training data for each trial.

Table 12.5 Variability in % of tumor volumes

	ssFCM	ISG	k-nn
Intra-observer	6	6	9
Inter-observer	4	17	5

The differences in volume estimates for multiple training sets chosen by one or more observers are indicated by the two rows in Table 12.5. This experiment indicates that, for the cases studied, ssFCM and ISG are less sensitive to the choice of training data by a single observer than the k-nn rule, and when more than one observer extracts training data, ISG becomes very unstable, but ssFCM and k-nn are better.

Herndon et al. (1996) approach gray and white matter volume estimation with MR images in a completely different manner. Their study uses only T1 images. First, six experts created crisp classifications for each pixel in one of four tissue classes, CSF, WM, GM or fat. Two estimates by each expert for each image were averaged, yielding a crisp range of T1 (msec) values for each class. The six experts each did this for 9 images (3 patients, 3 slices per patient), so the group provided a collective set of 54 crisp classifications of the four tissue classes in real T1 images.

Then, the relative agreement of the 54 experts was aggregated graphically, producing empirical distributions for each tissue class. These were subsequently fit by exponential membership functions with experimentally determined parameters, leading to an analytical model that attaches a possibilistic label vector to each pixel in an image. The components of this label vector are interpreted by Herndon et al. (1996) as estimates of the *voxel fraction* of tissue in each represented class. Multiplication of the pixel label vector by voxel volume (slice thickness times pixel area) produces an estimate of the volume of each tissue *for each pixel.* Summation of these estimates over any ROI in the T1 image produces a per slice estimate of ROI tissue volume. Finally, volume estimates over the same spatial ROI in adjacent slices are added, yielding an overall estimate of volume of each tissue for the slice set.

Herndon et al. (1996) report that in most of the cases tested using the fuzzy classifier with sets of slices with thicknesses of 1, 3 or 5 mm, the difference between computed and actual volumes was less than 5%. Predictably, their best results (generally less than 2% error) occurred for sets of 1 mm slices.

Hata et al. (1997, 1998) discuss a 3D approach that starts with segmentation of MR slices, followed by aggregation into volumes, and finally 3D rendering for visualization based almost *entirely* on rule-based processing. Measured and derived numerical features are fed directly into a fuzzy rule base which produces a membership function valued on the entire image (Section 12.8). Thresholding this function crisply segments each 2D slice in the input data. Structural knowledge of the relevant human anatomy empowers the rules that subsequently aggregate slice regions into 3D volumes.

The data used by Hata et al. (1997, 1998) to set up their model are 8 bit MR data in three planes (TR, TE, flip angle) for 124 contiguous 1.5 mm thick slices, resulting in a data set with $124 \times 256 \times 256$ *voxels* having gray levels in the set $\{0,\ldots,255\}$. The 1997 paper considered segmenting this data into three ROIs (whole brain, WM and GM); the 1998 paper extends the model to another set of four new ROIs (left cerebral hemisphere, right cerebral hemisphere, cerebellum, and brain stem).

Hata et al. allude to testing the model on a set of 36 similar MR voxel data sets, and state that they obtain an average error rate of 2.7%. However, it is not clear from their discussion what labels or ground truth enable them to make error rate calculations - they say only that comparisons to "genuine" ROIs are made. An example of 3D rendering for visualization of each of the ROIs is given.

Hata et al. (1998) also describe the application of their model to CT data of a human foot. The data for this application were 512×512 images with gray levels in the set $\{0,\ldots,2048\}$. Sets of 64 contiguous 3.44 mm thick slices thus gave $64 \times 512 \times 512$ voxels. Visualization imagery is presented to illustrate the viability of this technique, but no quantitative evaluation is offered.

The work of Hata et al. (1997, 1998) is most like our track Su in Figure 12.5, again with the added layer needed for three dimensional data. Perhaps the most interesting aspect of this work is its nearly total reliance on fuzzy input-output systems, instead of the more usual technique of combining such a system with more conventional types of processing such as done by Li et al. (1993) or Chang et al., (1995).

Delliapiane (1997) offers several examples of the fuzzy isovolume method. One MR slice of a normal patient shows an open path of best connectivity between two points within a CSF region. This is used to illustrate the basic idea: assign crisp labels to all the pixels (or voxels) along maximally connected paths chosen by thresholding the connectivity measure.

A second example is given that uses real 3D magnetic resonance tomography data from a patient with a brain tumor. Examples of two isovolumes (for the ventricle and brain) based on the voxel data are given. Perhaps more interestingly, Dellipiane (1997) also uses this data to demonstrate that 2D information - say, the location of an isoregion in a single slice, is better found by restricting the results of real 3D processing to the 2D slice domain.

Udupa and Samarasekera (1996) illustrate their fuzzy connected components approach to volume rendering by growing object (bone and soft tissue) regions from seed voxels selected from three dimensional CT data from a patient's knee. Udupa and Samarasekera (1996) also describe and illustrate two MR based studies.

In the first application, the clinical aim is to identify and visually render regions of constriction, narrowing or stenosis in MR angiography. The second application illustrates the use of fuzzy connectedness for segmentation and volume computations of various tissue structures (WM, GM and ventricle) in MR images of patients with multiple sclerosis lesions in the human brain. This study was based on upwards of 400 scenes (collected on site at the University of Pennsylvania). Segmentations and rendered volumes look quite reasonable to us, and since the authors *are* radiologists, performance evaluation utilizes clinically knowledgeable human experts. The work of this group exhibits very strong ties between the mathematical, computational and clinical aspects of image processing in medicine, and follows a pretty non-traditional line. It is well worth reading.

Bloch (1995, 1996, 1997) has done extensive work describing how various properties of fuzzy morphology definitions can be used as tools for data fusion and decision making in fuzzy set frameworks. In Bloch (1995), families of fuzzy mathematical morphology operators for erosion, dilation, opening, and closing are investigated. A thorough comparison of six definitions of morphology and fuzzy sets is presented in terms of logic and decision theory.

In Bloch (1996), a classification scheme is developed for the main operators used in numerical data fusion theories to combine information from multiple sensors. In this work, three classes of operators are defined based on properties such as decisiveness and ability to handle conflicting information. For example, context independent constant behavior operators do not consider external information and have the same behavior whatever the values of the information to be combined. Alternatively, context independent variable behavior operators depend on the values of the variables to be combined (e.g., behaving one way if both values are low). Finally, context dependent operators depend on global knowledge or measures (e.g., reliability) of the sources to be combined. Synthetic examples are provided to demonstrate how probability and Bayesian inference, fuzzy sets, possibility theory, MYCIN-like systems, and Dempster-Shafer evidence theory fit into the proposed classification. Criteria for determining the choice of operator is also described.

All the above concepts are utilized in Bloch's approach to three-dimensional reconstruction of blood vessels to assist vascular lesion interpretation tasks (Bloch, 1997). Fuzziness is incorporated in four different ways (i.e. segmentation, modelling imprecision, mathematical morphology and data fusion), without a priori geometrical model information. The approach fuses information from digital angiographic and echographic data to make a final binary decision, solving possible contradictions between the two modalities. Redundancy in the combined data from two orthogonal X-ray angiographic projections (which provide longitudinal information and overall 3D geometry of the vessel), and a series of endovascular echographic (EE) slices (high resolution vessel cross sections) helps to reduce imprecision and uncertainty in the final model. Segmentation of the EE images is based on fuzzy classification and mathematical morphology, whereas the digital angiographic images are segmented based on dynamic tracking of vessel centerlines and contours. During model reconstruction, fuzzy dilation is used to handle the spatial imprecision of the detected contours. The data is then fused using a fuzzy operator, and a binary decision on the contour is based on a 3D watershed algorithm which connects maximal membership points. Example images from a dog aorta are provided.

12.13 CONCLUSIONS AND DISCUSSION

Some (non-neural) fuzzy methods for segmentation and edge detection in medical images have been reviewed. The examples given suggest that fuzzy models are useful for medical image analysis, but that we still have a long way to go. To conclude, we first offer some remarks about improving the techniques mentioned in this chapter.

Segmentation by clustering, mathematical or human validation; human tissue labeling (Track US_A, Figure 12.5)

The better unsupervised techniques are still too slow. Improving speed via parallelization and optimization will improve their competitiveness with, e.g., neural networks, which appear most often in track Su, but can also appear as clustering networks in this track. There are a number of methods for speeding up FCM. Of these, perhaps the most promising is offered by reducing the size of the image it is applied to, either by random sampling (Sankar and Pal, 1994) or by preprocessing steps with fuzzy rules before clustering (Chang et al., 1995).

Development of dynamic cluster validity analysis also shows promise. Unsupervised methods need better ways to specify and adjust c, the number of tissue classes sought by the algorithm. A recent study of 23 validity indices of all three types - crisp, fuzzy and probabilistic - concluded that mathematical validation for mixtures of normal distributions was generally unreliable (Bezdek et al., 1997b).

The VGC method described here seems useful for a limited objective, but in general, the main problem persists: if c is unknown, is there an automatic way to deduce that your algorithm has found an interpretation of the image that humans will find agreeable? Many experimental studies of countless validity indices with various data sets offer little cause for optimism.

Our opinion is that much research needs to be done before much reliance can be placed on mathematical validation. Human evaluation and labeling will probably continue to be the most popular and certainly the most reliable method for systems using the track US_A approach, but this is subjective, time-consuming (and therefore expensive), and susceptible to errors introduced by human factors such as fatigue.

Initialization of unsupervised iterative methods such as FCM is a third important area of research. Much effort has been invested in ways to initialize unsupervised methods at points in the solution space that don't lead to local trap states (the FCM($\mathbf{V}_{M3}$) method discussed in Section 12.6, for instance). This is a pervasive problem for all gradient based techniques, and can be very serious in the application domain discussed here. The error functions that are being optimized have very large numbers of unknowns, so there will be a concomitant increase in the number of local solutions that make an algorithm happy but offer little to medical practitioners. One way to presumably circumvent this problem is to optimize with genetic algorithms (Hall et al., 1999), but this class of techniques still carries too much computational overhead to be practical for image analysis.

Segmentation by clustering; rule-based validation and tissue labeling (Track US_B, Figure 12.5)

The most desirable situation is that the evolution of some form of track US_B will eventually lead to a library of pre-labeled prototypical images which can be used to train a generic classifier, thereby eliminating intervention by humans to supply tissue labels for clusters formed by algorithms in this track. However, variations in real images from location to location in the same patient, from patient to patient, and from sensor to sensor make this possibility rather remote at present.

Image understanding groups have long since recognized the need for fusion of low- and mid-level data and information. One of the hallmarks of track US_B is the use of hybrid models that fuse low-level, pixel-based processing with mid-level rules, facts, heuristics and rules of thumb in knowledge-based systems. This is the trend in both medical and non-medical image understanding efforts. The papers reviewed here support this trend, and we believe that track US_B will ultimately yield a truly unsupervised design for (limited objective) medical image analysis.

At the input end, knowledge-based labeling with well established rules can dramatically reduce the number of unlabeled pixels (windows, cases, etc.) that need to be classified. And at the output end, rules can be used to "clean up" unsupervised outputs and then label the final tissue classes.

We believe that research in this area will grow, as will the achievement of dynamic cluster validity using cluster merging and splitting rules. These rules will be both mathematically and teleologically based. Aggregation of evidence using Sugeno-type fuzzy integrals should be of great help in this area.

Supervised segmentation; operator derived training data and tissue labels; classification of remaining pixels (Track Su, Figure 12.5)

Experience and careful documentation of many case studies in a particular area (such as segmentation of MR imagery) are needed to make any real progress in this area. The instability of supervised techniques to training sets across different sensors, patients and even within MR slices for the same patient might be improved by the generation of more globally representative training sets.

The growth industry in track Su is probably in neural networks and their fuzzy relatives, a topic that we purposely eschewed in this chapter. A careful survey of network based approaches to computational medical image analysis would be a welcome companion to our survey (and might be extensive enough to warrant its own volume in this handbook!).

However, without expert rules, we feel that supervised learning for segmentation of medical images - with or without the new network approaches that are nowadays so fashionable - has little chance to become the backbone of computational image analysis in fielded systems that are reliable enough to be accepted by the health care industry.

Mammography

The impact of automated computer interpretation systems which preprocess mammogram images is well established (Pisano and Shtern, 1994). Image processing steps such as segmentation and edge detection lead to image enhancements that will significantly ease the workload of practicing radiologists. The underlying premise of our review in this area is that fuzzy models have been successfully applied to a wide variety of other medical applications (e.g., brain tumor detection and volume estimation from magnetic resonance imagery) with excellent results. However, each data domain and application has unique peculiarities which necessitate adjustments to or even replacement of existing methods. Experiments reported so far indicate potential success for computational mammography with fuzzy models, but much needs to be done. The TS4 edge detection paradigm (or some relative of it) offers a rich platform for interactive experiments that can be done with practicing clinicians; we believe that this avenue will be explored and improved.

The use of wavelets for feature extraction in mammographic studies has increased dramatically in the last five years or so. Beyond the usual seduction of a new buzzword, however, there seems to be a very good reason for interest in this technique. Unlike global procedures such as the fast Fourier transform, wavelets offer a way to examine image substructure for local details. Since practicing clinicians do just this (focus their attention on regions of interest within the field of view), we believe that wavelet based features will become a mainstay in some areas of computational mammography (e.g., in the detection of microcalcifications).

Some final thoughts

The general accuracy of computational techniques that perform well needs to be further investigated across different imaging devices, types of patient problems, and medical communities. Perhaps the largest single impediment to real success in this domain is the lack of a very large (e.g., 50,000 sets of patient slices for MR studies of one affliction), well-documented database. The DDSM database, if constructed as advertised, offers a fairly good basis for mammography research, but still falls far short of what we believe to be an adequate number of cases for training and testing. There are many other important areas in medicine that share imaging sensors as a common thread (e.g., cancer of the cervix, prostate cancer, bone cancer, heart disease, lung disorders, etc.). The construction of such databases, that can be shared worldwide, is difficult and expensive. On the other hand, the reward - delivery to the health care industry of an economical means of vastly improved diagnosis, treatment and recovery procedures - is well worth the cost.

The classification of suspicious areas in a medical image should, when possible, be compared quantitatively to ground truth patient information, in terms of the number of false positives and false negatives. Furthermore, error rates only tell part of the story. They should always be benchmarked against visual assessment by practicing clinicians. The development of useful instruments for solicitation from and interaction between medical personnel and computer scientists is an area for careful research that is often ignored. For example, in mammography, useful judgmental functions for radiologists include assessment of the enhancement quality of the breast skin line and breast substructure, the sufficiency of the detail level within and outside these areas, and the level of differentiation between normal and abnormal tissues. However, providing outputs for these assessments and user friendly communication channels to convey them are non-trivial issues, and this aspect of medical image analysis is usually given scant attention by computer professionals. More work needs to be done to involve qualified human factors personnel throughout the system development process in the design of instruments that provide such feedback.

In terms of the time and space complexity of algorithms, the modification of spatial and intensity resolution has important consequences. The impact of varying these on system resources and the quality and utility of enhancements (contrast and structural detail) from whatever algorithms are used must be accounted for during investigations based on medical sensor data. Loss of resolution to satisfy computational constraints can degrade digital representations so much that the processing is medically useless.

Finally, fusion of outputs (multiple classifiers) is very fashionable nowadays, and rightfully so. Information of different types certainly resides in different sensors and processing streams. Multiple fuzzy models can be integrated to essentially cast a weighted vote for "significant" areas within an image, and we believe that the fuzzy integral will be useful for this purpose. The use of fuzzy integrals (cf. Krishnapuram and Lee, 1992, Bezdek et al., 1999) for combining confidences about pixels and pixel windows has enjoyed wide success in other application domains, and thus seems appropriate here as a useful means for fusion of multiple output streams. Kuncheva (1993) illustrates another way to increase the classification accuracy of medical decision support systems based on the use of fuzzy neurons in an aggregation network. Kuncheva's model (and many other neural network models that we have not covered in this chapter) does not explicitly concern medical imagery, but we suspect that methodologies of this kind will be useful in the domain of computational medicine.

REFERENCES

American College of Radiology (ACR), (1995). *Breast imaging reporting and data system (BI-RADS)*, 2nd edition, Reston, VA, American College of Radiology.

Apple Computer, Inc. (1997). *LaserWriter 16/600 PS product information*, Apple Computer, Inc.

Ball, G. and Hall, D.A. (1967). A clustering technique for summarizing multivariate data, *Behav. Sci.*, 12, 153-155.

Bensaid, A., Hall, L.O., Bezdek, J.C. and Clarke, L. (1996b). Partially supervised clustering for image segmentation, *Pattern Recognition*, 29(5), 859-871.

Bensaid, A., Hall, L.O., Bezdek, J.C., Clarke, L., Silbiger, M., Arrington, J. and Murtaugh, R. (1996a). Validity-guided (re)clustering with applications to image segmentation, *IEEE Trans. Fuzzy Systems*, 4(2), 112-123.

Bezdek, J.C. and Pal, S.K. (1992). *Fuzzy Models for Pattern Recognition*, IEEE Press, Piscataway, NJ.

Bezdek, J. C., Chuah, S. and Leep, D. (1986). Generalized k-nearest neighbor rules, *Fuzzy Sets and Systems*, 8(3), 237-256.

Bezdek, J.C., Hall, L.O., Clark, M.C., Goldgof, D.B. and Clarke, L.P. (1997a). Medical image analysis with fuzzy models, *Stat. Methods in Medical Research*, 6, 191-214.

Bezdek, J. C., Hall, L.O., and Clarke, L.P. (1993). Review of MR image segmentation techniques using pattern recognition, *Med. Physics*, 20, 1033-1047.

Bezdek, J.C., Keller, J., Krishnapuram, R. and Pal, N. R. (1999). *Fuzzy models and algorithms for pattern recognition and image processing*, Kluwer, Norwell, MA.

Bezdek, J. C., Li, W.Q., Attikiouzel, Y. A. and Windham, M.P. (1997b). A geometric approach to cluster validity for normal mixtures, *Soft Computing*, 1, 166-179.

Bezdek, J. C., Chandrasekhar, R. and Attikiouzel, Y. A. (1998b). A geometric approach to edge detection, *IEEE Trans. Fuzzy Systems*, 6(1), 52-75.

Bloch, I. (1996). Information combination operators for data fusion: A comparative review with classification. *IEEE Trans. on Systems, Man, and Cybernetics-Part A: Systems and Humans*, 26(1), 52-67.

Bloch, I. and Maitre, H. (1995). Fuzzy mathematical morphologies: A comparative study. *Pattern Recognition*, 28(9), 1341-1387.

Bloch, I., Sureda, F., Pellot, C. and Herment, A. (1997). Fuzzy modelling and fuzzy mathematical morphology applied to 3D reconstruction of blood vessels by multimodality data fusion, in *Fuzzy Information Engineering: A Guided Tour of Applications*, ed. D. Dubois, H. Prade and R.R. Yager, John Wiley and Sons, NY, 93-110.

Bombardier, V., Jaulent, M.-C., Bubel, A. and Bremont, J. (1997). Cooperation of two fuzzy segmentation operators for digital substract angiograms analysts. *Proc. of the Sixth IEEE International Conference on Fuzzy Systems*, 2, 1057-1062.

Boudraa, A. E., Champier, J., Damien, J., Besson, J. E., Bordet, J. C., and Mallet, J.J. (1992). Automatic left ventricular cavity detection using fuzzy ISODATA and connected-components labeling algorithms, *Proc. 14th IEEE Conf in Medicine and Biology*, CH3207-8, Piscataway, NJ, 1895-1896.

Boudraa, A. E., Mallet, J. J., Besson, J. E., Bouyoucef, S. E. and Champier, J. (1993). Left ventricle automated detection method in gated isotropic ventriculography using fuzzy clustering, *IEEE Trans. Med. Imaging*, 12(3), 451-465.

Boujeema, N., Stamon, G. and Lemoine, J. (1992a). Toward fuzzy models for visual perception, *SPIE Proc. 1818, Visual Communication and Image Processing*, Bellingham, WA, 1271-1281.

Boujeema, N., Stamon, G. and Lemoine, J. and Petit, E. (1992b). Fuzzy ventricular endocardium detection with gradual focusing decision, *Proc. International Conference of the IEEE EMBS,* 14, Piscataway, NJ, 1893-1894.

Boujemaa, N. and Stamon, G. (1994). Fuzzy modeling in early vision application to medical image segmentation, in *Proc. 7th INt'l Conf. on Image Analysis and Processing*, ed. S. Impedovo, World Scientific Press, Singapore, 649-656.

Brandt, M. E. and Kharas, Y. F. (1993). Simulation studies of fuzzy clustering in the context of brain magnetic resonance imaging, *Proc. 3rd Int'l. Conf. on Industrial Fuzzy Control and Intelligent Systems,* IEEE Press, Piscataway, NJ, 197-203.

Brandt, M. E., Bohan, T. P., Kramer, L. A. and Fletcher, J. M. (1994). Estimation of CSF, white and gray matter volumes in hydrocephalic children using fuzzy clustering of MR images, *Computerized medical imaging and graphics,* 18(1), 25-34.

Brandt, M. E., Fletcher, J. M. and Bohan, T. P. (1992). Estimation of CSF, white and gray matter volumes from MRIs of hydrocephalic and HIV-positive subjects, *Proc. SimTec/WNN*, IEEE Press, Piscataway, NJ, 643-650.

Brzakovic, D., Luo, X. M. and Brzakovic, P. (1990). An approach to automatic detection of tumors in mammograms, *IEEE Trans. on Medical Imaging,* 9(3), 233-241.

Cannon, R.L., Dave, J. and Bezdek, J.C. (1986). Efficient implementation of the fuzzy c-means clustering algorithms, *IEEE Trans. PAMI,* 8(2), 248-255.

Carazo, J.M., Rivera, F.F., Zapata, E.L., Radermacher, M. and Frank, J. (1990). Fuzzy sets-based classification of electron microscopy images of biological macromolecules with an application to ribosomeal particles, *Jo. Microscopy*, 157(2), 187-203.

Chandrasekar, R (1996). *Systematic Segmentation of Mammograms*, PhD Thesis, U. of W. Australia, Perth, 1996.

Chang, C. W., Hillman, G.R., Ying, H., Kent, T. and Yen, J. (1994). Segmentation of rat brain MR images using a hybrid fuzzy system, *Proc. NAFIPS/IFIS/NASA'94*, San Antonio, 55-59.

Chang, C. W., Hillman, G.R., Ying, H., Kent, T. and Yen, J. (1995). Automatic labeling of human brain structures in 3D MRI using fuzzy logic, *Proc. CFSA/IFIS/SOFT '95*, World Scientific, 27-34.

Chang, C.W., Hillman, G.R., Ying, H. and Yen, J. (1995). A two stage human brain MRI segmentation scheme using fuzzy logic, *Proc. 1995 FUZZ-IEEE*, IEEE Press, Piscataway, NJ, 649-654.

Clark, M., Hall, L.. Goldgof, D., Clarke, L., Velthuizen, R. and Silbiger, M. (1994). MRI segmentation using fuzzy clustering techniques: integrating knowledge, *IEEE Engineering in Medicine and Biology Magazine*, 13(5), 730-742.

Clark, M., Hall, L.O., Goldgof, D.B., Velthuizen, R., Murtagh, F.R., and Silbiger, M.S. (1998). Automatic tumor segmentation using knowledge-based techniques, *IEEE Trans. on Medical Imaging*, 17(2), 187-201.

De La Paz, R., Berstein, R., Hanson, W. and Walker, M. (1986). Approximate fuzzy c-means (AFCM) cluster analysis of medical magnetic resonance image (MRI) data - a system for medical research and education, *IEEE Trans. Geosci. and Remote Sensing*, GE25, 815-824.

de Oliveira, M.C. and Kitney, R. I. (1992). Texture analysis for discrimination of tissues in MRI data, *Proc. Computers in Cardiology*, IEEE press, Piscataway, NJ, 481-484.

Dellepiane, S. (1991). Image segmentation: Errors, sensitivity and uncertainty, *IEEE EMBS*, 253-254.

Dellepiane, S. (1997). The active role of 2-D and 3-D images: semi-automatic segmentation, *Contemporary perspectives in three-dimensional biomedical imaging*, ed. C. Roux and J. L. Coatrieux, IOS press, 165-189.

Dellepiane, S., Fontana, F. and Vernazza, G. L. (1996). Nonlinear image labeling for multivalued segmentation, *IEEE Trans. Image Processing*, 5(3), 429-446.

Di Gesu, V. (1994). Integrated fuzzy clustering, *Fuzzy Sets and Systems*, 68, 293-308.

Di Gesu, V. and Romeo, L. (1994). An application of integrated clustering to MRI segmentation, *Pattern Recognition Letters*, 731-738.

Di Gesu, V., De La Paz, R., Hanson, W.A. and Berstein, R. (1991). Clustering algorithms for MRI, in *Lecture notes for medical informatics*, K.P. Adlassing, B. Grabner, S. Bengtsson and R. Hansen, eds., Springer, Berlin, 534-539.

Green, W. B. (1989). *Digital image processing: a systems approach*, Van Nostrand Reinhold, New York, NY, 2nd edition.

Gustafson, E. E. and Kessel, W. (1978). Fuzzy clustering with a fuzzy covariance matrix, *Proc. IEEE CDC*, 761-766.

Hall, L. O. (1996). Learned fuzzy decision rules vs. decision trees in classifying microcalcifications in mammograms, *SPIE Proc. on Applications of Fuzzy Logic Technology III*, 2761, ed. B. Bosacchi and J. C. Bezdek, SPIE Press, Bellingham, WA, 54-61.

Hall, L. O., Ozyurt, I. B. and Bezdek, J. C. (1999). Clustering with a genetically opitimzed approach, *IEEE Trans. on Evolutionary Computation*, 3(2). 103-112.

Hata, Y., Hirano, S., and Kamiura, N. (1998). Medical image granulation by fuzzy inference, *Proc. 1998 NAFIPS Conference* ed. L. Hall and J. C. Bezdek, NAFIPS Press, Pensacola, FL, 188-192.

Hata, Y., Kobashi, N., Kamiura, N. and Ishikawa, M. (1997). Fuzzy logic approach to 3D magnetic resonance image segmentation, *Infor. Proc. in Medical Imaging*, Lecture notes in Comp. Science, v.1230, 387-392.

Herndon, R. C., Lancaster, J. L., Toga, A. W. and Fox, P. T. (1996). Quantification of white and gray matter volumes from T1 parametric images using fuzzy classifiers, *J. Mag. Res. Imaging*, 6(3), 425-435.

Hillman, G. R., Chang, C.W., Ying, H., Kent, T.A. and Yen, J. (1995). Automatic system for brain MRI analysis using a novel combination of fuzzy rule-based and automatic clustering techniques, *SPIE Proc. Med. Imaging 1995 : Image Processing*, v2434, ed. M. H. Lowe, Bellingham, WA, 16-25.

Hume, A., Thanisch, P. and Hartswood, M. (1996). On the evaluation of microcalcification detection algorithms, *Proc. Third International Workshop on Digital Mammography*, 71-72.

Huo, Z., Giger, M. L., Olopade, O. I., Wolverton, D. E., Zhong, W., Tahoces, P. G., Narvid, S. I., and Doi, K. (1996). Computer-aided diagnosis: breast cancer risk assessment from mammographic parenchymal pattern in digital mammograms, *Proc. Third International Workshop on Digital Mammography,* 87.

Hyman, T.J., Kurland, R.J., Levy, G.C. and Shoop, J.D. (1989). Characterization of normal brain tissue using seven calculated MRI parameters and a statistical analysis system, *Mag. Res. in Med.*, 11, 22-34.

Jain, A. K. and Flynn, P. J. (1996). Image segmentation by clustering, in *Advances in image understanding*, ed. K. Bowyer and N. Ahuja, IEEE Computer Society Press, Los Alamitos, CA., 65-83.

Jain, R., Kasturi, R. and Schunck, B. G. (1995). *Machine Vision*, McGraw-Hill, NY.

Just, M. and Thelen, M. (1988). Tissue characterization with T1, T2 and proton density values: Results in 160 patients with brain tumors, *Radiology*, 169, 779-785.

Keller, J., Gray, M. and Givens, J. (1985). A fuzzy k-nearest neighbor algorithm, *IEEE Trans. SMC,* 15, 580-585.

Keller, J., Krishnapuram, R., Gader, P.D. and Choi, Y.S. (1996). Fuzzy rule-based models in computer vision, in *Fuzzy Modeling: Paradigms and Practice*, ed. W. Pedrycz, Norwell, MA, 353-374.

Keller, J.M., Moore, J. and Gader, P. (1998). A fuzzy logic approach to detector scoring, *Proc. 1998 NAFIPS*, ed. J. C. Bezdek and L.O. Hall, Pensacola Beach, FL, 339-344.

Kopans, D. B. (1996). The potential impact of computer-aided diagnosis on clinical mammography, *Proc. Third International Workshop on Digital Mammography*, 35.

Kovalerchuk, B., Triantaphyllou, E., Ruiz, J.F., and Clayton, J. (1997). Fuzzy logic in computer-aided breast cancer diagnosis: Analysis of lobulation, *Artificial Intelligence in Medicine*, 11(1), 75-85.

Krishnapuram, R. and Keller, J. (1993). A possibilistic approach to clustering, *IEEE Trans. Fuzzy Systems*, 1(2), 98-110.

Krishnapuram, R. and Lee, J. (1992). Fuzzy-set-based hierarchical networks for information fusion in computer vision, *Neural Networks*, 5, 335-350.

Kuncheva, L. I. (1993). An aggregation of pro and con evidence for medical decision support systems, Comput. Biol. Med., 23(6), 417-424.

Levin, D.N, Hu, X., Tan, K.K., Galhotra, S., and Pellizzari, C. (1989). The brain: integrated three-dimensional display of MR and PET images, *Radiology*, 172, 783-789.

Li, C., Goldgof, D. and Hall, L. (1993). Knowledge-based classification and tissue labeling of MR images of human brains, *IEEE Trans. Med. Imaging*, 12(4), 740-750.

Li, H. D., Kallergi, M., Clarke, L. P., Jain, V. K. and Clark, R. A. (1995). Markov random field for tumor detection in digital mammography, *IEEE Transactions on Medical Imaging*, 14 (3), 565-576.

Liang, Z. (1993). Tissue classification and segmentation of MR images, *IEEE EMBS Magazine*, 12(1), 81-85.

Lo, S.-C. B., Lin, J.-S., Li, H., Hasegawa, A., Tsujii, O., Freedman, M. T. and Mun, S. K. (1996). Detection of clustered microcalcifications using fuzzy classification modeling and convolution neural network, *SPIE Proc. 2710, Medical Imaging: Image Processing*, 8-15.

Menhardt, W. and Schmidt, K.H. (1988), Computer vision on magnetic resonance images, *Pattern Recognition Letters,* 8, 73-85.

Morrison, M. and Attikouzel, Y. (1994). An introduction to the segmentation of Magnetic Resonance images, *Aust. Comp. Jo.*, 26(3), 90-98.

Namasivayam, A. and Hall, L. O. (1995), The use of fuzzy rules in classification of normal human brain tissues, *Proc. ISUMA-NAFIPS'95*, 157-162.

Pedrycz, W. (1985). Algorithms of fuzzy clustering with partial supervision, *Patt. Recog. Letters*, 3, 13-20.

Pisano, E. D. and Shtern, F. (1994). Image processing and computer aided diagnosis in digital mammography: A radiologist's perspective, in *State of the Art in Digital Mammographic Image Analysis,* ed. Bowyer, K. W. and Astley, S., Singapore, 280-291.

Raman, S.V., Sarkar, S. and Boyer, K.L. (1991). Tissue boundary refinement in magnetic resonance images using contour-based scale space matching, *IEEE Trans. on Med. Imag*. 10(2), 109-121.

Rosenfeld, A. (1984). The fuzzy geometry of image subsets, *Patt. Recog. Letters*, 11, 311-317.

Roubens, M. (1982). Fuzzy clustering algorithms and their cluster validity, *Eur. Jo. Oper. Res.*, 10, 294-301.

Sameti, M. and Ward, R. K. (1996). A fuzzy segmentation algorithm for mammogram partitioning, in *Digital Mammography '96*, ed. Doi, K., Giger, M.L., Nishikawa, R.M., and Schmidt, R.A., Elsevier Science Publishers, 471-474.

Sameti, M., Ward, R. K., Palcic, B., and Morgan-Parkes, J. (1997). Texture feature extraction for tumor detection in mammographic images, *Proc. IEEE Pacific Rim Conference on Communications, Computer and Signal Processing*, 831-834.

Shankar, B.U., and Pal, N.R. (1994). FFCM: An effective approach for large data sets, *Proc, 3rd Int'l. Conference on Fuzzy Logic, Neural Nets and Soft Computing*, IIZUKA, Fukuoka, Japan, 331-332.

Strickland, R. N. and Theodosiou, T. (1998). Fuzzy system for detecting microcalcifications in mammograms, *Proc. SPIE Applications and Science of Neural Networks, Fuzzy Systems and Evolutionary Computation*, 3455, eds. B. Bosacchi, J. C. Bezdek and D. Fogel, SPIE Press, Bellingham, WA, 317-327.

Sutton, M. A. and Bezdek, J. C. (1997). Enhancement and analysis of digital mammograms using fuzzy models, *SPIE Proc. 3240, 26th AIPR Workshop: Exploiting New Image Sources and Sensors*, SPIE Press, Bellingham, Washington, 179-190.

Udupa, J. K. and Samarasekera, S. W. (1996). Fuzzy connectedness and object definition: theory, algorithms and applications in image segmentation, *Graphical Models and Image Processing*, 58(3), 246-261.

Vaidyanathan, M., Clarke, L.P., Velthuizen, R. P., Phuphanich, S., Bensaid, A. M., Hall, L. O., Bezdek, J. C., Greenberg, H., Trotti, A. and Silbiger, M. (1995). Comparison of supervised MRI segmentation methods for tumor volume determination during therapy, *Mag. Res. Imaging*, 13(5), 719-728.

Vannier, M.W., Pilgram, T.K., Speidel, C.M., Neumann, L.R., Rickman, D.L. and Schertz, L.D. (1991). Validation of magnetic resonance imaging (MRI) multispectral tissue classification, *Computerized Medical Imaging and Graphics*, 15(4), 217-223.

Velthuizen, R. P., Hall, L. O., Clarke, L. P. and Silbiger, M. L. (1997). An investigation of mountain method clustering for large data sets, *Patt. Recog.*, 30(7), 1121-1135.

Wang, T. C. and Karayiannis, N. B. (1997). Compression of digital mammograms using wavelets and learning vector quantization, in *SPIE Proc. 3030, Applications of Artificial Neural Networks in Image Processing II,* SPIE Press, Bellingham, WA., 44-55.

Woods, K.S. (1994). *Automated Image Analysis Techniques for Digital Mammography*, Ph.D. dissertation, U. of South Florida, Tampa.

Xiaoping, X., Alperin, N., Levin, D.N., Tan, K.K. and Mengeot, M. (1991). Visualization of MR angiographic data with segmentation and volume rendering techniques, *J. Mag. Res. Imag.* 1(5).

Xie, X.L. and Beni, G. A. (1991). Validity measure for fuzzy clustering, *IEEE Trans. PAMI*, 3(8), 841-846.

Yager, R. R. and and Filev, D.P. (1994). Approximate clustering by the mountain method, *IEEE Trans. Syst., Man and Cyberns.*, 24(8), 1279-1283.

Zugaj, D. and Lattuati, V. (1998). A new approach of color image segmentations based on fusing region and edge segmentation outputs, *Patt. Recog.*, 31(2), 105-113.

Zurada, J. (1992). *Introduction to Artificial Neural Systems*, West Publishing, St. Paul, MN.

MAMMOGRAPHIC DATABASES REFERENCES

Bowyer, K., Kopans, D., Kegelmeyer, W. P., Moore, R., Sallam, M., Chang, K. and Woods, K. (1996). The digital database for screening mammography, *Proc. Third International Workshop on Digital Mammography,* 58. Electronic contact information includes e-mail (ddsm@bigpine.csee.usf.edu) and a URL address (http://marathon.csee.usf.edu/Mammography/Database.html).

Mammographic Image Analysis Society (MIAS), Digital Mammogram Database (1994). Electronic contact information includes e-mail (mias@sv1.smb.man.ac.uk) and a URL address (http://s20c.smb.man.ac.uk/services/MIAS/MIAScom.html).

Mascio, L. N., Frankel, S. D., Hernandez, J. M. and Logan, C. (1996). Building the LLNL/UCSF digital mammogram library with image ground truth, *Proc. Third International Workshop on Digital Mammography,* 57. Electronic contact information includes e-mail (mammo-db-help@llnl.gov) and a URL address (http://www-bio.llnl.gov/bbrp/healthcare/Projects/MamLibDetail.html).

Nijmegen Digital Mammogram Database. Electronic contact information includes the e-mail address (nico@mbfys.kun.nl) and the world wide web URL address (http://marathon.csee.usf.edu/Mammography/Nijmegen/README.1.)

Nishikawa, R. M., Johnson, R. E., Wolverton, D. E., Schmidt, R. A., Pisano, E. D., Hemminger, B. M., and Moody, J. (1996). A common database of mammograms for research in digital mammography (UC/UNC database), *Proc. Third International Workshop on Digital Mammography,* 59. This database is not currently on-line, and no electronic contact information is available (as of December 1998).

Washington University Digital Mammography Database (1997). Electronic contact information includes e-mail (jafein@artsci.wustl.edu) and a URL address (http://www.erl.wustl.edu/mammo/digital2.html).

III Management

13 STRATEGIC PLANNING

Miroslawa Lasek

Abstract: Strategic planning is the process in which organisations undertake such actions as identyfying its business objectives, selecting the acceptable means to achieve them and initiating the necessary courses of action and allocation of resources. All these actions are unstructured, ill-defined with uncertainty and impreciseness of the data. This chapter considers some concepts of using fuzzy sets to support actions of strategic planning, i.e. fuzzy portfolio analysis, hierarchical structures of fuzzy ratings and fuzzy approach in strategic planning of CIM implementation. Software tools: expert systems that help to use fuzzy concepts in strategic planning are also considered.

13.1 INTRODUCTION

In the *Encyclopedia of Banking and Finance* planning is defined as a major function of management, and a process in which organizations establish their goals and objectives (what the organization wants to do and why) and select the means of attaining those goals and objectives (when, where, and how to do them). At the level of strategic planning certain activities are accomplished. These include: (i) analysis of organization's environments (internal and external) and study trends on macro- and micro-level, (ii) analysis of alternative goals and objectives, (iii) fixing the selected goals and objectives, and (iv) elaborating the policy alternatives and resource utilization.

The above mentioned activities are complex, unstructured, ill-defined processes, incorporating multiattribute comparisons, uncertainty and impreciseness of the data. Strategic planning's activities are made in the presence of macroeconomics shocks and disturbances, firm- and industry-specific schocks and disturbances, internal schocks and disturbances, and refer to real marketing and production side, financial side, as well as to organizational side of the firm - what is illustrated in Fig. 1 (according to Oxelheim and Wihlborg (1991)). In the last years, to cope with strategic planning's activities, approaches have been searched which would be more sensitive to the uncertainty of the problems (see Wilson 1994). Fuzzy sets seem to be an appropriate way to deal with the kind of uncertainty and imprecision in the problems that strategic planning must address (Homburg (1994)). Decisions in strategic planning involve choice actions as well as information processing, particularly of human expertise which involves the processing of linguistic information (see e.g. Robson (1997)). To cope with processing of such information, fuzzy databanks and fuzzy query languages can be used as described in e.g. Bhattacharjee and Mazumdar (1998), Chiang *et al.* (1997), Dubois and Prade (1996), Kraft and Petry (1997), Nakajima *et al.* (1993), Yazici and Koyuncu (1997).

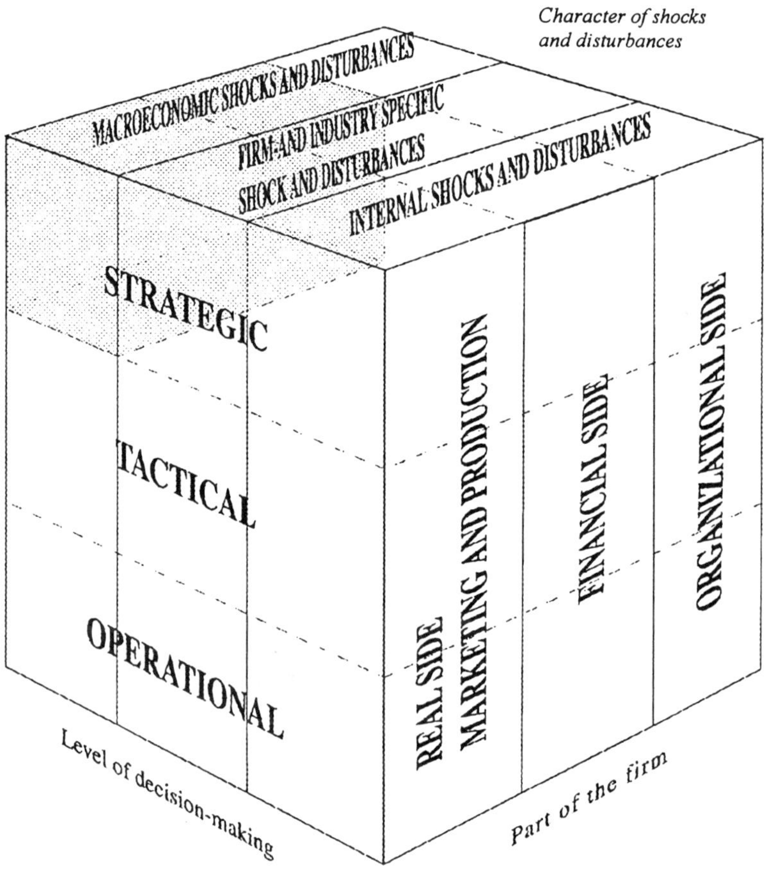

Figure 1 Corporate decision making under uncertainty, according to Oxelheim and Wihlborg (1991)

In this chapter some of the ideas of using fuzzy sets in strategic planning are presented. The first section of the chapter deals with fuzzy portfolio analysis in strategic planning. In such an portfolio analysis strategic planning decisions are considered not as an act of choice, but rather as information processing process in which solution space and goals are determined. The second part of the chapter concentrates on multi-criteria decisions in strategic planning. Hierarchical structures of fuzzy ratings are built, and then AHP procedure and „what-if" analysis with fuzzy sets and fuzzy arithmetic are applied. The third section provides fuzzy approach in strategic planning of CIM implementation with intensive processing of linguistic information. It is a good exemplary description of a specific strategic planning application of fuzzy sets. The last part of the chapter refers to software tools: expert systems that help to use fuzzy concepts in strategic planning, to make it easy and convenient.

13.2 FUZZY PORTFOLIO ANALYSIS

Classical portfolio analysis in strategic planning was used to describe the structure of a corporation by the positions of *Strategic Business Units* (*SBU's*) in a 2-dimensional portfolio matrix and try to find strategies to either keep this "portfolio" balanced or to aim at a balanced portfolio (see e.g. Burnes (1996), Robson (1997)). The classical portfolio matrix, which is primarily cash flow oriented, is shown in the Fig. 2. The main disadvantages in using the above described portfolio analysis in strategic planning are: two dimensionality - insufficient to describe adequately the strategic position of an *SBU*, difficulties with data collection and aggregation for factors which are combinations of many aspects, definition in portfolio matrices broad strategic categories which are much too rough to really define operational strategies to them, lack of the method for modelling and consideration of uncertainty: at the utmost what is done is to attach uncertainty factors to estimates and next to aggregate them together with the data in a rather heuristic and arbitrary way.

In Zimmermann (1989) *multiple strategic portfolios* under uncertainty are proposed to overcome the above mentioned disadvantages. In those strategic portfolios *multidimensional portfolio matrices* are used. In matrices, the locations of *SBU's* are defined by vectors but for visualisation - two dimensionality is used. For example let us consider the four dimensions:

1. *Attractiveness of Technology*
2. *Position of Technology*
3. *Attractiveness of Market*
4. *Competitive Position.*

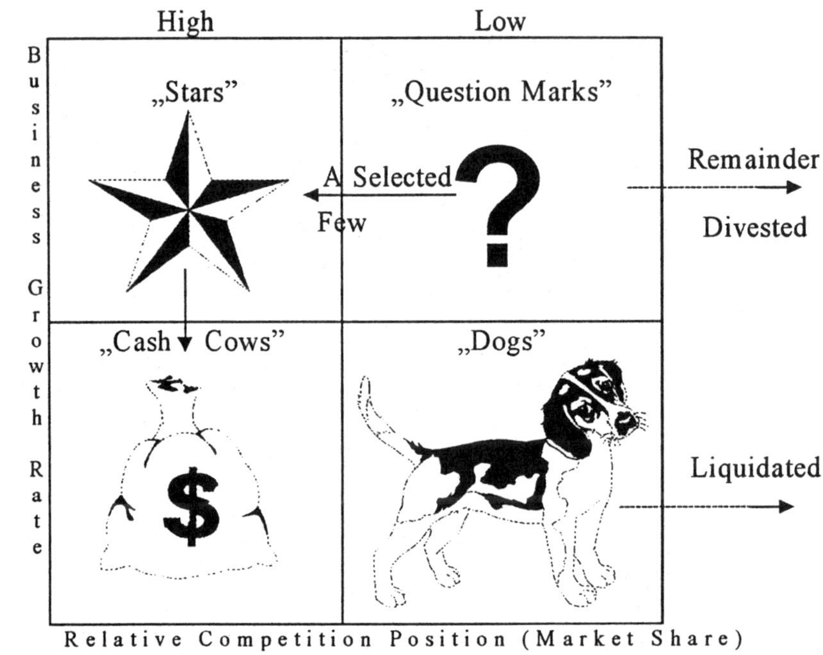

Figure 2 Portfolio Matrix, according to Zimmermann (1989)

We can consider a *Market-Technology-Portfolio*, in which the four above mentioned dimensions are discussed as shown in Fig. 3. A four dimensional matrix is obtained by combining two 2-dimensional portfolio matrices: the first for two dimensions: *Attractiveness of Technology* and *Position of Technology*, and the second for the last two dimensions: *Attractiveness of Market* and *Competitive Position*. Each of this 2-dimensional matrices consists of nine strategic categories by having 3 "fuzzy" intervals - *low, medium, high* - on each axes, and hence 9 strategic positions. Therefore, the combined matrix contains 9 x 9 = 81 strategic positions, although graphically only the 2-dimensional matrices are shown. Strategy assignment is made on the basis of the vectorially described position of a *SBU*.

Such concepts/categories as *Attractiveness of Technology*, *Position of Technology*, *Attractiveness of Market*, and *Competitive Position*, used in strategic planning to determine or more accurately measure the position of an *SBU* in the portfolio matrix, are very complex factors for which the volume of information that has to be collected, is extremely large. In such a case a proper structuring of the collected information is particularly important because of *informational uncertainty* (Biswas (1997), Chiang *et al.* (1997), Zimmermann and Gutsche (1991), Zimmermann (1989)). This uncertainty is caused by the high number of determinants that have to be taken into account when determining the locations of *SBU's* on the different axes of the portfolio matrices.

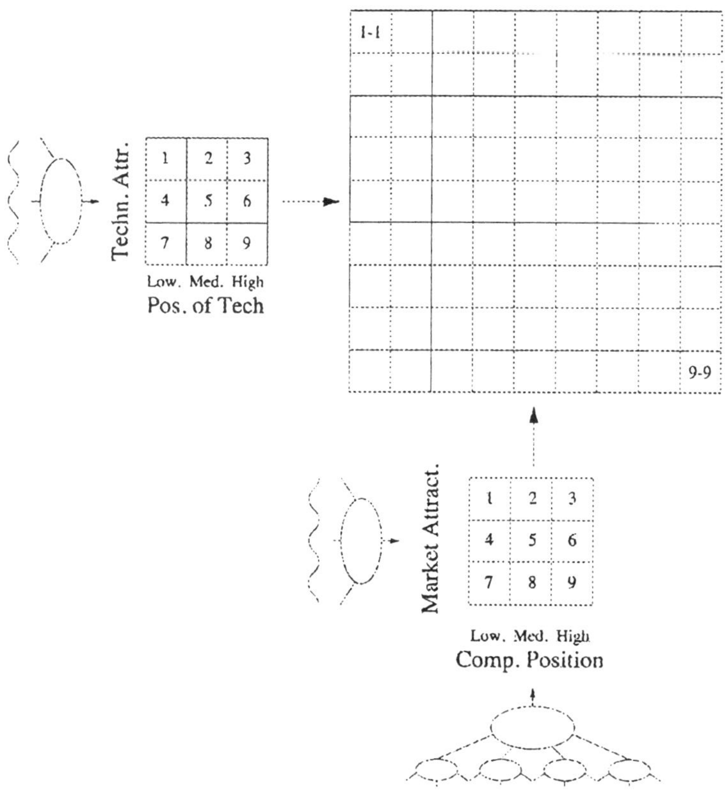

Figure 3 Market-Technology Portfolio, according to Zimmermann (1989)

Due to their complexity and in order to cope with the *informational uncertainty,* "dimensions" can be described by trees of subcriteria and categories. For example, such a tree (for *Attractiveness of Technology*) is shown in Fig. 4. It is assumed that the ratings of the leaves (lowest subcategories) in the considered problems are human evaluation categories and such categories are aggregated to measure a criterion on the top. The hierarchy of concepts can be aggregated step by step using fuzzy operations until the top concept of the hierarchy is attained. To make practical use of the model it is allowed that the subcategories are of unequal importance for the respective super category, and that the description of categories of each level may partly contain the same attributes.

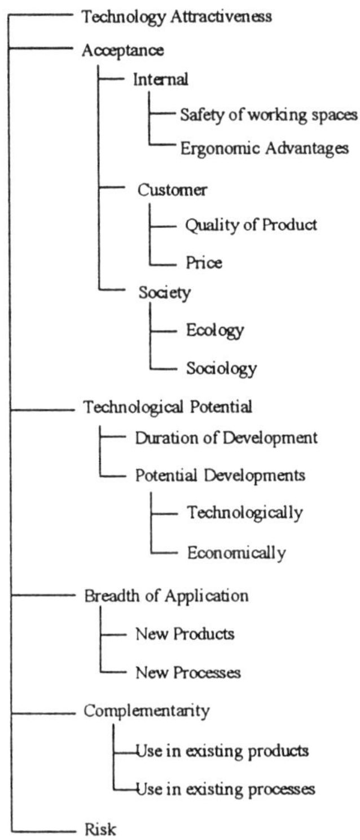

Figure 4 Evaluation Tree, according to Zimmermann (1989)

The uncertainty associated with the input of the information at the leaves of the tree includes evidential, *lexical (linguistic) uncertainty* and *stochastic uncertainty*. Human beings usually are not willing or able to distinguish at that level different kinds of uncertainty as is stated in Zimmermann (1989). They use *linguistic expressions*, which can be considered as *values of linguistic variables*. For example we can define the linguistic variable "degree of achievement", which may take values "not at all", "little", "medium", "considerably", "full". These values, which are *linguistic terms* can be represented using fuzzy sets. In Fig. 5 terms of "degree of achievement" are represented by trapezoidal membership functions.

If we use fuzzy sets, the formal connectives of fuzzy set theory can be applied in the aggregation procedure. For example, depending on the context, e.g. algebraic mean, the minimum or the maximum operator can be chosen.

In the *classical portfolio analysis* it is assumed that we have numerical data as input data, that is that all nodes in trees of categories are represented as numbers, which are used to define the vector for the first level of the tree and the location in the portfolio matrix. In the output, the *SBU's* are represented in the matrices by circles. Their diameters represent either profit or sales (see Fig. 6).

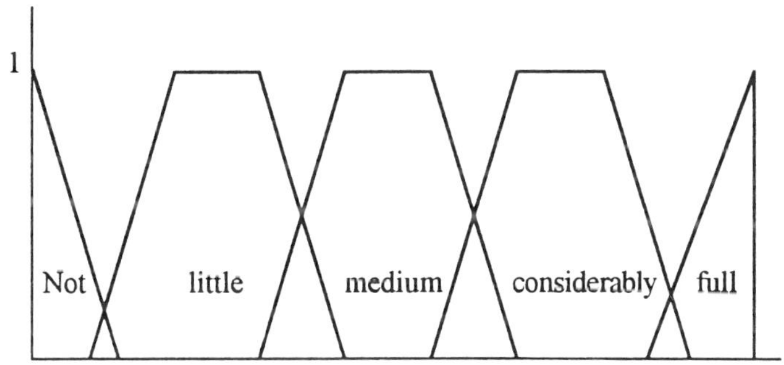

Figure 5 Terms of „degree of achievement", according to Zimmermann (1989)

In the *fuzzy portfolio analysis*, where the input data are linguistic terms, the *SBU's* are represented in the portfolio matrices by rectangles (see Fig. 7). The edges of those rectangles represent α-cuts of the linguistic variables computed for the roots of the trees. The size of the rectangles represents the uncertainty of the location of the *SBU*. The information for the first-level nodes is represented either by the fuzzy sets computed for each of these nodes or linguistically using linguistic approximation to these fuzzy sets (Zadeh (1975; 1976)).

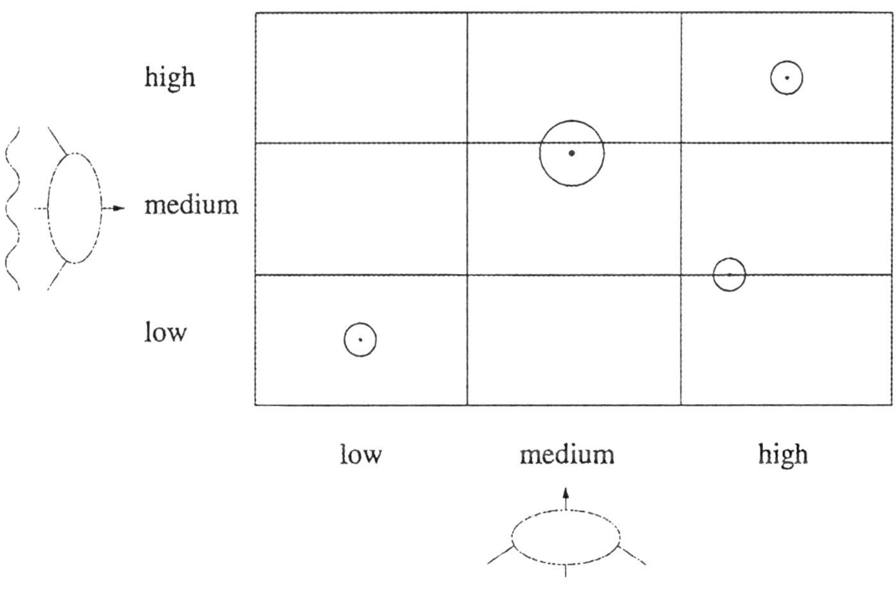

Figure 6 Portfolio Matrix with numerical input, according to Zimmermann (1989)

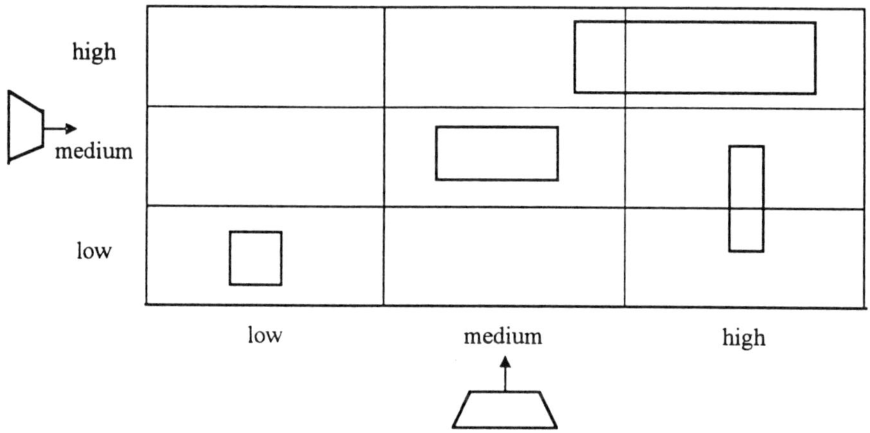

Figure 7 Portfolio Matrix with linguistic input, according to Zimmermann (1989)

13.3 HIERARCHICAL STRUCTURES OF FUZZY RATINGS

Strategic planning involves simultaneous consideration of various goals, subgoals and strategies, which are changed according to situations and problems. Analysis of strategic goals of enterprises is extremely complex, unstructured problem with only uncertain and imprecise data available. To cope with this problem a methodology, integrating the Analytic Hierarchy Process (AHP) procedure, "what-if" analysis, fuzzy sets and fuzzy arithmetic can be proposed (see Lasek (1992)).

Using this methodology, a decision maker can arrange strategic goals, subgoals, and strategies hierarchically, evaluate them or apply the Analytic Hierarchy Process (AHP) to determine the priorities for them. He can obtain the hierarchical structure of their ratings, and accomplish a "what-if" analysis to investigate the effect of a change in hierarchy or in ratings. Due to the impreciseness of the data, fuzzy numbers and fuzzy variables are used to evaluate or set priorities to goals, subgoals and strategies, and hierarchical structures are obtained with fuzzy ratings.

The main element of the methodology is the *concept of hierarchical structure of ratings*. At the top of such a hierarchical structure the most general goal or objective of the firm is placed - the so called overall purpose, for instance the "*overall competitive position of the firm*", the "*well-being of the company*" or the "*performance of the firm*". This complex purpose is decomposed into subgoals or subobjectives at the lower level of the hierarchy. For "*overall competitive position of the firm*" we can have for example subgoals: "*market size and share*" (to enlarge market size and share by an increase in percentage of the total market and reputation on the market) and "*future development*" (to ensure future development by the product's up-to-dateness and its flexibility towards market changes). Each of them, in turn, is decomposed into a set of subgoals at the next lower level. The process of the hierarchical structuring is continued down to the most specific subgoals or strategies placed at the lowest level of the hierarchy. For the example of "*overall competitive position of the firm*" at the lowest level of the hierarchy we can consider *strategies*: *introducing to production one of the new brands* of an existing products, that would help the firm to achieve the two subgoals "*market size and share*" and "*future development*", thus enabling the firm to realise its main strategic goal.

The general overview of the proposed methodology is shown in Fig. 8. The methodology comprises four steps: from building the hierarchical structure (step 1) through calling and performing the evaluation algorithm (step 2) until presenting the hierarchy with ratings to the user (step 3) and conducting „what-if" analysis (step 4).

The steps of the methodology are as follows.

Step 1. Building the hierarchical structure. The user suggests the overall purpose of the firm and arranges it hierarchically into goals, subgoals and strategies.

Step 2. Calling the evaluation algorithm. One from among two different algorithms can be chosen according to the kind of relationships between strategies, subgoals, goals and overall purpose. In the case of functional dependencies, the judgements are input in the lower levels of the hierarchy. Ratings in higher levels are computed in a down-top fashion by performing the algebraic operations pointed out by the user. As an example of the hierarchy with functional dependencies serves the hierarchy with the net profit at the top as the most general objective, gross profit and income tax at the first lower level, income from sale, trade tax and total costs at the second lower level, and volume of sale, price of product, variable costs and fixed costs at the lowest level. From the ratings of the lower levels given by the user, the ratings in higher levels can be computed as follows:

income of sale = *volume of sale* x *price of product*,
total costs = *variable costs* + *fixed costs*,
gross profit = *income from sale* - *trade tax* - *total costs*,
net profit = *gross profit* - *income tax*.

In the case of preference relations, the user can only set priorities (weights) of goals, subgoals and strategies. In the methodology, a multiple-attribute process, developed by Saaty (1980) (new publications see e.g. Kwiesielewicz (1998), McCauley-Bell and Badiru (1996), Olson (1996), Triantaphyllou and Lin (1996)) and called *Analytic Hierarchy Process* (*AHP*), is used. *AHP* is applied to establish the priorities of the goals, subgoals and strategies within each level of the hierarchy. This is accomplished by assessing importance ratios between goals, subgoals and strategies in pairwise comparison matrices. In the matrices, each pair of elements from a given level of the hierarchy is evaluated separately. The evaluation is done by the judgement of the degree to which one element of the pair dominates the other with respect to each of the elements in the next higher level.

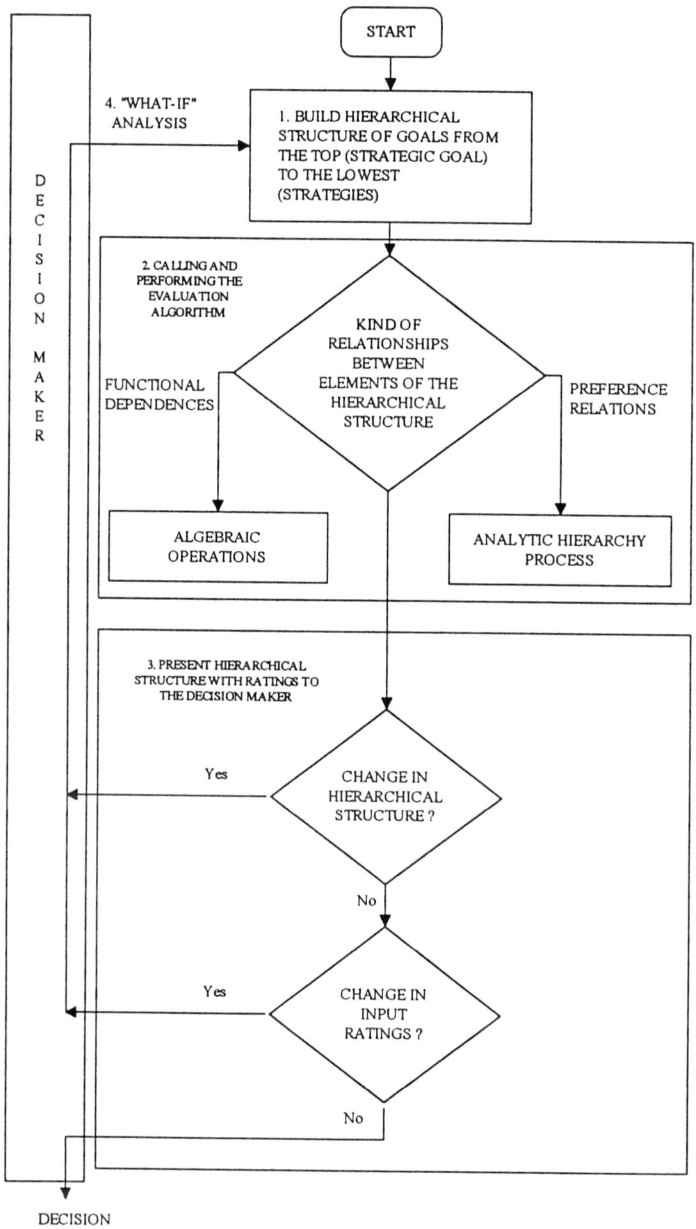

Figure 8 An overview of the proposed methodology, according to Lasek (1992)

Saaty proposed a scale for making the judgements. To calculate the priorities from pairwise comparisons in described methodology, due to its simple algebraic formula, the method of the geometric mean was used, which was one of the methods suggested by Saaty and used by others. When pairwise comparisons are made for all levels in the hierarchy, *AHP* is used to obtain priorities of lowest level elements with respect to the element on the top of the hierarchy - i.e. final ratings for analysing hierarchy. As an example of such a hierarchy to which this procedure is applied, the hierarchy can serve with *well-being of a firm* at the top, subgoals such as *production rate*, *current ratio*, *cumulative net profit after taxes*, *monetary inventory value* at the lower level and actual, *real strategies of activity of the firm* at the lowest level (see Kivijarvi and Tuominen (1989)).

In both cases of functional dependencies and of preference relations, the decision makers can use imprecise, qualitative terms to express their evaluations or preferences. These terms are (i) linguistic quantities like *low*, *good*, *high*, *acceptable*, *medium*, *excellent*, *excessive* when functional dependencies are considered or (ii) linguistic intensity of importance like *equal importance* of elements, *weak*, *essential/strong*, *very strong/demonstrated* or *absolute importance* of one element over another, which are used in *Saaty's priority scale* and which are commonly applied in other applications of *AHP*, when preference relations are considered.

The qualitative terms are defined using fuzzy sets and then fuzzy ratings are computed using the methods of performing arithmetical operations over fuzzy sets. Real numbers as evaluations, priorities or ratings can also be used and are then treated as a special case of fuzzy numbers with the grade of membership equal to one.

In the literature a number of algorithms has been proposed to perform algebraic operations (addition, subtraction, multiplication or division) for fuzzy quantities. In prototype software, the method of performing algebraic operations was based on *Zadeh's extension principle*. This method was chosen because of its computational simplicity and the ease of implementation. Other methods can also be used, e.g. fuzzy arithmetic as proposed by Buckley, Dubois and Prade or Laarhoven and Pedrycz (see Kwiesielewicz (1998), Ribeiro (1996), Triantaphyllou and Lin (1996); see also special issue on fuzzy arithmetic in *Fuzzy Sets and Systems* vol 91, 1997).

To rank fuzzy ratings of various strategies, an algorithm which allows to compare fuzzy sets is applied. A number of methods has been suggested in the literature for the comparison of fuzzy sets (see e.g. Cheng (1998), Fortemps and Roubens (1996), Grzegorzewski (1998), Peneva and Popchev (1998), Ribeiro (1996)). In the described methodology the concept of *maximizing set* $M(S)$ is used (see e.g. Borgulya (1997)):

$M(S) = \{(\mu_M(r), r)\}, \quad r \in S,$

where

$M(r) = r/r_{max}, \; r_{max} = sup\ S,$

and S is the union of the supports of the fuzzy sets associated with all ratings.

In the procedure of comparison of fuzzy ratings: (i) a maximizing set is built, (ii) fuzzy sets are formed which are the intersections of fuzzy ratings and the maximizing set, (iii) a fuzzy optimal alternative is built as a fuzzy set in which a maximum grade of membership from the prior formed sets is assigned to each of the alternatives, (iv) alternatives are ranked on the basis of their grades of membership in the set representing the fuzzy optimal alternative.

In Lasek (1992) a simple example is given to illustrate fuzzy operations in the methodology.

Step 3. Presenting the hierarchy that contains ratings to the user.

Step 4. Conducting a "what-if" analysis using the information obtained at the *step 3*. The users can change either the hierarchical structure or only the input ratings in it and observe the effect of changes in hierarchy or ratings on the entire hierarchy.

Let us consider a simple illustrative problem. To make the reader's investigation easier it deals with only a narrow, limited dimension of strategic planning. The user of the methodology (and software) can consider the problem of strategic planning in the range which he recognizes to be necessary in real situation. He can include even the whole range of strategic planning, building his own appropriately expanded hierarchy to analysis.

Let us assume that the top management wishes to increase the overall competitive position of an enterprise on the market. Its chief executives propose three major strategies which can be undertaken to achieve this strategic goal. The three proposed strategies are to add one of the new brands A, B or C of an existing product to its production.

According to the proposed methodology, the decision maker starts the analysis with decomposing the above mentioned strategic goal - increase of the overall competitive position of a firm - into subgoals and strategies. A possible hierarchy given by the decision maker for the analysis of one of the strategies is shown in Figure 9. This hierarchy consists of three levels. At the top of the hierarchy lies the most general goal - an increase of the overall competitive position of the firm. This is decomposed into two subgoals: (1) to enlarge market size and share by an increase in percentage of the total market and reputation on the market, (2) to ensure future development by the product's up-to-dateness and its flexibility towards market changes. At the lowest level of the hierarchy there are strategies: introducing to production one of the new brands A or B or C. These strategies are selected by the managers as strategies that would help the firm to achieve the two subgoals (1) and (2), thus enabling the firm to realize its main strategic goal.

After the hierarchy has been built, AHP is used to set priorities among the subgoals and among the strategies. The user provides subjective value judgements in the pairwise comparison matrices. Making judgements in this example, fuzzy terms used to define the intensity of importance in Saaty's priority scale are applied. They are: „equal importance", „weak importance", „essential or strong importance", „very strong or demonstrated importance", „absolute importance".

These fuzzy terms are defined in this example as follows:

Equal importance of both elements = {(1.0, 1.0)},

Weak importance of one element over another = {(0.5, 1.0), (0.75, 2.0), (1.0, 3.0), (0.75, 4.0), (0.5, 5.0)},

Essential or strong importance of one element over another = {(0.5, 3.0), (0.75, 4.0), (1.0, 5.0), (0.75, 6.0), (0.5, 7.0)},
Very strong or demonstrated importance of one element over another = {(0.5, 5.0), (0.75, 6.0), (1.0, 7.0), (0.75, 8.0), (0.5, 9.0)},
Absolute importance of one element over another = {(0.5, 9.0), (1.0, 10)}.

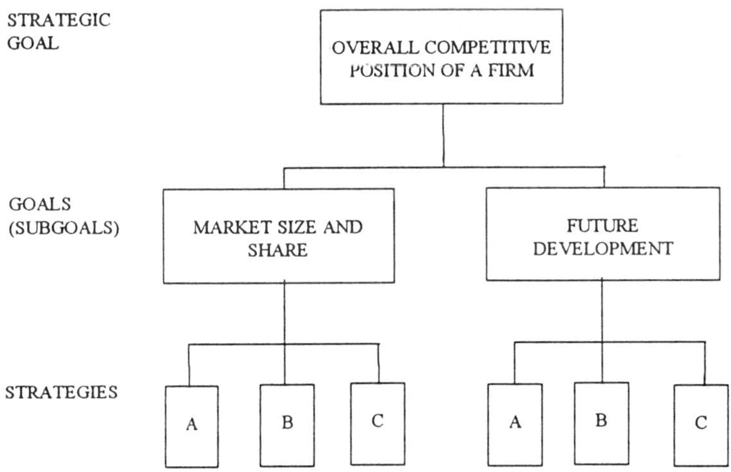

Figure 9 Hierarchy for the selection of strategy, according to Lasek (1992)

Using the above presented terms, decision maker is asked to make pairwise judgements for the two subgoals selected in the example. Judgements involve assigning the priority of one subgoal over the other in the terms defined above. The decision maker answers the question: which of the two subgoals is more important and how much more important is it in relation to the overall competitive position of the firm. The matrix of pairwise comparisons of the subgoals given by the decision maker is shown in Table 1.

Table 1. Pairwise judgement matrix

Subgoal	Market size and share	Future Development
Market size and share	Equal Importance	Essential or Strong importance
Future Development	Reciprocal of Essential or strong importance	Equal Importance

The next step is to rate each of the competing strategies with reference to each of the two subgoals. Based on the judgements of the decision maker the ratings are obtained here by pairwise comparisons of all three competing strategies and done in pairwise comparison matrices. The matrices for the comparison of strategies A, B, C with respect to each of the two subgoals are presented in Table 2.

Table 2. Comparison of strategies with respect to two subgoals

Market size and share	A	B	C
A	Equal Importance	Very strong Importance	Absolute importance
B	*	Equal Importance	Essential importance
C	*	*	Equal importance

Future Development	A	B	C
A	Equal Importance	Weak Importance	*
B	*	Equal Importance	*
C	Very strong Importance	Absolute Importance	Equal importance

* Reciprocal in row *j*, column *i* of the rating assigned in row *i*, column *j*.

For each matrix in Tables 1 and 2, the vector of priorities is obtained by computing the geometric mean of each row of the matrix and normalizing the resulting numbers. The calculated priority vectors are shown in Table 3. These vectors represent the priority indices of each subgoal, i.e. „market size and share" and „future development", with respect to the overall competitive position of a firm (column vector v_1), and priority indices of the three strategies A, B, C with respect to each of the subgoals (column vectors v_2 and v_3).

Table 3. Fuzzy priority vectors

Overall competitive Position of a firm	V_1
Market size and share	{(0.5, 0.5374), (1.0, 0.8333), (0.75, 0.7310), (0.75, 0.8953), (0.5, 1.2539)}
Future Development	{(0.5, 0.1173), (1.0, 0.1667), (0.75, 0.1492), (0.75, 0.1695), (0.5, 0.2736)}
Market size And share	**V_2**
A	{(0.5, 0.5995), (1.0, 0.7796), (0.9375, 0.7642), (0.8750, 0.7826), (0.5, 0.9974)}
B	{(0.5, 0.1169), (1.0, 0.1691), (0.9375, 0.1658), (0.9375, 0.1854), (0.5, 0.2490)}
C	{(0.5, 0.0409), (1.0, 0.0513), (0.9375, 0.0503), (0.8750, 0.0515), (0.5, 0.0742)}
Future Development	**V_3**
A	{(0.5, 0.0806), (1.0, 0.1451), (0.8750, 0.1334), (0.8750, 0.1468), (0.5, 0.2321)}
B	{(0.5, 0.0455), (1.0, 0.0619), (0.75, 0.0562), (0.8750, 0.0627), (0.5, 0.1116)}
C	{(0.5, 0.5966), (1.0, 0.7930), (0.75, 0.7518), (0.8750, 0.8024), (0.5, 1.0400)}

To obtain the priorities of the strategies with respect to the overall competitive position of the firm, the matrix with vectors v_2 and v_3 as columns is multiplied by the column vector v_1. The obtained results are given in Table 4.

Table 4. Fuzzy priorities of strategies A, B, C with respect to the overall competitive position of the firm

Overall competitive position of the firm	
A	{(0.5, 0.3316), (1.0, 0.6738), (0.9375, 0.6610), (0.875, 0.6741), (0.5, 1.3142)}
B	{(0.5, 0.0681), (1.0, 0.1512), (0.9375, 0.1485), (0.9375, 0.1649), (0.5, 0.3427)}
C	{(0.5, 0.0919), (1.0, 0.1750), (0.9375, 0.1741), (0.8750, 0.1751), (0.5, 0.3776)}

In order to choose the best strategy, final fuzzy ratings of strategies A, B and C, as presented in Table 4, are compared. The maximizing set is formed as follows:

$M(S)$ = {(0.2523, 0.3316), (0.5127, 0.6738), (0.5030, 0.6610), (0.5129, 0.6741), (1.0, 1.3142), (0.0518, 0.0681), (0.1151, 0.1512), (0.1130, 0.1485), (0.1255, 0.1649), (0.2608, 0.3427), (0.0699, 0.0919), (0.1332, 0.1750), (0.1325, 0.1741), (0.1332, 0.1751), (0.2873, 0.3776)}.

The intersection of each fuzzy rating and maximizing set is then obtained as

w_1 = {(0.2523, 0.3316), (0.5127, 0.6738), (0.5030, 0.6610), (0.5129, 0.6741), (0.50, 1.3142)},

w_2 = {(0.0518, 0.0681), (0.1151, 0.1512), (0.1130, 0.1485), (0.1255, 0.1649), (0.2608, 0.3427)},

w_3 = {(0.0699, 0.0919), (0.1332, 0.1750), (0.1325, 0.1741), (0.1332, 0.1751), (0.2873, 0.3776)}.

The fuzzy optimal alternative is

{(0.5129, A), (0.2608, B), (0.2873, C)},

and after normalizing it we get

$$\{(1.0, A), (0.5085, B), (0.5601, C)\}.$$

This result indicates that the strategy A is the most desirable strategy (best strategy), whereas B is the worst.

The hierarchical structure that the decision maker has built in the first step of his analysis is filled with the computed ratings and presented to him. The decision maker may decide to continue the analysis, and then study what will be the effect of a change in hierarchy or judgements over the entire system. Such „what-if" analysis can give him more insights into the interdependencies of different strategies, subgoals and strategic goal of the firm.

13.4 A FUZZY LINGUISTIC APPROACH IN STRATEGIC PLANNING OF *CIM* IMPLEMENTATION

To be competitive in the future, companies have to consider in their strategic planning the problem of implementation of new technologies such as *CAD*, *CAM*, *FMS*, *MRPII*, *JIT* and to apply them in the way which can ensure integrated information processing leading to the so called *Computer Integrated Manufacturing* (*CIM*).

Wilhelm and Parsaei (1991) presented the problem of how a fuzzy linguistic approach can help the decision makers in strategic planning of the phased implementation of a *CIM* strategy. A heuristic algorithm is proposed to describe the capability of available technologies to satisfy a common set of *CIM* goals of strategic interest for an organization (e.g. *high product quality*, *low product cost*, *good asset utilization*, *short product development lead times*, *minimum inventories*, *flexibility*), and next to select a technology from those available to be used at each phase in the implementation of the *CIM* plan.

In the proposed approach it is attempted to assess effectiveness of each advanced technology in meeting strategic *CIM* goals in terms of natural language. To accomplish this, two *fuzzy linguistic variables* are defined: *X="Importance"* and *Y="Capability"*. By the use of the variable *"Importance"* - the importance associated with each of a set of goals common to all enabling technology is measured, and by the use of the variable *"Capability"* - the capability of each technology to meet the strategic *CIM* goals of the organization.

The primary values of the two fuzzy linguistic variables are defined as fuzzy sets on the universe of discourse [0,1]. For variable *"Importance"* the terms are given: *critical*, *important*, *unimportant*, for variable *"Capability"* - *superior*, *average*, *poor*.

New variable's values can be formed from the primary values using operators such as *"not"*, *"and"*, *"or"*, and *linguistic hedges*, e.g. *indeed*, *more or less*, *above*, *below*. In Wilhelm and Parsaei (1991) an example is given illustrating the use of operators and hedges. That is when we assume that x=*Not important*, and y=*Above average*, it is stated in Wilhelm and Parsaei (1991) that:

$$f_{M(x)} = 1 - f_{M(IMPORTANT)}, \text{ and}$$

$$f_{M(y)} = 1 - \left[f_{M(AVERAGE)} \right]^{1.5}$$

where $f_{M(x)}(u_x)$ is the compatibility function of u_x in $M(x)$; $M(x)$ is the semantic rule for associating a meaning with each variable name and can be given as fuzzy subset:

$$M(x) = \{x, f_{M(x)}(u_x)/u_x \in U_x\}.$$

For x and y the following fuzzy relational equation holds (Wilhelm and Parsaei (1991)):

$$x_i = r_{ij} \circ y_j,$$

where:
x_i - value of *Importance* of a *CIM* goal i,
y_j - value of *Capability* of an enabling technology j,
r_{ij} - a linguistic assessment of the extent to which a *CIM* goal i can be achieved if enabling technology j is selected for implementation,
$\circ$ - a composition operator defined such that the compatability function of $r_{ij} \circ y_j$ is determined from

$$f_{M(r \circ y)}(u_x, u_y) = \bigvee_{u_y} \left[f_{M(r)}(u_x, u_y) \wedge f_{M(y)}(u_y) \right],$$

for each u_x, and $\wedge$ is minimum, $\vee$ is maximum operator.

In the above equation, values of x_i and of y_j are assumed to be given by the decision maker. The goal is to find r_{ij}, given x_i and y_j Procedure for solving fuzzy equations is used as follows and r_{ij} is found as:

$$r_{ij} = x_i^{-1} ⓐ\, y_j$$

where
x_i^{-1} - the inverse, or transpose, of x_i,
ⓐ - a compositional operator defined such that the membership function of

$$x_i^{-1} ⓐ\, y_j$$

is determined from

$$f_{M\left(x^{-1} \ \alpha \ y\right)} = \wedge \left[f_{M(x)}^{-1}(u_x) \ \ \alpha \ \ f_{M(y)}(u_y) \right],$$

α - a compositional operator defined such that the elements of its compatibility function are determined as follows:

$$\left[f_{M(x)}^{-1}(u_x) \ \ \alpha \ \ f_{M(y)}(u_y) \right] = \begin{cases} 1 & \text{if } f_{M(x)}^{-1}(u_x) \le f_{M(y)}(u_y), \\ f_{M(y)}(u_y) & \text{if } f_{M(x)}^{-1}(u_x) > f_{M(y)}(u_y). \end{cases}$$

For each of the $j=1, 2,..., n$ enabling technologies, the intersection of the $i=1, 2,..., m$ binary fuzzy relations r_{ij} over the common set of m *CIM* goals is formed. This intersection is given as a fuzzy relation $\Re_j = \cap i\, r_{ij}$ with a function:

$$f_{\Re_j}(u_x, u_y) = \underset{r_{ij}}{\wedge} f_{M(r)}(u_x, u_y) \quad over \quad i = 1,2,\ldots,m$$

$$= \underset{u_x, u_y}{Min} \left[f_{M(r)}(u_x, u_y) \right].$$

Next x^* - an "ideal" value of X, e.g. importance of implementing the overall *CIM* strategy to the future of the company is established. This "ideal" value can be e.g. *indeed critical.*

In the next step the fuzzy relational equation $y_j^{-1} = \Re_j \textcircled{\alpha}\ x^{*-1}$ for each technology $j=1,2,...,n$ is solved, so that function of y_j^{-1} is taken as:

$$f^{-1}_{M(y)} = \wedge \left[f_{M(\Re)}(u_x, u_y) \;\alpha\; f^{-1}_{M(x^*)}(u_x) \right].$$

Thus the function of *Capability* of each of the technologies over all *CIM* goals is obtained. Linguistic meanings to those functions can be assigned and the "best" can be selected. It is possible to compare each function with the function for y^* = *indeed superior* and next selected the "best" as the "nearest" to it. To do this the relative *Hamming distance* of each function to function for y^* can be computed:

$$\delta_j = \frac{1}{s} \sum_{u_y} \left| f_{M(y)}(u_y) - f_{M(y^*)}(u_y) \right| \quad for \quad j = 1,2,\ldots,n,$$

where:

s - number of elements in the universe of discourse u_y,

$f_{M(y)}(u_y)$ - compatability function for technology j,

$f_{M(y^*)}(u_y)$ - compatability function for y^*.

The technology j for which the above calculated δ_j is a minimum, that is technology which has the minimum *Hamming distance* can be selected as the best.

13.5 FUZZY EXPERT SYSTEMS IN STRATEGIC PLANNING

Fuzzy models and methods have been used in a large number of commercial applications of the technology of expert systems (see e.g. Grabot and Caillaud (1996), Graham (1991), McCauley-Bell and Badiru (1996), Turksen (1992), Yeung and Tsang (1998)). Some of the commercial products known as the "expert system shells" are enough robust to cope with uncertainty and to be used to solve problems from the strategic planning domain (see e.g. Mayer *et al.* (1994), Mechler (1995), Rommelfanger (1994)).

System *ESP* is an expert system aimed at solving problems in the strategic planning area (Zimmermann (1989)). The basic structure of *ESP* is shown in Fig. 10.

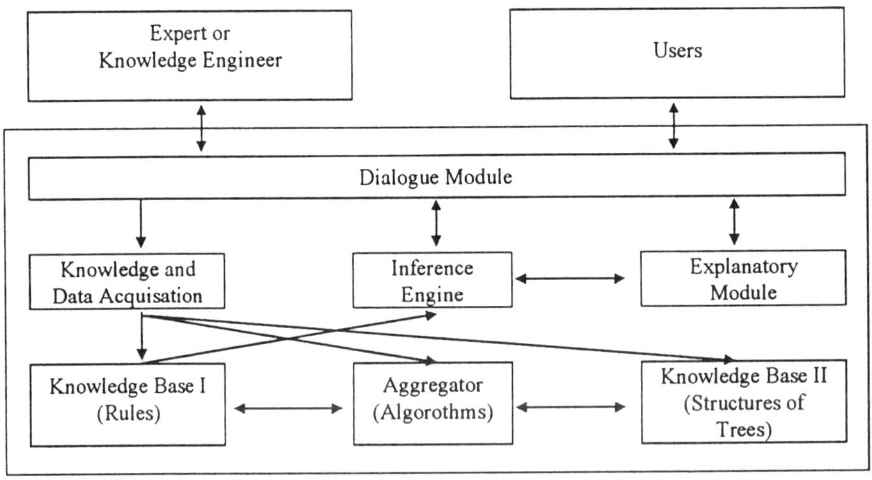

Figure 10 Structure of ESP, according to Zimmermann (1989)

ESP consists of two knowledge bases: *Knowledge Base I* and *Knowledge Base II*. *Knowledge Base I* contains rules which assign basic strategy recommendations for locations of *SBU* in multi-dimensional portfolio matrices and detailed supplementary recommendations to profiles of first-levels of trees (portfolio matrices and trees are understood as it is described in part 13.2 of this chapter). *Knowledge Base II* contains tree structures (free definable) which determine the location of an *SBU* in the different dimensions of the multi-dimensional matrix. The algorithm that is called *Aggregator* computes their values and characteristic values for the linguistic values for all nodes of the trees on the basis of available structural knowledge and on the basis of data that are introduced for each terminal leaf by the user. The information which is provided by the *Aggregator* is used for the visual presentation of two-dimensional matrices and profiles and also supports the explanatory module.

ESP could be considered as a second generation expert system which works with shallow knowledge (*Knowledge Base I*) as well as with deep knowledge (*Knowledge Base II*).

One of the new software tools that offers methods which can be used in solving problems in strategic planning area is DataEngine of Management Intelligenter Technologien GmbH, Aachen. DataEngine contains fuzzy methods such as (DataEngine (1997)):

- Fuzzy Rule Based Inference,
- Fuzzy C-Means Clustering,
- Fuzzy Kohonen Neural Network.

Structure of DataEngine is illustrated in Fig. 11.

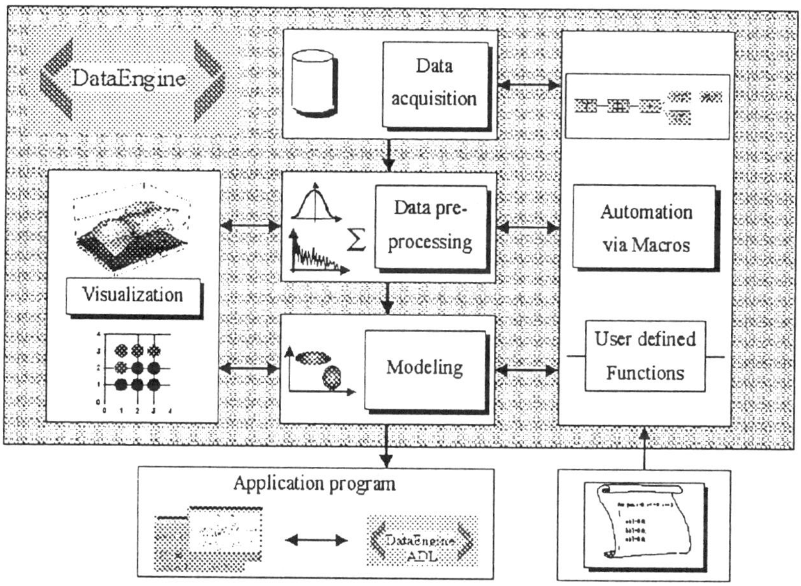

Figure 11 Structure of DataEngine, according to DataEngine Manual (1997)

DataEngine has been applied in strategic planning in scenario analysis, i.e. in the analysis of „possible pictures of the future called scenarios" (Mißler-Behr (1996)). Fuzzy C-Means Clustering method was used in this case. Applying this method it is possible to classify objects in regards of many criteria to classes. In

Fuzzy C-Means Clustering each object is not assigned to exactly one class, but with different degrees of membership to different classes. The aim of using Fuzzy C-Means Clustering in scenario analysis was to answer two questions: (1) what is the equivalent number of classes to represent the total scenario space ?, (2) what are the representative scenarios of these classes ?.

Algorithm on the whole as described in DataEngine (1997), consists of four steps (it is possible to use modified variants; DataEngine uses a slightly modified version). In the first step the membership values μ_{iK} of the k objects x_k to each of the i clusters, $i=1, ..., c$ and $k=1, ..., K$, are initialized (for example randomly) such that:

$$\sum_{i=1}^{c} \mu_{ik} = 1 \qquad \forall k = 1,\cdots,K \quad and \quad \mu_{ik} \in [0,1] \quad \begin{matrix} \forall i = 1,\ldots,c \\ \forall k = 1,\ldots,K \end{matrix}$$

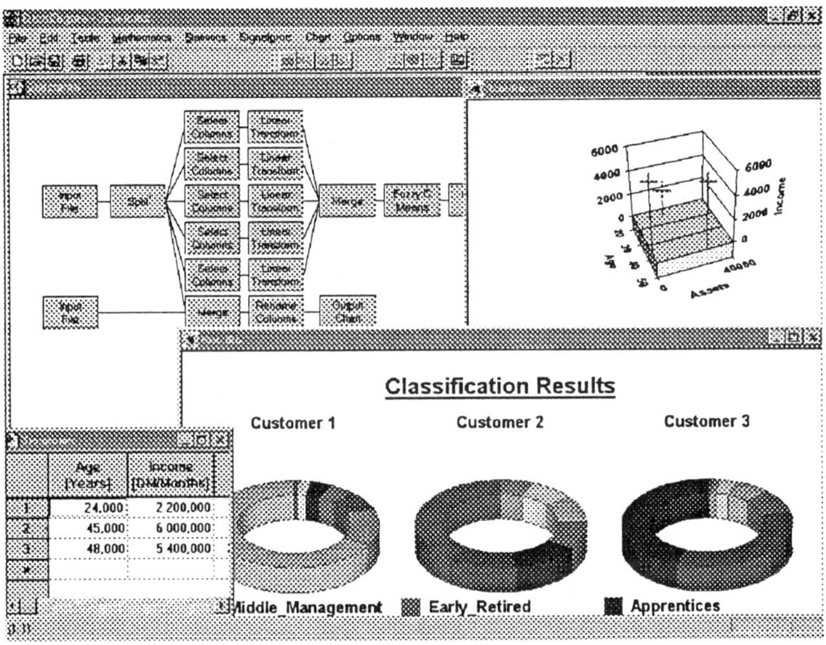

Figure 12 Screenshot of DataEngine

In the second step, using membership values μ_{ik} (as determined in the previous step), the cluster centers v_i are calculated:

$$v_i = \frac{\sum_{k=1}^{K} (\mu_{ik})^m \times x_k}{\sum_{k=1}^{K} (\mu_{ik})^m}, \qquad \forall i = 1, \ldots c$$

In the next, third step of the method, new membership values μ_{ik}^{new}, using received foregoing cluster centers v_i, are calculated:

$$\mu_{ik}^{new} = \frac{1}{\sum_{j=1}^{c} \left(\frac{\|v_i - x_k\|}{\|v_j - x_k\|} \right)^{\frac{2}{m-1}}}, \qquad \begin{array}{l} \forall i = 1, \ldots, c \\ \forall k = 1, \ldots K \end{array}$$

In step four we test if $\|\mu^{new} - \mu\| > \varepsilon$. If $\|\mu^{new} - \mu\| > \varepsilon$, let $\mu = \mu^{new}$ and we go to step 2. If the distance between two successive membership matrices μ is less or equal convergence threshold ε algorithm stops.

To calculate the vector distances Euclidean measure is employed.

The exponent m constitutes the degree of fuzziness of the results. If $m \to 1$, the results become more and more „unfuzzy". If $m \to \infty$, the objects tend to have membership values to clusters bound for the reciprocals of the number of classes $\frac{1}{c}$.

As a result of the algorithm we obtain the membership values of the individual objects to clusters and cluster centers that characterize individual clusters (classes).

Fuzzy C-Means Clustering method was applied to the analysis of scenarios of possible development of the regional bank in Germany (Mißler-Behr (1996)). Eighty nine various scenarios, each written with 37 features were analysed using Fuzzy C-Means. The clusters of scenarios were formed. The method facilitated considerably the analysis of so called total scenario space, more objective selection of the scenarios and even allowed to find possibly further scenarios.

13.6 SUMMARY

Although strategic planning is an important area in business and management practice and involves complex, unstructured, ill-defined activities with processing uncertain and imprecise information it has received so far relatively little attention as the domain of fuzzy sets applications. Results considered in the literature and cases reported in this chapter indicate that fuzzy sets and methods are very powerful tools for strategic planning. However, further research is required in order to take full advantage from the potential of using fuzzy sets in strategic planning.

References

Bhattacharjee, T.K. and Mazumdar, A.K. (1998). Axiomatisation of fuzzy multivalued dependencies in a fuzzy relational data model, *Fuzzy Sets and Systems*, 96, 343-352.

Biswas, T. (1997), *Decision-making under Uncertainty*, Macmillan, New York.

Borgulya, I. (1997). A ranking method for multiple-criteria decision-making, *International Journal of Systems Science* 28, 905-912.

Burnes, B. (1996). *Managing Change. A strategic Approach to Organizational Dynamics*, Pitman Publishing, London.

Cheng, Ch.-H. (1998). A new approach for ranking fuzzy numbers by distance method, *Fuzzy Sets and Systems*, 95, 307-317.

Chiang, D.-A., Chow, L.R. and Hsien, N.-Ch (1997). Fuzzy information in extended fuzzy relational databases, *Fuzzy Sets and Systems*, 92, 1-20.

DataEngine (1997): (1) Tutorials and Theory (2) Overview and User Manual, MIT GmbH, Aachen.

Dubois, D. and Prade, H. (1996). Semantics of quotient operators in fuzzy relational databases, *Fuzzy Sets and Systems*, 78, 89-93.

Encyclopedia of Banking and Finance (1991), McGraw-Hill.

Fortemps, Ph. and Roubens, M.. (1996). Ranking and defuzzification methods based on area compensation, *Fuzzy Sets and Systems*, 82, 319-330.

Grabot, B. and Caillaud, E. (1996). Imprecise Knowledge in Expert Systems: A Simple Shell, *Expert Systems with Applications*, 10, 99-112.

Graham, I. (1991). Fuzzy logic in commercial expert systems - Results and prospects, *Fuzzy Sets and Systems*, 40, 451-472.

Grzegorzewski, P. (1998). Metrics and orders in space of fuzzy numbers, *Fuzzy Sets and Systems*, 97, 83-94.

Homburg, O. (1994). Integration of Fuzzy-Sets in the Strategic Corporate Planning Process, in: *Proceedings Second European Congress on Intelligent Techniques and Soft Computing EUFIT'94, Aachen*, vol. 3, 1622-1626.

Kivijarvi, H. and Tuominen, M. (1989). Computer-based multi-attribute simulation of production and financial strategies, *Engng. Costs and Production Econ.*, 17, 331-341.

Kraft, D.H. and Petry, F.E. (1997). Fuzzy information systems: managing uncertainty in databases and information retrieval systems, *Fuzzy Sets and Systems*, 90, 183-191.

Kwiesielewicz, M. (1998). A note on the fuzzy extension of Saaty's priority theory, *Fuzzy Sets and Systems*, 95, 161-172.

Lasek, M. (1992). Hierarchical structures of fuzzy ratings in the analysis of strategic goals of enterprises, *Fuzzy Sets and Systems*, 50, 127-134.

Mayer, A., Mechler, B., Schlindwein, A. and Wolke R. (1994). *Fuzzy Logic. Einführung und Leitfaden zur praktischen Anwendung mit Fuzzy-Shell in C++*, Addison-Wesley, Bonn.

McCauley-Bell, P. and Badiru, A.B. (1996). Fuzzy Modeling and Analytic Hierarchy Processing to Quantify Risk Levels Associated with Occupational Injuries. Part I: The Development of Fuzzy -Linguistic Risk Levels, Part II: The Development of a Fuzzy Rule-Based Model for the Prediction of Injury, *IEEE Transactions on Fuzzy Systems*, 4, 124-138.

Mechler, B. (1995). *Intelligente Informations-systeme. Fuzzy Logic und Neuronale Systeme zur Unterstützung von Management-entscheidungen*, Addison-Wesley, Bonn.

Mißler-Behr, M. (1996). Applying Fuzzy Classification in Strategic Planning, in: *Proceedings Fourth European Congress on Intelligent Techniques and Soft Computing EUFIT'96, Aachen*, vol. 2, 1302-1306.

Nakajima, H., Sogoh, T. and Arao, M. (1993). Development of an efficient fuzzy SQL for a large scale fuzzy relational database, in: *Proceedings 5th IFSA World Congress, Seoul*, Korea, 517-530.

Olson, D.L. (1996). *Decision Aids for Selection Problems*, Springer-Verlag, New York.

Oxelheim, L. and Wihlborg, C.G. (1991), Corporate Strategies in a Turbulent World Economy, *Management International Rewiev*, 31, 293-315.

Peneva, V. and Popchev, I.. (1998). Comparison of clusters from fuzzy numbers, *Fuzzy Sets and Systems*, 97, 75-81.

Ribeiro, R.A. (1996). Fuzzy multiple attribute decision making: A review and new preference elicitation techniques, *Fuzzy Sets and Systems*, 78, 155-181.

Robson, W. (1997). *Strategic Management & Information Systems*, Financial Times Pitman Publishing, London.

Rommelfanger, H. (1994). *Fuzzy Decision Support-Systeme. Entscheiden bei Unschärfe*, Springer-Verlag, Heidelberg.

Saaty, T.L. (1980). *The Analytic Hierarchy Process*, McGraw-Hill, New York.

Triantaphyllou, E. and Lin, Ch.-T. (1996). Development and Evaluation of Five Fuzzy Multiattribute Decision-Making Methods, *International Journal of Approximate Reasoning*, 14, 281-310.

Turksen, I.B. (1992). Fuzzy expert systems for IE/OR/MS, *Fuzzy Sets and Systems*, 51, 1-27.

Wilhelm, M.R. and Parsaei, H.R. (1991). A fuzzy linguistic approach to implementing a strategy for computer integrated manufacturing, *Fuzzy Sets and Systems*, 42, 191-204.

Wilson, I. (1994). Strategic Planning isn't Dead - is Changed, *Long Range Planning*, 27, 12-24.

Yazici, A. and Koyuncu, M. (1997). Fuzzy object-oriented database modeling coupled with fuzzy logic, *Fuzzy Sets and Systems*, 89, 1-26.

Yeung, D.S. and Tsang, E.C.C. (1998). A Multilevel Weighted Fuzzy Reasoning Algorithm for Expert Systems, *IEEE Transactions on Systems, Man, and Cybernetics*, 28, 149-158.

Zadeh, L.A. (1975). The concept of a linguistic variable and its application to approximate reasoning. Parts 1 and 2, *Inform. Sci.* 8 (1975) 199-249, 301-357. Part 3, *Inform. Sci.* 9 (1976) 43-80.

Zimmermann, H.-J. and Gutsche, L. (1991). *Multi-Criteria Analyse. Einführung in die Theorie der Entscheidungen bei Mehrfachzielsetzungen*, Springer-Verlag, Berlin Heidelberg.

Zimmermann, H.-J. (1989). *Strategic Planning, Operations Research and Knowledge Based Systems*, in: J.-L. Verdegay and M. Delgado, The Interface between Artificial Intelligence and Operations Research in Fuzzy Environment, 253-274.

14 DECISION AND PLANNING IN RESEARCH AND DEVELOPMENT

Brigitte Werners
Richard Weber

Abstract: The importance of research and development has been increasing steadily during the last decades and is still growing. As a consequence, methods and tools which can support the R&D planning and decision process also become more numerous and sophisticated. This main chapter gives an overview over important methods and reviews the literature in this area. Special attention is given to the modeling of different kinds of uncertainty using fuzzy logic which is particularly important in R&D planning. A broad spectrum of fuzzy approaches is used to support in research and development decision making.

14.1 INTRODUCTION

The management of research and development can be described as a complex, multi-phase process. After observing the environment in order to recognize chances and analyzing the strategic posture, research and development projects have to be identified and defined. They also must be evaluated and promising projects are to be singled out and implemented. The management of the projects and possible modifications of project descriptions on the basis of additional observations follows. Generally, the resource requirements of the suggested projects exceed the existing capacities. Hence, project selection is one of the crucial problems of research and development management. This particularly since existing interdependences and synergics have to be taken into consideration. A special feature of this area is the high degree of uncertainty concerning possible environmental developments and also concerning the successful termination of research and development activities.

Engineering design as a part of the development process also contains several uncertainties and needs adequate tools.

In chapter 2 strategic planning of research and development is briefly treated. Some decision support systems are introduced which base on fuzzy methods and advise in strategic R&D planning.

The main focus of our considerations in chapter 3 will be on the selection of R&D projects and programs by mathematical programming models of which a representative basic model will be shown. Then, uncertainty modeling and the use of fuzzy set theory to represent uncertainties in R&D planning models are described. It will also be shown how the suggested approaches can be integrated into an interactive decision support system.

R&D project management contains the scheduling of activities in a project. These aspects are considered in more detail in chapter 4. Fuzzy network techniques are introduced which support project control. Here, too, fuzzy mathematical programming approaches are useful to consider scarce resources and optimize development time.

Chapter 5 is mostly concerned with some engineering aspects in R&D management. An application of fuzzy set theory for information management in simultaneous engineering is introduced. Many applications in engineering design and construction exist which use linguistic variables and other fuzzy logic approaches, some of them are shortly described.

14.2 STRATEGIC PLANNING OF RESEARCH AND DEVELOPMENT

Strategic planning of R&D generally has to be part of the strategic planning process during which the long-term development has to be determined. A detailed diagnosis of environmental influences and the strategic posture is essential for the development of R&D strategies. Due to the uncertainty of the future and the fact that most problems are semi-structured especially fuzzy methods are promising and have already been developed.

STRATASSIST is a general decision support system for strategic planners which uses fuzzy sets technology. It is developed by Green Hall (1993) and helps a decision maker to formulate an appropriate strategy for his or her firm after analyzing the respective crisp and vague data using the concept of Michael Porter (1980, 1985).

Lasek (1992) has implemented a decision support system to analyze strategic goals of enterprises. It assists in the formulation of strategic goals and subgoals of a firm and in the selection of strategies to achieve these subgoals and goals. For that purpose it contains a hierarchical evaluation procedure, a "what-if" hierarchical analysis model, and a fuzzy ratings approach. Before introducing or further investigating a new technology it is important to know its public acceptance. For instance, Onisawa et al. (1986) use fuzzy measure analysis to determine the structure of public attitude towards the use of nuclear energy.

A considerably large amount of research has been done on the topic of handling multiple criteria problems using fuzzy sets. Many of the resulting methods are appropriate in the context of research and development planning problems. Strategic planning approaches, for example, combine goals and evaluate strategies on the basis of fuzzy sets. One commonly used technique for analyzing the current and future business position is the business portfolio approach. Originally, a business portfolio of Strategic Business Units (SBU) is evaluated and visualized in a two-dimensional portfolio matrix with the dimensions "relative competition position" and "business growth rate" (Hedley 1992). Dependent on the position of the different Strategic Business Units in this matrix, strategies are recommended to keep the portfolio balanced. Of particular importance in strategic R&D planning are similar tools to evaluate the technology portfolio of a firm (Pfeiffer, Dögl 1997).

Some of the major problems of such approaches are (Zimmermann 1996, p. 192):

Dimensionality: Two dimensions are insufficient to describe the strategic position of a SBU.

Data collection and aggregation: A lot of factors have to be considered and aggregated to get evaluations with respect to the two dimensions of the portfolio. This is usually done intuitively or using scoring models.

Strategy assignment: Only broad strategic categories have been defined to which strategies are assigned. The knowledge and experience of experts is not included so that no detailed strategies can be determined.

Modeling and consideration of uncertainty: The different kinds of uncertainty which are particularly important in the area of strategic planning are hardly considered in classical approaches.

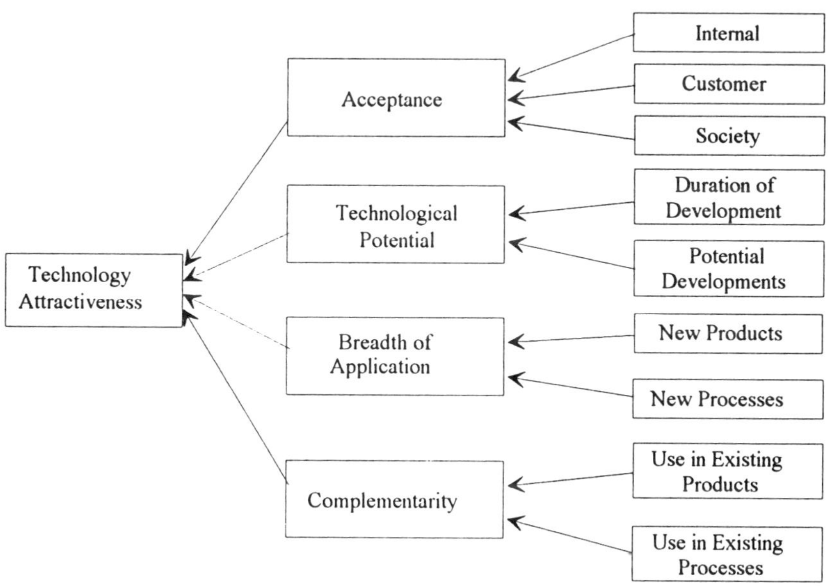

Figure 14.1 Criteria tree for technology attractiveness (Zimmermann 1996, p.195)

Management of uncertainty is a very important component in expert systems (Zadeh 1983). For example ESP, an Expert System for Strategic Planning (Zimmermann, 1989), improves classical approaches by introducing fuzzy methods into portfolio evaluations. A further development can be found in Werners (1993). In addition to the market oriented consideration of the SBUs, the technological positions are evaluated in this portfolio approach with respect to technology attractiveness and technological strength of the firm. A combined analysis with a business portfolio leads to valuable additional information.

Each dimension of the portfolio matrix is defined by a criteria hierarchy of subcriteria and strategic factors. Fig. 14.1 shows a part of the tree for the dimension technology attractiveness.

For each business unit under consideration the expert gives his or her evaluation with respect to each factor. The expert can use linguistic terms for a qualitative evaluation which are defined for respective linguistic variables. An example with trapezoidal membership functions is given in the following figure 14.2.

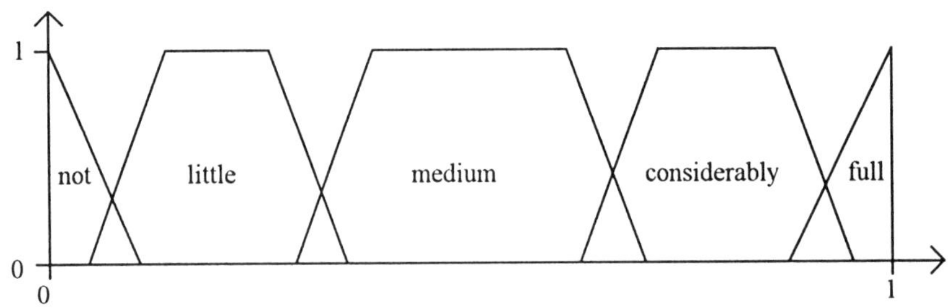

Figure 14.2 Terms of "degree of achievement" (Zimmermann 1996, p. 196)

These linguistic ratings have to be aggregated adequately. Zimmermann suggests the γ-operator which is empirically tested and well suited to model compensation of different criteria (Zimmermann, Zysno 1980). Another compensatory operator is the "Fuzzy and" suggested by Werners (1984, 1988). This operator is very easy to use in calculations especially in combination with trapezoidal membership functions.

The fuzziness of the position of a technology or of a business unit as a result of the uncertainty in evaluation can be represented in the fuzzy portfolio model. An example is given in fig. 14.3. Here, the hatched part symbolizes the uncertainty in positioning the strategic business unit. Choosing a certain membership degree $\alpha \in [0, 1]$ the respective α-level cut can be drawn, too. That is the crisp portfolio model with degree α.

Even more informative is a combined consideration of product and process technologies of a strategic business unit or a comparison of the positioning with respect to market criteria in contrast to technological criteria. In addition to the evaluation of a business unit as a whole with respect to technological criteria, the necessary product and production technologies first are considered autonomously and then the consequences are combined. This allows to develop strategies in more detail and to analyze synergies explicitly.

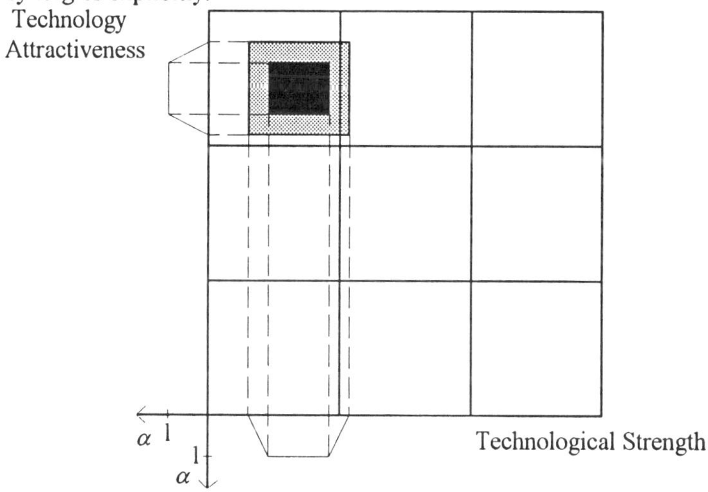

Figure 14.3 Fuzzy technology portfolio (Werners 1993, p. 300)

If there are several component technologies relevant for a product the technological position of each technology is determined and then the positioning of the product resp. the belonging SBU is deduced (Pfeiffer, Dögl 1997, p. 424). Here, the varying importances of the technologies for a SBU and the kind of aggregation can be modeled using fuzzy methods. For example, in the following figure the technological positions of the three component technologies A, B, and C of a product are given together with the resulting position of the product for the considered SBU. If the rather unattractive component technology A can be substituted by the highly attractive technology X, the position of the future product will change positively. This is demonstrated by the position of the SBU_{FUTURE} in the technology portfolio with chosen degree α (fig. 14.4). If the attractive technology X can be integrated into several products the replacement of the respective unattractive component technologies by X changes the technological position of all those products. The analysis of this kind of synergistic effects can be supported by a model which contains information of component technologies. Here, additionally, fuzzy dependences can be modeled.

A similar evaluation can be done considering the process technologies of the product and its components. The interdependences between product, process, and material technologies on one hand and the strategic business units on the other hand are modeled using fuzzy rules and linguistic variables. Comparing the positions of product technologies and process technologies gives additional advice on necessary strategic consequences. Those information are the basis for the fuzzy expert system for strategic technology planning StraTeP as suggested by Werners (1993).

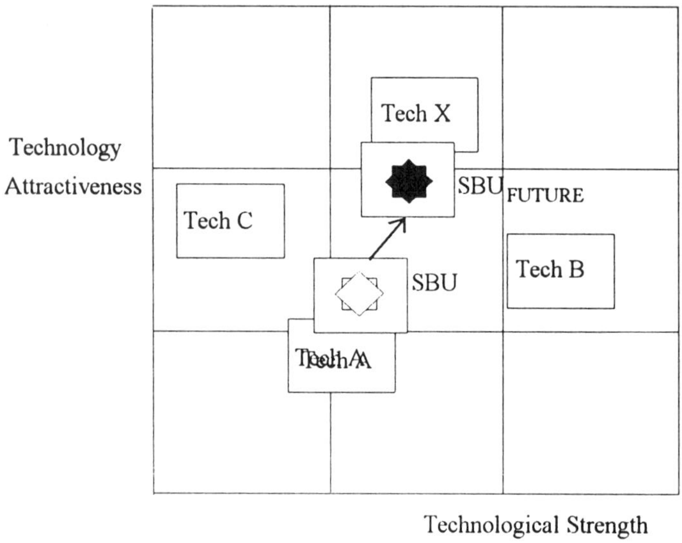

Figure 14.4 Substitution of component technology A by X (Werners 1993)

The analysis and selection of a technology transfer strategy is characterized as a multi-criteria decision making problem in an ill-structured situation. Chang, Chen (1994) suggest a fuzzy algorithm to determine the most appropriate technology transfer strategy. As an application in biotechnology they describe exemplarily the selection of the transfer strategy for the hepatitis B vaccine technology in more detail. One of the tasks of the Development Center for Biotechnology in Taiwan is to transfer technology to commercial applications and should be supported by an adequate system. A committee of four decision makers has been formed to determine the most appropriate technology transfer strategy. After screening, the following selection criteria were considered:

- technological availability,
- market potential,
- policy support, and
- management ability.

Commonly successful strategies for transferring R&D results to industry have been: purchasing, i.e. firms buy R&D; working with an industrial partner in a joint venture, both partners share the required resources; licensing; cooperative R&D. There is no sharp boundary between the linguistic evaluations of innovations with respect to the criteria and the available market data which are imprecise and fuzzy due to time, space, measurement error, or uncertain environment. The degree of appropriateness of a certain strategy is not exactly measurable but described using terms like highly appropriate, sometimes appropriate. Chang, Chen (1994) suggest a fuzzy based hierarchical multi-person multi-criteria algorithm to evaluate the appropriateness of strategies and to choose the best one. The fuzzy appropriateness indices for the alternative strategies contain information about the rating and, additionally, about its uncertainty. That renders it a valuable tool for decision support in such an ill-structured problem area.

The above depicted applications demonstrate that especially fuzzy multi-criteria approaches and fuzzy reasoning systems can assist in strategic R&D problem solving.

14.3 SELECTION OF R&D PROJECTS AND PROGRAMS

Decision models for R&D planning support the decision maker in composing the best possible R&D program from suitable R&D projects (Liberatore, Titus 1983). To this end an appropriate objective function has to be optimized subject to constraints that have to be satisfied by this program (Brockhoff 1994). Therefore, an R&D program planning problem can be expressed as a mathematical programming model in which the objective function, which is possibly vector-valued, is optimized subject to existing constraints. Objective functions as well as constraints can be linear or nonlinear. Nonlinear objective functions and/or constraints are, for instance, used to model project interdependences by a quadratic objective function (Fox et al. 1984) or to model resource allocation for R&D projects (e.g. Bard et al. 1988). Most of the models which can be found in literature, however, use linear functions to model interdependences of different projects. We will focus our attention essentially on this type of models because efficient algorithms exist for their solution.

Aside from R&D planning models which consider simultaneously different relationships ranking methods and other methods from the area of multi-attribute decision making have been suggested in literature (Anandalingam, Olsson 1989). Drawing profiles and using check lists (Souder, Mandakovic 1986) or the use of scoring methods are, for instance, suggested in this context. Other authors suggest the use of the analytic hierarchy process (Saaty 1990) for R&D project selection (Liberatore 1987). One disadvantage of these approaches is that an "optimal" R&D project selection is not achieved because there is no simultaneous consideration of existing constraints. Therefore, project interdependences cannot be considered properly.

Extensive surveys of classical approaches and mathematical programming models for R&D planning are given by Baker, Freeland (1975), Booker, Bryson (1985), and Liberatore, Stylianou (1995), respectively. The application of mathematical R&D planning models in reality has been investigated empirically and it has been found that the frequency of use of such methods is increasing (e.g. Watts, Higgins 1987, Lee et al. 1986). This can be credited to a certain extent to the fact that management comprehends to an increasing degree the limited human ability of information processing and hence, accepts the support by computers more than in the past. In addition, theories, methods, and systems which facilitate the modeling and processing of subjective factors have been advanced. The empirical investigations mentioned above agree in stating an increased requirement of user friendly systems for the decision support of those responsible for R&D management. Hence, a further consideration of these approaches seems to be promising. We will shortly introduce premises, criteria, interdependences, and uncertainties which are relevant in models for R&D program planning.

Premises of models

We shall first briefly describe the essential premises which are made in most of the existing decision models for R&D program planning:

Decision alternatives: Frequently, it is assumed that the possible decision alternatives are well-known and that the information needed for their description exists completely at the beginning of the planning process. The number of the considered decision alternatives influences the decision process. If only a few decision alternatives are taken into account, they are normally explicitly enumerated. If, on the other hand, a large number of R&D projects have to be considered or if different projects exist in a continuous space, the definition of decision alternatives is done implicitly by defining a feasible set via constraints. If necessary, formulations and specific methods can be used to ensure integer values of the decision variables.

Kind of projects: The majority of published models assumes that the selection of R&D projects refers to a set of new projects which have not yet been started (e.g. Czajkowski, Jones 1986, Steele 1988). A study of Raelin and Balachandra (1985) investigates in addition criteria which can be used to decide about continuing or stopping running R&D projects. Based on this publication, Bard et al. (1988) have developed an R&D program planning model in which running projects and their termination is considered explicitly. This can be achieved by introducing additional variables and constraints for the different projects. The optimal budget allocation to existing projects can also be supported by mathematical models (e.g. Lootsma et al. 1990).

Results of projects: With respect to the respective result, projects can be classified into those which lead to new or improved products (product innovation) and those which lead to new or improved production methods (process innovation) (Baker et al. 1986). This difference is relevant for the relationships to be considered and it should, therefore, be contained in the model. In the following, we will show an approach for the selection of R&D projects which lead to product innovations and for which the simultaneous determination of projects for new products has to be modeled. A more detailed model which includes sales strategies can be found in Weber et al. (1990).

Time of decision: The solutions of R&D program planning models consider decisions at various points of time. Most of the models one can find in literature support a decision only in one period (e.g. Czajkowski, Jones 1986). In these models the decision concentrates on the selection of projects available at one specific point of time. For those projects which are not optimal at that moment a later consideration is not possible. Approaches which consider decisions also at later points of time can be differentiated in models of fixed and models of flexible planning (Brockhoff 1994). While the models of fixed planning determine the decision for all periods up to a certain planning horizon, models of flexible planning only provide decisions definitely for the first period considering for all consecutive time intervals alternative plans which are geared to the respective environmental scenario.

Criteria

In deciding about R&D projects, generally several criteria have to be considered simultaneously (e.g. Danila 1985). Frequently, in addition to quantitative or quantifiable criteria, non-quantifiable criteria have to be taken into consideration when deciding about R&D projects, for example the scientific potential which can be achieved by executing an R&D project, new or improved technical options, insight into new areas of expertise and catching up with the newest technological know-how.

In addition to all criteria of corporate strategy which, in general, are difficult to quantify, Brockhoff (1994) mentions the following classification of relevant criteria for R&D projects:

- financial criteria,
- marketing criteria,
- R&D related criteria.

Financial criteria normally mentioned in the literature are return on investment, sales, net present value, profit (Brockhoff, Chakrabarti 1988). Marketing criteria one often finds are market share, sales growth, number of potential customers. R&D related criteria are, for instance, the expected number of successful projects or the number of projects which are contained in the R&D program of an enterprise.

In order to be able to integrate multi-dimensional and partially qualitative criteria into mathematical programming models more recent approaches suggest to aggregate those criteria for each project to a cardinal evaluation. These values are then used to model a one-dimensional linear objective function, which is applied to determine the optimal R&D program. For the investigation of the weights which are used to aggregate the different criteria the analytic hierarchy process by Saaty is suggested by several authors (e.g. Bard 1986, Liberatore 1987). The hierarchical structuring of the decision process, which is necessary for the use of Saaty's method, is particularly useful for the decision maker. Several objective functions can be handled simultaneously by models based on the theory of fuzzy sets. Here, common membership functions are determined which express feasibility aspects as well as degrees of optimality (Werners 1987, Zimmermann 1988).

Interdependences

Research and development within one enterprise is interrelated to various other functional areas of a corporation. These interdependences are particularly strong between research and development on one side and marketing and production on the other side (Horwitch, Thietart 1987, Weber et al. 1990). In addition, there are relevant project interdependences, internal ones as well as external ones which are relevant for the R&D selection decision (Gear, Cowie 1980). External interdependences are, for instance, the relationships between R&D projects and external influences such as social and economical aspects. These are, generally, not modeled explicitly in R&D program planning models.

Internal project interdependences are very often easier to model quantitatively and, therefore, they can be included more easily into mathematical models. These interdependences are classified into the three groups technical interactions, resource interactions, benefit interactions (Czajkowski and Jones 1986).

When composing the R&D program of a corporation from single R&D projects, certain conditions exist which require either the common execution of specific projects or which exclude the simultaneous realization of some projects. These technical interactions can normally be modeled mathematically by using logical variables.

If several projects need the same scarce resources, these projects have to be considered simultaneously in the selection process. This is done using mathematical programming models with budget or capacity constraints. Sometimes certain synergy effects may exist if two or more specific projects are selected, i.e. when executing these projects simultaneously less of the resources are used than if the projects are executed consecutively.

Benefit interactions can also occur when two or more R&D projects are executed simultaneously. This means that a different result is achieved when executing two projects simultaneously than if the results of these two projects has been added. These synergy effects also have to be integrated into the model.

Uncertainty

Uncertainties connected with R&D projects result from quite different sources and they have to be modeled respectively. One of the main problems, especially when considering R&D projects, is the uncertainty of the project result. This uncertainty concerns the magnitude of the result as well as the occurrence of a positive result per se. In addition, very often the necessary resources are not known exactly and existing interdependences are partially uncertain (Czajkowski, Jones 1986). In some approaches at least partial aspects of existing uncertainties have been modeled. In this respect two classes of models can be distinguished, those basing on probability theory and those using the theory of fuzzy sets. Those of the first class use stochastic concepts. In general, it is assumed that all states of nature which can possibly happen in the future are known together with their prospective probabilities, which may have been estimated subjectively. The ways in which these data are included in mathematical models depend on whether the uncertainties refer to the results or to the resources, and on the risk related attitude of the decision maker.

Approaches on the basis of fuzzy set theory make it possible to include uncertainties of different kinds. This pertains, in particular, to uncertainties of the description of phenomena and requirements. Uncertainties can, generally, be modeled linguistically rather than numerically which facilitates the communication between the formal system and the user. A survey of these approaches is contained in Zimmermann (1987). Dias (1988) suggests a method to select from R&D project appraisals for which linguistic values are used for each criterion.

Mathematical programming and R&D

In the following a decision model based on mathematical programming is exemplarily presented which supports the planning of the R&D program of an enterprise and which includes most of the aspects mentioned so far. We start with a basic model. Similar models are part of many suggestions which can be found in the literature. This approach will then be extended by including the modeling of uncertainty.

Premises and goal

The problem is considered to compose optimally an R&D program which itself consists of several different single programs each of which results in a new product. These programs again consist of several R&D projects needing certain resources. Let there exist n different R&D projects to choose from. From these projects m different R&D programs shall be composed. It is possible to assign one project to several programs at the same time. Figure 14.5 indicates such a structure assuming that there are $m=5$ programs x_j with net present value NPV_j composed of $n=10$ different projects y_i with their required financial means B_i.

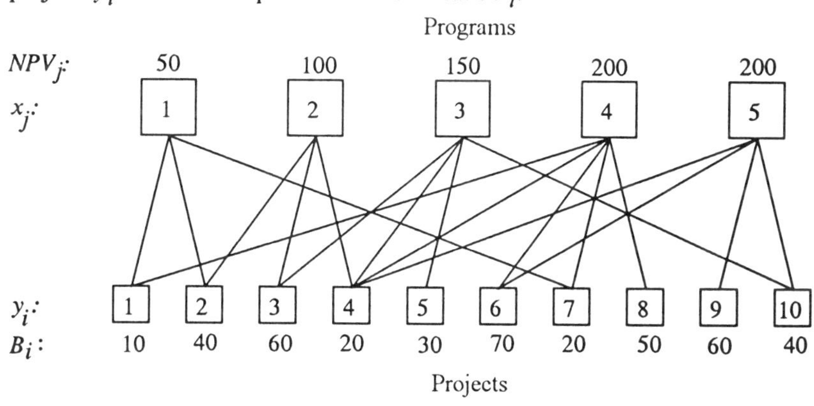

Figure 14.5 Project - program relationships

The model will not consider explicitly projects which have already been selected at an earlier point of time, i.e. the model will only support decisions concerning projects which have not been subject to a decision before. To each R&D program we assign a new product of which the cash flow is known. Hence, each program can be evaluated by the net present value of the positive and negative cash flows originated by this program. For the time being, additional criteria are not considered. The following objective function can then be formulated:

Objective function of the decision model

$$\sum_{j=1}^{m} NPV_j * x_j \rightarrow max \qquad (1)$$

where NPV_j is the net present value of program j and x_j is a binary decision variable defined as follows:

$$x_j = \begin{cases} 1 & \text{if program } j \text{ is executed} \\ 0 & \text{else} \end{cases}$$

Constraints of the decision model

Technical interactions contained in the model describe the fact that certain projects are necessary for the completion of a certain program. When deciding to execute a certain program it has to be ensured that all R&D projects are selected which are necessary for this program. A possible interrelationship between programs and projects is depicted in fig. 14.5 and is modeled by the following constraint:

$$NTN_j * x_j - \sum_{i=1}^{n} TN_{ij} * y_i \leq 0 \qquad j = 1,\dots,m \qquad (2)$$

Here, the binary decision variable y_i is defined as:

$$y_i = \begin{cases} 1 & \text{if project } i \text{ is selected} \\ 0 & \text{else} \end{cases}$$

The coefficients are defined as follows:

$$TN_{ij} = \begin{cases} 1 & \text{if project } i \text{ is technically necessary for program } j \\ 0 & \text{else} \end{cases}$$

$$NTN_j := \sum_{i=1}^{n} TN_{ij}$$

NTN_j is the number of projects which are technically necessary for program j.

Resource interactions: When planning the R&D programs of an enterprise certain capacity constraints generally have to be considered. These constraints concern the financial budget as well as resources of different kinds, for instance, the number of available persons in the area of R&D (Frame 1983). The type of modeling is identical for all types of resources. Therefore, in our model we shall exemplarily include only the financial capacities. If the requirement of the decision maker is "The cumulated expenses for the selected R&D projects must not exceed the available R&D budget" then the following constraint has to be included:

$$\sum_{i=1}^{n} B_i * y_i \leq B \tag{3}$$

where B = available budget for R&D
B_i = required financial means for the realization of project i.

Figure 14.6 depicts exemplarily the optimal solution of the basic model for the parameters shown there and assuming a budget of $B=120$. The optimal solution is the execution of program 2 containing projects 2, 3, and 4. The resulting net present value is 100.

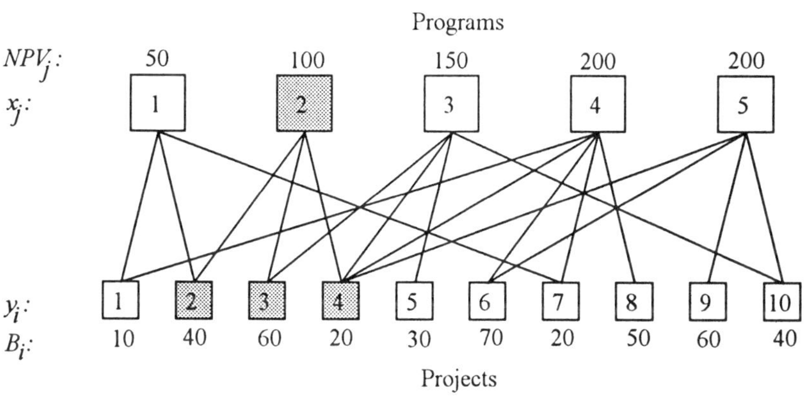

Figure 14.6 Optimal solution of the basic model

If there is a "synergistic" resource interdependence of two projects i and h in the sense that they use the amount B_{ih} less if they are executed simultaneously, then this situation is modeled by including constraint (4).

$$\sum_{i=1}^{n} B_i * y_i \leq B + \frac{1}{2} \sum_{i=1}^{n} \sum_{h=1}^{n} B_{ih} * int_{ih} \tag{4}$$

Here, B_{ih} has to be determined by the decision maker. The variables and coefficients are defined as follows:

$$int_{ih} = \begin{cases} 1 & \text{if } y_i = y_h = 1, \text{ i.e both projects have been selected} \\ 0 & \text{else} \end{cases}$$

This is derived by adding the following inequalities:

$$y_i \geq int_{ih}$$
$$y_h \geq int_{ih}$$
$$y_i + y_h - 1 \leq int_{ih}$$

Benefit interactions: In analogy to the resource interactions on the input side there might be synergy effects on the output side (Schmidt 1993). These benefit interactions lead to a higher or a lower net present value if two specific programs are executed simultaneously. In this case the above mentioned objective function (1) has to be modified as follows

$$\sum_{j=1}^{m} NPV_j * x_j + \frac{1}{2} \sum_{j=1}^{m} \sum_{k=1}^{m} APV_{jk} * x_{jk} \rightarrow max \qquad (5)$$

Here x_{jk} is a variable which is 1 if both programs j and k are selected, i.e. if x_j and x_k both have the value of 1. APV_{jk} is the additional net present value if programs j and k are realized simultaneously.

Uncertainty in R&D models

To model uncertainty basically different theories exist. The appropriate one has to be chosen context dependently. Of particular importance are the stochastic approaches and those on the basis of fuzzy set theory. In the following they shall be considered in more detail in the context of R&D planning.

Stochastic approaches

If the coefficients of the objective function are not deterministic and if the decision maker knows the corresponding probability distributions, one may use the expected contributions of the different R&D programs as coefficients of the objective function (e.g. Fox et al. 1984). This approach, however, assumes a risk-neutral decision maker and hence, this type of modeling is not always adequate. To allow for risk aversion the variance of the respective probability distributions can also be considered as a part of the objective function. If the decision maker is capable to state the consequences of his or her decisions as complete utility functions together with the respective probability distributions, then the goal is to maximize the respective utility of an R&D program (e.g. Madey, Dean 1985; Booker, Bryson 1985). It is, however, very questionable whether such assumptions are realistic.

Sometimes, the R&D projects have stochastic components for example budget requests. To use the mean values is not adequate because depending on the factual requirements some of the constraints may be violated, which leads to unfeasibility of the solution. To avoid this, chance-constrained approaches may be used. These models for R&D program planning have, for instance, been suggested by Czajkowski and Jones (1986). In all of these cases, however, it is assumed that the decision maker knows the probability distributions of the random variables or at least the respective moments of the distributions.

Fuzzy approaches

Some of the uncertainties which occur in R&D program planning can be modeled by using results of fuzzy set theory. In the following the above mentioned basic assumptions are used. A basic model for decision making in a fuzzy environment was suggested by Bellman and Zadeh (1970). In contrast to the classical model of a decision under certainty, the authors suggest a symmetrical model in which the goals as well as the constraints are defined as fuzzy sets. This approach can also be applied to linear programming models (e.g. Zimmermann 1978).

First it is assumed that the decision maker when planning his or her R&D program does not necessarily want to maximize the objective function but he or she is already satisfied with reaching certain aspiration levels. These aspiration levels may even be considered of different importance or strictness. If a level z should be reached, this can be modeled using fuzzy goals according to one of the following:

$$\sum_{j=1}^{m} NPV_j * x_j \gtrsim z \tag{6}$$

$$\sum_{j=1}^{m} NPV_j * x_j + \frac{1}{2} \sum_{j=1}^{m} \sum_{k=1}^{m} APV_{jk} * x_{jk} \gtrsim z \tag{7}$$

Using such a fuzzy approach other or additional goals can similarly be modeled and considered simultaneously. Suggestions for suitable and constraint-dependent aspiration levels z can be established by using interactive models (see Werners 1987).

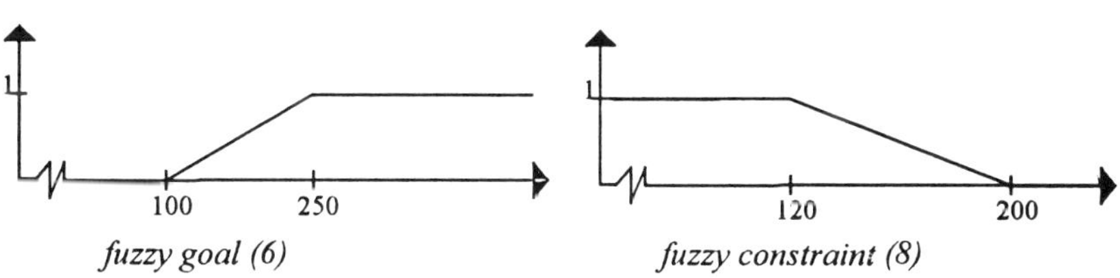

Figure 14.7 Linear membership functions for a goal and a constraint

In figure 14.7 the former solution value 100 is the lower bound of the requested total net present value. It is assumed that the decision maker is fully satisfied with each resulting *NPV* above 250.

The capacity constraints can now also be relaxed and modeled as fuzzy constraints if, for instance, the decision maker requests that "the expenses for research and development should, if possible, not exceed the planned R&D budget". This is modeled by the following constraint:

$$\sum_{i=1}^{n} B_i * y_i \lesssim B \tag{8}$$

Fig. 14.7 shows a membership function where the decision maker allows to exceed the budget of 120 up to a maximum of 200. All other constraints are treated similarly. The aggregation of the different fuzzy constraints of the model for R&D program planning can then be performed by using the logical 'and'. It is also possible to use a hierarchical compensatory aggregation, which is appropriate to model compensation (Werners 1988).

Here, it is assumed that a violation of the budget of 120 is tolerated up to 200 dependent on positive consequences for the objective function value. Then the maximizing solution shown in figure 14.8 results. A net present value of 200 is achieved with a budget of 170.

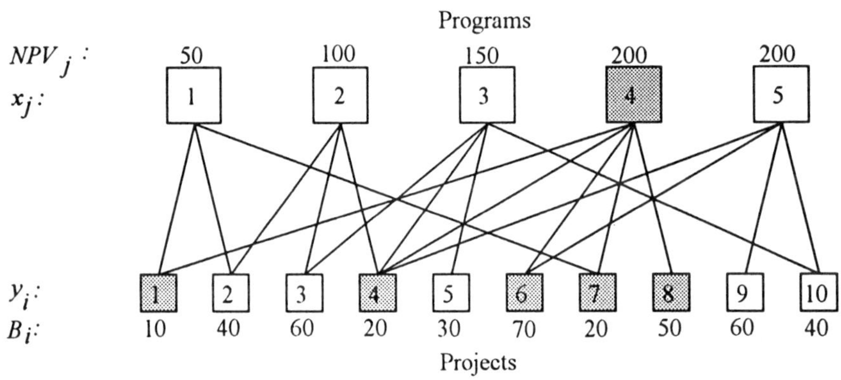

Figure 14.8 Maximizing decision for the fuzzy R&D model

Decision support systems for R&D program planning

Known solutions of the models for R&D program planning are especially useful for the decision maker if he or she obtains additional information which offers new insights into the problem structure. On this basis he or she may be in a position to modify the model. If fuzzy modeling approaches are used which are more flexible with respect to the aggregation operators used they can also be built into the model.

Support of the decision maker can be further improved by integrating models for R&D program planning into an interactive decision support system. To this end, the decision support system MOPS can be used, which was suggested by Werners (1984). In this system several objective functions are considered and the use of subjective evaluations and fuzzy formulations is possible. These aspects were considered as essential for R&D program planning. MOPS offers in particular

- extensive analytical decision support by using flexible models on the basis of the theory of fuzzy sets,
- ways to easily modify the models by entering new parameters, and
- efficient analytical evaluations by determining equivalent models which can be solved by available efficient standard software.

Modeling is supported in so far as the system only requires the planner to formulate crisp and fuzzy constraints, and crisp or variable goals. If the user does not provide the membership function characterizing the goal, an appropriate membership function which depends on the restrictions is suggested by the system. This membership function characterizes the fuzzy set "maximize the goal under fuzzy constraints". It would, for instance, be possible to provide the membership function for the objective function "net present value of the program" subject to flexible "budgetary constraints".

In order to facilitate a progressive articulation of problem information, interactive modifications of the fuzzy models are possible. They can be implemented by the decision maker very easily and they are supported by the system giving additional information. In addition to linear membership functions those nonlinear membership functions can also be considered for which suitable transformations to linear crisp models exist.

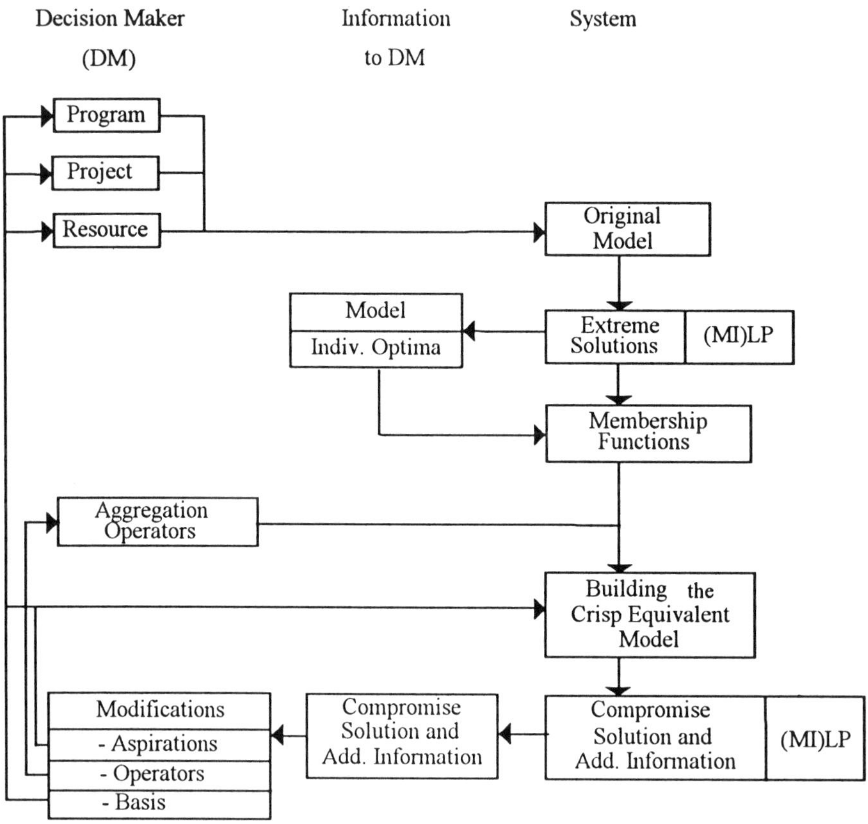

Figure 14.9 Structure of the decision support system MOPS

Figure 14.9 depicts the rough structure of the decision support process by MOPS (Werners 1984). On the left-hand side, information is shown which is entered into the system by the decision maker. It consists primarily of information about the goals and constraints and of the membership functions as far as they are known. To determine extreme solutions, the system uses a commercially available linear programming (LP) resp. mixed integer linear programming (MILP) code after the models have been formulated accordingly. These extreme solutions are then presented to the decision maker. If he or she accepts one of them as the best compromise, the process stops. The decision maker also determines the way in which goals and constraints have to be aggregated. This can also be performed using a hierarchical structure. The computer then formulates the respective compromise model by considering the hierarchical levels sequentially. Depending on the operator used for aggregation it finds the compromise solution by using an LP or MILP code and provides this solution together with other additional information to the decision maker. If he or she accepts this solution as the best compromise, the procedure stops. Otherwise, it suggests possible modifications. If the decision maker modifies his or her aspirations in a feasible way, the system computes directly a new compromise solution and presents it to the decision maker.

In this phase of the decision process the decision maker gains further insights into the complex interrelationships of the selection of R&D projects. Hence, the planner gets the possibility to investigate interactively the influence of different formulations of constraints and goals and to modify his or her aspirations accordingly.

Additional facilities of support for the planner should be generated which, in particular, reduce inhibitions to formulate and use mathematical programming models. In addition, interfaces to data base systems should be provided in order to be able to use existing project and resource information easily. Conceivable is a context dependent preparation of the data in which, for instance, projects, programs and resources are described according to their characteristic parameters. Typical relationships such as e.g. the objective function can generally be fixed and provided such that the adequate mathematical modeling can be automated to a certain degree. In the process shown in figure 14.9 this corresponds to an additional system's support of the partial analysis shown on the left-hand side, which so far has to be performed by the user. Further support can be expected from future developments of structured modeling suggested by Geoffrion (1989), in which model structures can be handled in a similar way as the processing of data.

Very promising are methods of artificial intelligence in this context. They can e.g. be used to formulate interdependences which are typical for an enterprise or modeling heuristics of experts, which can then be used to improve existing compromise solutions or modify constraints or objective functions automatically. This pertains also to the use of frames and semantic nets in which inheritance of properties can be used to simplify the modeling process. It seems conceivable that descriptions of versions of projects which only differ to minor parts can be described together with their expected results.

For the acceptance of the use of mathematical methods it is also important that these procedures are computationally efficient. For integer programming, which is particularly relevant for the planning of research and development programs, new heuristic approaches are very promising. In this context simulated annealing and TABU-Search (Glover, Greenberg 1989) are certainly of particular interest. Additionally, the further development of inexpensive and powerful hardware technology will presumably lead to an increasing user acceptance of mathematically based approaches.

14.4 R&D PROJECT MANAGEMENT IN UNCERTAIN ENVIRONMENTS

Another important issue in research and development is management and scheduling of a program resp. a project after it has been chosen (Kort 1998). Some of the frequently used instruments for project management are the common network techniques critical path method (CPM), metra potential method (MPM), and program evaluation and review technique (PERT) (see f.i. Lockyer, Gordon 1992).

These techniques consider the classical problem of organizing a complex project which has been broken down into tasks or activities. Such techniques are not only valuable as mathematical tools for planning and controlling the schedule of special projects, they also proved to be of great importance for the total process of team building and functioning in project management (Graham 1985). With these instruments especially well structured projects can be analyzed and the relevant tasks are scheduled. Additionally, there exist instruments for capacity and cost analysis and optimization. Usually, these techniques base on a graph theoretical model of the tasks and their technological dependences. Graph theoretical algorithms are used to determine earliest and latest starting dates for the activities and the minimal duration of the overall project. R&D projects are characterized by different kinds of uncertainty which concern the necessity of activities, the structure, and the duration of the activities among others. Very often, available information concerning an R&D project is vague, highly aggregated, and changes over time. The standard technique PERT handles uncertain time durations using a specific stochastic approach, uncertainties concerning the structure have been modeled using GERT (for an overview see f.i. Pagnoni 1990). Both approaches use stochastic models and they are not sufficiently well suited to model other kinds of uncertainty. Therefore, in the following we will describe some approaches of fuzzy project network analysis appropriate for R&D project management considering fuzzy durations and fuzzy structure.

One frequently cited approach is the calculation of a network analysis with fuzzy duration estimation in Dubois, Prade (1988, pp. 60). The authors consider a project where the tasks are represented by the arrows of a directed graph and the events are represented by the nodes. If the durations of the different tasks are poorly known in advance they must be estimated subjectively. This can quite naturally be done by using fuzzy numbers resp. fuzzy intervals. Before describing some fuzzy approaches we will explain the fundamentals in project planning for crisp data in an activity-to-arrow representation.

We consider a directed graph $G = (N, A)$ where N is the set of nodes i, A is the set of arrows (i, j) between the nodes i and j. If the duration of each task (i, j) is given by a crisp value d_{ij} then the earliest dates t_j for the events represented by nodes can be calculated as follows:

$$t_j = \begin{cases} max \left\{ t_i + d_{ij} \,\middle|\, i \in P_j \right\} & P_j \neq \varnothing \\ t_0 & P_j = \varnothing \end{cases}$$

where $P_j = \left\{ i \,\middle|\, (i,j) \in A \right\}$ is the set of predecessors of j, t_0 is the starting date of the project.

The earliest finishing date of the project is obtained by $t_n = max \left\{ t_i \,\middle|\, i \in N \right\}$, the earliest finishing date of the last event n.

Accordingly, the latest dates T_i of the events represented by nodes can be calculated:

$$T_i = \begin{cases} min \left\{ T_j - d_{ij} \mid j \in S_i \right\} & S_i \neq \emptyset \\ T_n & S_i = \emptyset \end{cases}$$

where $S_i = \left\{ j \mid (i,j) \in A \right\}$ is the set of successors of i, T_n is in general set equal to t_n.

Using the results of these formulas the earliest starting time ES_{ij} and the latest starting time LS_{ij} of each activity *(i, j)* can be derived by

$$ES_{ij} = t_i, \quad LS_{ij} = T_j - d_{ij}$$

The difference of earliest and latest starting time is the slack time of the task, during which the task may be delayed without the project going beyond its earliest finishing date. It is called total float, *TF*. A critical path is a set of succeeding activities from the start to the final event for which each activity has a total float of zero. An event is critical if the difference between latest and earliest date is zero. Each event on a critical path is critical.

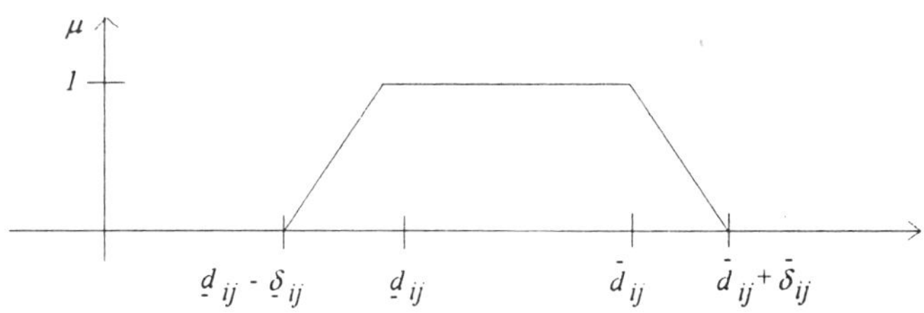

Figure 14.10 Fuzzy duration $d_{ij} = (\underline{d}_{ij}, \bar{d}_{ij}, \underline{\delta}_{ij}, \bar{\delta}_{ij})_{LR}$

Fuzzy durations

This above approach has been generalized to model uncertainty relating to the durations of activities with fuzzy sets. Using fuzzy trapezoidal intervals in *LR* representation to model uncertain durations, $d_{ij} = (\underline{d}_{ij}, \bar{d}_{ij}, \underline{\delta}_{ij}, \bar{\delta}_{ij})_{LR}$, results in efficient calculations of earliest and latest dates for the events.

The following standard operations can easily be performed as a result of the extension principle with arbitrary fuzzy intervals $M = (\underline{m}, \overline{m}, \underline{\alpha}, \overline{\alpha})_{LR}$ and $N = (\underline{n}, \overline{n}, \underline{\beta}, \overline{\beta})_{LR}$ (Dubois, Prade 1988, p. 50):

- the exact calculations

$$M \oplus N = (\underline{m} + \underline{n},\ \overline{m} + \overline{n},\ \underline{\alpha} + \underline{\beta},\ \overline{\alpha} + \overline{\beta})_{LR}$$

$$M \ominus N = (\underline{m} - \overline{n},\ \overline{m} - \underline{n},\ \underline{\alpha} + \overline{\beta},\ \overline{\alpha} + \underline{\beta})_{LR}$$

- and the global approximations

$$\tilde{max}\,(M,N) = (\ max(\underline{m},\underline{n}),\ max(\overline{m},\overline{n}),\ max\,(\underline{m},\underline{n}) - max(\underline{m} - \underline{\alpha}, \underline{n} - \underline{\beta}),$$
$$max(\overline{m} + \overline{\alpha}, \overline{n} + \overline{\beta}) - max(\overline{m},\overline{n})\ )_{LR}$$
$$\tilde{min}\,(M,N) = (\ min(\underline{m},\underline{n}),\ min(\overline{m},\overline{n}),\ min(\underline{m},\underline{n}) - min(\underline{m} - \underline{\alpha}, \underline{n} - \underline{\beta}),$$
$$min(\overline{m} + \overline{\alpha}, \overline{n} + \overline{\beta}) - min(\overline{m},\overline{n})\ )_{LR}.$$

The following example in fig. 14.11 shows a fuzzy network model of a project with durations of activities which are fuzzy numbers in linear *LR* representation with $\tilde{2} = (2,2,1,1)_{LR}$, $\tilde{3} = (3,3,1,1)_{LR}$, $\tilde{4} = (4,4,1,1)_{LR}$.

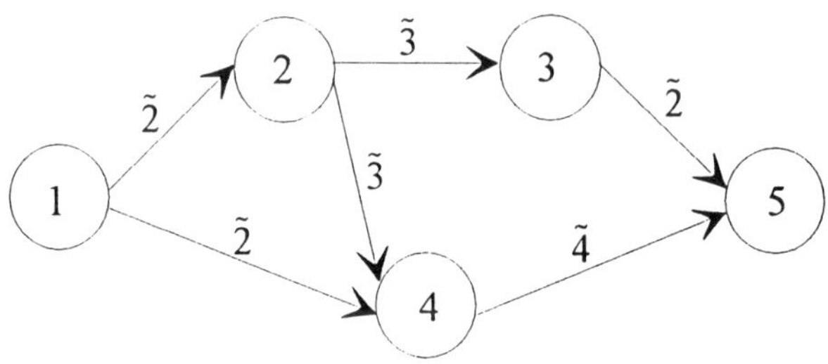

Figure 14.11 A project network with fuzzy durations

The earliest dates of the nodes are given in the following table. To determine the latest dates of all events the latest date of the last node $\tilde{T}_n$ has to be fixed. Here, $\tilde{T}_n = (11,11,1,1)$ is chosen.

Node i	Earliest date $\tilde{t}_i$	Latest date $\tilde{T}_i$
1	(0,0,0,0)	(2,2,4,4)
2	(2,2,1,1)	(4,4,3,3)
3	(5,5,2,2)	(9,9,2,2)
4	(5,5,2,2)	(7,7,2,2)
5	(9,9,3,3)	(11,11,1,1)

A modified Gantt chart as suggested by Geidel (1990) can be used to illustrate the resulting schedule of the project activities each with fuzzy earliest starting date and fuzzy earliest finishing date, see fig. 14.12. An alternative visualization technique for a fuzzy schedule is a Gantt chart with an optimistic and a pessimistic schedule as is used by Hapke et al. (1994).

The results demonstrate that the possible deviations of durations on a path accumulate and therefore the uncertainty of the earliest finishing times increase with an increasing number of activities on the path. The more activities are on a path the larger is the spread of the interval of the possible earliest finishing times and the more unprecise are the planning results. Updating the calculation during the realization of the project results in an increasing precision of the planning dates.

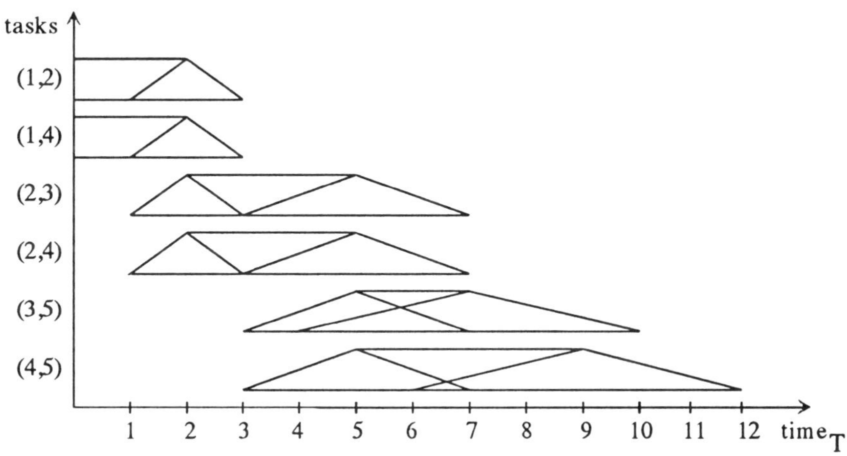

Figure 14.12 Modified Gantt chart

A specific problem in fuzzy project planning concerns the calculation of the latest dates of the nodes. Generally in classical CPM planning, the latest possible date of the sink node is set equal to the earliest date of that node. In the fuzzy context there are a lot of imprecisions in determining $\tilde{t}_n$. Therefore, Dubois, Prade (1988, p. 63) suggest to define $\tilde{T}_n$ independently of $\tilde{t}_n$ as a kind of required project terminal date.

In the example we have chosen $\tilde{T}_5 = (11,11,1,1)$, in contrast to $\tilde{t}_5 = (9,9,3,3)$. The calculation of the latest project starting event yields $\tilde{T}_0 = (2,2,4,4)$. This result does not seem to be very plausible. It is the consequence of using the subtraction operator $\ominus$.

The interval $\left[\tilde{t}_i, \tilde{T}_i\right]$ becomes a fuzzy interval bounded by fuzzy quantities. As a consequence the determination of a critical node becomes rather uncertain according to how the fuzzy intervals $\tilde{t}_i$ and $\tilde{T}_i$ overlap (fig. 14.13).

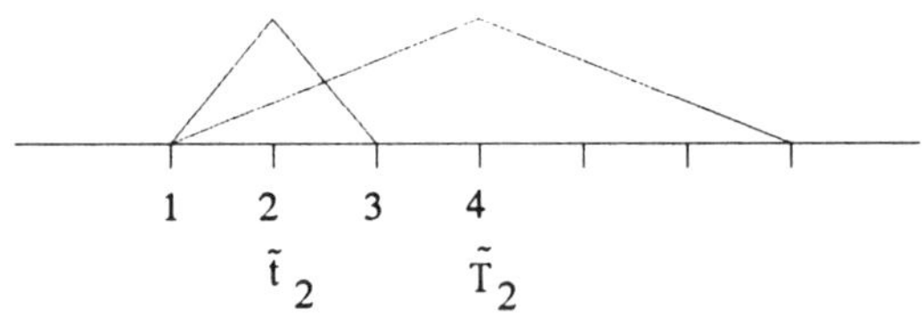

Figure 14.13 Bounds of a fuzzy slack interval

Rommelfanger (1994) suggests a modified approach to calculate the latest dates because in fuzzy set theory the subtraction as defined above is not the inversion of the addition. The latest date of a node has to be determined so that all succeeding tasks can be realized without any delay of the fuzzy earliest finishing date of the project. Therefore, Rommelfanger uses the following approach to calculate the latest dates for the nodes:

$$\tilde{T}_i = \begin{cases} \min\limits_{j \in S_i} \left\{ \tilde{X}_{ij} \mid \tilde{X}_{ij} \oplus \tilde{d}_{ij} = \tilde{T}_j \right\} & \text{for } S_i \neq \emptyset \\ \tilde{t}_n & \text{for } S_i = \emptyset \end{cases}$$

Here as well, the determination of $\tilde{X}_{ij}$ is rather easy for fuzzy trapezoidal intervals

$$\text{with} \quad \tilde{d}_{ij} = (\underline{d}_{ij}, \overline{d}_{ij}, \underline{\delta}_{ij}, \overline{\delta}_{ij})_{RL} \quad \text{and} \quad \tilde{T}_j = (\underline{T}_j, \overline{T}_j, \underline{\tau}_j, \overline{\tau}_j)_{RL}$$

$$\tilde{X}_{ij} = (\underline{T}_j - \underline{d}_{ij}, \overline{T}_j - \overline{d}_{ij}, \underline{\tau}_j - \underline{\delta}_{ij}, \overline{\tau}_j - \overline{\delta}_{ij})_{RL}$$

Using this formula the latest date of the last event can be set equal to the fuzzy earliest date and the calculated results are much more plausible:

Node	Earliest date	Latest date
1	(0,0,0,0)	(0,0,0,0)
2	(2,2,1,1)	(2,2,1,1)
3	(5,5,2,2)	(7,7,2,2)
4	(5,5,2,2)	(5,5,2,2)
5	(9,9,3,3)	(9,9,3,3)

In our example the events 1, 2, 4, and 5 are critical. With this approach for all critical events the earliest date is equal to the latest date. To determine the total float *TF* of the activities Rommelfanger (1994) suggests the solution of the following fuzzy equality: $\tilde{T}F_{ij} + \left(\tilde{t}_i + \tilde{d}_{ij}\right) = \tilde{T}_j$. In the example, the fuzzy slack time of activity (2,3) is calculated by $\tilde{T}F_{23} + ((2,2,1,1) + (3,3,1,1)) = (6,6,3,3)$. The result $\tilde{T}F_{23}$ $= (1,1,1,1)$ seems to be rather natural in this context. Here, we only consider the total float of an activity. The determination and interpretation of different types of slack in a fuzzy context is discussed in detail in Rabetge (1991, pp. 163).

There might be problems in determining $\tilde{X}$ respectively $\tilde{T}F$ if the spread of the result is smaller than that of one component of the addition. Then a generalized calculation scheme is suggested for $\tilde{X}_{ij} = (\underline{x}_{ij}, \overline{x}_{ij}, \underline{\xi}_{ij}, \overline{\xi}_{ij})$ in the following way (Rommelfanger 1994):

$$\underline{x}_{ij} = min\ (\underline{T}_j - \underline{d}_{ij}, \overline{T}_j - \overline{d}_{ij}) - max\ (\overline{\delta}_{ij} - \overline{\tau}_j,\ 0)$$

$$\overline{x}_{ij} = \overline{T}_j - \overline{d}_{ij} - max\ (\overline{\delta}_{ij} - \overline{\tau}_j,\ 0)$$

$$\underline{\xi}_{ij} = max\ (0,\ \underline{\tau}_j - \underline{\delta}_{ij} - max\ (0,\ (\underline{T}_j - \underline{d}_{ij}) - (\overline{T}_j - \overline{d}_{ij})) - max\ (\overline{\delta}_{ij} - \overline{\tau}_j,\ 0))$$

$$\overline{\xi}_{ij} = max\ (0,\ \overline{\tau}_j - \overline{\delta}_{ij})$$

Resulting from a fuzzy time schedule the possibility of meeting a specified crisp project completion time t^0 can be calculated (McCahon 1993):

$$Poss\left(\tilde{t}_n\ is\ t^0\right) = \sup_{t \in S_{\tilde{t}_n}} \left[\ min \left\{ m_{\tilde{t}_n}(t),\ m_{t^0}(t) \right\}\right] = \begin{cases} \mu_{\tilde{t}_n}(t) & if\ t^0 \in S_{\tilde{t}_n} \\ 0 & else \end{cases}$$

where $S_{\tilde{t}_n}$ is the support set of $\tilde{t}_n$, μ_{t0} is the characteristic function of set $\{t0\}$.

In addition to the generalized CPM approach described above which uses an activity-on-arrow representation, activity-on-node representations with uncertain durations are objects of interest in R&D project scheduling, too. Geidel (1990) has developed such a fuzzyfied metra potential method (MPM) with minimal and maximal time distances between different tasks. Extensions of a fuzzy PERT approach additionally including cost factors are developed by Chanas, Kamburowski (1981) and Mon et al. (1995).

Fuzzy structure

The above described approaches can be improved by considering uncertainties concerning the structure. Uncertain structure is of particular importance in R&D projects where for instance an experiment might be successful, unsuccessful, or successful to a certain degree. A well known stochastic approach to handle uncertainties concerning the necessity of activities is GERT. This approach is especially schemed for technological forecasting of complex systems. But in R&D applications often forecasted values of parameters are indefinite themselves. Thus, additional fuzzy models to represent nodes and/or times of activities and appropriate efficient algorithms should be developed to support project management. A first approach FUZNETIC is suggested by Itakura, Nishikawa (1984) who use fuzzy sets to model the possibility that an activity is undertaken once its start node is realized and to model the duration of an activity.

In R&D projects often alternative kinds of execution of a task are possible. Control nets which contain choice modules are suited to model such situations. During the execution of the control net decisions about which alternative to choose are necessary. Pagnoni (1990, pp. 185) suggests to use fuzzy attributes to express the desirability of the alternative executions. Then optimal executions or outranking of executions of choice schemata are deduced from a set of fuzzy operation attributes by determining the overall degree of desirability.

Cost and resource planning

Another possible way to determine the earliest starting and finishing times of the activities in a crisply defined project is to formulate a mathematical programming model and to solve it by using exact mathematical programming tools or heuristics. This approach for classical project scheduling problems is of great importance if resources have to be scheduled in addition to the determination of times. Often, the project completion time can be reduced with additional costs. Then the goal 'minimize overall completion time' and the goal 'minimize costs' are in conflict with each other. DePorter, Ellis (1990) suggest a fuzzy mathematical programming approach to handle multiple objectives in mathematical programming models for the scheduling of project activities. Accordingly, the interactive fuzzy mathematical programming system MOPS (Werners 1984) described in part 3 can be used to schedule activities considering multiple objectives. With this system additionally vaguely known resource requirements can be handled. An approach for simultaneous multiple criteria R&D project selection and scheduling using fuzzy logic is also suggested by Coffin, Taylor (1996).

Hapke et al. (1994) suggest a generalization of the classical approach which is included in their fuzzy project scheduling system for software development. They describe in detail the application of this system to a software project management problem where the number of system designers, user interface designers, and programmers and their allocation are to be planned in the uncertain environment of a development project. The decision support system calculates the consequences on the earliest project finishing time employing alternative numbers of engineers, for example. If the project due date has been fixed in advance the system can be used to minimize the possibility of overstepping the required completion time. This system assists during the development process. Whenever an activity is completed the system determines the new values for the following activities. Therefore, the closer the end of the project is the lower is the degree of uncertainty for the remaining dates.

14.5 SIMULTANEOUS ENGINEERING, CONFIGURATION, DESIGN

The above described project scheduling methods and especially the network techniques are useful to plan and control new product development processes in detail. Nowadays, the entire development process from idea to market entry attracts growing attention. This can be seen for instance in the extensive literature on simultaneous resp. concurrent engineering (Eversheim et al. 1996). To control the development process often phased review processes are used in practice which divide the process into different stages (Cooper 1994). A new stage can be started if the preceding stage is finished successfully which means that all conditions at a review point or gate are fulfilled. The new product models today feature parallel processing which makes the planning task more complex. The overlapping of stages is impossible in the above model with crisp gates. Therefore, some changes are necessary. One very important change is to consider fuzzy gates as suggested by Cooper (1994). Fuzzy gates allow to support conditional and situational decisions, e.g. whether the next task can be started even if not all information are on hand. Thus, a project can move ahead without loss of time whereas the missing parts of the preceding tasks are completed. This can result in a shorter project development time.

The actual grade of ripeness of a certain aggregate which is part of a product under development is an important information during the development process. Usually, different kinds of data are available, some of them are subjective and some objective. Additionally, the interdependences between the evaluations with respect to different facts can not be represented adequately by a uniform mathematical expression or by a stochastic approach. For instance, the department for commercial vehicles development of a wellknown German automobile producer decided to model the available expert knowledge with fuzzy rules and linguistic variables and hence got an efficient instrument to constantly evaluate and document the actual grade of ripeness (v. Altrock 1993, p. 142-145). The system is in common use and helps to improve the development process.

Simultaneous engineering

A further reduction of development process durations can be achieved by parallizing the tasks or activities as far as possible. Simultaneous Engineering focuses on the parallel execution of project and process development to reduce the time between the first idea about a new product and its introduction into the market (Eversheim et al. 1997). Several departments within a company are involved in this process and usually the respective tasks are performed sequentially. Usually, during the development of a new product information has to be passed from one department to a succeeding department after the complete and detailed information is obtained which is necessary for further processing. Generally, however, an early information interchange between the different involved teams can improve the development process and is the basis for parallizing activities. Hence, the adequate processing of incomplete information has the potential to shorten the process of new product development. The information which is accessible at an early stage is, however, uncertain. Therefore, fuzzy techniques are well suited to improve simultaneous engineering.

Here, we will shortly present an approach for the treatment of uncertain information within the product development process as suggested by Eversheim et al. (1994, 1996, 1997). Any information is regarded consisting of two components: the information-content itself and the reliability of this content with additional context information. Both parts can contain uncertainties. The uncertainties on a content level are handled using an object-oriented model where linguistic variables are integrated to represent vagueness. The context dependent description of uncertainties bases on the concept of "process of maturing" of information during the product development process. To determine the information-maturity the four attributes

- expected frequency for changes,
- expected date of the next change,
- expected extent of the next change, and
- possible reasons for the next change

are defined and have to be evaluated by fuzzy values or linguistic variables. Fuzzy query techniques are then used to offer a user-friendly access to uncertain development information. Introducing this method, it will be possible to break the sequential and document oriented information flow by modeling uncertain information and creating small required information units. Hence, an early information pass on will be possible which meets the demands of an integrated product and process development.

Configuration and design

Configuration and design are parts of the development process where fuzzy techniques have been used intensively. As explained by Wood et al. (1992) engineering design can be viewed as a goal-directed process with the stages synthesis, evaluation, and decision or as a mapping of the functional description of a process to a form description. In both cases an evaluation and selection from different alternatives is necessary where the context and the attributes of design alternatives are described only approximately or imprecisely. Several approaches exist to handle design imprecision. We will present some applications of fuzzy approaches in engineering design.

Zimmermann, Sebastian (1994) consider the class of so called "composition type problems". Specific problems of that class are e.g. configuration problems, planning, and scheduling, tasks which are all present in R&D management. The suggested model to solve composition type problems consists of a knowledge model and a requirement model. The domain knowledge in the particular area is represented by the knowledge model. This knowledge is formulated using conceptual hierarchies, where appropriate techniques from the field of artificial intelligence, like e.g. rule based systems and frame hierarchies are used. The requirements model contains functional requirements for the respective goal system, global and local constraints. Here, techniques from the field of classical operations research, like e.g. optimization models, multi-criteria decision models, and simulation models are employed. A more detailed description can be found in Sebastian (1994).

In most design applications it becomes obvious that in both models different kinds of uncertainty have to be considered in order to find adequate solutions. Therefore, many authors have suggested to integrate fuzzy techniques into the above mentioned models. Engineering design calculations with fuzzy parameters have been developed by Wood et al. (1992). Fuzzy approaches to solve multi-criteria problems have been suggested e.g. by Knosala, Pedrycz (1992) who especially consider design processes in mechanical engineering. Borisov, Naglis (1985) suggest a fuzzy sets approach to support the layout planning of an industrial robot. Here, the fuzzy sets and possibility theory approach is used to evaluate the different alternatives in the planning phase with respect to multiple criteria considering ordinal or vague information. Multi-attribute fuzzy decisions in construction strategies are considered by Eldukair, Ayyub (1992). Darzentas et al. (1994) suggest a fuzzy reasoning system as decision aid for designers. The selection of the best material is also part of the design process. Wang, Chang (1995) propose a fuzzy sets multiple criteria decision making approach to select the most suitable tool steel materials for a specific manufacturing application under fuzzy environment.

14.6 CONCLUSION

The above shortly or extensively described approaches demonstrate the high potential of fuzzy sets theory in helping to improve the research and development process.

Because vague, uncertain and linguistically expressed information, and expert knowledge are of special importance in research and development planning and decision making, corresponding methods are necessary for an adequate decision support. Those have been and can further be developed using results of the theory of fuzzy sets and possibility theory. Different methods have shown to be of special importance in this context. First, the different fuzzy multi-criteria methods have to be mentioned which can be used on each planning level, for strategic planning as well as for operational or tactical planning, for engineering design as well as for simultaneous engineering. Additionally, fuzzy reasoning and fuzzy expert systems are successfully applicable in different stages and for the entire development process. By integrating fuzzy techniques for knowledge processing into knowledge management systems (e.g. Abecker et al. 1998) the efficiency of decision and planning in R&D could be further improved. If additional requirements can be quantified as for instance in project selection under limited resources fuzzy mathematical programming systems can efficiently be used. Also fuzzy project planning methods are of special importance in vague planning situations as occur in decision and planning in research and development.

REFERENCES

Abecker, A.; Bernardi, A.; Hinkelmann, K.; Kühn, O.; Sintek, M. (1998): Toward a Technology for Organizational Memories, in: IEEE Intelligent Systems, May/June 1998, pp. 40-48.

Altrock, C. von, (1993): Fuzzy Logic - Technologie, Band 1, München.

Anandalingam, G.; Olsson, C.E. (1989): A multi-stage multi-attribute decision model for project selection, in: European Journal of Operational Research 43, pp. 271-283.

Baker, N.; Freeland, J. (1975): Recent advances in R&D benefit measurement and project selection methods, in: Management Science 21, No. 10, pp. 1164-1175.

Baker, N.R.; Green, S.G.; Bean, A.S. (1986): The need for strategic balance in R&D project portfolios, in: Research Management 29, pp. 38-43.

Bard, J.F. (1986): Evaluating space station applications of automation and robotics, in: IEEE Transactions on Engineering Management 33, No. 2, pp. 102-111.

Bard, J.F.; Balachandra, R.; Kaufmann, P.E. (1988): An interactive approach to R&D project selection and termination, in: IEEE Transactions on Engineering Management 35, No. 3, pp. 139-146.

Bellman, R.E.; Zadeh, L.A. (1970): Decision-making in a fuzzy environment, in: Management Science 17, No. 4, pp. 141-164.

Booker, J.M.; Bryson, M.C. (1985): Decision analysis in project management: An overview, in: IEEE Transactions on Engineering Management 32, No. 1, pp. 1-9.

Borisov, A.; Naglis, L. (1985): Multi-criteria choice of alternatives in an expert system for computer-aided design of industrial robot installations, in: Fuzzy Sets and Systems 16, pp. 93-101.

Brockhoff, K. (1994): Forschung und Entwicklung: Planung und Kontrolle, München, Wien, 4., rev. ed.

Brockhoff, K.; Chakrabarti, A.K. (1988): R&D/marketing linkage and innovation strategy: Some West German experience, in: IEEE Transactions on Engineering Management 35, No. 3, pp. 167-174.

Chanas, S.; Kamburowski, J. (1981): The use of fuzzy variables in PERT, in: Fuzzy Sets and Systems 5, pp. 11-19.

Chang, P.-L.; Chen, Y.-C. (1994): A fuzzy multi-criteria decision making method for technology transfer strategy selection in biotechnology, in: Fuzzy Sets and Systems 63, pp. 131-139.

Coffin, M.A.; Taylor, B.W. III (1996): Multiple criteria R&D project selection and scheduling using fuzzy logic, in: Computers and Operations Research 23, No. 3, pp. 207-220.

Cooper, R.G. (1994): Perspective: Third-generation new product processes, in: J. Prod. Innov. Manag. 11, pp. 3-14.

Czajkowski, A.F.; Jones, S. (1986): Selecting interrelated R&D projects in space technology planning, in: IEEE Transactions on Engineering Management 33, No. 1, pp. 17-24.

Danila, N.V. (1985): Evaluating pharmaceutical R&D, in: Mar, B.W., Newell, W.T. and Saxberg, B.O. (eds.): Managing High Technology, Amsterdam, pp. 25-40.

Darzentas, J.; Darzentas, J.; Spyrou, T. (1994): Fuzzy reasoning and system thinking in a decision aid for designers, in: EUFIT '94, Second European Congress on Intelligent Techniques and Soft Computing, Aachen, pp. 1609-1618.

DePorter, E.L.; Ellis, K.P. (1990): Optimization of project networks with goal programming and fuzzy linear programming, in: Computers and Industrial Engineering 19, pp. 500-504.

Dias, O.P. jr. (1988): The R&D project selection problem with fuzzy coefficients, in: Fuzzy Sets and Systems 26, pp. 299-316.

Dubois, D.; Prade, H. (1988): Possibility Theory: An Approach to Computerized Processing of Uncertainty, New-York.

Eldukair, Z.A.; Ayyub, B.M. (1992): Multi-attribute fuzzy decisions in construction strategies, in: Fuzzy Sets and Systems 46, pp. 155-165.

Eversheim, W.; Bochtler, W.; Laufenberg, L. (1994): Methods and models for integrated modelling of product and processes, in: Production Engineering 1/2, pp. 173-176.

Eversheim, W.; Derichs, T.; Roggatz, A.; Zimmermann, H.-J. (1996): Management of uncertain information in simultaneous engineering, in: Sebastian, H.-J., Antonsson, E.K. (eds.): Fuzzy Sets in Engineering Design and Configuration, Boston, London, Dordrecht, pp. 233-250.

Eversheim, W.; Roggatz, A.; Zimmermann, H.-J.; Derichs, T. (1997): Information management for concurrent engineering, in: European Journal of Operational Research 100, No. 2, pp. 253-265.

Fox, G.E.; Baker, N.R.; Bryant, J.L. (1984): Economic models for R and D project selection in the presence of project interactions, in: Management Science 30, No. 7, pp. 890-902.

Frame, J.D. (1983): Quantitative indicators for evaluation of basic research programs/projects, in: IEEE Transactions on Engineering Management 30, No. 3, pp. 106-111.

Gear, T.E.; Cowie, G.C. (1980): A note on modeling project interdependence in research and development, in: Decision Science 11, pp. 738-748.

Geidel, J. (1990): Project scheduling with fuzzy data, in: Methods of Operations Research 62, pp. 339-347.

Geoffrion, A.M. (1989): Computer-based modeling environments, in: European Journal of Operational Research 41, pp. 33-45.

Glover, F.; Greenberg, H.J. (1989): New approaches for heuristic search: A bilateral linkage with artificial intelligence, in: European Journal of Operational Research 39, pp. 119-130.

Graham, R. J. (1985), The role of network techniques in team building for project management, in: Dean, B.V. (ed.), Project Management: Methods and Studies, Amsterdam, pp. 163-171.

Green Hall, N. (1993): A Fuzzycalc implementation of STRATASSIST - A DSS for strategic planners, in: EUFIT '93 - First European Congress on Fuzzy and Intelligent Technologies, Aachen, September 7-10, pp. 547-551.

Hapke, M.; Jaszkiewicz, A.; Slowinski, R. (1994): Fuzzy project scheduling system for software development, in: Fuzzy Sets and Systems 67, pp. 101-117.

Hedley, B. (1992): A Fundamental Approach to Strategy Development, (Reprinted from Long Range Planning), in: Hahn, D.; Taylor, B. (eds.): Strategische Unternehmensplanung, 6. ed., Heidelberg, pp. 176-202.

Horwitch, M.; Thietart, R.A. (1987): The effect of business interdependencies on product R&D-intensive business performance, in: Management Science 33, No. 2, pp. 178-197.

Itakura, H.; Nishikawa, Y. (1984): Fuzzy network technique for technological forecasting, in: Fuzzy Sets and Systems 14, pp. 99-113.

Kort, P.M. (1998): Optimal R&D investment of the firm, in: OR Spektrum 20, No. 3, pp. 155-164.

Knosala, R.; Pedrycz, W. (1992): Evaluation of design alternatives in mechanical engineering, in: Fuzzy Sets and Systems 47, pp. 269-280.

Lasek, M. (1992): Hierarchical structures of fuzzy ratings in the analysis of strategic goals of enterprises, in: Fuzzy Sets and Systems 50, pp. 127-134.

Lee, J.; Lee, S.; Bae, Z.-T. (1986): R&D project selection: behavior and practice in a newly industrializing country, in: IEEE Transactions on Engineering Management 33, No. 3, pp. 141-147.

Liberatore, M.J. (1987): An extension of the analytic hierarchy process for industrial R&D project selection and resource allocation, in: IEEE Transactions on Engineering Management 34, No. 1, pp. 12-18.

Liberatore, M.J.; Stylianou, A.C. (1995): Expert support systems for new product development decision making: A modeling framework and applications, in: Management Science 41, No. 8, pp. 1296-1316.

Liberatore, M.J.; Titus, G.J. (1983): The practice of management science in R&D project management, in: Management Science 29, No. 8, pp. 962-974.

Lockyer, K.; Gordon, J. (1992): Critical Path Analysis and other Project Network Techniques, 5. ed., repr., London.

Lootsma, F.A.; Mensch, T.C.A.; Vos, F.A. (1990): Multi-criteria analysis and budget reallocation in long-term research planning, in: European Journal of Operational Research 47, pp. 293-305.

Madey, G.R.; Dean, B.V. (1985): Strategic planning for investment in R&D using decision analysis and mathematical programming, in: IEEE Transactions on Engineering Management 32, No. 2, pp. 84-90.

McCahon, C.S. (1993): Using PERT as an approximation of fuzzy project-network analysis, in: IEEE Transactions on Engineering Management 40, No. 2, pp. 146-153.

Mon, D.-L.; Cheng, C.-H.; Lu, H.-C. (1995): Application of fuzzy distributions on project management, in: Fuzzy Sets and Systems 73, pp. 227-234.

Onisawa, T.; Sugeno, M.; Nishiwaki, Y.; Kawai, H.; Harima, Y. (1986): Fuzzy measure analysis of public attitude towards the use of nuclear energy, in: Fuzzy Sets and Systems 20, pp. 259-289.

Pagnoni, A. (1990): Project Engineering, Computer-Oriented Planning and Operational Decision Making, Berlin, Heidelberg.

Pfeiffer, W.; Dögl, R. (1997): Das Technologie-Portfolio-Konzept zur Beherrschung der Schnittstelle Technik und Unternehmensstrategie, in: Hahn, D.; Taylor, B. (eds.): Strategische Unternehmensplanung - strategische Unternehmensführung, 7., rev. ed., Heidelberg, pp. 407-435.

Porter, M.E. (1980): Competitive Strategy, New York.
Porter, M.E. (1985): Competitive Advantage, New York.
Rabetge, C. (1991): Fuzzy Sets in der Netzplantechnik, Wiesbaden.
Raelin, J.A.; Balachandra, R. (1985): R&D project termination in high-tech industries, in: IEEE Transactions on Engineering Management 32, No. 1, pp. 16-23.
Rommelfanger, H.J. (1994): Network analysis and information flow in fuzzy environment, in: Fuzzy Sets and Systems 67, pp. 119-128.
Saaty, T.L. (1990): The Analytic Hierarchy Process - Planning, Priority Setting, Resource Allocation, 2nd ed., Pittsburgh, PA 1990.
Schmidt, R.L. (1993): A model for R&D project selection with combined benefit, outcome, and resource interactions, in: IEEE Transactions on Engineering Management 40, No. 4, pp. 403-410.
Sebastian, H.-J. (1994): Intelligente Entscheidungsunterstützung bei der Systemkonfigurierung, in: Werners, B., Gabriel, R. (Hrsg.): Operations Research, Reflexionen aus Theorie und Praxis, Festschrift zum 60. Geburtstag von Hans-Jürgen Zimmermann, Heidelberg, pp. 275-304.
Souder, W.E.; Mandakovic, T. (1986): R&D project selection models, in: Research Management 29, pp. 36-42.
Steele, L.W. (1988): Selecting R&D programs and objectives, in: Research Technology Management 31, pp. 17-36.
Wang, M.-J.J.; Chang, T.-C. (1995): Tool steel materials selection under fuzzy environment, in: Fuzzy Sets and Systems 72, pp. 263-270.
Watts, K.M.; Higgins, J.C. (1987): The use of advanced management techniques in R&D, in: OMEGA International Journal of Management Science 15, No. 1, pp. 21-29.
Weber, R.; Werners, B.; Zimmermann, H.-J. (1990): Planning models for research and development, in: European Journal of Operational Research 48, pp. 175-188.
Werners, B. (1984): Interaktive Entscheidungsunterstützung durch ein flexibles mathematisches Programmierungssystem, München.
Werners, B. (1987): Interactive multiple objective programming subject to flexible constraints, in: European Journal of Operational Research 31, pp. 342-349.
Werners, B. (1988): Aggregation models in mathematical programming, in: Mitra, G. (ed.): Mathematical Models for Decision Support, Berlin, pp. 295-319.
Werners, B. (1993): Unterstützung der strategischen Technologieplanung durch wissensbasierte Systeme, Aachener Beiträge zu den Wirtschaftswissenschaften, Aachen.
Wood, K.L.; Otto, K.N.; Antonsson, E.K. (1992): Engineering design calculations with fuzzy parameters, in: Fuzzy Sets and Systems 52, pp. 1-20.
Zadeh, L.A. (1983): The role of fuzzy logic in the management of uncertainty in expert systems, in: Fuzzy Sets and Systems 11, No. 3, pp. 199-277.
Zimmermann, H.-J. (1978): Fuzzy programming and linear programming with several objective functions, in: Fuzzy Sets and Systems 1, pp. 45-55.
Zimmermann, H.-J. (1987): Fuzzy Sets, Decision Making, and Expert Systems, Boston.
Zimmermann, H.-J. (1988): Interactive decision support for semi-structured mathematical programming problems, in: Mitra, G. (ed.): Mathematical Models for Decision Support, Berlin, pp. 307-319.
Zimmermann, H.-J. (1989): Strategic planning, operations research and knowledge based systems, in: Verdegay, J.-L., Delgado, M. (eds.): The Interface between Artificial Intelligence and Operations Research in Fuzzy Environment, Köln, pp. 253-274.
Zimmermann, H.-J. (1996): Fuzzy Set Theory - and Its Applications, 3rd ed., Boston, Dordrecht, London.

Zimmermann, H.-J.; Sebastian, H.-J. (1994): Fuzzy design - Integration of fuzzy theory with knowledge-based system-design, in: Proceedings of The Third IEEE Conference on Fuzzy Systems, June 26 - June 29, 1994, Orlando, pp. 352-357.

Zimmermann, H.-J.; Zysno, P. (1980): Latent connectives in human decision making, in: Fuzzy Sets and Systems 4, pp. 37-51.

15 PRODUCTION PLANNING AND SCHEDULING: FUZZY AND CRISP APPROACHES

I.B. Türkşen
M.H. Fazel Zarandi

Abstract: We survey some of the main Crisp and Fuzzy (Computational) Intelligence works, which have been developed for the solution of the real world production planning and scheduling problems. Development of better production planning and scheduling can help many industries, since timely deliveries and good quality of products are important for customers, as well as the special goals of manufacturers, e.g., the reduction of the inventory costs, improvement of the utilization of the productive capacity, synchronization of the different sections of the plants, etc. However, attaining an optimal scheduling of production in different industries is very hard, because of the conflicts among different goals. The examples of such conflicting goals are: the optimization of the production rate, production quality, intra-order production consistency, etc. Within the last thirty years or so, it has been found that fuzzy set theory incorporated with fuzzy cluster analysis is a suitable tool to analyse and diagnose these sorts of complicated problems.

15.1 INTRODUCTION

Generally, the applications of the fuzzy set and fuzzy logic systems fall into: (i) fuzzy decision analysis systems, (ii) fuzzy control systems, and more generally (iii) the complex systems that attempt to support or replace the human experts. The fuzzy systems approach demonstrates many advantages in real world applications (Tűrkşen and Berg, 1991, Slany, 1994, and Jang and Gulley, 1995). They may be summarized as:

1. Fuzzy systems are conceptually easy to understand: the mathematical concepts behind fuzzy reasoning are analogous to human like reasoning. What makes fuzzy systems effective is their naturalness of approach and not their far reaching complexity

2. Fuzzy systems are flexible: with any given system, it is easy to manage it or layer more functionality on top of it without starting again from scratch.

3. Fuzzy systems are tolerant of imprecise data: everything is imprecise if you

look close enough, but more than that, most things are imprecise even on careful inspection. Fuzzy reasoning builds this understanding into the process.

4. Fuzzy systems can model most nonlinear functions of arbitrary complexity: one can create a fuzzy system to match any set of input-output data. This process is made particularly easy by adaptive techniques such as adaptive neuro-fuzzy inference systems.

5. Fuzzy systems can be built on top of the experience of experts: in direct contrast to neural networks, which take training data and generate opaque, impenetrable models, fuzzy system approach lets you stand on the knowledge of experts who already understand a given system.

6. Fuzzy systems can be blended with conventional control techniques: fuzzy systems don't necessarily replace conventional control methods. In many cases fuzzy systems augment them and simplify their implementation.

7. Fuzzy systems are based on natural language: the basis for fuzzy systems are the basis for human communication. This observation underpins many of the other statements about fuzzy systems.

8. Fuzzy systems provide better communication between the experts and the managers: fuzzy systems can help the model builders communicate their complex results to management in an appropriate natural context.

With these advantages in mind, the organization of this paper is as follows: In Section 2, we briefly discuss the management of imprecision and fuzzy production planning and scheduling. Section 3 is devoted to the explanation of fuzzy expert systems for production planning and scheduling. In Section 4, we present knowledge base systems and examples of fuzzy rules in prototype studies of production planning and scheduling. In Section 5, aggregate production planning and scheduling is discussed from the perspective of classical and fuzzy set theory. In this section, we also discuss the major difficulties of the classical production planning and scheduling models, and demonstrate some well known classical detailed scheduling problems, as well as some well known fuzzy models which are developed for production planning and scheduling problems. In Section 6, we discuss detailed fuzzy production planning and scheduling systems. Similarities between fuzzy systems and Just-In-Time systems is presented in Section 7. Finally, in Section 8, we discuss the application of fuzzy cluster analysis in production planning and scheduling.

15.2. MANAGEMENT OF IMPRECISION

Management of imprecision (Zadeh, 1977) is a major concern of most knowledge engineers involved in the design of the production planning, inventory control, and scheduling in many industries. Whenever managers are to interact in an intelligent way with this environment, they have to reach out for a new kind of reasoning based on imprecise knowledge. The major characteristics of the imprecise knowledge are as follows:

1. It contains uncertainties: experts and/or sensors cannot and do not usually measure precisely the exact values of their domain of concern.

2. It includes errors: some minor measurement errors and/or some gross judgmental errors are embedded in data acquisition.

3. It is almost always incomplete: ignorance is a well recognized component of knowledge imparted from experts and managers. Input-output data analyses may not always reveal knowledge about the total domain of concern.

Studies that investigate certain aspects of the management of imprecision in production planning and scheduling, can be found in the works of Mamdani (1977), Kickert (1978), Holmbald and Ostergaard (1982), Rinks (1982), Whalen and Schott (1983, 1985), Holsapple and Winston (1986), Tűrkşen (1986), Tűrkşen (1989), Tűrkşen and Zhong (1990), and Zimmermann (1991).

Production planning and scheduling is concerned with the determination of the production level in a specific horizon for various production systems under the categories of simple, batch, or continuous job activities. It aims to provide an overall guidance for the effective allocation of productive resources. In such cases, the aggregate means that the planning is conducted at a gross level to meet the total demand from all products that share the same limited resources of a dedicated facility. All of these systems are complex and the optimization and control of them are often interactable. In contrast, the detailed means that the jobs are assigned to production units in specific order known as sequences or schedules.

Traditional aggregate production planning, inventory control, and scheduling models are mainly occupied with deterministic or stochastic and probabilistic methods. However, managers usually state their goals, objectives and operative constraints, and the like, in regards to natural language expressions and linguistic variables. These expressions and variables are often imbued with uncertainty, ambiguity, vagueness, and with imprecision. This is an inescapable reality despite great efforts of the scientists and engineers to deem definitions and measurements as precise as possible.

In some of the existing models for production planning, inventory control, and scheduling, researchers have used some of the stochastic methods. Naturally, stochastic models provide valuable insights into a system's behavior under specific assumptions and identify useful guidelines. However, these systems are not often implemented. Many managers often do not understand the basic assumptions of such models. On the other hand, such models are either too complex or the precise information required to determine the values of the model parameters cannot be obtained or are not available. Thus, such models are, by default, inadequate to help managers to cope with the natural behavior of many real-life systems (Tűrkşen and Berg, 1991).

In fuzzy production planning and scheduling the main challenge is to find suitable realization of operations with the intersection of fuzzy constraints for the satisfaction (with a degree) of the overall requirements. In mathematical foundations of the fuzzy theory, there exist enough flexibility subject to certain axioms. In fuzzy systems, it is not difficult to merge hard and soft constraints. Moreover, the main source of difficulty for constructing reasonable production planning and scheduling stems from the conflicting nature of the criteria and goals. In classical scheduling methods, to reduce the complexity of the problem, researchers often consider only a small subset of the goals, or only one criterion.

However, in real applications, it is reasonable to find viable compromises between conflicting objectives and goals, by determining a degree of satisfaction amongst degrees of importance associated with objectives and goals. Fuzzy approach provide tools to satisfy constraints to a certain degree and to take into account the relative importance in a format easily understood by human experts. This gradual satisfaction can help the scheduler to find approximately good solutions and to simplify the complexity of the production planning and scheduling problems. In general, as stated by Slany (1994), "the constructing of a fuzzy schedule will involve establishing a knowledge base of empirical mapping functions, fuzzy rules and weighting factors which express the preferences of decision makers and state of knowledge, such that the degree of constraint satisfaction of any given schedule can be computed, and finally searching for maximal satisfaction of the fuzzy goals and constraints."

15.3 FUZZY EXPERT SYSTEMS

In general, as stated by Kusiak and Chen (1988), an expert system is a computer system which can solve the problems by using human expertise and knowledge of the system environment. Feigenbaum (1982) defines an expert system as "an intelligent computer program that uses knowledge and inference procedures to solve problems that are difficult enough to require significant human expertise for their solution. The knowledge necessary to perform at such a level, plus the inference procedures used can be thought of as a model of the expertise of the best practitioners of the field. The knowledge of an expert system consists of facts and heuristics. The facts constitute a body of information that is widely shared, publicly available, and generally agreed upon by experts in the field. The heuristics are mostly private, rarely discussed rules of good judgment (rules of plausible reasoning, rules of good guessing) that characterize the domain of expert level decision making. The performance level of an expert system is primarily a function of the size and quality of the knowledge base that it possesses."

15.3.1 MODULES OF AN EXPERT SYSTEM

The main modules of a classical expert system are as follows:

1. **Knowledge Base**: The knowledge base represents the knowledge related to the problem domain. The provision of an appropriate structure and representation of the knowledge are very important factors. A subset of a knowledge base is the fact base, which contains both symbolic information and numeric information. This subset enables the expert system to handle qualitative as well as quantitative information. The rule bases which are part of the knowledge base contain the heuristics. They are used by domain experts to diagnose or predict or control, etc., a system behavior. The main distinction between the conventional database structures and a knowledge base is that in the former the rules exploit the facts, whereas in the latter the facts trigger the rules. A production rule has the form of **IF-THEN**, where **IF** part is the antecedent for conditions, and **THEN** part is the consequent for conclusions or actions.

2. **Inference Mechanism**: The inference mechanism uses the knowledge base and the rule base for inference and derivation of decisions for given observed facts. An inference mechanism may be data driven as the case of forward chaining or it may be goal driven as in the case of backward chaining.

3. **Knowledge Acquisition**: This module involves the adaptation of the expert system to a specific domain. A knowledge engineer who is familiar with the expert system technology and problem domain can handle the knowledge acquisitions, whose goal is extracting and organizing the knowledge and expertise embedded in the problem domain.

4. **User Interface**: Mitchie (1980) states that for interacting with an expert

system, three possible ways exists as follows:

- Consultation, where the user is in the position of a client.
- Building the model, where the user is in the position of a tutor.
- Education, where the user is in the position of a pupil harvesting the knowledge base.

Fuzzy expert system models are similar to the classical expert system models, where input spaces are mapped to an output space. A typical expert system consists of (i) an inference engine which uses the domain knowledge as well as information gained from a problem to provide an expert solution, (ii) a knowledge base which contains the expert domain knowledge for problem solving, and (iii) working memory. A fuzzy expert system uses a collection of fuzzy membership functions and rules for knowledge representation and reasoning with observed system state data.

In the past two decades, expert systems have been implemented in production planning and scheduling frequently. A list of different applications of expert systems in this area was presented by Kumara *et al.* (1986). De and Whinston (1986) explained the advantages of the natural language understanding, planning and scheduling, and integrated these in a manufacturing system. Kempf (1985) studied several expert system models which have been developed for manufacturing systems. He stated that artificial intelligence can offer the most practical opportunities for solving the complex problems of manufacturing systems. Parunak (1987) reviewed the application of Distributed Artificial Intelligence (DAI) in scheduling and control. Applications of rule based systems in production and manufacturing systems have also been presented by Rayson (1985). Some other references in application of AI in p Halpin and Noah (1986), Orsini (1986), Gaines (1987), Kusiak and Chen (1988), Kusiak (1992), and Zweben and Fox (1994). Generally, approximate reasoning process is executed with the following steps in a fuzzy expert system:

1. **Fuzzification**: the degree of the memberships for each rule's premise is determined by matching the linguistic terms with the actual values of the input variables.

2. **Inference**: the conjunctive combination of antecedent memberships is executed and projected to the conclusion part of each rule to determine the degree of membership on its consequent.

3. **Composition**: all fuzzy subsets determined in step 2 above for each rule's output variable are combined to form a single fuzzy subset for each output variable.

4. **Defuzzification**: the overall output fuzzy set is converted into a crisp number on the base variable domain.

In the approximate reasoning procedure outlined above, the inference module is the heart of every fuzzy knowledge base system. The most important methods of inferencing are known as follows:

1. **Compositional Rule of Inference (CRI)**: This method was presented by Zadeh (1973) as a generalization of the classical basic inference method. It is known as "Generalized Modus Ponens (GMP)" which is the extension of classical Modus Ponens in fuzzy theory.

2. **Approximate Analogical Reasoning (AAR)**: This method was developed by Tűrkşen and Zhong (1988). In this method, first one computes pattern matching scores between the antecedents of the rules and the observation using similarity measures, and then modifies the consequents of the rules by the pattern matching scores. A variation of this method considers only those rules where the pattern matching scores of the rules are above a certain threshold (Tűrkşen and Lucas, 1991).

3. **Interpolative Reasoning (IR)**: Zadeh (1973) proposed this alternative method. A particular interpretation of this method is known as Mamdani schema (1974).

15.4 KNOWLEDGE-BASED SYSTEMS

Informally, approximate reasoning is the process by which a possible imprecise conclusion is deduced from a collection of imprecise premises. The structure of a possible knowledge-based three essential components as follows (Tűrkşen, 1988):

1. Factual knowledge or observed system state or facts:

$$X_1 \text{ isr } A_1^* \text{ AND } X_2 \text{ isr } A_2^* \text{ AND } \dots \text{ AND } X_n \text{ isr } A_n^*$$

where "isr" is a short form for "is contained in", "is similar to", "is compatible to", etc. (Zadeh, 1995, and Tűrkşen, 1996).

A short hand notation for the observed system state is usually expressed as:

$$A^* = A_1^* \text{ AND } A_2^* \text{ AND } \dots \text{ AND } A_n^*$$

The user of an expert system inputs the factual knowledge and chooses the linguistic terms for observed states of the system.

2. Rule-base or expert rules or, for short, rules:

$$\textbf{IF } X_1 \text{ isr } A_1 \text{ AND } X_2 \text{ isr } A_2 \text{ AND } \ldots \text{ AND } X_n \text{ isr } A_n, \\ \textbf{THEN } Y \text{ isr } B$$

in short hand notation:

$$A \rightarrow B$$

where,

$$A = A_1 \text{ AND } A = A_2 \text{ AND } \ldots \text{ AND } A_n$$

for the multi-input and single-output, MISO, system.

During the knowledge acquisition phase, the domain experts provide these rules or alternately these rules could be extracted from input-output data via fuzzy cluster analyses techniques (Tűrkşen and Fazel Zarandi, 1997). The linguistic terms and their membership values and/or functions are also specified in such rules either by experts or by other automated techniques based on input-output data analyses.

3. Expert system advice or response:

$$Y \text{ isr (should be) } B^*$$

After the inference subsystem operates on the rules and given facts in accordance with an inference scheme, the expert system response is provided to a user. Some examples of the rules in a fuzzy rule-base are as follows:

- "**IF** heat is *low* AND pressure is *high* **THEN** valve is" (Mamdani, 1975).

- "**IF** the percentage of chemical element E in heat H_1 is *less than* in heat H_2, **THEN** E-compatibility (percentage of element E) of H_1 preceding H_2 is *medium*" (Slany, 1994).

- "**IF** the current inventory level is *much higher* than the preferred level AND the deviation is decreasing at *medium* rate,
THEN the production rate is to be *moderately* slowed down" (Türkşen, 1985).

- "**IF** the sales forecast for period t is *high* and the inventory for period $t - 1$ is *not high* and the labour force for period $t - 1$ is *high*,
THEN the production for period t is (should be) *high*" (Rinks, 1982).

- "**IF** the demand variability is *low* AND the service level is to be *high* AND the gross profit is to be *high*,
  THEN the safety stock level should be *medium*" (Tűrkşen, 1988).

- **IF** Carbon isr 0.05-o.13\% AND Magnesium isr 0.20-0.90\% AND Phosphorus isr 0.005-0.01\% AND Sulfur isr 0.005-0.01\% AND Silicon isr 0.07-0.12\% AND Aluminium isr 0.03-0.05\% AND Columbium isr 0.0004-0.03\% AND Titanium isr 0.001-0.007\%AND Calcium isr 0.0002-0.003\% AND Nitrogen isr 0.002-0.005\% AND Tundish Start isr 31.67-45.84 tons AND Tundish End isr 64.19-66.26 tons AND Average Casting Speed isr 1.37-1.74, m/min, AND Maximum Casting Speed isr 0.80-1.75, m/min
  AND Minimum Casting Speed isr 1.16-1.76, m/min,
  THEN Mixed Zone Tonnage isr 4.25-9.88 tons" (Tűrkşen and Fazel Zarandi, 1997).

In the final rule, for example, Carbon, Magnesium, Phosphorus, Sulfur, Silicon, Aluminium, Columbium, Titanium, Calcium, Nitrogen, Tundish Start, Tundish End, Average Casting Speed, Maximum Casting Speed, and Minimum Casting Speed, are the input variables and Mixed Zones' tonnage is the output variable in the continuous casting of a steel plant.

15.5 AGGREGATE PRODUCTION PLANNING AND DETAILED SCHEDULING

As stated above, aggregation production planning is concerned with determining the production level for a specific planning horizon and for various kinds of production systems such as a simple job production, batch production, or continuous production. Production planning and scheduling system, as stated by Kusiak and Chen (1988), determines the efficiency and productivity of the manufacturing systems. Shaw and Whinston (1986) state that the following capabilities are desirable for a suitable expert system of production planning and scheduling:

1. Support on-line decisions,
2. Schedule operations dynamically,
3. Coordinate manufacturing resources,
4. Synchronize processes for different jobs,
5. Monitor the execution of plans.

In this section, first, we demonstrate some basic aspects of classical production planning and scheduling. Then we will review some well known classical AI models for production planning and control. Finally, we will review some well known fuzzy models for production planning and scheduling.

15.5.1 DETAILED SCHEDULING PROBLEMS

Detailed scheduling, as stated by Pinado (1995), concerns the allocation of limited resources to tasks over time. It is a decision making process with the aim of optimizing one or more objectives. In Continuous Casting of an Iron and Steel Industry, for example, the resources are: Ladel, Tundish, Strands, Tourching, etc., and the tasks are: the operations of pouring heats from ladles, flying tundish, mould changes, cooling, cutting, etc. In general, the detailed scheduling problems are classified into general job shop scheduling, flow shop scheduling, and open shop scheduling. While scheduling is referred to timing of the arrivals and departures of the jobs in the shops, sequencing is referred to the order in which jobs enter into shops for desired processes or services. Traditionally, the performance measures in scheduling are completion time (C_i), i.e., the time at which processing of job J_i finishes, flow time (F_i), i.e., the time that job spends in the workshop, mean flow time (F), maximum completion time (C_{max}), lateness (L_i), i.e., the difference between the completion time and due date of job J_j, tardiness (T_i), i.e., max $\{L_i, 0\}$, and their variations, etc.

Dorn and Froeschl (1993) also categorized the scheduling problems as follows:

1. Mathematical-analytical methods: These methods which are based on operations research, can be theoretically sound and yielding optimal solution. However, these methods encounter many limitations, especially in the era of complexity and real world applications.

2. Knowledge based methods: These methods try to represent all constraints of a scheduling problem explicitly, even if some of them are only vaguely known.

It should be noted that in the above classification, there exist a distinction between the constraints in mathematical programming and the constraints in artificial intelligence cases. In the mathematical programming, constraints are mathematical equalities or inequalities, whereas in the artificial intelligence, which is based on logic programming, the constraints represent knowledge that can be updated and manipulated in many ways.

Comprehensive studies about classical scheduling are found in the works of Conway *et al.* (1967), Elmaghraby (1968), Baker (1974), Coffman (1975), Rinnooy Kan (1976), French (1982), Baker (1995), and Pinado (1995).

In classical scheduling, the methods and theories are based on strong mathematical theories. However, these methods have not been very successful in realistic applications. The main problems with these methods, as stated by Dorn and Slany (1994), are as follows:

1. The algorithms are too complex for real world applications.
2. The methods demand exact knowledge about durations and technical constraints.
3. The effort to formalize a new scheduling problem is considerable.
4. The models consider only one or two criteria.
5. Only linear relations among parameters are considered.
6. No antagonistic knowledge can be modeled.

Obviously, the shop scheduling problem is a hard problem both theoretically and practically. The combinatorial characteristics of these problems make them very complex. As an example, in general job shop scheduling with n jobs and m machines, there exist $(n!)^m$ possible solutions. Consider if we have only 5 jobs and three machines, then there exists $(5!)^3 = 1728000$ possible configurations. Now, consider that we have 30 jobs in flow shop environment, and we want to schedule these jobs for the aim of minimizing the tardiness. For this purpose, we must check $30! = 2.6525286 \times 10^{32}$ possible solutions with a complete enumeration technique to find the optimal solution or solutions. Thus, when one attempts to find the optimal schedule with a search method, the complexity of the problem becomes very high. In other words, most of the scheduling problems fall into the class of NP-hard problems. These types of problems, as stated by Garey and Johnson (1979), are "... just as hard as a large number of the other problems that are widely recognized as being difficult and that have been confounding the experts for years." As a result, all of these problems are intractable. At times branch and bound and dynamic programming technique may be used for certain small and moderate size problems. Zadeh (1973) states that "... as the complexity of a system increases, our ability to make precise and yet significant statements about the behavior diminishes until a threshold is reached beyond which precision and significance (or relevance) become almost mutually exclusive characteristics. It is in this sense that precise quantitative analyses of the behavior of humanistic systems are not likely to have much relevance to the real world societal, political, economic, and other types of problems which involve humans either as individuals or in groups". Based on this, he suggested linguistic methods and more recently "computation with words" (Zadeh, 1995) instead of quantitative methods and proposed the following approaches:

1. Use of so-called linguistic variables in place of or in addition to numerical variables;

2. Characterization of simple relations between variables by conditional fuzzy statements;

3. Characterization of complex relations by fuzzy algorithms.

Slany (1994) also states that fuzzy sets and fuzzy logic can provide a systematic framework for the representation, application, propagation and relaxation of imprecise constraints, and evaluation of schedules with respect to vaguely defined goals.

15.5.2 CLASSIFICATION OF SCHEDULING PROBLEMS

The classification scheme $\alpha|\beta|\gamma$, suggested by Graham *et al.* (1979), is generally used to identify detailed scheduling problems, where α represented the machine environment (e.g., F_3 for flow shop with 3 machines), β indicates the job characteristics (e.g., preemptions, no-wait, ready times, etc.), and γ indicates the objective function (e.g., make-span, maximum lateness, tardiness, etc.). For example, $F_2|$ no-wait $|C_{\max}$ refers to flow shop with 2 machines where delays between the machines are not permitted and the criterion is minimization of the maximum completion time ($C_{\max}$).

A simple flow shop as stated by Baker (1974), Conway *et al.* (1967), French (1982), and Pinado (1995) is a special case of job shop that consists of "*M*" machines which are arranged in series, and "*n*" jobs which must be processed on all machines in the same order. More complicated flow shops, as shown in Figure 1, are: skip flow shops, reentrant flow shops, compound flow shops, and finite queue flow shops.

Generally, in flow shops, the optimal schedule need not be a permutation schedule. Nevertheless, the complexity of these systems is very high. Over the past four decades, many attempts have been made by the researchers to formulate and solve the n-jobs/M-machines/flow shop problems. In literature, we can not find any easy algorithm which provides an optimal solution for flow shop problems, except for 2-machine for $F_2||C_{\max}$ special cases of 3-machine problems for $F_3|\min p_{i1} \geq \max p_{i3} \geq \max p_{i2} \mid F_{max}$ or $C_{\max}$, where p_{ij} is the processing time of job i on machine j (Johnson, 1954), and Gilmore and Gomory's (1964) algorithm for $F_2|$ no-wait $|C_t$, where C_t is cycle time.

While for developing heuristics for flow shops, integer programming, and branch-and-bound techniques have been implemented more frequently, these techniques are not effective on even medium-size problems. Coffman (1976), Rinnooy Kan (1976), and Lenstra *et al.* (1977) stated that all flow shop problems, as well as job shop problems, with exception of one or two cases, are NP-

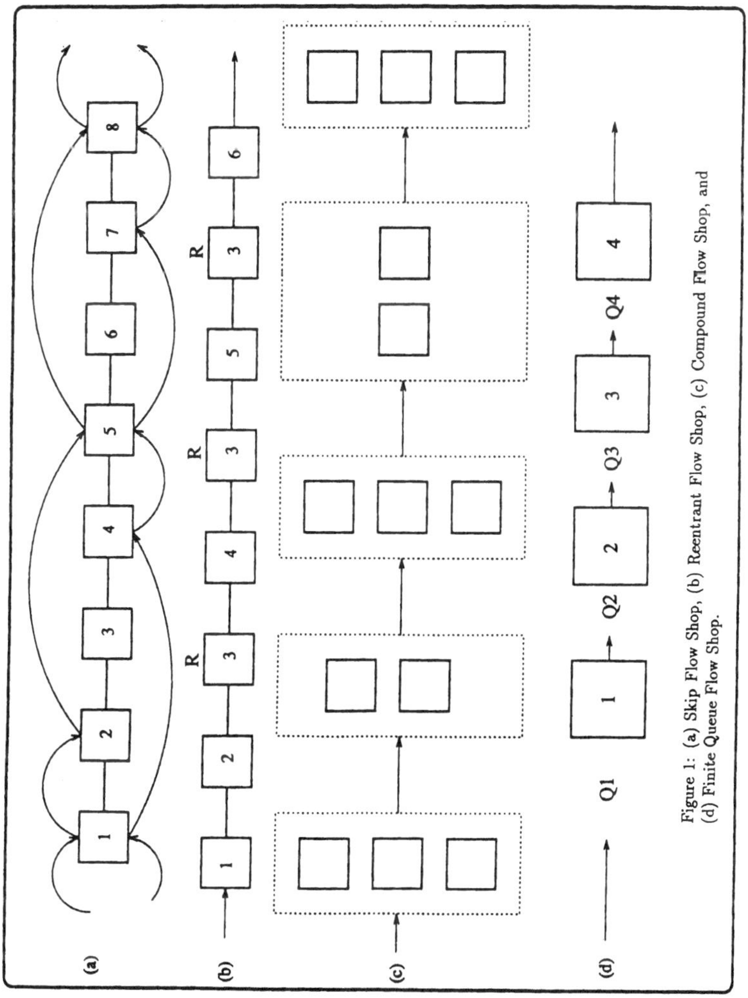

Figure 1: (a) Skip Flow Shop, (b) Reentrant Flow Shop, (c) Compound Flow Shop, and (d) Finite Queue Flow Shop.

complete. For this reason King (1979) stated that "... there seems little point in seeking optimal solutions to the very large scheduling problems of the kind frequently encountered in many practical situations, they are certain, at least for the foreseeable future, to be beyond the available computational capabilities. For these kinds of problems, it would appear that we must concentrate on constructing heuristic polynomial time algorithms that seek *approximate solutions that are satisfactory even though they may not be optimal*." Some of the well-known heuristics for flow shop scheduling are the works of Page (1961), Palmer (1965), Campbell *et al.* (1970), Gupta (1971), Bonney and Gundry (1976), Bestwick and Hastings (1976), King and Spachis (1980), and Dudek *et al.* (1992).

15.5.3 NO-WAIT AND BLOCKING PRODUCTION PLANNING AND SCHEDULING

The flow shop problems are the simplest scheduling problems, however, they are hard problems. In the no-wait flow shop scheduling, a job must be of a job and the start time of its first operation, is equal to the sum of the processing times of all its operation.

A class of problems which have close relation with no-wait systems are scheduling with blocking. Blocking occurs when production of a job must remain on a machine after finishing its process until the subsequent machine becomes ready to accept the job. Kamoun *et al.* (1993) states that theoretically any blocking production scheduling, for two-machines problem, can be transformed into a no-wait schedule, without changing the objective function value. The no-wait and blocking production planning and scheduling can occur for the following reasons:

1. The characteristic of the production environment is no-wait. In a steel plant, for example, delays between operations result in cooling that makes the operation prohibitively difficult. Such situation also occur in plastic molding and silverware production systems, pharmaceutical industries, chemical and petrochemical industries, aluminum manufacturing, etc.

2. There is not any intermediate buffer or in-process inventory or storage available between machines or workstations.

3. In the Just-In-Time systems, one of the main goals is to eliminate or decrease the work-in-process inventories, as well as the finished product inventories. Since this system is a highly coordinated manufacturing process, it can be interpreted as no-wait systems.

4. In some service industries, the cost of waiting for the process is very high for customer satisfaction. These kinds of systems are also a kind of no-wait systems.

5. In sequence dependent environments, the no-wait system increases the processing time, if the processing time of the job on the subsequent machine or workstation is an increasing function of the operation and its waiting time in the current and previous machines or workstations.

All no-wait scheduling problems, except $F_2|$ no-wait $|C_t$, with any criterion are NP-hard. For example, $F_4|$ no-wait $|C_{max}$ (Papadimitriou and Kanellakis, 1980), $F_2|$ no-wait $| L_{msx}$ (Rock, 1984), $F_2|$ no-wait $|\Sigma C_j$ (Rock, 1984), $F_3|$ no-wait $|C_{max}$

(Rock, 1984), $F_2|$ no-wait, $m_1 = 2\ |C_{max}$ (Sriskandarajah and Ladel, 1986), $F_2|$ min $C_t\ |\ W_t$ (Matsuo, 1990), $F_2^g\ |$ no-wait $|C_{max}$ (Sahni and Cho, 1979), are NP-hard in strong sense, where, F_m, m = 1,2,3,4 is flowshop with m machines, C_{max} is maximum completion time, C_t is cycle time, and L_{max} is maximum lateness. Thus, essentially most no-wait systems have heuristic solutions. More details of no-wait and blocking flowshop systems can be found in the works of Sriskandarajah and Goyal (1988), Sriskandarajah (1993), and Hall *et al.* (1997). Since all of the no-wait and blocking systems are very hard problems, and all heuristic algorithms which have been developed to this date produce approximate solution, to develop fuzzy system models for them are potential domains of research. Fuzzy models are generally and usually more appropriate for such systems than classical heuristic methods, since the real world decision making problems often contain more than one criterion and information and data for such systems are usually imprecise.

15.5.4 EXAMPLES OF CLASSICAL EXPERT SYSTEMS

In this part, we briefly review some of the well known AI models for production planning and scheduling.

Bruno *et al.* (1986) implemented expert systems for production planning and scheduling of the parts in a flexible manufacturing system. In this case, the environment is a batch production system, where each batch is composed of 100 to 200 parts. They used a dynamic priority scheme for the generation of release times for batches. The constraints of the system are: due dates, capacity of the queue, the fixture vacancy, and the maintenance of the machines. As shown in Figure 2, in this approach, the scheduling system is composed of two subsystems as: expert system for schedule generation, and simulation system for the evaluation of the schedules.

The first subsystem is written in OPS-5, and the second subsystem is written in Fortran-77. They demonstrated the following conclusion for their approach:

1. Non-procedural programming increases flexibility and maintainability of the planning and scheduling systems.
2. Heuristic knowledge can be introduced to direct decision making.
3. Algorithmic knowledge can be easily integrated as external routines called by OPS-5 statement.

O'Connor (1984) developed two expert systems for scheduling and capacity planning, named as ISA and IMACS. In ISA, which is a rule based expert system, the customer orders are scheduled. ISA can generate a loading strategy for difficult orders, and is able to cluster the orders into two groups as follows:

1. Material shortage group.
2. Credit line insufficient group.

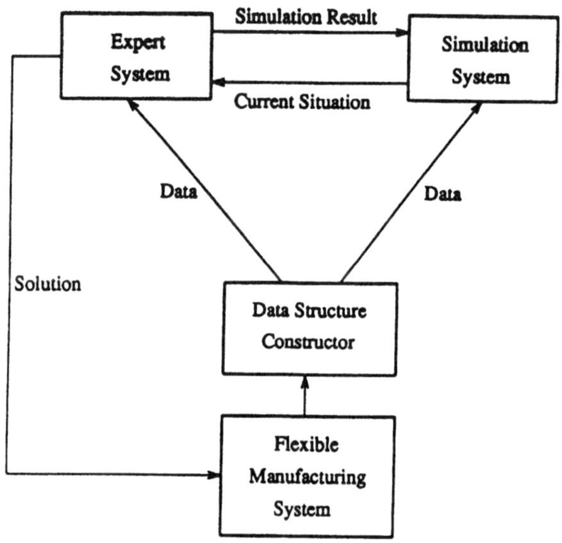

Figure 2: Bruno *et al* Expert System for Batch Production.

The second expert system (IMACS) is for capacity planning. It is able to generate schedules for the inventory control, capacity expansion, etc.

Subramanyan and Askin (1986) developed an expert system for the scheduling of an Flexible Manufacturing System (FMS). In this research, they determined the following statuses for FMS: (1) System status, (2) Machine status, and (3) Job status. In this research, the first and second statuses have the following characteristics: (a) Heavily loaded, (b) Moderately loaded, and (c) Under loaded. For the third status, they determine the following situations: (i) Critically waiting for the machine in the queue, (ii) Moderately late, and (iii) Normal situation.

O'Connor (1984) presented a hierarchy with three levels for experts system decision making as follows:

1. The rules representing various object statuses at different levels, coded in PROLOG.
2. Rule acquisition by a job shop simulation, and composition of the knowledge acquisition from the domain experts.
3. Simulation for expert planning and scheduling.

Ben-Arieh (1986) presented an expert scheduling system for an assembly line. The system is divided into two subsystems as Level I and II. In the Level I subsystem, he presented two algorithmic knowledge as follows:

1. The production rules are written in PROLOG, for checking the conditions of calling computing routines for decision making.
2. The computation routines are written on PASCAL, for the calculation of the costs of the candidate solutions.

The process of the schedule generation is controlled in Level II subsystem. In this subsystem, the simulation is written in SLAM II. Software provides the inference for human controllers.

Shaw and Whinston (1985) used a rule based distributed artificial intelligence (DAI) of Parunak (1987) for production planning and control of a cellular flexible manufacturing system. In their study, they used Petri-net approach for modeling the processes.

Newman and Kempf (1985), and Thesen and Lei (1986) also applied expert systems for production planning and scheduling of the robotic manufacturing system.

Two large scale artificial intelligence scheduling systems were developed in Carnegie-Mellon University. The first one, ISIS is a hierarchical model for constraint directed scheduling of a job shop which was developed by Fox (1983). ISIS can combine hierarchical search, constraint directed reasoning, and scheduling. ISIS was developed for job shop scheduling problems, and was implemented in Westinghouse Electric Corporation plant that produces turbine blades. In ISIS, the fundamental knowledge that affects the scheduling are as follows: (i) Organizational goals, (ii) Physical limitations, (iii) Causal restrictions, (iv) Availability, and (v) Performance.

This approach is based on Schema Representation Language (SRL), which is a frame based knowledge representation language. In this investigation, a frame is a structured piece of knowledge. The configuration of a frame in this approach is as follows:

OPERATION
Next Operation
Previous Operation
Machines
Operators
Durations

In this research, many constraints such as due dates, resources, costs, and the like, are considered in the process of schedule generation which uses a frame structure. The constraint of the product length, for example, is expressed as follows:

PRODUCTION-LENGTH-CONSTRAINT
Constraint-Of: product-length-requirement
Instance required-constraint:
RELAXATION-TYPE: required
PREDICATE: product-length-predicate

ISIS has a hierarchical structure with three levels. The system uses the constraint-directed approach to search the state space for the determination of alternative schedules. Newman and Kempf (1985), Nardi (1985), Miller (1984), and Papas (1986) have discussed some aspects of ISIS.

In ISIS, a constraint-directed search in the systems' state is implemented for the determination of the alternative schedules. There are three levels for the performance of tasks as follows:

- **LEVEL 1**: Selection of orders to be scheduled.
- **LEVEL 2**: Capacity analysis and determination of starting and finished time for the selected orders.
- **LEVEL 3**: Scheduling of all necessary resources for orders selected.

In ISIS, a pre-search process is implemented for the examination of the constraints and determination of the scheduling directions. The search is started by a null schedule for generating alternative partial schedules. The system creates new states for the other operations when an operation is known. It rates the states in the search space based on the set of related constraints and previous states. The system uses the weighted average of the constraints for the determination of the rates of the states with multiple constraints. The weights of the constraints are selected according to the importance of the constraints. A weight may be static or dynamic, based on the relationship between the pre-determined goal state and the current state. Then, the analysis is continued to examine the validity of the schedule generated. The constraints may be relaxed in order to obtain a feasible schedule. ISIS can restart the search when the constraints are not met in a current schedule. In such cases, it may change the search strategy form forward to backward chaining or vice versa.

The second one, CALLISTO was developed by Fox and Smith (1985). It is a project scheduling system that uses a forward dispatch approach and improved versions of the heuristics provided by Morton and Smunt (1986).

Morton and Smunt (1986) applied the expert system in a flexible manufacturing system (FMS) called "PATRIACH". The system contains for levels as follows:

- **LEVEL 1**: Strategic Planning: Which is a manual decision support system.
- **LEVEL 2**: Capacity Planning: Which is a semi-automatic decision support system.
- **LEVEL 3**: Scheduling: Which uses OR for generating the heuristic solutions.
- **LEVEL 4**: Dispatching: Which mixes the solution of the LEVEL 3 with simple rules of thumb and occasional human help.

In this approach, the main constraints of the scheduling in LEVEL 3 are: resource utilization and due dates, and in LEVEL 4 they are: (a) time that the operation is actually available, (b) temporary interruptions such as tool unavailability, and (c) interventions of a human operator.

Yu (1986) developed an expert system to implement Operations Research (OR) techniques, called Automatic Reasoning Operations Research Applications (AURORA). In this expert planning system, pattern matching method is used for finding the problem patterns. A linear programming approach is applied after the pattern is matched. The model and its related linear programming can be transferred by a meta-planner to an automatic programmer. Then, by getting the suitable code from the automatic programmer, the system tries to solve the problem. AURORA was used in air force mission planning, as well as the cargo loading problem of NASA space shuttle programs.

As stated by Fox (1983), two areas of research in artificial intelligence potentially impacts the scheduling problem: search planning and constraint analysis. Research in planing aspect of scheduling in artificial intelligence can be traced back to LT and GPS. Both LT and GPS were developed by Newell and

Simon (1963). In LT a heuristic search is introduced for problem solving in system states, and transfer of states by operators. The system can decompose the problem into subproblems by backwards approach from the goals. In GPS the focus is on the most relevant operators in achieving the goal state for the reduction of the search space.

15.5.5 FUZZY EXPERT SYSTEMS FOR AGGREGATE PRODUCTION PLANNING AND SCHEDULING:

Traditional aggregate production planning, inventory control, and scheduling models, as stated in the previous section, are mainly occupied with deterministic or stochastic and probabilistic methods. However, management usually states the goals, objectives and operative constraints, and the like, in regards to natural language expressions and linguistic variables. These expressions and variables are often imbued with uncertainty, ambiguity, vagueness, and with imprecision. This is an inescapable reality despite great efforts of the scientists and engineers to deem definitions and measurements as precise as possible.

Most of the existing models for production planning, inventory control, and scheduling have used a kind of stochastic method to deal with random uncertainty. Of course stochastic models provide valuable insights into a system's behavior under specific assumptions, i.e., random behavior, etc., and identify useful guidelines. However, these systems are not often implemented. Many managers often do not understand the basic assumptions of these models. On the other hand, these models are either too complex or the precise information required to determine the values of the model parameters cannot be obtained or are not available. Thus, such models are, by default, inadequate to help managers to cope with the natural behavior of many real-life systems (Tűrkşen and Berg, 1991).

As stated by Tűrkşen (1988), the control of complex production processes often involves the implementation of the available knowledge of experts on certain problems. Such expert knowledge is usually articulated in the form of linguistic rules of a natural language. For example, in a production-inventory system, suppose one of the production rules specified by an expert is as follows:

> "**IF** the current inventory level is *much higher* than the preferred level but the deviation is decreasing at a *medium* rate (say due to fairly high demand), **THEN** the production rate is to be *moderately* slowed down", (Tűrkşen, 1985).

In such cases, the rule-based decision making is stated with the linguistic variables, as shown in the example given above. In general, expert production planning and scheduling systems are mainly rule-based systems (Steffen, 1986). Hence, fuzzy set theory, which admits representation and computation with words, should be used in modeling decision processes and formulating actions with multi-valued controllers. The reasons for the representation of managerial knowledge by linguistic rules is becoming more and more acceptable in the design of expert systems for production planning and inventory control as follows:

1. It has been shown that linguistic rule based system models determine outputs which are as compatible as crisp (non-fuzzy) system models proposed by operations research and/or automatic control methodologies, (Rinks, 1982).
2. Some real-life application systems, such as supervisory process control systems, rely much more directly on human interaction for their inputs and for the interpretation of their outputs, (Rouse, 1972). Controllers of such systems quite often tend to communicate with fuzzy linguistic variables, (Zadeh, 1977).
3. In most cases, the imprecision is forced upon the experts by the fact that there are no units of measurement for the attributes of a complex system and no quantitative criteria for representing the values of such attributes, Zadeh, 1983).

Fuzzy expert systems have the potential basis for modeling humanistic systems. They are able to integrate the imprecisions into the knowledge representation and can utilize the imprecise knowledge in the inferencing mechanism in approximate reasoning approach. In such a case, a fuzzy expert system is the integration of artificial intelligence, problem domain, and fuzzy sets theory. In the steel industry, for example, the inventory, scheduling, capacity, and resource considerations and their coordination necessitate aggregate planning. In this industry, batch production characteristics (in iron-making and steel making), sequence dependency of the processes as well as no-waiting requirements, make the problem extremely complicated. In such cases, fuzzy aggregate production planning and control knowledge with representation based on fuzzy cluster analysis and fuzzy inference become more suitable, since the classical models have serious difficulties for implementation in such complicated industrial situations.

Rinks (1982) was the first who attempted to use fuzzy expert systems approach associated with an aggregate production planning. He illustrates how simple heuristic rules used by managers may closely approximate the results of optimization techniques for the aggregate production planning and scheduling problems. In his model, he used the point valued linguistic variables to transform reasonable rules in his model to transform reasonable rules of thumb into an operational model. Then, he applied Compositional Rule of Inference to deduce a recommendation based on observed system state and the rules in the rule base.

Yao and Tűrkşen (1983), and Tűrkşen and Yao (1984) developed a model of multi-valued controllers in production systems. In this research, the connectives and the implication involved in the multi-valued controllers are represented through their normal forms. They state that the approach has some advantages as follows:

1. It is a natural extension, i.e., fuzzification of the boolean normal forms.
2. It treats the connectives and the implications as fuzzy entities, i.e., generates Type II fuzziness which captures additional fuzziness introduced by the combination of fuzzy concepts in statements (decision rules).
3. The relation $A_o\ (A \rightarrow B) = B$ is preserved under certain regularity conditions of A and B.

In this and later studies conducted by Tűrkşen and his colleges (1986, 1989, 1993, 1995, 1996), the knowledge representation of the controllers are formulated in the form of a set of normal form based rules. In such systems the inference methods are also executed with Type II fuzziness for any observed system state, where the best fitted set of rules are chosen for the execution. Finally, it is stated that this formulation avoids the arbitrary aggregation of the decision rules and guarantees the best use of the expert knowledge.

Tűrkşen (1988) also developed several prototypes based on approximate reasoning with interval-valued fuzzy sets for production planning and inventory control. He stated that with this system the inference engine which provide a framework that incorporates imprecise knowledge as a bounded quantity produces robust results. Moreover, in this model, the observed system states did not have any corresponding rules that exactly matched any of the rules in the rule base.

Tűrkşen and Zhong (1988) proposed an Approximate Analogical Reasoning Schema (AARS) which exhibits the advantages of fuzzy sets theory and analogical reasoning in expert systems development. In this research, they developed two algorithms based on similarity measures and modification functions for both point-valued and interval-valued AARS. Then, they implement this system for a prototype of an aggregate production planning environment with 27 rules and run the program in the Personal Consultant Plus (PC^+). In this system the independent variable are: sales forecast for period $t(F_t)$, inventory level at the end of period t-1 (I_{t-1}), and work force level at the end of period t-1 (W_{t-1}), and the decision variables are: production rate for period t (P_t), and work force level or change in work force level for period $t(W_t)$ or (Dw_t). An example of the rules is as follows:

> "**IF** F_t is *High*, AND I_{t-1} is *High*, AND W_{t-1} is *Low*,
> **THEN** P_t should be *Average* and DW_t should be *Rather-High*."

They compared the proposed AARS model with other available models developed for the same industry such as company's actual model, Linear Decision Rules (LDR), (Holt *et al.* 1955), and Rink's point-valued fuzzy set method (1982), etc. The results of the comparisons show that fuzzy system models were just as good or, at times, better than their crisp variables.

Tűrkşen (1988) proposed an approximate reasoning framework based on the developments in interval-valued fuzzy set representation and inference (Tűrkşen, 1986) and implemented it on the aggregate production planning and inventory safety stock level control. The framework of the model is essentially designed to accomplish the following:

1. An interval valued representation of rules and observed system facts;
2. A search for the rules closest to the observed system facts based on a similarity measures;
3. An inference based on the implication of the rules found in (2);
4. An advice based on the linguistic approximation of the result found in (3).

The sample of the rules generated is as follows:

> "**IF** the demand variability X_1 is to be A_1, **AND** the service level X_2 is to be A_2,
> **THEN** the safety stock level Y is (should be) B".

Where the generic linguistic variables A_1, A_2 and B can take on linguistic variables such as *high (H)*, *medium (M)*, and low (L). Thus, the above rule may be restated as follows:

> "**IF** the demand variability X_1 is M, AND the service level X_2 is to be H,
> **THEN** the safety stock level is (should be) H."

In other research, Tűrkşen (1989) develops an algorithm based on approximate reasoning and interval-valued fuzzy representation and inference for production planning. In this research, he shows that despite the resistance from the established deterministic and probabilistic researchers, fuzzy reasoning has gained recognition in aggregate production planning and inventory control, scheduling, operation research, etc. His proposal to represent the linguistic connectives, the implications and generalized Modus Ponens with the normal forms, provide a convenient and workable structure for control of production and inventory problems.

Tűrkşen and Berg (1991) developed an approximate reasoning framework for development of an expert system prototype to aid management in planning inventory capacities. They attempted to show how operation research (OR) and approximate reasoning (AR) can be synthesised for the solution of real-life problems in the area of knowledge-based systems. That is, they state that OR methodologies can generate valuable insights into the understanding of the problem domain and AR is used for providing a framework for restructuring the insight gained form OR to real-world problems where there is insufficient information to identify parameters of the system at hand or where such data are not available and the environmental conditions are uncertain. The parameters of the system in this research are as follows: (1) Demand Rate (λ), (2) Mean Demand Size (μ^{-1}), (3) Failure Rate (θ), and (4) Repair Rate (σ^{-1}).

An example of the rules in this research is as follows:

> "**IF** the demand arrival rate, λ, is *low*,
> AND the mean demand size, μ^{-1}, is *medium*,
> AND the failure rate of the machine, θ, is *low*,
> AND the machine repair rate, σ^{-1}, is *high*,
> **THEN** in order to have a *high* fraction of satisfied demand, ρ, increase N (inventory level) *moderately*."

Here, 21 rules were developed for a real-life industry and the simulation were done with both Max-Min operators and Bold Union/Intersection operators. The OR expert's assessments and the real system behavior via model analysis are expressed in a natural language setting with linguistic terms of linguistic variables. Thus, this approach provides a framework for the flexible expressions in human judgements. Moreover, approximate reasoning based on fuzzy logic can handle the linguistics, i.e., semantics, uncertainties and imprecision. As a result, it appears that the merger of operations research and approximate reasoning is inevitable with the advance of fuzzy expert systems. It shows that approximate reasoning can provide aggregation of the human knowledge as well as user friendliness.

Tűrkşen and Fazel Zarandi (1997) discuss the fuzzy cluster analysis together with an integrated fuzzy systems modeling, in general, and its application in batch production systems, such as continuous casting of steel plants, in particular. This will be discussed further in section 8.

15.6 DETAILED FUZZY PRODUCTION PLANNING AND SCHEDULING

Bensana *et al.* (1988) developed a fuzzy scheduling approach for job shop problems by implementing the concepts of artificial intelligent, fuzzy set theory and traditional operational research. In their model, there exist *K* jobs which must be processed on a set of *M* machines. The system, as shown in Figure 3, consists of two planning modules and one supervisor module.

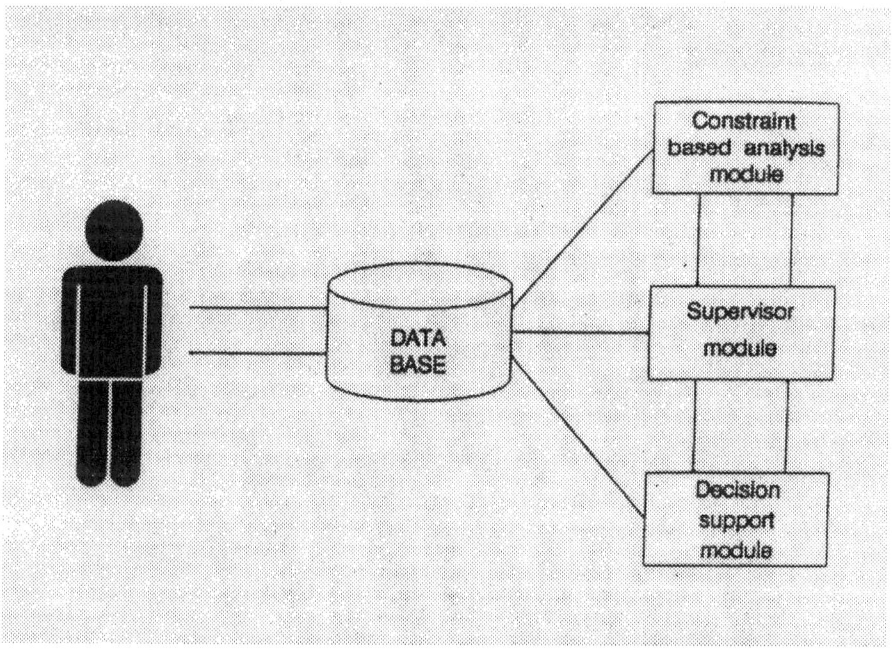

Figure 3: The structure of OPAL for fuzzy Job Shop Scheduling

1. **Constraint-base analysis module:** This module is related to the sequence of jobs, schedule in progress, and the time constraints for the processing of jobs. It generates new precedence constraints by subsequent systematical comparison of the existing precedence constraints. The system generates three states as: (a) success (i.e., there is a feasible and hence a complete schedule is derived), (b) failure (i.e., there exist no feasible schedule, due to conflicting precedence constraints), and (c) wait (i.e., the schedule in process is incomplete).

2. **Decision-support module:** This module generates new precedence constraints, by selecting a subset of the set of all unordered pairs of operations and choosing one element of this subset to form a new precedence

specific machines, specific operations, temporal location, influence on the quality of the schedule, or influence on the resolution speed. The grades of membership of the unfixed pairs of operations in the sets "which" are defined by those criteria may be fuzzy." When more than one criterion is used in selection, the system implements min-operator between criterion. In the next step, the fuzziness of an element is fixed to generate a new precedence constraint. The result of this step is a collection of IF-THEN rules. These rules may differ by their origin and their range of application. The efficiency of the rules depends on the goals and the state of completion of the schedule, where, each rule has a grade of membership function with respect to the different goals.

In this system, a schedule is constructed by adding the precedence constraints among the operations. The constraint-based analysis module is stopped and a new edge is generated for the development of a schedule graph and the current schedule is stored. Then, the decision support module generates a new precedence constraint to the graph of schedule. This time constraint-based analysis module checks the consequent precedence constraints. Here, when the explored path leads to a failure state, back-tracking can occur. If the system can not find a feasible solution for scheduling problem, it is necessary to modify the data to recover a feasible solution.

Hintz and Zimmermann (1989) applied fuzzy set theory in production planning and control of flexible manufacturing system (FMS), where the system consists of highly automated work stations linked by a computerized material handling system. In this system, the jobs can follow diverse routes through the machines. In this case, the criteria are minimization of the flow time, maximization of the machine utilization, and meeting the jobs' due dates. In general, these objectives are in conflict with each other. They decomposed the problem into master scheduling, tool loading, release time scheduling, and machine scheduling subproblems. For master scheduling, they implemented fuzzy linear programming. A heuristic algorithm was developed for tool loading problem. Finally, by using approximate reasoning (AR) they solve the release time scheduling and machine scheduling.

In the first part, where short-term production planning is determined, the objective function is to balance the machines utilization with the following constraints:

1. Parts can only be processed when they are released from earlier production stages.
2. They have to meet the given due dates in order to match the following operations and assembling requirement.

3. The capacity of the FMS must not be exceeded.
   They are classified into appropriate groups, because the machines may partially be substituted by each other.
4. There is only a limited number of (expensive) fixtures and pallets available.

They modeled the first and the second constraints as fuzzy constraints, and the third and the forth constraints as crisp constraints. Then, they used fuzzy linear programming (FLP) to solve the problem.

In this approach, AR is implemented to formulate the third and forth criteria. They stated that AR is a good tool to model such complex systems where the resolution of conflicting goals of scheduling would be needed. The linguistic variables are slack time (critically-short, short) and waiting time (short, medium, long) and the criterion is the due date urgency (urgent, not-urgent). Some examples of the rules are as follows:

IF waiting time is *long* AND slack time is *critically-short*
THEN the due date criterion urgent (1.0).
IF waiting time is *medium* AND slack time is *critically-short*
THEN the due date criterion is urgent (0.8).

IF waiting time is *short* AND slack time is *critically-short*
THEN the due date criterion is urgent (0.6).

IF waiting time is *long* AND slack time is *short*
THEN the due date criterion is urgent (0.5).

IF waiting time is *medium* AND slack time is *short*
THEN the due date criterion is urgent (0.2).

IF waiting time is medium AND slack time is *short*
THEN the due date criterion is not-urgent (0.7).

Obviously, each of these rules is formed by a fuzzy combination of the slack time, waiting time to imply the due date criteria. The configuration of their proposed AR is shown in Figure 4.

This approach is developed in a general simulation software for FMS. It is shown that the method dominated the classical priority scheduling with respect to all three criteria.

Holtz and Desonki (1981) developed a fuzzy mathematical model for determining optimal maintenance cycles in electrical power plants. They stated that "... stochastic models have been used before, but because of the very low frequency of breakdowns it seemed that a model based on frequentistic arguments was not appropriate". The crisp mathematical model in this research is as follows:

$$\min C = C(T_1, T_2, \ldots, T_N) \tag{1}$$

$$= \sum_{j=1}^{N} (x_{1j} T_j + x_{2j} + \frac{x_{3j}}{T_j}); \; i = 1.2,3; \; j = 1,2,\ldots,N \tag{2}$$

$$s.t. \quad \sum_{j=1}^{N} (x_{1j} T_j + y_{2j} + \frac{y_{3j}}{T_j}) \leq M_h \tag{3}$$

$$\sum_{j=1}^{N} (z_{1j} T_j + z_{2j} + \frac{x_{3j}}{T_j} \leq B_{\max} \tag{4}$$

$$T_j \geq 0 \tag{5}$$

where, C is total cost, T_j is the cycle times of maintenance operations, x_{ij} is the coefficient of the crisp cost function, y_{ij} is the coefficient of the manpower requirement function, z_{ij} is the coefficient of the breakdown function, M_h is the number of man-hour available for maintenance per year, B is the number of breakdowns per year, and $B_{\max}$ is the maximum of acceptable breakdowns per year.

Then, they used the concepts of the fuzzy sets theory and defined the optimal decision as

$$T_{j,opt.} = T_j \circ \mu_{Tj} = \max_{j} \min_{i} \mu_i(T_j) \tag{6}$$

In this study, they presented two types of membership functions: linear membership function, and nonlinear two-parameter membership function. For the first strategy the membership functions are as follows:

$$\mu_C(T_j) = \frac{1}{2}\left\{[1+sgn(C_L-C)]+[1+sgn(C-C_L)]\cdot\left(\frac{C_U-C}{C_L-C_U}\right)\right\} \quad (7)$$

$$\mu_{Mh}(T_j) = \frac{1}{2}\left\{[1+sgn(M_h-M_{hL})]+[1+sgn(M_L-M_{hL})]\cdot\left(\frac{M_{h_U}-M_h}{M_{h_L}-M_{h_U}}\right)\right\} \quad (8)$$

$$\mu_B(T_j) = \frac{1}{2}\left\{[1+sgn(B_L-B)]+[1+sgn(B_U-B_L)]\cdot\left(\frac{B_U-B}{B_L-B_U}\right)\right\} \quad (9)$$

where, C_L and C_U and lower bound and upper bound of the total costs, M_{h_L} *and* M_{h_U} C is the total cost, are lower bound and upper bound of number of man-hour available for maintenance per year, and B_L and B_U are lower and upper bounds of the number of the breakdowns per year.

For defining the nonlinear membership function, they developed the following model:

$$\mu_C(T_j) = \frac{1}{2}\left\{[1+sgn(C_L-C)]+\frac{1+sgn(C-C_L)}{1+\left(\frac{1}{b_1}-1\right)\left|\frac{C-C_L}{c_1}\right|}\right\} \quad (10)$$

where, b_1 and c_1 are used as means of better fitting of the membership function to the real world situation. Other membership functions were developed similar to the above model. They used min-operator for intersection and showed that the model enhanced the maintenance scheduling of the real world electrical power plant.

McCahon and Lee (1992) used fuzzy set theory for developing a model for job sequencing for a flow shop, where "there is no need to force the manager to give a single precise number nor a probability distribution to represent the processing time". In this paper, the processing times are determined by a natural language in expressions of fuzzy numbers. In this study, there exist *n* jobs that should be processed through *m* workstations, with the aims of minimization of fuzzy makespan and fuzzy mean flow time. These performance measures are used for comparing the alternatives, as well as the interpretation of the impact of the fuzzy processing times on job completion time, flow time and makespan. They modified the Campbell *et al.* (1970) classical algorithm that accepts fuzzy processing times. Then, they used the Lee-Li (1988) fuzzy number comparison method for comparing the fuzzy numbers, whenever such comparison had to be made during the above sequencing algorithm. They developed deterministic sequences, and then calculated fuzzy makespans and fuzzy job mean flow times,

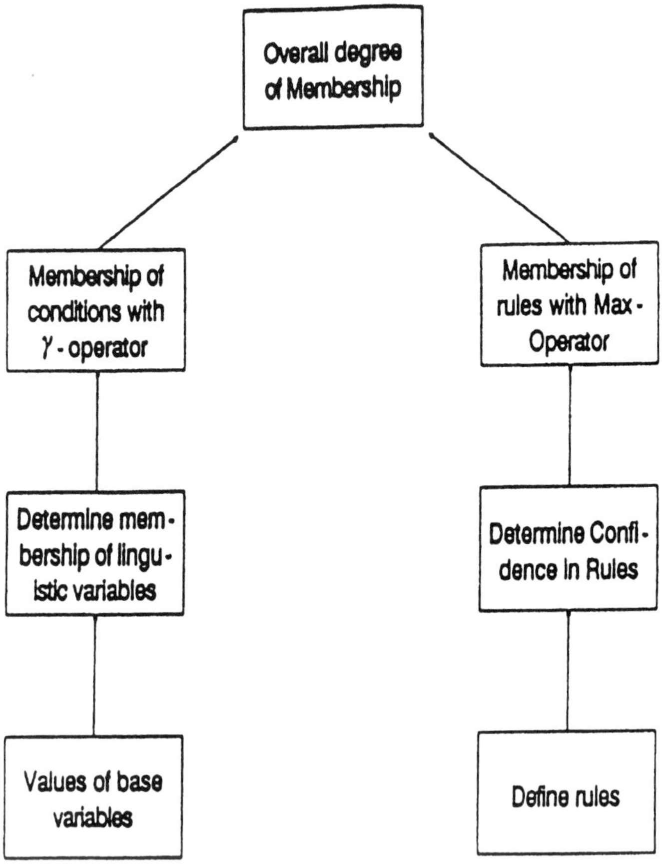

Figure 4: Components of AR for FMS.

by using fuzzy arithmetic. This research shows that through the use of possibility theory, the manager can get a better understanding of how the processing times spread the effect of the job mean flow time and makespan, and thus can plan the activities of the shop accordingly.

Hung *et al.* (1995) presented a fuzzy Longest Processing Time (LTP) algorithm for the aim of scheduling. In this research, they assume that (a) more than two machines are available, (b) the tasks are not preemptive, (c) the number of the possible execution times for each task is finite, and (d) the discrete membership function for each tasks is known. Based on the above assumptions, they developed an algorithm with 12 steps as a discrete fuzzy LTP scheduling algorithm, where, the inputs are n tasks each with a discrete membership function for its processing time and a set of m processors, and the output is a schedule with a membership function that determines its completion time.

Dubois *et al.* (1995) developed a constraint based method for job shop scheduling problems. This approach can account for the flexibility of temporal constraints and the uncertainty of the durations of the operations. They demonstrate the set of solutions to the job shop problems by an implementation of fuzzy sets, where the membership functions reflects the performance of the system. In this research, the uncertainty of the operation durations is qualitatively described by possibility distributions. They formulated a simple job shop scheduling with framework of the constraint satisfaction approach in artificial intelligence, and adapted a combinatorial search method for the solution of the problem.

Han *et al.* (1994) applied the concepts of fuzzy sets theory in a generalized single machine with fuzzy due dates and controllable machine speed, where the main criterion is minimization of the maximum lateness. In this paper, there is only one machine which n jobs need to be processed on the machine. In this research, the membership function of the jobs are described as the degree of satisfaction with respect to their related completion time. Here, since the speed of the machine can be changeable for different jobs, the objective is the determination of the optimal job sequence according to the machine speeds. The objective function of the model is expressed as the minimization of the total sum of costs associated with job sequences, machine speeds and dissatisfaction with respect to completion time of jobs. For this purpose, they developed a polynomial time algorithm and tested their algorithm with a small size problem.

Petrovic and Pupin (1994), discuss the complexity of the inventory control, where it encounters various sources of uncertainty and imprecision. They, developed a method based on fuzzy knowledge-based approach for the aim of the solution of the problem of order quantity determination, where the demand, lead time, and inventory level are uncertain. In this research, uncertain demands are demonstrated by fuzzy numbers, and vaguely defined relations between them are modeled by fuzzy IF-THEN rules. They presented several examples for verification of their proposed approach, and compared them with the traditional probabilistic approaches and showed the capability of fuzzy rule-based system for solving these kinds of the complex production planning and inventory control system.

Grabot and Geneste (1994), implemented fuzzy sets theory and fuzzy systems in multiple criteria decision making, where, they build aggregated rules which allows to obtain a compromise between satisfaction of several criteria. In this research, when the criteria of the performance is changed with the evolution of the production environment, the aggregated rules are parametrized in order to model the respective influence of the elementary rules which they are composed of. They showed that this approach is more appropriated for real world job shop problems than the classical SPT or DISPATCHING approaches.

Custodio *et al.* (1994) used fuzzy decision making in the production planning and scheduling. In this research, a short-range planning and scheduling problem is addressed by using nonclassical approach by the support of fuzzy sets theory. They used a hierarchical structure with three decision level as HIGHER, MIDDLE, and LOWER. In this research, each level is responsible for a different production problem with a different time scale. In each level, the needed task is from one level to the next by a heuristic method. This heuristic is able to solve the short-range planning and scheduling problems with non-stationary policy. In this paper, the higher decision level determines safety stock levels, which can compensate for future resource failures, and at the medium level loading rates are computed.

Ishibuchi *et al.* (1994) presented a model for fuzzy flow shop scheduling problems, where the due dates of the jobs are fuzzy. In this research, the membership functions of the due dates correspond to the degree of satisfaction of a completion time. The main criterion in this paper is maximization of the minimum degree of satisfaction over given jobs. They used several local search algorithms such as multi-start descent, simulated annealing and taboo search algorithms for such problems. For the comparison of the performance of each algorithm they used computer simulations on randomly generated test problems. They showed by simulation that the performance of their proposed fuzzy model is better than the classical algorithms for fuzzy flow shop scheduling problems.

In summary, the models briefly reviewed above show that fuzzy sets theory and fuzzy systems have gained recognition in many areas, especially in decision making, aggregate production planning, inventory control, production systems, scheduling, etc. In production planning and scheduling of no-wait, blocking, and JIT systems, where the classical crisp models encounter serious difficulties, fuzzy systems models are "good" candidates as shown by the results of many researchers as demonstrated by the research reviewed above.

15.7 FUZZY SYSTEMS FOR JUST IN TIME (JIT) PRODUCTION PLANNING AND SCHEDULING

Planning and Scheduling:

Since Japanese invention of JIT system, many companies tried to implement this system in their production planning and manufacturing systems all around the world. In the real world applications, JIT systems have shown that they are able to raise product quality, reduce the completion time of a throughput in a shop, increase system flexibility, and increase the role of human experts in the production systems. While, before the advent of JIT, it was reasonable to have large work-in-process (WIP) buffers as well as finished product stocks to absorb scheduling errors and to decouple complex systems, today, as stated by Morton and Pentico (1993) ".. it is widely accepted that these comfortable buffers need to be progressively eliminated." Their reasons for this expression are as follows:

1. The increasing complexity and rapid obsolescence of products.
2. More rapid learning to eliminate process flaws.
3. Customer desires for much shorter lead times with easier order revision during that lead time.
4. More flexible reaction to desired changes in product mix.
5. More flexible reaction to emergency floor problems.
6. More rapid awareness of variation in product quality and its exact source.

15.7.1 PUSH AND PULL SYSTEMS

While the push systems assume a completely customized order that must traverse the entire shop after an order is placed, the pull systems assume some standardized products for the delivery of customer orders from inventories with little lead time (Morton and Pentico, 1993). JIT systems with the aids of kanban and conwip are systems in which raw materials, work in process inventories, and final orders are delivered to the next stage exactly when needed. Morton and Pentico (1993) presented the procedures that triggers customer orders in the push and pull systems as follows:

1. Push systems

   (a) Creation of paperwork to create the order from scratch.
   (b) Negotiation of a due date with the customer.
   (c) Decision when to release the job order to shop floor.
   (d) Dynamic calculation of priorities.
   (e) Priorities to push the job through the stages.
   (f) Delivery of a complete job to a customer.

2. Pull systems

   (a) Ship from final inventory.
   (b) Instruct the final machine to take existing WIP and finish a new unit for final inventory.
   (c) Replace final machine orders with WIP from previous machine, which in turn replace orders with WIP from previous machine, and so on.

Obviously, without large work-in-process inventories to decouple the stages of the system, the human expert ought to struggle to adapt to the larger system directly. In such cases, human scheduler encounters greater difficulties with respect to rapid system changes, which reduces the buffers, and increases the system complexities. However, humans have a broad base for common sense; intuitive and poorly understood thinking abilities; and the ability to strategically summarize the knowledge. These abilities taken together often work to allow experts to react quickly and accurately to glitches and problems. On the other hand, definition of the criterion in this environment is very difficult. It is very difficult to quantify some objectives such as customer satisfactions which are often expressed qualitatively which require promptness. Moreover, shops are usually encountered with some objectives that are quite in conflict with each other. Maximization of the shop throughput over some time periods and customer satisfaction of desired quality and minimization of current out-of-pocket costs are often in conflict with each other. Fuzzy sets theory and fuzzy systems can combine different types of objectives to reach reasonable solutions without resorting to the complicated mathematical operations.

There exist many similarities between the objectives of fuzzy systems and Just-In-Time systems.

1. The objective of both fuzzy production planning and scheduling systems and JIT systems is to enhance the accuracy of the systems by considering the human behavior. Thus, it is necessary to evolve, develop and integrate many concepts and techniques and to begin changing the culture of the manufacturing systems environments.

2. It is not difficult to embrace both JIT and fuzzy system concepts within the overall concept of any manufacturing system's goals.

3. Both fuzzy systems and JIT systems try to create an organization in which every employee from top to bottom is working toward making that organization the best in its particular field. It is based upon the idea that each person is the expert in his/her own job (with a degree). By harnessing the collective thinking power, allowing the creation of job knowledge by every one in the organization, the productivity, as well as the proficiency of the systems and organizations can be improved very rapidly.

15.8 CLUSTER ANALYSIS IN PRODUCTION PLANNING AND SCHEDULING

Grouping the objects, based on their similarities, is a fundamental mode of better understanding and learning. Jain and Dubes (1988) stated that "... the act of sorting similar things into categories is one of the most primitive and common pursuits of men. Our languages consists of words which help us to recognize and discuss the different types of events, objects and people we encounter. Each noun is a label used to describe a class of things which have certain striking features in common. We need to be able to classify our activities and surroundings simply in order to make life manageable, since it would be impossible to treat everything we encounter as unique." Clustering methods are more prevalent than the other methods for grouping or classifying the objects. These methods, as stated by Hartigan (1975), Romesburg (1984), and Backer (1995) can help to:

1. Formulate hypotheses concerning the origin of a population (e.g. evaluation studies);
2. Describe the sample in terms of typology (e.g., market analysis);
3. Predict the future behavior of population types (e.g., modeling economic prospects);
4. Optimize a functional process (e.g., information retrieval);
5. Assist identification (e.g., diagnosing);
6. Measure the different effects of treatments on classes within the population (e.g., analysis of variance).

In general, clustering is a method for grouping and classifying a number of similar objects together, based on a measure of similarity or distance between the pairs of objects. Here, the objects are grouped in a cluster in such a way that the similarity between two objects within a group ought to be large and the similarity between two groups ought to be small. In other words, in clustering, the distance within a group must be small and the distance between groups must be large. Thus, in cluster analysis, the inputs of the system are the set of objects and their similarities or distances, and the output of the system is a number of clusters or family partitions of the object set. Clustering, as stated by Pal and Bezdek (1994), "is also called unsupervised learning, the word learning referring here to learning the correct labels (and possibly vector prototypes or quantizers) for good subgroups in the data."

In contrast to the manual grouping process, clustering techniques, as stated by Jain and Dubes (1988), offer several advantages. They can apply a specified objective criterion consistently to form the groups, i.e., different individuals may not always identify the same clusters in data. Moreover, a clustering algorithm can form the groups in a fraction of time required by a manual grouping, particularly if a long list of descriptors of features is associated with each object. Clustering also can reduce the computational complexity of many complex problems, especially the various decision-making algorithms in pattern recognition.

In conventional clustering methods it is assumed that an object can belong to one and only one cluster. In real world the separation of the clusters is fuzzy. Of course fuzzy sets theory, by allowing the algorithm to assign a membership function to the "c" clusters, offers many advantages over conventional clustering. Fuzzy clustering technique has many advantages in real applications. Two obvious advantages of this technique are (Yoshinari *et al.* 1993): (a) an automatic formation of fuzzy sets and fuzzy relations (i.e. the method is unsupervised learning approach where they do not require any initial knowledge about the structure in the data), and (b) the fuzzy models can take full advantages of the structure-free approach, i.e., the "hidden" structure (Tűrkşen and Fazel Zarandi, 1997) becomes reveal within the process of clustering.

Clustering methods have been used in different areas of industrial engineering and operational research. In classical production planning and scheduling, for reducing the size of the problem, a kind of stringing or clustering or chaining have been implemented. Partitioning in scheduling, lot sizing, group technology, production smoothing in JIT systems, etc., are samples of partitioning which are frequently used in classical production planning and scheduling. Moreover, fuzzy clustering have been implemented in several areas in the past decade or so. Actually, this method is very useful for modeling different manufacturing systems such as cyclic systems, Just-In-Time systems, batch production systems, periodic production systems, group technology, etc. It has also close relation with sequencing and scheduling problems with different criteria. In production planning and scheduling of hot mill in steel production, for example, one goal is stringing different customer orders in such a way that we can attain the maximization of the throughputs of the mill as well as minimization of the inventory and the maximization of the customer satisfactions (minimization of the tardiness and maximization of the quality of the products). For this purpose, fuzzy clustering is a suitable tool for the above mentioned stringing operations.

In a recent real life scheduling of continuous casting operations, fuzzy clustering and modeling have been used rather successfully. Tűrkşen and Fazel Zarandi (1997) present a fuzzy system model through fuzzy cluster analysis for the prediction of tardiness of continuous casting in a steel plant. In this research, they used both identity norm matrix and covariance norm matrix. They investigate and analyse complex interactions of variables affecting a scheduling system and its performance measure under current operating conditions to expose the *hidden rules* that are operating under day-to-day implementation of the company procedures by the schedulers of the company. The input variables of the system are: *priority, grade transition penalty, customer orders' width, minimum width of slab, maximum width of slab, quality, customer orders' weight, number of slabs needed for customer orders, and weight of slabs.* The output of the system is *tardiness of customer delivery due date.* The results of fuzzy clustering approach for this data set are four rules which demonstrate fuzzy relationships between antecedents, i.e., inputs and consequent, i.e., output. In the next stage, they use parametric approximate reasoning schema in inference mechanism, and tune input output membership functions. The results of the research show that the mean error of the system with covariance norm matrix is rather less than that of with identity norm matrix.

15.8.1 PARTITIONING IN CLASSICAL PRODUCTION PLANNING AND SCHEDULING

In classical production planning and scheduling, it is possible to reduce the size of the problem by implementing a kind of stringing or clustering or chaining. Here, we mention some of the models developed for this purpose.

Consider that there exists n jobs that we want to partition them into c strings or clusters, each cluster contains n_i, $i = 1, 2, ..., c$ jobs. Let p_{ij} be the processing time of job i in cluster j and $P_i = \sum_{j=1}^{n_i} p_{ij}$ be the total processing time. For minimization of the mean job flow time ($\bar{F}$), French (1982) presented the following theorem.

Theorem (French, 1982): For the problem $n|1|\bar{F}$ with required strings of the jobs whose processing order is fixed in advance and whose processing cannot be preempted once started, the optimal schedule(s) are such that

$$\frac{P_{i(1)}}{n_{i(1)}} \leq \frac{P_{i(2)}}{n_{i(2)}} \leq \frac{P_{i(3)}}{n_{i(3)}} \leq \ldots \leq \frac{P_{i(c)}}{n_{i(c)}} \tag{11}$$

where n_i is the number of jobs in cluster i and n is total number of jobs, $\bar{F}$ is mean job flow time, and $i(1)$th cluster is scheduled first, and $i(2)$th cluster is scheduled second, etc. (The proof is easy with interchanging two adjacent cluster. For more detail please refer to French (1982).)

In scheduling of parallel machines, Morton and Pentico (1993) also used a kind of clustering method.

15.8.2 GROUP TECHNOLOGY

Cellular manufacturing systems is a kind of layout in which machines are grouped into a cell. In this case, as stated by Stevenson (1995), "... grouping are determined by the operations needed to perform work for a set of similar items, or part families, that require similar processing." The advantages of this grouping or clustering are faster throughput time, less material handling, less work-in-process inventories, and less setup times. The grouping or clustering process for cellular manufacturing systems is called group technology. In this technique, items with similarities in either design characteristics (i.e. size, shape,

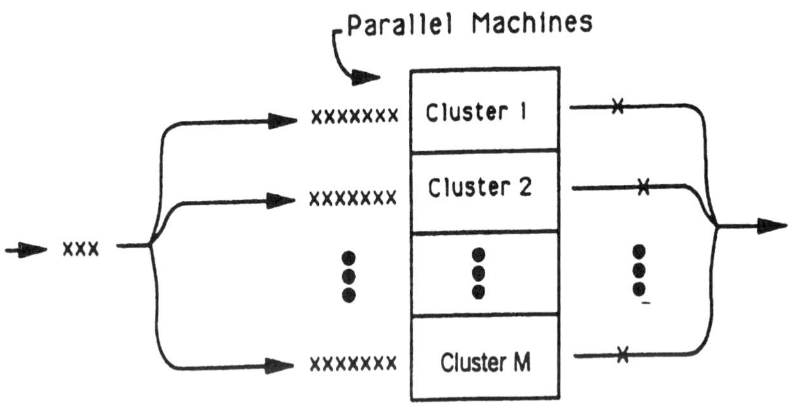

Figure 5: Clustering in Parellel Machines.

and functions), or manufacturing characteristics (i.e., type of sequence of operations required) are grouped or clustered into part families. Of course, design characteristics and manufacturing characteristics are often correlated. In classical manufacturing systems there exist some heuristic methods, such as ranking method and bounded energy method for modeling these systems.

Kusiak (1992) discuss the advantages of the group technology (GT) for the reduction of the lead time, work-in-process, labour, tooling, rework and scrap materials, setup time, delivery time, and paper works, in manufacturing systems. He stated that GT can be applied to manufacturing systems in two ways as follows.

1. **Logical:** where the machines are dedicated to part families but their positions in a factory are not altered.

2. **Physical:** where the different machines are grouped in cells, and each cell is dedicated to part families to exploit flow shop efficiency.

He emphasizes the capability of the cluster analysis for real world application which is more practical and simple for GT, and classify this approach as clustering by using matrix formulation and clustering by using integer programming formulation. He presents two algorithms for clustering for machine-part problems called: cluster identification algorithm, and branching algorithm. GT also is discussed by McAuley (1972), King (1980), Chan and Milner (1982), Kusiak (1985), Askin and Subramanian (1987), Kusiak (1990), Kusiak and Cheng (1990).

However, most of these methods are not applicable in real world problems and are very time consuming. Moreover, the methods based on hard clustering are not able to consider the overlaps between different clusters or groups, while in real world problems these overlaps often exist, hence the real models are fuzzy.

15.8.3 LOT SIZING

Lot sizing, where it is necessary to sequence a number of jobs, which are the same or similar, in a group that has zero or minimum set up costs between the members of the batches, is a kind of clustering method. In production and inventory management, where the main goal is minimizing the set up costs, sum of ordering costs, and holding costs, lot sizing plays an important roll. Some classical methods of lot sizing are lot-for-lot ordering, economic order quantity, fixed-period ordering, part period ordering, and Wagner-Within (1958) method.

15.9 CONCLUSIONS

In this survey, first we described the classical methods of production planing and scheduling and application of Artificial Intelligence (AI) in this area. Then, we discussed the application of fuzzy sets theory and fuzzy logic in production planning, inventory control, and scheduling. Moreover, we reviewed the well known AI and Fuzzy Artificial Intelligence (FAI) methods which have been developed for production planning and scheduling. The study shows that although the classical models provide valuable insights into the behavior of the systems under specific assumptions and identify useful guidelines, they have serious difficulties for implementation, in real world industries. These models which are based on classical mathematics pose the following drawbacks:

1. Most of the models are not understandable by the managers.
2. They are too complex.
3. They need the exact knowledge which is not often available.
4. They only consider one or two criteria and are implemented in small scale problems.
5. Most of them are not flexible.
6. The complexity of these systems are too high and most of them are located in NP-hard class of problems.

In contrast, Computational Intelligence models which are based on *IF-THEN* rules and are more similar to the common sense logic, and are more acceptable and understandable for the researchers and managers. For this reason, during the past two decades many researchers and industries have concentrated on developing models, systems, and packages based on Computational Intelligence for production planning and scheduling. The survey also showed that fuzzy sets theory in cooperation of Computational Intelligence can help production planners and schedulers for the development of more appropriate models and systems for real world production planning and scheduling. Fuzzy models developed for this purpose have demonstrated several advantages as follows:

1. They are not as complex as classical models and are more understandable by the managers due to the linguistic and qualitative nature of fuzzy rules.
2. They are suitable for solving complex problems.
3. They are very flexible and can support Just-In-Time (JIT) systems.
4. They can improve the quality of the production planning and scheduling activities.
5. They can use imprecise or semantically uncertain information.
6. Since they are able to combine with the conventional models, the cost of improving the systems by fuzzy systems are marginal.

It is also found that fuzzy clustering methods are more appropriate for aggregate production planning and control, and scheduling. In particular, there are advantages in the analysis of current operating systems, and their diagnosis. They can also be used in prediction and forecasting of operations under current operating rules.

15.10 REFERENCES

Anderberg, M.R., 1973, *Cluster Analysis for Applications*, Academic Press, Inc, New York.

Askin, R., and Subramanian, S., 1987, A cost based heuristic for group technology

Backer, E., 1978, *Cluster Analysis by Optimal Decomposition of Induced Fuzzy Sets*, Delft University Press, Delft.

Backer, E., 1995, *Computer-assisted Reasoning in Cluster Analysis*, Prentice Hall.

Baker, K.R., 1974, *Introduction to Sequencing and Scheduling*, Wiley, New York.

Baker, K.R., 1995, *Elements of Sequencing and Scheduling*, Amos Tuck School of Business Administration, Dartmouth College, Hanover.

Ball, G.H., and Hall, D.J., 1964, Some fundamental concepts and synthesis procedures for pattern recognition preprocessors, *International Conference on Microwaves, Circuit Theory, and Information Theory*, September, Tokyo.

Barone, J., Filev, D.P., and Yager, R.R., 1993, Mountain method-based fuzzy clustering: Methodological considerations, *Technical Report # MII-1320*, Machine Intelligent Institute, Iona College, New Rochelle, NY.

Bauman, E., Dorofeyuk, E., and Filev, D., 1990, *Fuzzy Identification of Nonlinear Dynamical Systems*, Proceedings International Conference on Fuzzy Logic and Neural Nets, Japan.

Bellman, R., and Zadeh, L.A., 1970, Decision making in a fuzzy environment, *Management Science*, Vol. 17, pp. 141-164.

Ben-Arieh, D., 1986, Knowledge-based control system for automated production and assembly, in: A. Kusiak (ed), *Modeling and Design of Flexible Manufacturing Systems*, Elsevier, New York, pp. 347-368.

Bensana, E., Bel, G., and Dubois, D., 1988, OPAL: A multi-knowledge-based system for industrial job-shop scheduling, *Inter. J. Product. Res.*, Vol. 26, pp. 795-819.

Bestwick, P.F., and Hastings, N.A.J., 1976, A new bound for machine scheduling, *Opl. Res. Q.*, Vol. 27, p 479.

Bezdek, J. C., 1981, *Pattern Recognition with Fuzzy Objective Function Algorithms*, Plenum Press, New York.

Bezdek, J. C., 1973, *Fuzzy Mathematics in Pattern Classification*, Ph.D. Thesis, Applied Math. Center, Cornell University, Ithaca.

Bezdek, J.C., 1974, Numerical Taxonomy with Fuzzy Sets.

Bezdek, J.C., Coray, C., Gunderson, R., and Watson, J., 1981, Detection and Characterization of Cluster Substructure. I. Linear Structure: Fuzzy c-Lines, *SIAM J. Appl. Math.*, Vol. 40.

Bezdek, P.C., and Pal, S.K., 1992, *Fuzzy Models for Pattern Recognition*, Plenum Press, New York.

Bonney, M.C., and Gundry, S.W., 1976, Solutions to the constrained flow shop sequencing problem, *Operational Research Quarterly*, Vol. 27, pp. 869-883.

Bruno, B., Elia, A., and Laface, P., 1986, A rule-based system to schedule production, *IEEE Computer*, Vol. 19, No. 7, pp. 32-40.

Campbell, H.G., Dudeck, R.A., and Smith, M.L., 1970, A heuristic algorithm for the n-job, M-machine sequencing problem, *Management Science*, Vol. 16/B, pp. 630-637.

Chan, H.M., and Milner, D.A., 1982, Discrete clustering algorithm for group formation in cellular manufacturing, *Journal of Manufacturing Systems*, Vol. 1, No. 1, pp. 65-74.

Coffman, E.G. (ed.), 1975, *Computer and Job Shop Scheduling Theory*, Englewood Cliffs, New Jersey, Prentice Hall.

Coffman, E.G., 1976, *Computer and Job Shop Scheduling Theory*, Wiley, New York.

Conway R., Maxwell, W., and Miller, L., 1967, *Theory of Scheduling*, Addition-Weseley, Reading, MA.

Cormack, R.M., 1971, A review of classification, *Journal of the Royal Statistical Society*, Seties A, Vol. 134, pp. 321-367.

Custodio, L.M.M., Sentierio, J.J.S., and Bispo, C.F.G., 1994, Production Planning and Scheduling using a Fuzzy Decision System, *IEEE Transactions Robotics and Automation*, Vol. 10, No. 2, pp. 160-168.

Day, N.E., 1969, Estimating the components of a mixture of two normal distributions, *Biometrika*, Vol. 56, pp. 463-474.

De, S., and Whinston, A.B., 1986, A framework for integrated problem solving in manufacturing, *IIE Transactions*, September, pp. 286-297.

Deluca, A., and Termini, S., 1972, A Definition of a Nonprobabilistic Entropy in the Setting of Fuzzy Sets Theory, *Infor. Control*, Vol. 20, pp. 301-312.

Devijver, P.A., and Kittler, J., 1982, *Pattern Recognition: a Statistical Approach*, Prentice Hall, Englewood Cliffs, NJ.

Dorn, J., and Froeschl, K.A., 1993, *Scheduling of Production Processes*, Ellis Horwood Series in Artificial Intelligence, Ellis Horwood Limited.

Dorn, J., and Slany, W., 1994, A Flow Shop with Compatibility Constraints in a Steelmaking Plant, in *Intelligent Scheduling*, Zweben, M., and Fox, M., (ed.), pp. 629-654.

Dubes, R., and Jain, A.K., 1980, Clustering methodologies in exploratory data analysis, In *Advances in Computers*, Vol. 19 (M.C. Yovits, ed.), Academic Press, Inc., New York, pp. 113-215.

Dubois, D., and Prade, H., 1985, A review of fuzzy sets aggregation connectives, *Information Sciences*, Vol. 36, pp. 85-121.

Dubois, D., Fargier, H., and Parade, H., 1995, Fuzzy constraints in Job-Shop scheduling, *Journal of Intelligent Manufacturing*, Vol. 6, No. 4, pp. 215-234.

Duda, R.O., and Hart, P.E., 1973, *Pattern Classification and Scene Analysis*, John Wiley & Sons, Inc., New York.

Dudek, R.A, Panwalkar, S.S., and Smith, M.L., 1992, The lessons of Flowshop Scheduling Research, *Opns. Res.*, Vol. 40, pp 7-13.

Elmaghraby, S.E., 1968, The machine scheduling problem-review and extensions, *Naval Research Logistics Quarterly*, Vol. 15, pp. 205-232.

Emami, M.R., Tűrkşen, I.B., and Goldenberg, A.A., 1996, Fuzzy Clustering in Fuzzy Modeling, *IEEE Transactions on Fuzzy Systems*, (submitted).

Emami, M.R., Tűrkşen, I.B., and Goldenberg, A.A., 1996, A Unified Parameterized Formulation of Reasoning Process in Fuzzy Modeling and Control, *Fuzzy Sets and Systems*, (submitted).

Engelke, H., Grotrian, J., Scheuing, C., Schmackpfeffer, A., Schwarz, W., Soft, B., and Tomann, J., 1985, Integrated manufacturing modeling system, *IBM Journal of Research and Development*, Vol. 29, No. 4, pp. 343-354.

Everitt, B.S., 1974, *Cluster Analysis*, John Wiley & Sons, New York.

Everitt, B.S., 1979, Unresolved problems in cluster analysis, *Biometrics*, Vol. 35, pp. 169-181.

Feigenbaum, E.A., 1982, *Knowledge engineering for the 1980's*, Computer Science Department, Stanford University.

Fox, M.S., and Mostow, D.J., 1977, Maximal Consistent Interpretations of Errorful Data in Hierarchically Modeled Domains, *Fifth International Joint Conference on Artificial Intelligence*, Karlsruhe, West Germany.

Fox, M.S., 1979, On Inheritance in Knowledge Representation, *Proceedings of the Sixth International Conference on Artificial Intelligence*, Tokyo, Japan.

Fox, M.S., 1981, The Intelligent Management System: An Overview, Technical Report, Robotics Institute, Carnegie-Mellon, Pittsburg, PA.

Fox, M.S., Lowenfeld S., and Kleinosky, P., 1983, Techniques for Sensor Based Diagnosis, *Proceedings of the International Joint Conference on Artificial Intelligence*, Karlsruhe, West Germany.

Fox, M.S., 1983, *Constraint-directed Search: A Case Study of Job-shop Scheduling*, Ph.D. thesis, Carnegie-Mellon University, Pittsburg, PA, December, Technical Report CMU-R1-TR-83-22.

Fox, M.S., and Smith, S.F., 1984, ISIS: A knowledge based system for factory scheduling, *Expert Systems Journal*, Vol.1, No.1, pp. 25-49.

Fox, M.S., 1984, Artificial intelligence in the factory of the future, *Proceedings of the ACM 12th annual Computer Science Conference*, Philadelphia, pp. 29-38.

Fox, M.S., 1986, Industrial Applications of Artificial Intelligence, *Robotics*, Vol. 2, pp. 301-311.

French, S., 1982, *Sequencing and Scheduling: An introduction to the mathematics of the Job Shop*, Ellis Horwood, Chichester.

Friedman, H.P., and Rubin, J., 1967, On some invariant criteria for grouping data, *Journal of the American Statistical Association 62*, pp. 1159-1178.

Fu, K.S., 1982, *Syntactic Pattern Recognition and Application*, Prentice Hall, Englewood Cliffs, NJ.

Fukanaga, K., and Narendra, P.M., 1975, A branch and bound algorithm for computing k-nearest neighbors, *IEEE Transactions on Computers C 24*, pp. 750-753.

Fukunyama, Y., and Sugeno, M., 1989, A new method of choosing the number of clusters for the fuzzy *c*-Means method, in *Proc. 5th Fuzzy Systems Symposium* (in Japanese), pp. 247-250.

Gabbay, H., 1979, Multistage production planning, *Management Science*, Vol. 25, No. 11, pp. 1138-1148.

Gaines, B.R., 1987, Expert system in integrated manufacturing: Structure, development and applications, in: A. Kusiak (ed), *Artificial Intelligence: Computer Integrated Manufacturing*, IFS, Kempston, Bedford, UK and Springer, New York.

Garey, M.R., and Johnson, D.S., 1979, Computers and Intractability: A Guide to the Theory of NP-Completeness, Bell Telephone Laboratories, U.S.A.

Gilmore, P.C., and Gomory, R.E., 1964, Sequencing a one-state variable machine: a solvable case of the travelling salesman problem. *Operation Research*, Vol. 12, pp. 655-679.

Gordon, A.D., and Henderson, J.T., 1977, Algorithm for Euclidean sum of squares classification, *Biometrics*, Vol. 33, pp. 355-362.

Grabot, B., and Geneste, L., 1994, Dispatching rules in scheduling: a fuzzy approach, *International Journal of Production Research*, Vol. 32, No. 4, pp. 903-915.

Gunderson, R., 1978, Application of Fuzzy ISODATA Algorithm to Star Trecher Pointing Systems, in: *Proc. 7th Triennial World IFAC Congress*, Helsinki, pp. 1319-1323.

Gupta, J.N.D., 1971, A functional heuristic algorithm for the flow shop scheduling problem, *Operational Research Quarterly*, Vol. 22, pp. 39-47.

Gupta, M.M., and Kaufmann, A., 1988, *Fuzzy Mathematical Models in Engineering and Management Science*, North Holland Publishers: Amsterdam.

Hall, N.G., and Sriskandarajah, C., 1996, A Survey of Machine Scheduling Problems with Blocking and No-wait in Process, *Operations Research*, Vol. 44, No. 3, pp. 510-525.

Hall, N.G., Kamoun, H., and Sriskandarajah, C., 1997, Scheduling in Robotic Cells: Classification, Two and Three Machine Cells, *Operations Research*, Vol. 45, No. 3, pp. 421-439.

Halpin, S.M., and Noah, W.W., 1986, Adaptive user interface for planning and decision aids in C^3 I systems, *IEEE Transactions on Systems, Man, and Cybernetics*, Vol. 16, No. 6, pp. 909-918.

Han, S., Ishii, H., and Fujii, S., 1994, One machine scheduling problem with fuzzy due dates, *European Journal of Operational Research*, Vol. 79, No. 1, pp. 1-12.

Hartigan, J.A., 1975, *Clustering algorithms*, John Wiley, New York.

Hintz, G.W., and Zimmermann, H.J., 1989, A method to control flexible manufacturing systems, *EJOR*, pp. 321-334.

Holmbald, L.P., and Ostergaard, J.J., 1982, *Control of a Cement Kiln by Fuzzy Logic*, in: M. Gupta and E.Sanchez, Eds, Fuzzy Information and Decision Process, North Holand, Amsterdam.

Holsapple, C.W., and Whinston, A.B., 1986, *Manager Guide Expert Systems*,

Dow Jones-Irwin, Homewood, IL.

Holt, C.C., Modigliani, F., and Simon, H., 1955, A linear decision rule for production and employment scheduling, *Management Sci.*, Vol.2, pp. 1-30.

Holtz, M., and Desonki, 1981, Fuzzy model fur instanhaltung, *Unscharfe Modellbildung und Steuerung*, IV, 54-62, Karl-Marx-Stadt.

Hou, E., Ansari, N., and Ren, H., 1994, A genetic algorithm for multiprocessor scheduling, *IEEE Trans. on Parallel and Distributed Systems*, Vol. 5, No. 2, pp. 113-120.

Hung, T.P., Huang, C.H., and Yu, K.M., 1995, A Fuzzy LPT Algorithm for Scheduling, *IEEE*, pp. 2588-2592.

Hundal, T.S., and Rajgopal, J., 1988, An extension of Palmer's heuristic for flow shop scheduling problem, *International Journal of Production Research*, Vol. 26, pp. 1119-1124.

Ishibuchi, H., Yamamoto, N., Misaki, S., and Tanaka, H., 1994, Local search algorithms for flow shop scheduling with fuzzy due dates, *International Journal of Production Economics*, Vol. 33, No. 1-3, pp. 53-66.

Jain, A.K., and Dubes, R.C., 1988, *Algorithms for Clustering Data*, Prentice Hall, Inc., Englewood Cliffs, New Jersey.

Jang, J.S.R., and Gulley, N., 1995, *Fuzzy Logic Toolbox for Use with MATLAB*, The MathWorks, Inc., U.S.A.

Johnson, S.M., 1954, Optimal two-and three-stage production schedules with set up times included, *Naval Research logistics Quarterly*, Vol. 1, pp. 61-68.

Kamoun, H., Hall, N.G., and Sriskandarajah, C., 1993, Scheduling in Robotic Cells: Heuristics and Cell Design, *Working Paper # 93-08*, Department of Industrial Engineering, University of Toronto.

Kandel, A., 1986, *Fuzzy Mathematical Techniques with Application*, Addison-Wesley, Reading, MA.

Kaufmann, A., 1973, *Theory of Fuzzy Subsets*, Academic Press, New York.

Kempf, K.G., 1985, Manufacturing and artificial intelligence, *Robotics*, Vol. 1, No. 1. pp. 3-12.

Kickert, W.J.M., 1978, *Fuzzy Theories on Decision-making*, Leiden, Boston, London.

King, J.R., 1980, Machine-Component Group Formation in Production Flow Analysis, *International Journal of Production Research*, Vol. 18, No. 2, pp. 213-232.

King, J.R., and Spachis, 1980, Heuristics for flow shop scheduling, *International Journal of Production Research*, Vol. 18, pp. 345-357.

Kumara, S.R.T., Joshi, S., Kashyap, R.L., Moodi, C.L., and Chang, T.C., 1986, Expert systems in industrial engineering, *International Journal of Production Research*, Vol. 24, No. 5, pp. 1107-1125.

Kusiak, A., 1985, The part families problem in flexible manufacturing systems, *Annals of Operations Research*, Vol. 3, pp. 279-300.

Kusiak, A., and Chen, M., 1988, Expert systems for planning and scheduling manufacturing systems, *European Journal of Operational Research*, No. 34, pp. 113-130.

Kusiak, A., 1990, *Intelligent Manufacturing Systems*, Prentice-Hall, Englewood Cliffs, NJ.

Kusiak, A., and Cheng, C-H., 1990, A Branch and Bound Algorithm for Solving the Group Technology Problem, *Journal of Manufacturing Systems*, Vol. 6, No. 2, pp. 117-124.

Kusiak, A., 1992, Group Technology, in: Kusiak, A. (ed.), *Intelligent Design and Manufacturing*, John Wiley & Sons Inc., New York, pp. 289-302.

Lance, G.N., and Williams, W.T., 1967, A general theory of classificatory sorting strategies, *Computer Journal*, Vol. 10, pp. 271-277.

Lee, E.S., and Li, R.J., 1988, Comparison of fuzzy numbers based on the probability measure of fuzzy events, *Computers and Math. with Applications*, Vol. 15, pp. 887-896.

Lei, L., and Wang, T.J., 1994, Determining Optimal Cyclic Hoist Schedules in a Single-Hoist Electroplating Line, *IIE Transaction*, Vol. 26, No. 2, pp. 25-33.

Lenstra, J.K., Rinnooy Kan, A.H.G., and Brucker, P., 1977, Complexity of machine scheduling problems, *Ann. Discrete Math*, Vol. p. 343.

Lenstra, J.K., 1977, *Sequencing By Enumerative Methods*, Mathematisch Centrum, Amsterdam.

Mamdani, E.H., and Assilian, S., 1974, Application of fuzzy algorithms for control of simple dynamic plant, *Proc. Inst, Elec, Eng.*, Vol. 121, pp. 1585-1588.

Mamdani, A., 1977, Applications of the fuzzy set theory to control systems: a survey, in: M. Gupta and E. Sanchez, Eds., *Fuzzy Information and Decision Processes*, North-Holand, Amsterdam.

Matsuo, H., Shang, J.S., and Sullivan, R.S., 1991, A Crane Scheduling in a Computer Integrated Manufacturing Environment, *Management Science*, Vol. 37, pp. 587-606.

McAuley, J., 1972, Machine Grouping for Efficient Production, *Production Engineer*, February, pp. 53-57.

McCahon, C.S., and Lee, E.S., 1992, Fuzzy Job Sequencing for a Flowshop, *EJOR*, Vol. 62, pp. 294-301.

McQueen, J.B., 1967, Some methods of classification and analysis of multivariate observations, *Proceedings of Fifth Berkeley Symposium on Mathematical Statistics and Probability*, pp. 281-297.

Michalski, R.S., 1980, Knowledge acquisition through conceptual clustering: theoretical framework and algorithm for partitioning data into conjunctive concepts, *Int. Journal of Policy Analysis and Information Systems*, Vol. 4.

Miller, R.K., 1984, *Artificial Intelligence Applications for Manufacturing*, SEAI Technical Publications, Madison, GA.

Mitchie, D., 1980, Expert Systems, *Computer Journal*, Vol. 23, pp. 369-376.

Morton, T.E., and Smunt, T.L., 1986, A planning and scheduling system for flexible manufacturing, in A Kusiak (ed), *Flexible Manufacturing System: Method and Studies*, North-Holland, Amsterdam, pp. 151-164.

Morton, T.E., and Pentico, 1993, *Heuristic Scheduling System: with Application to Production Systems and Project Management*, Wiley Inc, New York.

Nakanishi, H., Tűrkşen, I.B., and Sugeno, M., 1993 A Review and Comparison of Six Reasoning Methods, *Fuzzy Sets and Systems*, Vol. 57, No. 3, pp. 257-294.

Nardi, B., 1985, Westinghouse unveils ISIS-II factory order system, *Intellinews*, Vol. 1, No. 4, pp. 1-2.

Naylor, A.W., and Maletz, M.C., 1986, The manufacturing game: A formal approach to manufacturing software, *IEEE Transactions on Systems, Man and Cybernetics*, Vol. 16, No. 3, pp. 321-334.

Newman, P.A., and Kempf, K.G., 1985, Opportunistic scheduling for robotic machine tending, *The second Conference on Artificial Intelligence Applications*, Miami Beach, FL, Dec. 11-13, pp.168-173.

Newell, A, and Simon, H.A., 1956, The Logic Theory Machine: A Complex Information Processing System, *IEEE Transactions on Information Theory*, IT-2, No. 3, pp. 61-79.

Newell, A., and Simon, H.A., 1963, GPS: A program that simulates human thought, in: Computers and Thought, E. Feigenbaum and j. Feldman (Eds), New York, McGraw-Hill Co.

O'Connor, D.E., 1984, Using expert systems to manage change and complexity in manufacturing, in: W. Reitman (ed.), *Artificial Intelligence Applications for Business*, Ablex Norwood, NJ, pp. 149-158.

Orsini, J-E., 1986, Artificial Intelligence: A way through the strategic planning crisis, *Long Range Planning*, Vol. 19, No. 4, pp. 71-77.

Page, E.S., 1961, An approach to the scheduling of jobs on machines, *Machine Scheduling Problems*, Martinus Nijhoff, The Hauge.

Pal, N.R., and Bezdek, J.C., 1994, On Cluster Validity for the Fuzzy *c*-Means Model, *IEEE Trans. Fuzzy Systems* (submitted).

Palmer, D.S., 1965, Sequencing jobs through a multi-stage process in the minimum total time - A quick method of obtaining a near optimal, *Operational Research Quarterly*, Vol. 16, pp. 101-107.

Papadimitriou, C.H., and Kanellakis, P.C., 1980, Flowshop scheduling with limited temporary storage, *J. Assoc. Comput. Mach.*, Vol. 27, pp. 533-549.

Papas, P., 1986, ISIS: A project in review, *Symposium on Real Time Optimization in Automated Manufacturing Facilities*, National Bureau of Standards, Gaithersburg, MD, January.

Parunak, H.V.D., 1987, Distributed AI Systems, in: A. Kusiak (ed.), *Artificial intelligence: Computer Integrated Manufacture*, IFS, Kempston, Bedford, UK and Springer, New York.

Petrovic, D., and Pupin, S., 1994, Fuzzy knowledge-based approach to treating uncertainty in inventory control, *Computer Integrated Manufacturing Systems*, Vol. 7, No. 3, pp. 147-152.

Pinado, M., 1995, *Scheduling: Theory, Algorithms, and Systems*, Prentice Hall, Englewood Cliffs, New Jersey.

Rayson, P.T., 1985, A review of expert systems principles and their role in manufacturing systems, *Robotica*, Vol. 3, pp. 279-287.

Rinks, D.B., 1982, A heuristic approach to aggregate production scheduling using linguistic variables Methodology and applications, in: Yager, R.R., Ed., *Fuzzy Sets and Possibility Theory: Recent Developments*, Pergamon Press, New York, pp. 562-581.

Rinnooy Kan, A.H.G., 1976, *Machine Scheduling Problems: Classification, Complexity and Computations*, Nijhoff, The Hague.

Rock, H., 1984, The three machine no-wait flow shop problem is NP-hard, *J. Assoc. Comput. Mach.*, Vol. 31, pp. 336-345.

Rock, H., 1984, Some new results in flow shop scheduling, *Z. Oper. Res.*, Vol. 28, pp. 1-16.

Romesburg, H.C., 1984, *Cluster Analysis for Researchers*, Lifetime Learning Publications, Belmont, CA.

Rouse, W.B., 1972, Cognitive Source of Suboptimal Human Prediction, Ph.D. Thesis, Massachusetts Institute of Technology.

Ruspini, H., 1969, A new approach to clustering, *Information and Control*, Vol. 15, pp. 22-32.

Ruspini, H., 1970, Numerical Methods for Fuzzy Clustering, *Inf. Sci.*, Vol. 2, pp. 319-350.

Sahni, S., and Cho, Y., 1979, Complexity of Scheduling Shops with No-wait In Process, *Math. O.R.*, Vol. 4, pp. 448-457.

Selim, S.Z., and Ismail, M.A., 1984, K-mean-type algorithm: a generalized convergence theorem and characterization of local optimality, *IEEE Transactions on Pattern Analysis and Machine Intelligence PAMI 6*, pp. 81-87.

Shannon, C.E., 1948, A Mathematical Theory of Communication, *Bell Syst. Tech. J.*, Vol. XXVII-3, pp. 379-423.

Shaw, M.J.P., and Whinston, A.B., 1985, Task bidding and distributed planning in flexible manufacturing, *The Second Conference on Artificial Intelligence Applications*, Miami, Beach, FL, December 11-13, pp. 184-189.

Shaw, M.J.P., and Whinston, A.B., 1986, Application of artificial intelligence to planning and scheduling in flexible Manufacturing, in A. Kusiak (ed.), *Flexible Manufacturing Systems: Methods and Studies*, North-Holland, Amsterdam, pp. 223-242.

Slany, W., 1994, *Fuzzy Scheduling*, Ph.D. thesis, Technical University Vienna, Vienna.

Sneath, P.H.A., and Sokal, R.R., 1973, *Numerical Taxonomy*, Freeman and Co., San Francisco.

Sriskandarajah, C., and Ladel, 1986, Some no-wait shops scheduling problems: complexity aspects, *European Jour. of Operational Research*, Vol. 24, pp. 424-438.

Sriskandarajah, C., and Goyal, S.K., 1988, No-wait Shops Scheduling: Computational Complexity and Approximate Algorithm, *OPSEARCH*, Vol. 25, No. 4, pp. 220-224.

Sriskandarajah, C., 1993, Performance of Scheduling Algorithm for No-wait Flowshops with Parallel machines, *European Journal of Operational Research*, Vol. 70, pp. 365-378.

Steffen, M.S., 1986, A survey of artificial intelligence-based scheduling systems, *Proceedings of Fall Industrial Engineering Conference*, Dec. 7-10, Boston, MA, pp. 395-405.

Stevenson, W.J., 1995, *Production/Operation Management*, Fifth Edition, IRWIN.

Subramanyam, S., and Askin, R.G., 1986, An expert system approach to scheduling in flexible manufacturing systems, in: A. Kusiak (ed.), *Flexible Manufacturing Systems: Methods and Studies*, North-Holland, Amsterdam, pp. 243-256.

Sugeno, M., and Yasukawa, T., 1993, A Fuzzy Logic Based Approach to Qualitative Modeling, *IEEE Transactions On Fuzzy Systems*, Vol., 1, No., 1, pp. 7-31.

Symon, M.J., 1981, Clustering criteria and multivariate normal mixture, *Biometrics*, Vol. 37, pp. 35-43.

Takada, T., Sanou, K, and Fukumara, S., 1995, A neural network system for solving an assortment problem in the steel industry, *Annals of Operations Research*, Vol. 57, pp. 265-281.

Takahashi, T., Kawasaki, S., Saegusa, M., Tamura, S., and Ebina, K., 1992, Scheduling for steelmaking process using a mathematical programming method, *KOBELCO Technology Review*, No. 13, pp. 13-16.

Terano, T., Asai, K., and Sugeno, M., *Fuzzy Systems Theory and its Applications*, Academic Press, San Diego, CA.

Thesen, A., and Lei, L., 1986, An expert system for scheduling robot in a flexible electroplating system with dynamically changing work loads, in: K.E., Stecke and R. Suri (eds.), *Flexible Manufacturing Systems: Operations Research Models and Applications*, Elsevier, New York, pp. 555-566.

Titterington, D.M., Smith, A.F.M., and Quarterly, U.E., 1985, *Statistical Analysis of Finite Mixture Distributions*, John & Sons. Inc., New York.

Tukey, J.W., 1977, *Exploratory Data Analysis*, Addison-Wesley, Reading, MA.

Tűrkşen, I.B., and Yao, D.D.W., 1984, Representation of Connectives in Fuzzy Reasoning, The View Through Normal Forms, *IEEE-SMC*, Vol. 14, No. 1, pp. 146-151.

Tűrkşen, I.B., 1985, Fuzzy sets and systems and their applications in production research, in: Bullinger, H.J. and Warnecke, H.J., (eds.), *Toward the Factory of the Future*, Springer, Berlin-New York, pp. 649-656.

Tűrkşen, I.B., 1988, Approximate Reasoning for Production Planning, *Fuzzy Sets and Systems*, Vol. 26, pp. 23-37.

Tűrkşen, I.B., and Berg, M., 1991, An Expert System Prototype for Inventory Capacity Planning: An Approximate Reasoning Approach, *International Journal of Approximate Reasoning*, Vol. 5, pp. 223-250.

Tűrkşen, I.B., and Zhong, Z., 1990, An Approximate Analogical Reasoning Schema based on Similarity Measures and Interval-valued Fuzzy Sets, *Fuzzy Sets and Systems*, Vol. 34, pp. 323-346.

Tűrkşen, I.B., 1989, Four Methods of Approximate Reasoning with Interval-Valued Fuzzy Sets, *International Journal of Approximate Reasoning*, Vol. 3. pp. 121-142.

Tűrkşen, I.B., and Lucas, C., 1991, A pattern matching inference model and its comparison with inference methods, *Proceedings of IFSA 91*, Brussels Belgium, July 7-12, pp. 231-234.

Tűrkşen, I.B., Ulguray, D., and Wang, Q., 1992, Hierarchical Scheduling based on Approximate Reasoning - A Comparison with ISIS, *Fuzzy Sets and Systems*, Vol. 46, pp. 349-371.

Tűrkşen, I.B., 1992, Fuzzy Expert Systems for IE/OR/MS, *Fuzzy Sets and Systems*, Vol. 51, pp. 1-27.

Tűrkşen, I.B., 1992, Interval Valued Fuzzy Sets and 'Compensatory And', *Fuzzy Sets and Systems*, Vol. 51, pp. 295-307.

Tűrkşen, I.B., 1993, Interval Valued Fuzzy Sets and Fuzzy Connectives, *Journal of Interval Computations*, Vol. 4, pp. 125-142.

Tűrkşen, I.B., 1996, Non-Specificity and Interval Valued Fuzzy Sets, (*Special Issue) Fuzzy Sets and Systems*, Vol. 80, pp. 87-100.

Tűrkşen, I.B., 1995, Type I and Interval-Valued Type II Fuzzy Sets and Logics, P.P. Wang (ed), *Advances in Fuzzy Theory and Technology*, Vol. 3, pp. 31-81.

Tűrkşen, I.B., 1996, Computational Intelligence: Extended Truth Tables and Fuzzy Normal Forms, *Proceedings of NATO ASI on Computational Intelligence*, Antalya, August 26 - September 4, to appear.

Tűrkşen, I.B., 1997, Type I and Type II Fuzzy System Modeling, *Fuzzy Sets and Systems: Special Issue on scheduling*, Vol. 19, 119-143.

Tűrkşen, I.B., and Fazel Zarandi, M.H., 1998, Fuzzy System Models For Aggregate Scheduling Analysis, *International Journal of Approximate Reasoning: Special Issue on Scheduling*, Vol. 19, 119-143.

Wagner, H.M., and Whitin, T.M., 1958, Dynamic Version of Economic Lot Size Model, *Management Science*, Vol., 5, pp. 89-96.

Whalen, T., and Schott, B., 1983, Decision support with fuzzy production systems, in P.P. Wang, (ed), *Advances in Fuzzy Set Theory and Applications*, pp. 199-216, Plenum Press, New York.

Whalen, T., and Schott, B., 1985, Alternative logics for approximate reasoning in expert systems: a comparative study, *International J. Man-Machine Stud.*, Vol. 22, pp. 327-346.

Wolfe, J.H., 1970, Pattern clustering by multivariate mixture analysis, *Multivariate Behavioral Research*, Vol. 5, pp. 329-350.

Xie, X.L., and Beni, G.A., 1991, Validity Measure for Fuzzy Clustering, IEEE Trans. PAMI, Vol. 3, No. 8, pp. 841-846.

Yager, R.R., and Filev, D.P., 1994, *Essentials of Fuzzy Modeling and Control*, John Wiley & Sons, Inc.

Yao, D.D.W., and Tűrkşen, I.B., 1983, On Multi-Valued Controllers in Production Systems, *Proc. of ICPR*, Windsor, Ontario, August 22-24.

Yoshinari, Y., Pedrycz, W., and Hirota, K., 1993, Construction of fuzzy models through clustering techniques, *Fuzzy Sets and Systems*, Vol. 54, No. 2, pp. 157-165.

Yu, D.H., 1986, *A System for Utilizing Operations Research Techniques in Knowledge Based Planning System*, Technical report, The Mitre Corporation, Bedford, MA.

Zadeh, L.A., 1965, Fuzzy Sets, *Information and Control*, Vol. 8, pp. 338-353.

Zadeh, L.A., 1973, Outline of a new approach to the analysis of complex systems and decision processes, *IEEE Trans. Syst., Man, Cybern.*, Vol. 3, pp. 28-44.

Zadeh, L.A., 1977, A Theory of Approximate Reasoning, Memorandum No. UCB/ERLM77/58.

Zadeh, L.A., 1979, Fuzzy sets and information granularity, in *advances in Fuzzy Set Theory and Applications*, Gupta, M.M., Ragade, R.K. & Yayger, R.R. (eds.), Amsterdam: North Holand, pp. 3-18.

Zadeh, L.A., 1983, The role of fuzzy logic in the management of uncertainty in expert systems, *Fuzzy Sets and Systems*, Vol. 11, pp. 199-227.

Zadeh, L.A., 1992, Fuzzy Sets and Fuzzy Logic: An overview. In Stuart C. Shapiro, editor, *Encyclopedia of Artificial Intelligence*, John Wiley & Sons, Inc. Vol. 1, pp. 507-508.

Zadeh, L.A., 1995, New Frontiers in Fuzzy Logic, *Proceedings of the Sixth International Fuzzy System Association World Congress*, Vol. 1, pp. 1-2.

Zimmermann, H.J., 1987, *Fuzzy Sets, Decision Making, and Expert Systems*, Kluwer: Dordrecht.

Zimmerman, H.J., 1991, *Fuzzy Set Theory and Its Application*, Kluwer Academic Publishers, Netherlands.

Zweben, M., and Fox, M.S., 1994, *Intelligent Scheduling*, Morgan Kaufmann Publishers, INC, San Francisco, California.

16 FUZZY SETS METHODOLOGIES IN ACTUARIAL SCIENCE

Richard A. Derrig
Krzysztof M. Ostaszewski

Abstract: Actuarial science encompasses all types of quantifications of risks under conditions of uncertainty for the purpose of setting insurance premiums and reserves. Traditional actuarial methodologies have been built upon probabilistic models, and are often driven by stringent regulation of the insurance business. Deregulation and global competition of the last two decades have opened the door for new methodologies, among them being fuzzy methods. Here we present the uses of fuzzy sets in areas such as: underwriting, risk classification, interest rates, ratemaking, valuation of premium and taxes. We also discuss new opportunities for expanding fuzzy sets methodologies in actuarial science.

16.1 WHY IS ACTUARIAL SCIENCE SO LATE IN JOINING THE FUZZY SCIENCE?

The area of actuarial science has been relatively late in joining the vast field of applications of fuzzy sets methodology. In order to explain why, we must first understand the nature of the actuarial problem. Actuaries are professionals who quantify risk for the purpose of deriving a price for insuring against uncertain events. One part of the work of an actuary is the evaluation of the premium collected by an insurance enterprise from its customers for the payments of future claims and benefits. Let us note that the same kind of problem applies to the gradual funding of any future liabilities, including pension liabilities, another field of activity for actuaries. The second major part of actuarial work is the establishment of reserves. Given that the future claim payments are uncertain, it is simply impossible for the premium payment stream of cash flows of an insurance enterprise to match exactly the stream of cash flows required to pay future claims. The difference in the present value of the two must either be positive, or receive a subsidy from the insurance firm in order to be positive. This statement is simply equivalent to the insurance firm being solvent. Reserves constitute that amount, which together with future premiums, will be sufficient for the discharge of all future claim and benefit obligations. Again, this concept extends to pension funding where it is referred to as accrued liabilities.

Actuaries face uncertainty due to unpredictability of various factors influencing future premiums, claims, benefits, and investments cash flows. Traditional actuarial science (e.g., Bowers et al., 1986) assumes deterministic interest rates and a probabilistic model for the distribution of future claims. In some countries, including the United States, this has lead to stringent regulation of the actual premium and reserve calculation process. Insurance is regulated at the state level in the U.S., with model national regulation written by the National Association of the Insurance Commissioners (Black and Skipper, 1994). This structure has produced a situation where the calculation of premiums and reserves for insurance is driven by legislation, sometimes specifically prescribing the methodology allowed. Indeed, until 1980, life insurance firms in the United States, for which the interest rate used in discounting future cash flows is of utmost importance, due to long term nature of their contracts, had to use one interest rate, prescribed by law effective at the inception of the contract. A change of interest rate would then require lobbying all state legislatures to change the law, not an easy task indeed. Since 1980, the interest rate effective for a given life insurance contract is a function of Moody's Investment Grade Corporate Bond Index, but still remains unchanged for the life of the contract. In addition to that, a large portion of the methodology is also prescribed by state statutes, including mortality tables. In the Commonwealth of Massachusetts premium rates for personal auto insurance are set by the Insurance Commissioner, based upon data and recommendations filed by the Auto Insurers Bureau (Derrig, 1993). These static methodologies developed in a world radically different economically from our current environment.

The Golden Age (Black and Skipper, 1994) of the U.S. insurers, the 1950s and 1960s, was characterized by nearly complete knowledge of claim related cash flows -- because of actuarial knowledge; and predictability of other cash flows (i.e., lapses, surrenders, new business, investment returns) -- because of an economic environment providing stability of those factors. One could say that Golden Age was the "quiet before the storm". Subsequent developments (Sametz, 1987), such as:

- Unprecedented levels of inflation, and unpredictability of inflation rate;
- Unprecedented levels of volatility of financial markets, especially interest rates;
- Unprecedented deregulation, consumerism, and competition;

all leading to greater efficiency in consumer behavior, disintermediation, and change in the industry position versus other financial institutions, resulted in the insurance industry experiencing what is common in those three factors, i.e., "the unprecedented". First and foremost was the unpredictability of cash flows, or even a complete makeover of the nature of those cash flows. For example (Tullis and Polkinghorn, 1992), annuities, which have been historically a relatively unimportant part of the life insurance industry, used primarily to provide an income stream after retirement, acquire new significance as savings vehicles through the use of single and flexible premium deferred annuities, and the recent extraordinary growth of variable annuities. In 1982, total annuity reserves of U.S. life companies exceeded life insurance reserves for the first time, and by the 1990s they reached twice the level of life reserves. The popularity of annuities and other investment-related products in the United States has been aided by the provisions of the Tax Reform Act of 1986 (Babbel and Stricker, 1987). Three major milestones in the recent history of the life insurance industry (Asay, Bouyoucos and Marciano, 1993) occurred:

- In the early 1980s, the short term interest rates were at the record highs, causing massive disintermediation as policyholders fled to higher yields;
- In the middle of the 1980s, there occurred a record decline in the level of nominal interest rates, resulting in refinancing and prepayments of a large portion of insurers portfolios;
- At the end of the 1980s, insurers pursuing higher yields were often caught taking too much credit risk in their investment portfolios.

The market nature of insurance products has changed as well. Ostaszewski (1998) points out that the historical *Paul v. Virginia* Supreme Court decision of 1867, which lead to state regulation of insurance, appeared to have been based on the perception of insurance as a private contract between two local parties (thus ... no interstate commerce in insurance, and no central federal government regulation). Even though there are no traded markets in insurance products, there has been a decisive move towards competitive pricing of insurance, with mortality protection becoming nearly a commodity, and with catastrophe futures markets under development.

The traditional vision of insurance and pension valuation in the United States, before the storms described here, called for nearly deterministic calculations based on the interest rate, mortality, and other methodology prescribed by legislation. The resulting role of an actuary was limited to efficient professional calculation following the guidelines. This was in striking contrast with other English speaking nations, such as United Kingdom, Canada, or Australia, which over time developed a different role for an actuary. The concept of Appointed Actuary, as created in those countries, called upon the actuarial professional to create a model based on assumptions and methodology created by that professional, in accordance with the standards of practice, but nevertheless a creation of the actuary, which was the basis for premium and insurance. This opening for creativity was badly needed in the United States.

It was finally provided by the 1990 Amendments to the Standard Valuation Law, which require that an Appointed Actuary in the United States not only should prove that the reserves and premiums of an insurance firm are calculated properly, using a legal methodology, but also that they indeed have proper economic meaning, i.e., are sufficient to discharge company's obligations to pay claims and benefits. As a result, it is now standard that insurance firms, and indeed other financial institutions, test their long term solvency (a process often referred to as *cash flow testing)*under a large number of economic scenarios of the future, mostly interest rates scenarios, but other factors are considered, too. In addition to that, new standards of practice in Canada call for the Appointed Actuary to provide Dynamic Solvency Testing analysis of the firm. This kind of testing not only investigates the development of the company under cash flow testing scenarios, but also provides sensitivity analysis, by inquiring about the economic value of the firm under changes to various input factors, such as interest rates, mortality, policy lapsation, epidemics, etc.

We see, therefore, that the traditional actuarial analysis, although based in the probabilistic methodology, was indeed quite immune to any invasion by fuzzy sets methodology. This was especially true of the situation in the United States, given the extreme form of inflexibility written into the pre-1980 Standard Valuation Law, and even its improved pre-1990 form.

It should then come as no surprise that the early applications of fuzzy sets methodology came from European scholars, and that the decade of 1990s is marked by a sudden increase in the interest in fuzzy sets applications in actuarial science.

16.2 UNDERWRITING

The earliest work known to us directly applying fuzzy sets methodology to actuarial science was by DeWit (1982) in which he pointed out that the process of insurance underwriting, i.e., the process of selection and evaluation of risks to be insured, is indeed fraught with uncertainty which may not be properly described by probability. For example, in group insurance, even though actual pricing of the insurance product is done based on concrete data, one cannot directly apply the conclusions to any group in which participation is voluntary. Traditionally, a minimum participation rate of 75% is a requirement for issuing a group policy, but this crisp boundary is by no means a definite cure for resolving the uncertainty of the insured participants. Furthermore, the age/sex composition of the insured group is generally needed to be either stable, or reach a steady-state eventually. What constitutes "stable" is subject to the underwriter's judgement, a fuzzy approach indeed. Other factors of similar nature include description of the industry, if the group is work-related, credit rating of the policyholder, and stability of the insurance provider (a group which changes the insurance carrier every year is generally a very poor risk). Finally, there is the classical actuarial problem of *credibility* (Fuhrer, 1993) of group experience. Typically a group seeking insurance will provide an insurer with information about its claims history. These data provide one basis for premium derivation. But is it prudent to base the pricing decision on such limited data? Should industry-wide, or country-wide experience be used instead? If the claims experience of the group is better, i.e., its payments are lower, than the corresponding industry or country, the group may not welcome such generalization. The standard approach to this problem is to assign a degree of credibility, a number between 0 and 1, to the premium derived based on the group data, and one minus that number to the premium derived from the industry or country experience. The weighted average of the two is then used as the premium for the group. One cannot avoid noticing a striking similarity between this process and the fuzzy sets methodology. Karwowski and Ostaszewski (1992, 1993) investigated development of fuzzy-based credibility measure in insurance.

The work of DeWit (1982) was followed by Erbach (1990, also see Erbach and Seah, 1993) who in 1987, together with his two colleagues, Douglas Holmes and Robert J. Purdy, working for a Canadian insurance firm, developed *Zeno*,a prototype life insurance automated underwriter using a mixture of fuzzy and other techniques. *Zeno* was intended to do final underwriting of the majority of individual life insurance cases. The prototype was carried far enough so that it proved feasible, at which point the company turned it over to the regular systems, and it was promptly abandoned in favor of traditional human judgement. One may indeed wonder if this was a proof of any weakness of fuzzy methodology, or a signal of the world not yet ready for innovation.

In the United States one area which begs for innovation in underwriting is nonmedical issues. Some small amount individual life products are issued without gathering any medical information about the prospects, just based on data in the simple application form. Better methods of distinguishing between acceptable and not acceptable risks in this area, which *Zeno* appeared to had been able to handle, could yield significant profits. We can only wait and see if such a breakthrough will indeed happen.

Lemaire (1990) expanded on the work of DeWit (1982) by suggesting fuzzy logic methodology for insurance underwriting in general, as well as fuzzy calculation of insurance premiums and reserves. The underwriting methodology underwent further refinement in the work of Young (1993), which published a specific algorithm for group health underwriting, utilizing fuzziness of rules such as mentioned above. Horgby, et.al. (1997) introduced fuzzy inference rules by generalized modus ponens as a means of underwriting mortality coverage for applicants with diabetes mellitus. Twenty seven medically related factors are represented as fuzzy input parameters to a fuzzy controller scheme with a center of area defuzzifyer to extract a crisp premium surcharge.

16.3 USING FUZZY ACTUARIAL PRESENT VALUES AND FUZZY ARITHMETIC

Lemaire's work (1990) included calculations of present values under fuzzy interest rates and fuzzy factors influencing future cash flows being discounted. Calculations of actuarial present values, i.e., expected values of present values of future random payments, are at the very core of actuarial science, especially when applied to life insurance and annuities. The pioneering work in this area, based on the standard rules of fuzzy arithmetic (see, e.g., Zimmerman, 1991) was done by Buckley (1986, 1987) and was an extension of the classical mathematics of finance. Calzi (1990) expanded on Buckley's ideas. Although traditional actuarial science as applied to life insurance assumed the interest rate to be constant and given for the entire duration of the contract, there is a wide recognition of the uncertainty of interest rates. In his discussion of the time value of money, Trowbridge (1989) put it very well: "The inexperienced actuary may tend to take an assumption about the time value of money as a given, and devote little or no attention to the appropriateness of the interest rate assumed. As he gains knowledge and experience, however, the actuary learns to differentiate between gross interest and net, before and after tax, nominal, effective, and real rates of interest, and internal rate of return. He gains a knowledge of the yield curve, the relationship between interest rates for different maturity periods. He recognizes that any specific interest rate has a basic component for time preference, and additional components for the possibility of default and the expectation of inflation. He knows that interest rate changes can affect assets and liabilities differently." Indeed, the last challenge identified by Trowbridge may be a unique field for deep investigations utilizing fuzzy set methodology. When valuing assets, one can usually determine their market values, or use market interest rates reasonably closely related to the risk of default of the entity providing asset cash flows. But on the liabilities side, do we expect the actuary to provide for the risk of default of the firm he or she works for? How would the clients of the insurance firm view it, if they learned that the firm is considering a possibility of defaulting on the promises of claim and benefit payments made to them? But on the other hand, is it reasonable to exclude one's default risk, while including it in assets of firms of similar risk profile that the firm holds among its investments? The fuzziness lurking beneath this problem calls for an innovative approach indeed.

Yet another statement of actuarial principles brings about the inherent vagueness of certain considerations of financial natures. Dicke *et al.* (1991) write "Actuaries are often called upon to place a value on future contingent cash flows related to the operations of a financial security system. Because the actuarial value is, in general, a random variable, it may be preferable to state the conditions under which the actuarial value may be expected to fall within a given range." This is a direct endorsement of probabilistic models. Such probabilistic models of interest rates in the insurance firm model have been proposed by Panjer and Bellhouse (1980), Frees (1990), Dufresne (1992), and others. But the immense complexity of stochastic models brings about calls for simplicity, at least in presentation of the analysis performed. One of the key challenges of modern insurance theory lies in the area of asset/liability management. This refers to the process of protecting the company, and its profits, in the environment of changing interest rates. One tool of asset/liability management is the concept of *duration,* defined as the logarithmic derivative of the price of a financial instrument (e.g., insurance firm asset or liability). A simple model of an insurance firm compares duration of assets and duration of liabilities, usually calling for the two to be equal, or close to each other, to match the sensitivity of assets and liabilities to changes in interest rates. Unfortunately, duration-based analysis has some pitfalls. The main one is represented by the so called *Short Straddle Model* of an insurance firm which proclaims that a typical firm with assets and liabilities of similar duration will lose its economic value under any change of interest rates (Babbel and Stricker, 1987). A much better view of the company situation is obtained if one can see how the economic value changes with changes in uncertain interest rates. Ostaszewski (1993) points out that traditional fuzzy finance models call for the uncertainty of interest rates to be represented by Figure 1 below:

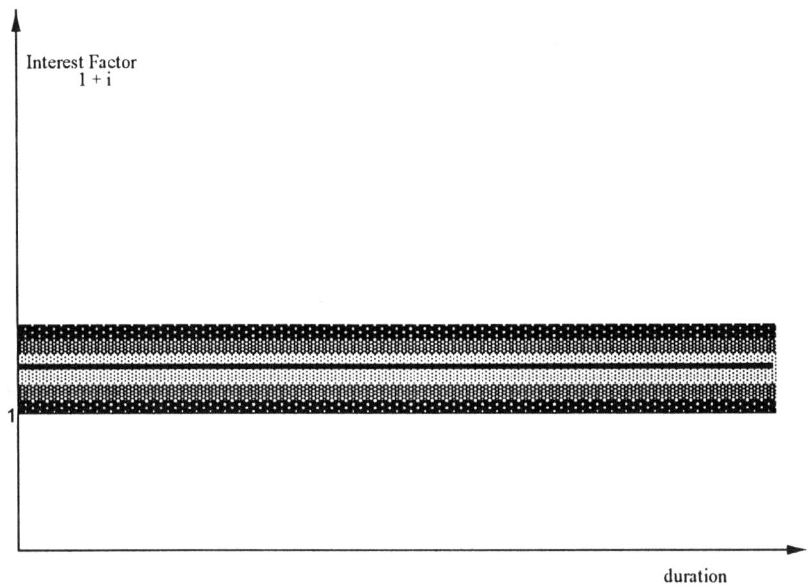

Figure 1 Uncertainty of Interest Rate with No Duration Variation

Figure 1, however, is not a realistic picture of interest rates. The near term interest rates are generally less uncertain than the long term returns which typically results in an upward sloping yield curve. Furthermore, one can receive a premium for undertaking the risk of a longer term investment. Thus a better picture of fuzzy interest rates may be the one presented below in Figure 2. The resulting "fuzzy yield curve" can then be used to calculate fuzzy present values, and can provide a very good representation of sensitivity to changes in interest rate levels.

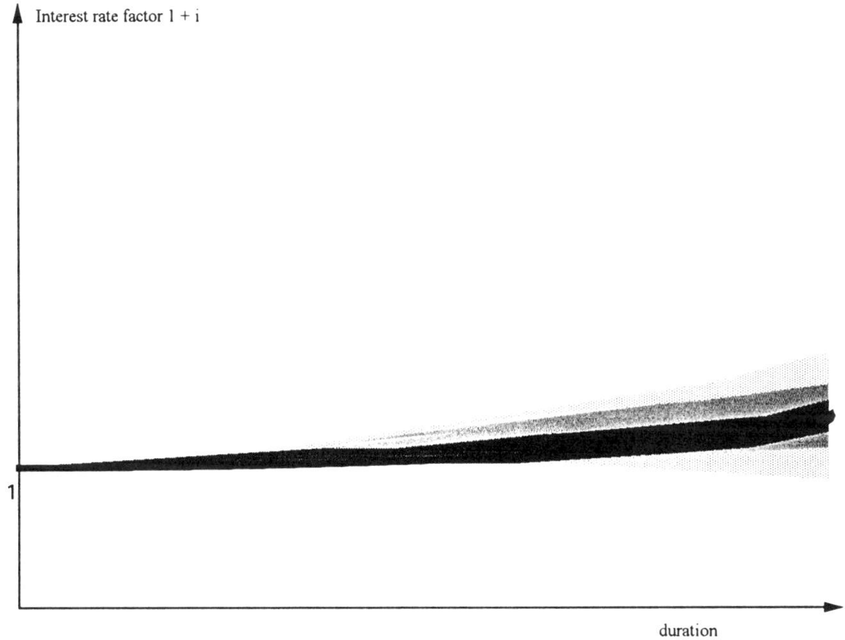

Figure 2 Uncertainty of Interest Rate with Duration Variation

We should properly note that the vast field of investment analysis, which is of natural interest to actuarial scientists, has been effectively "infiltrated" by fuzzy sets methodology, used mostly to create expert systems for the purpose of security selection and portfolio design. An example of such work is given by Wong, Wang, Goh and Quek (1992), but we should note that successful expert systems tend to remain proprietary, due to tremendous monetary incentives involved.

16.4 RISK AND CLAIM CLASSIFICATION

Ostaszewski (1993) pointed out that insurance risk classification often has to resort to rather vague and uncertain methods, or methods which are excessively precise -- as in a case of a person who may fail to classify as a preferred risk for life insurance application because of having body weight exceeding the stated limit by half a pound (this was also noted by Lemaire, 1990). Kandel (1982), writing from a different perspective, says: "In a very fundamental way, the intimate relation between the theory of fuzzy sets and the theory of pattern recognition and classification rests on the fact that most real-world classes are fuzzy in nature."

Ebanks, Karwowski, and Ostaszewski (1992) use measures of fuzziness to classify risks. In many situations, we do know in advance what characteristics a preferred risk possesses. Any applicant can be compared, in terms of measurements featured in the characteristics, to the "ideal" preferred risk, and then a membership degree can be assigned to each measurement. This produces a feature vector of fuzzy measurements describing the individual. By measuring the fuzziness of that individual as a preferred risk, we can determine a new classification.

Derrig and Ostaszewski (1995) use fuzzy clustering for risk and claim classification. They use the fuzzy c-means algorithm as discussed by Bezdek (1981). Let us illustrate this with a very simple example, originating from Lemaire (1990), and partly from Ostaszewski (1993). Suppose that four prospective insureds are defined by four characteristics, height, gender, weight, and resting pulse, and initially classified by sex. Given the following data:

Person 1: Height 175 cm, Gender 0 (male), Weight 92 kG, Resting Pulse 110;
Person 2: Height 185 cm, Gender 0 (male), Weight 75 kG, Resting Pulse 75;
Person 3: Height 160 cm, Gender 1 (female), Weight 55 kG, Resting Pulse 72;
Person 4: Height 150 cm, Gender 1 (female), Weight 90kG, Resting Pulse 100,

Persons 1 and 2 have a degree of membership in Cluster 1 of 1.00, and in Cluster 2 of 0.00, with persons 3 and 4 having exactly the opposite situation. We should note that the example given is intentionally exaggerated, but it does refer to the rather fundamental question of how to determine the nature of clusters of similar risks in insurance. If significantly different risks are insured under one rate, good risks tend to either underinsure, or leave in pursuit of a better deal. This process is commonly referred to as *adverse selection (Casualty Actuarial Society, 1990, p.35).* The end result is that the insurance firm is not only losing good customers, but it ends up insuring a group with significantly higher expected future claims. On the other hand, too fine of a classification results in increased costs, while crisp schemes produce situations such as a person 1 cm too short for a preferred rate ending up paying a significantly higher premium, because of this nearly nonexistent distinction from preferred risks. Ostaszewski (1993) points out that lack of actuarially fair classification is economically equivalent to price discrimination in favor of high risk prospects.

When the fuzzy c-means algorithm is applied to the above group of four risks, clustering gradually changes away from gender-based to a fuzzy partitioning combining inputs from all four factors (gender, weight, height, and esting pulse) used to measure risk. This is illustrated in the following figures.

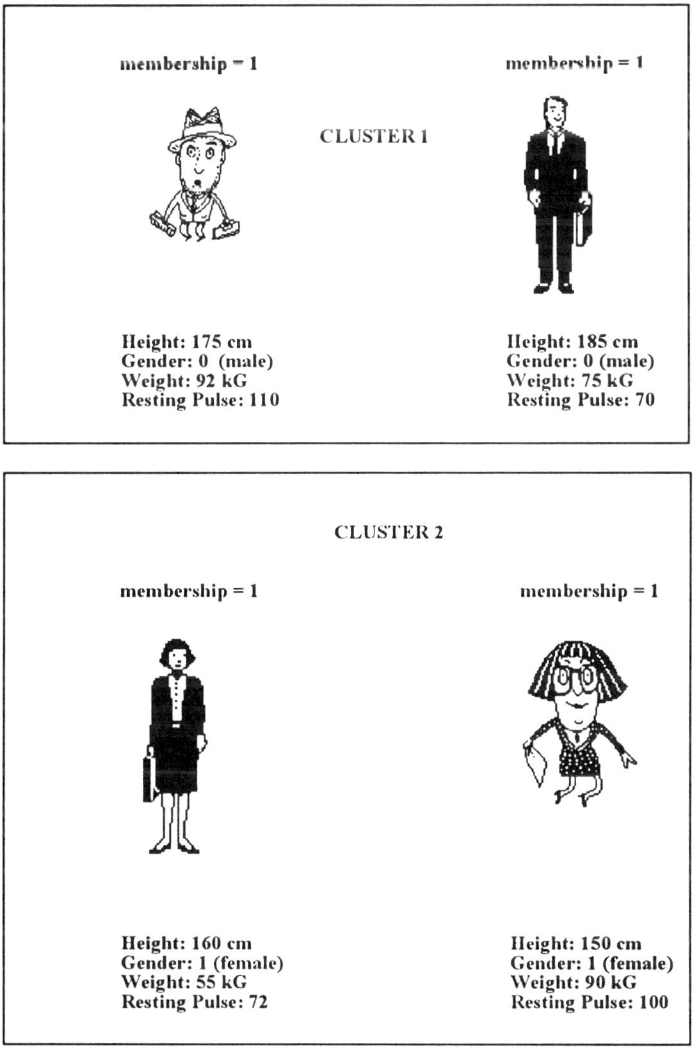

Figure 3 Initial Partition

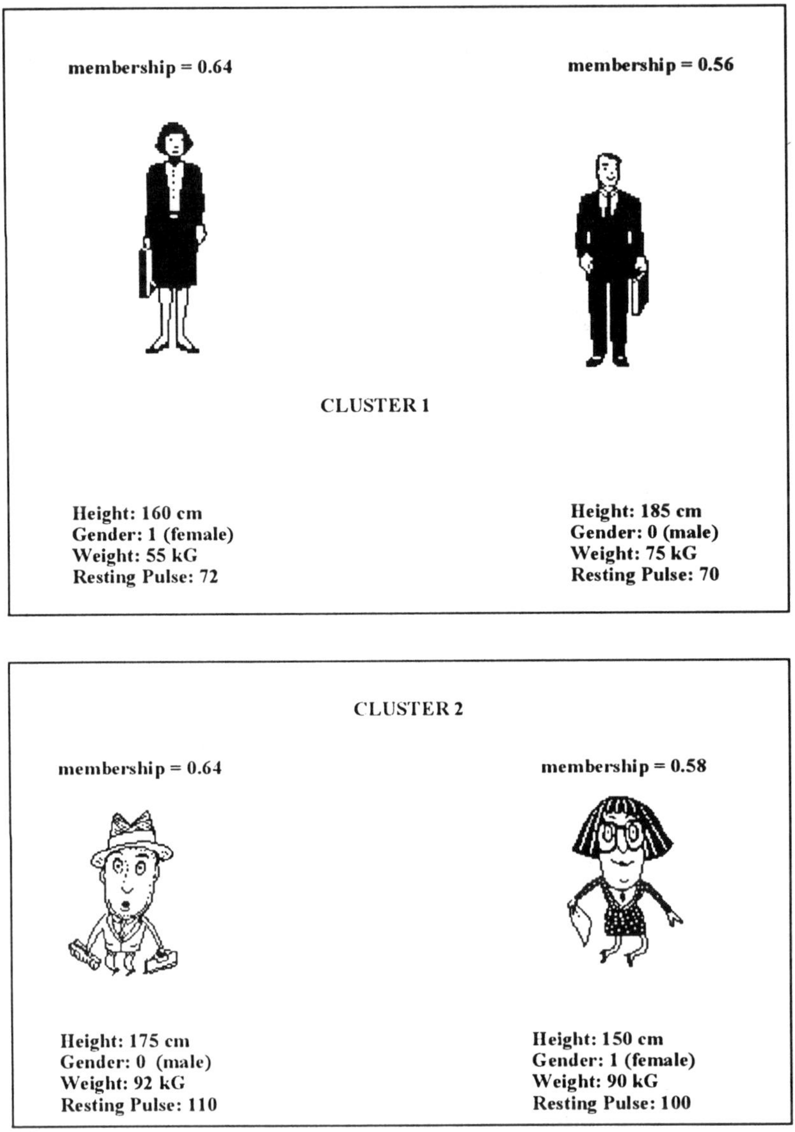

Figure 4 **Final Iteration Partition**

Derrig and Ostaszewski (1995) applied the fuzzy c-means algorithm to the problem of automobile territory rating in Massachusetts. As Conger (1987) describes "In Massachusetts, the past ten years have witnessed the evolution of an increasingly sophisticated system of methodologies for determining the definitions of rating territories for private passenger automobile insurance. In contrast to territory schemes in other states, which tend to group geographically contiguous towns, these Massachusetts methodologies have had as their goal the grouping of towns with similar expected losses per exposure, regardless of the geographic contiguity or non-contiguity of the grouped towns."

The methodology used in Massachusetts for arriving at town groupings results in pure premium indices for each of the 360 towns (or, more precisely, 350 towns and ten areas into which Boston is divided for automobile rating purposes). The indices, which are numbers relatively close to 1 (either greater than 1 or smaller, the indices represent expected losses in relation to those of the entire state expected losses) are then ordered and territories are created by analyzing such ordering. Since frequent switches from one territory to another are undesirable, numerous restrictions on moving towns from one territory to another exist. Also, capping is used, which restricts the maximum price movement of any town. Such difficulties in clustering warranted an investigation of fuzzy clustering. Resulting fuzzy clusters are much more flexible, as a town belonging partially to two territories could in the final assignment belong to the one of them which is more appropriate because of the regulatory limitations. It should be noted that although stability of territory assignment is desirable and convenient, the system of clustering towns into territories should meet the standard criterion for risk classification -- it should be responsive, to loss control (or lack thereof). Towns have an incentive to reduce their relative loss costs by maintaining their roads, law enforcement, safety engineering, and law enforcement, if those actions bring about lower premiums. If the system is not responsive, or slow to respond, the incentives are lost.

Derrig and Ostaszewski (1995) applied the fuzzy c-means algorithm for the 350 non-Boston towns, as the ten Boston towns are traditionally separated. The pure premium indices were calculated for the following coverages for all 350 towns: Bodily Injury Liability; Personal Injury Protection, Property Damage Liability, Collision, Comprehensive, and Combined. The data for the 1993 indices were used. The initial clustering was the actual 1993 territory assignment, i.e., there were 16 non-Boston territories. In addition to the above calculation, they also performed a calculation adding two more coordinates for each town -- its geographical coordinates (latitude and longitude) divided by the coordinates of Boston (the division is performed to adjust the scale for the numbers to other coordinates, which are all close to 1). By performing the algorithm on these coordinates they increased the chance of arriving at clusters which are not only actuarially similar, but also relatively close geographically. Recall that other states in the United States do use geographical proximity as an important factor in determining rating territories. The results of the work of Derrig and Ostaszewski (1995) were quite revealing. The traditional regulatory problem in Massachusetts has been the shifting of certain towns from one territory to another, and back. Fuzzy clustering showed that for those towns their membership in territories is indeed best described as fuzzy, and a more flexible regulatory approach may be needed for future decisions.

It should be noted that the work of Derrig and Ostaszewski (1995) contained some pioneering insight into use of fuzzy sets methodology to detect fraudulent claims in property-liability insurance. Cox (1995) devised fuzzy parameters to compare individual medical provider behavior to a peer group for the purpose of detecting "anomalous" behavior (better known as fraud and build-up) in health insurance. Insurance fraud is a very significant problem of the industry, especially in the presence of weak economic disincentives. Social insurance programs, such as Workers' Compensation, Medicare, and Social Security in the United States, have been often victimized by unscrupulous individuals submitting fraudulent claims. This detection has been traditionally done by special investigative units (SIU) in claim processing, (IRC, 1997) and requires significant time and money outlays, sometimes with only modest returns (Derrig and Weisberg, 1998). It should be frankly admitted that pursuit of every claim exaggerated by a few dollars does not make economic sense. Fuzzy clustering allows for distinguishing between grossly fraudulent claims and the so called claim build-up. The insurer is then free to investigate those suspicious claims which show the greatest threat of fraud. Detection of insurance fraud is still a relatively recent application of fuzzy sets methodology, and we believe that there are many promising opportunities ahead for it.

16.5 PROPERTY/CASUALTY INSURANCE PRICING

As we have already pointed out, one part of the job of an actuary is the determination of the cost of an insurance products. What is the cost of the insurance product? It is simply the cost of future covered claims, expenses and taxes. As simple as we attempt to make this definition, its practical application may not turn out to be so, given the quite fuzzy nature of the concepts of "covered" (what about insurance fraud and build-up?) claims, expenses, and taxes. Furthermore, once the cost is determined, we proceed to the price of an insurance product, which is the expected cost plus an expected profit margin. Since both the cost and profit margin are uncertain, there has to be some model of that uncertainty. It has traditionally been probabilistic. Derrig (1990) discusses the development of property/casualty insurance pricing in the United States.

The greatest challenge for property/casualty insurance lies in forecasting of claim costs, as well as in providing for fair profit and taxes. Cummins and Derrig (1993, 1997) proposed a fuzzy model for property/casualty insurance pricing. They began with claim costs trends. As we pointed out above, premium estimates are based upon expected claim costs for the period of coverage. Expected claim costs are based upon historical data, but must be developed to ultimate costs (i.e., we must provide for all already incurred costs of claims, and costs which may appear in the further processing of the claims), and must be trended to the policy period (i.e., must provide for expected changes in claim costs in the future caused, for example, by inflation). We see that the trends studied here are fuzzy, due to factors such as:

- Selection of historical data and data periods used in estimations;
- Forecasting models used, curve-fitting or econometric;
- Curves to be used, linear, exponential, and others;
- Statistical measures of fit, accuracy and bias;
- Reasonability of results, and
- Possible incompatibility of fit and reasonability.

Cummins and Derrig (1993) studied claim cost trends, and compared existing forecasting methods with respect to their forecasting accuracy, bias, and reasonability. Their main conclusion was that forecast methods that are nearly as accurate and unbiased may not produce expected claim costs that are nearly the same. They suggested assigning a membership degree to a method for its accuracy, bias and reasonableness separately. They then derived a composite fuzzy inference measure of the accuracy, bias, and reasonableness of a forecasting method. This produced much greater insight into the value of various methods than the commonly used methods comparisons of regression R-Squares or the preference of a company actuary.

Cummins and Derrig (1997) also provide examples of calculations of fuzzy insurance premiums for property/casualty insurance. They note that the premium calculation faces a first level uncertainty due to:

- cash flow magnitudes;
- cash flow patterns;
- risk free interest rates;
- risk adjustments, and
- tax rates.

These uncertainties are traditionally handled through probabilistic models and actuaries' judgement. But there is also second level uncertainty due to:

- historical data quality
- development methods
- trend or forecasting methods
- expense allocations
- surplus allocations
- capital market modeling, and
- insurance markets

By using, fuzzy parameters: losses, risk-free rate, risk-adjustment, and crisp parameters: flow patterns, tax rates, Cummins and Derrig (1997) derived fuzzy insurance premiums as generalizations of the crisp net present value pricing model (Myers and Cohn (1987), Derrig (1990). One interesting comment the authors have about their results is that the answers turned out to be "fuzzier" than they expected, indicating a degree of uncertainty which may not have been accounted for in traditional pricing methods.

Although presenting a prospective client with a fuzzy premium is not realistic, we should note that a fuzzy price may be a very valuable tool in assessing the range of premiums which needs to be considered, as well as in informing the management of the uncertainty of the premium calculation process.

16.6 FUZZY TAXES

Income taxes have a major effect on product pricing and insurance-investment portfolio management (Derrig, 1994). Derrig and Ostaszewski (1997) develop applications of fuzzy sets methodology to the management of the tax liability of a property/casualty insurance company. Myer's Theorem (1984) says that the risk-adjusted present value of the tax liability on investment income from a risky investment portfolio held by a corporation is

$$PV(\widetilde{Tr}_A) = \frac{Tr_F}{1 + r_F}$$

where r_A is the rate of return on the risky portfolio and r_F is the risk-free rate of return. The present value of the tax liability is independent of the investment strategy, and determined solely by the effective tax rate and the risk free rate. Derrig and Ostaszewski (1997) use fuzzy sets techniques to evaluate the effective tax rate as a fuzzy number, by considering fuzzy investment returns on a portfolio of government bonds, stocks, and within a context of a liability portfolio providing a tax shield.

For example, the effect of liability tax shield on the effective tax rates with fuzzy investment returns and liability shield are presented below in the Figure 5. We see that this work provides a tool for estimation of the effects of liabilities on taxes.

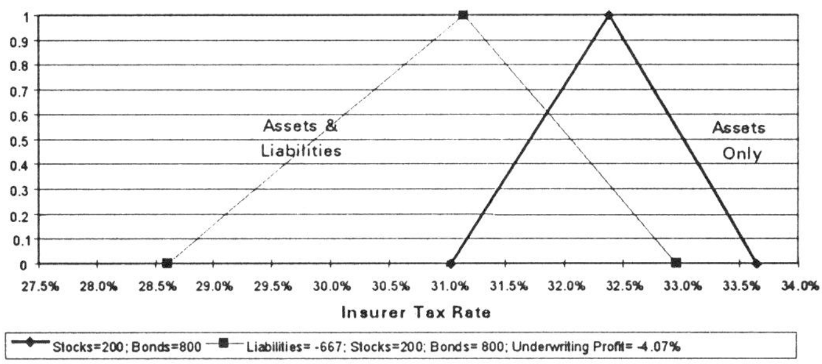

Figure 5 Fuzzy Investment Tax Rates: Effect of Liability Tax Shield

Figure 6 provides a picture of the effective tax rate under varying asset portfolio compositions. Cummins and Grace (1994) determined that property/casualty insurers in the United States perceive a yield advantage for longer maturity tax exempt bonds, implying the existence of a portfolio with an effective tax rate lower than 35 percent (the current corporate income tax rate in the United States). This can be justified only by a tax clientele effect -- a marginal buyer with a marginal tax rate of less that the insurers' 35% less their 5.1% minimum proration, alternative minimum tax rate, and capital gains income. But this perception of insurers is merely just that, a perception, and further studies are needed to determine if it is grounded in reality. The analysis of the effective tax rate under varying portfolio compositions, as in the Figure 6, suggests that small allocation shifts between bonds and stocks may have little effect on the expected tax rate.

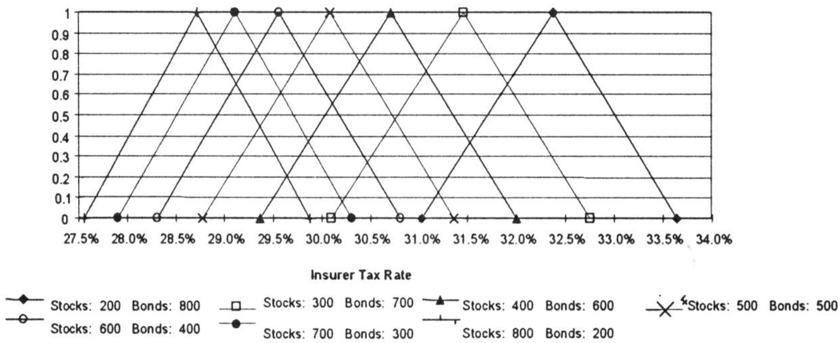

Figure 6 Fuzzy Investment Tax Rates with Selected Asset Mixes

Derrig and Ostaszewski (1995) also determine that the beta one company (i.e., a company with the same degree of risk as the whole market) has a fuzzier variation around the expected as the leverage of liability to surplus increases. This implies increasing uncertainty with increased leverage, a fact intuitively perceived by management, but now possibly quantified with fuzzy methodology, as in Figure 7.

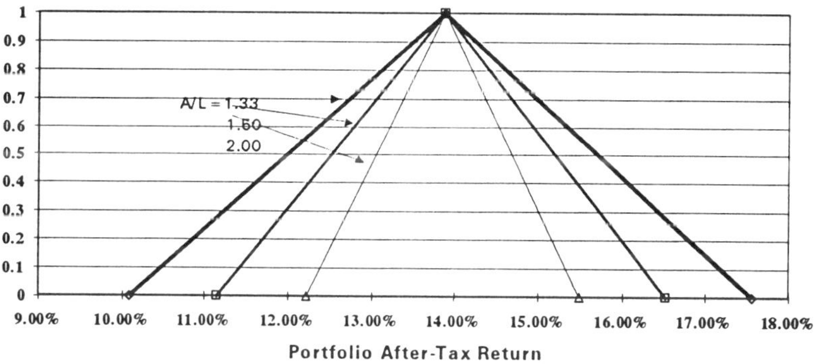

Figure 7 Fuzzy After-Tax Returns Effect of Levered Investments

16.7 THE FUTURE OF FUZZY SETS METHODS IN ACTUARIAL SCIENCE

Ostaszewski (1993) provided a broad overview of possible applications of fuzzy sets methodology in actuarial science. Recent developments do indicate that expert systems in the areas of underwriting and fraud detection hold the greatest promise in immediate applications. Indeed, such applications may be under development among the most competitive insurers.

But in the longer perspective, fuzzy sets can provide a great opportunity in risk classification, and in providing a new, fresh perspective on integrated management of an insurance firm. Some recent developments are also promising in combining fuzzy sets methodology with intervals of possibilities (Babad and Berliner, 1994), Kohonen's Self-Organizing Feature Map (Brockett, Xia, and Derrig, 1998), and neural networks. Operational control of premium price changes via fuzzy logic would begin adapting fuzzy controller methods in industrial processes to financial decisions. For example, Young (1996) codified several common actuarial concepts as fuzzy parameters in a rate changing controller. Prior rate changes, actual to expected ratios, change of business, and cancellations all combine in a modus ponens inference to produce a rate change indication for a line of business. Horgby (1998) discusses a similar fuzzy inference system for life insurance classification. Basic mortality classes are supplemented by fuzzy parameters such as overweight, high blood pressure and elevated cholesterol that combine to determine a high risk surcharge. Even though the entrance of fuzzy sets into actuarial science has been somewhat delayed, especially in the United States, we will probably see more progress in this area, especially in view of increasing competitiveness and globalization of the insurance industry.

REFERENCES:

Asay, Michael R., Peter R. Bouyoucos, and Anthony Marciano, (1993), "An Economic Approach to Valuation of Single Premium Deferred Annuities," *Financial Optimization,* edited by S.A. Zenios, Cambridge University Press, 101-131.

Babad, Y. and Berliner, B., (1994), "The Use of Intervals of Possibilities to Measure and Evaluate Financial Risk and Uncertainty," *4th AFIR Conference Proceedings, International Actuarial Association,* April 20-22, (Orlando), 14: 111-140.

Babbel, David F., and Robert Stricker, (1987), "Asset-Liability Management for Insurers", in: *Insurance Perspectives,* Goldman Sachs, May.

Bezdek, J.C., (1981), *Pattern Recognition with Fuzzy Objective Function Algorithms,* New York, Plenum Press.

Black, Jr., Kenneth, and Skipper, Jr., Harold, (1994), *Life Insurance* (Twelfth Edition), Prentice-Hall Simon & Schuster.

Bowers, N.L., Gerber, H.U., Hickman, J.C., Jones, D.C., and Nesbitt, C.J., (1986), *Actuarial Mathematics,* Society of Actuaries, Schaumburg, Illinois.

Brockett, P.L., Xia, X. and Derrig, R.A., (1998), "Using Kohonen's Self-Organizing Feature Map to Uncover Automobile Bodily Injury Claims Fraud," *Journal of Risk and Insurance,* 65-2:245-274.

Buckley, J.J., (1986), "Portfolio analysis using possibility distributions": *Approximate Reasoning in Intelligent System Decision and Control: Proceedings of the International Conference,* 69-76, January 8-10, edited by Sanchez, E., and Zadeh, L.A., Pergamon Press, Elmsford, New York.

Buckley, J.J., (1987), "The Fuzzy Mathematics of Finance," *Fuzzy Sets and Systems,* 21: 257-273.

Calzi, M.L., (1990), "Towards a General Setting for the Fuzzy Mathematics of Finance," *Fuzzy Sets and Systems,* 35:265-280.

Casualty Actuarial Society, (1990), *Foundations of Casualty Actuarial Science,* R&S Financial Printing, New York.

Conger, R.F., (1984), "The Construction of Automobile Rating Territories in Massachusetts", *Proceedings of the Casualty Actuarial Society,* May, 74(141):74.

Cox, Earl, (1995) "A Fuzzy System for Detecting Anomalous Behaviors in Healthcare Provider Claims", *Intelligent Systems for Finance and Business,* edited by S. Goonatilake and P. Treleven, John Wiley & Sons, 111-135.

Cummins, J. D. and Grace, Elizabeth, (1994), "Tax Management and Investment Strategies of Property-Liability Insurers," *Journal of Banking and Finance,* 18:43-72.

Cummins, J.D. and Derrig, Richard A., (1993), "Fuzzy Trends in Property-Liability Insurance Claim Costs," *Journal of Risk and Insurance,* September, 60:429-465.

Cummins, J.D. and Derrig, Richard A., (1997), "Fuzzy Financial Pricing of Property-Liability Insurance," *North American Actuarial Journal*, October, 1:21-44.

Derrig, Richard A. and Weisberg, Herbert I., (1998) "AIB PIP Claim Screening Experiment Final Report, Understanding and Improving the Claim Investigation Process", Massachusetts Fraudulent Claims Payment Filing, DOI Docket R98-41, Boston.

Derrig, Richard A. and Ostaszewski, Krzysztof M., (1997), "Managing the Tax Liability of a Property-Casualty Insurance Company", *Journal of Risk and Insurance*, 64-4:696-711.

Derrig, Richard A. and Ostaszewski, Krzysztof M., (1995), "Fuzzy Techniques of Pattern Recognition in Risk and Claim Classification, *Journal of Risk and Insurance*, 62-3:447-483.

Derrig, Richard A., (1994), "Theoretical Considerations of the Effect of Federal Income Taxes on Investment Income in Property-Liability Ratemaking," *Journal of Risk and Insurance,* 61-4:691-709.

Derrig, Richard A., (1993), "Price Regulation in U.S. Automobile Insurance, A Case Study of Massachusetts Private Passenger Automobile Insurance 1978-1990, *The Geneva Papers on Risk and Insurance*, 18:158-173.

Derrig, Richard A., (1990), "The Development of Property-Liability Insurance Pricing Models in the United States 1969-1989," *1st AFIR Conference Proceedings, International Actuarial Association,* May 23-26, 1990 (Paris), 4:237-264. Reprinted in *Casualty Actuarial Society Actuarial Forum,* Fall 1991, 19-46.

DeWit, G.W., (1982) "Underwriting and Uncertainty," *Insurance: Mathematics and Economics,* 1:277-285.

Dicke, A.A., Bergquist, W., Clancy, R.P., Miller, R.A. III, Panjer, Harry H., Peterson, D.M., Watson, C.B.H., and Luckner, Warren, (1991), *Principles of Actuarial Science,* by the Society of Actuaries Committee of Actuarial Principles, accepted by the Board of Governors of the Society of Actuaries in October.

Dufresne, Daniel, (1992), "On Discounting when Rates of Return are Random," *Proceedings of Twenty Fourth International Congress of Actuaries,* 1:27-43.

Ebanks, Bruce, Karwowski, Waldemar, and Ostaszewski, Krzysztof M., (1992), "Application of Measures of Fuzziness to Risk Classification in Insurance," *Computing and Information,* IEEE Computer Society Press, Los Alamitos, California, 290-291

Erbach, David W., (1990) "The Use of Expert Systems to do Risk Analysis," *Intelligent Systems in Business,* Edited by Richardson and DeFries, Ablex Publishing Co., New Jersey.

Erbach, David W., and Seah, Eric, (1993), Discussion of "The Application of Fuzzy Sets to Group Health Underwriting" Virginia R. Young, *Transactions of the Society of Actuaries* 45:585-587.

Frees, E.W., (1990), "Stochastic Life Contingencies with Solvency Considerations," *Transactions of the Society of Actuaries,* 42:91-129.

Fuhrer, C. S., (1993),"Credibility Theory," *Record of the Society of Actuaries* 19(1B):863-881.

Horgby, Per-Johan, (1998), "Risk Classification By Fuzzy Inference", *The Geneva Papers on Risk and Insurance Theory*, 23:63-82.

Horgby, Per-Johan, Lohse, Ralf, and Sittaro, Nocola-Alexander, (1997), "Fuzzy Underwriting: An Application of Fuzzy Logic to Medical Underwriting", *Journal of Actuarial Practice*, 5:1-79-105.

Insurance Research Council (1997), *Fighting Fraud in the Insurance Industry*, 2nd Edition., Wheaton, Il.

Kandel, A., (1982), *Fuzzy Techniques in Pattern Recognition,* John Wiley and Sons, New York.

Karwowski, Waldemar, and Ostaszewski, Krzysztof M., (1992), "An Analysis of Possible Applications of Fuzzy Set Theory to the Actuarial Credibility Theory," *Proceeding of the Annual Meeting of the North American Fuzzy Information Processing Society,* Puerto Vallarta, December.

Karwowski, Waldemar, and Ostaszewski, Krzysztof M., (1993), "Towards Fuzzy Actuarial Credibility," *Proceeding of the Annual Meeting of the North American Fuzzy Information Processing Society,* Allentown, Pennsylvania, August .

Lemaire, Jean, (1990) "Fuzzy Insurance," *ASTIN Bulletin,* 20:33-56.

Myers, Stewart C. and Cohn, Richard A., (1987),"A Discounted Cash Flow Approach to Property-Liability Insurance Rate Regulations," Chapter 3, 55-78 of J. D. Cummins and S. E. Harrington, editors, *Fair Rate of Return in Property-Liability Insurance,* Boston, Massachusetts, Kluwer-Nijhoff.

Myers, Stewart C., (1984), Paper prepared for the Massachusetts Automobile rate hearings, *Massachusetts Rating Bureau Filing for 1985 Automobile Rates,* Boston, Mass., December .

Ostaszewski, Krzysztof M., (1998), "Is Paul v. Virginia Dead?", *Fair Value of Insurance Liabilities,* I.T. Vanderhoof and E. Altman, editors, Kluwer Academic Press, Boston, 351-360

Ostaszewski, Krzysztof M., (1993), *An Investigation into the Possible Applications of Fuzzy Sets Methods in Actuarial Science,* Society of Actuaries, Schaumburg, Illinois.

Panjer, H.H., and Bellhouse, D.R., (1980), "Stochastic modelling of interest rates with applications to life contingencies," *Journal of Risk and Insurance,* 47:91-110.

Sametz, Arnold W., (1987), "The 'New' Financial Environment of the United States", in *Handbook of Financial Markets and Institutions,* edited by Edward I. Altman, 6th edition, John Wiley and Sons, New York.

Trowbridge, C.L., (1989), *Fundamental Concepts of Actuarial Science,* Actuarial Education and Research Fund, Schaumburg, Illinois.

Tullis, Mark A., Polkinghorn, Philip K., (1992), *Valuation of Life Insurance Liabilities,* Second Edition, Actex Publications, Winsted, Connecticut.

Wong, F.S., Wang, P.Z., Goh, T.H., and Quek, B.K., (1992), "Fuzzy Neural Systems for Stock Selection," *Financial Analysts Journal,* January/February, 48(1):47-52.

Young, Virginia R., (1996), "Insurance Rate Changing: A Fuzzy Logic Approach", *Journal of Risk and Insurance,* 63:461-484.

Young, Virginia R., (1994), "The Application of Fuzzy Sets to Group Health Underwriting," *Transactions of the Society of Actuaries* 45:551-584.

Zimmermann, H.J., (1991), *Fuzzy Set Theory and its Applications,* Second Edition, Kluwer Academic Publishers, Boston, Massachusetts.

IV Behavioral, Cognitive and Social

17 FUZZY SET THEORY AND APPLICATIONS IN PSYCHOLOGY

Michael Smithson

Gregg C. Oden

Abstract: While psychologists have known for some time that people use categories with blurred edges and gradations of membership, the field has been slow to take up ideas from fuzzy set theory and fuzzy logic. Despite critiques and controversies, by the late 80's fuzzy sets achieved a degree of legitimacy in psychology. This chapter reviews recent developments and the current state of fuzzy set theory and applications in psychology. Beginning with the gradient thesis, it surveys the uses of fuzzy sets in the operationalization and measurement of psychological concepts. The focus then shifts to modeling and data analytic techniques, and ends with a review of psychological theories and frameworks that use fuzzy sets. Because of space limitations, many topics and important works have had to be treated with more brevity than we would have wished, but we have striven for a balanced account of this diverse and still rapidly growing literature.

17.1 INTRODUCTION

Psychologists have long known that although human beings break up their world into categories, they often use categories with blurred edges and gradations of membership. Indeed, psychology inherited this idea fairly directly from a long line of natural philosophers' discourses which date back to Aristotle's observations about the uses of concepts such as color. Nevertheless, psychology has been less than wholeheartedly enthusiastic in adopting fuzzy set theory or fuzzy logic. Within the first decade after the publication of Lotfi Zadeh's 1965 paper on fuzzy sets, cognitive researchers in psychology such as Rosch (1973), Hersh & Caramazza (1976), and Oden (1977b) published experimental investigations comparing aspects of fuzzy set theory with human performance. The results, although not entirely consistent or conclusive, provided a measure of support for some versions of fuzzy set theory. At the same time, psychologists continue to be ambivalent as to whether fuzzy set theory should be considered to be a descriptive or normative account of graded categories.

By the early 80's, several conceptual and theoretical critiques of fuzzy sets from the psychological community had appeared. Fuzzy set theory was alleged to suffer from theoretical incoherence or paradoxes, mismatches with human intuition, and a lack of measurement foundations. Psychologists of judgment and decision making remained largely unconvinced that fuzzy sets could offer anything that was not already handled by subjective probability and utility.

Despite the critiques and controversies, by the late 80's there was growing evidence that fuzzy sets had achieved a degree of legitimacy in psychology. Fuzzy-set-based theories of perception (e.g., Oden & Massaro 1978 and sequelae) and memory (Massaro, Weldon, & Kitzis, 1991) appeared, fuzzy sets were utilized to extend conventional rating scales and solve measurement problems (Hesketh et al. 1988, Smithson et al. 1983, Wallsten et al. 1986, Zwick et al. 1988), and they formed the basis for new data-analytic techniques (Smithson & Knibb 1986, Smithson 1989). A book (Smithson 1987) and an edited anthology (Zetenyi 1988) devoted primarily to fuzzy set applications in psychology signaled its arrival in the field. In general, fuzzy theories have come to be accepted as a regular part of the psychological landscape; researchers are prepared to consider them on the basis of their individual characteristics rather than demanding a prior justification of the use of the notions of fuzzy sets and fuzzy truth values on which they are founded.

17.2 GRADED CATEGORIES AND FUZZY CONCEPTS

One of the most obvious applications of fuzzy set theory is to problems of categorization or classification where the categories seem to be blurred or have awkward 'borderline' cases. In areas such as personality assessment, diagnosis of disorders, and occupational counseling, arguments have been advanced that concepts and categories would be better understood as consisting of fuzzy or overlapping sets rather than discrete, mutually exclusive ones. Despite the apparent utility of fuzzy set theory for aiding this understanding, its adoption has not been widespread.

It is worth bearing in mind that in psychology there is a long-standing tradition of treating discrete constructs as if they are mutually exclusive crisp *sets* and continuous ones as if they are quantified *scales*. This is important for two reasons. One is that scale construction and classical test-theory are well-developed and firmly entrenched in psychology. The second is that psychologists are accustomed to analyzing scaled data with statistical techniques that until recently they thought of as entirely divorced from the techniques used for analyzing categorical data. These traditions, combined with the usual reluctance to abandon the Law of the Excluded Middle (e.g., can someone be depressed and not depressed at the same time?), have meant that psychologists usually turn first to scale construction rather than fuzzy sets when dealing with constructs that are graded.

The major exceptions to this tendency fall mostly into two groups. One is where psychologists have adopted the concept of prototypicality, and thereby sets containing objects that are more or less similar to the prototypical members of those sets. The other involves using fuzzy set concepts to clarify or extend a theoretical framework. We will survey these two groups in that order.

Many applications of the first kind simply use the notion of a fuzzy set to invoke gradations of membership and overlapping categories. Serban et al. (1987) provide a typical example of this kind of discussion regarding Borderline Personality Disorder. Broughton (1990) claims that prototypicality and fuzzy sets are useful for organizing research and assessment in personality psychology, particularly for refining personality assessment instruments and improving abnormal diagnosis. He also states that prototypes provide a better account of how people actually think about person variables and behavior. Horowitz and Malle (1993) review the concept of depression from a psychotherapeutic standpoint and argue that because it lacks external referents and is inherently subjective, depression (along with other psychological states) may be thought of as a fuzzy concept. Waterhouse et al. (1989) make similar claims about autism, as does Burisch (1993) concerning burnout.

A somewhat more sophisticated version of this kind of application uses prototypes and fuzzy sets to explore whether crucial psychological concepts have multiple prototypes or other kinds of internal structure. An early example is Smithson et al.'s (1983) investigation into perceptions of different types of small-scale helping (prosocial) behavior. The authors developed an empirically based taxonomy of helping using a combination of fuzzy set and multidimensional scaling approaches, which in turn yielded a pre-theoretical organization of that field and resolutions of what had been regarded as conflicting findings. Russell and Bullock (1986) argued that controversy over assessment of ability to recognize facially expressed emotion could be reconciled when natural language concepts of emotion are thought of as overlapping and fuzzy, rather than as crisp and mutually exclusive. Subjects produced reliably graded responses which also indicated that individual expressions belonged to more than one category. More recently, Fehr and Russell (1991) have argued that the concept of "love" has an internal structure and fuzzy borders, and that subtypes of love can be reliably ordered from better to poorer examples of love.

While these examples all involve static frameworks, a few similar applications have occurred in dynamical theories that have heretofore used crisp 'stages' to describe dynamics. Moore et al. (1991) argues, for instance, that stage-theories of child development should recognize that stages have fuzzy boundaries and that a child may belong partly to more than one stage at a time.

Extensions or developments of existing theoretical frameworks in psychology via fuzzy set theory are comparatively rare, but may well become more common in the future. An interesting example is the work by Hesketh and her colleagues (Hesketh et al. 1988, 1990, 1992, 1993). They have used fuzzy rating-scales to test hypotheses derived from theories of work adjustment, career choice, career compromise, and occupational counseling. Most of these hypotheses had remained untested for lack of conceptual development and operationalization. More recently, Smithson and Hesketh (1998) have used fuzzy set theoretic machinery to extend the Holland 'interest category' approach to assessing person-environment fit in occupational psychology. In a similar vein, Allinger et al. (1993) have applied a fuzzy approach to decision making to problems of personnel assessment and selection.

Applications of fuzzy sets by Foddy and Smithson (1989, 1996) to status characteristics theory (SCT) have performed similar functions to those outlined above. SCT provides a theory of how people process status-bearing information about one another to make judgments on matters such as another person's ability, and decisions such as whether to defer to another person or not. Like the career choice theories extended by Hesketh and her colleagues, SCT traditionally treats status characteristics as if they are crisp sets. Foddy and Smithson have attempted to show that a fuzzy reconceptualization and extension of SCT enables it to account for phenomena (e.g., double standards) that its crisp version cannot accommodate, and that measuring degrees of possession of characteristics yields payoffs in predictive power for SCT.

There have been other less 'successful' attempts to import fuzzy sets into existing theoretical frameworks. A typical example is social categorization theory, which has embraced the concept of prototypicality but not fuzzy set machinery. Capozza and Nanni's (1986) article is among the few in this area that use fuzzy sets to motivate hypothesis testing or theoretical developments. Researchers and theorists in this area apparently have deemed scales, prototypes, and measures of similarity sufficient for their needs.

Lastly, it is clear that only a restricted range of fuzzy theoretic concepts and techniques have been utilized in most applications to existing psychological frameworks. Most such applications simply use grades of membership, usually in conjunction with prototypes and measures of similarity to prototypes. Some use fuzzy intersection and union, but few use fuzzy negation, fuzzy logic, or other kinds of fuzzy inference. Possibility theory remains almost entirely unused, despite a long-standing interest in possibilism in psychology (see Smithson's 1988 review and programmatic suggestions). This state of affairs may change, particularly now that fuzzy logic is more often associated with neural nets and genetic algorithms and the development of so-called 'intelligent' systems (see below).

17.3 MEASUREMENT AND DATA ANALYSIS

There is the potential for a rich dialog between fuzzy sets and psychology on issues of measurement. Although fuzziness is not considered a fundamental aspect of measurement in psychology (measurement error, for instance, is handled by probabilistic accounts), some researchers have tried to represent fuzziness explicitly in their measurement procedures. Others have attempted to apply fuzzy sets to scale development problems and even to qualitative data analysis. There are also developments in measurement that have a fuzzy 'spirit' to them even though their proponents have not drawn a connection with fuzzy sets. This section will examine these issues and developments.

First, however, we will briefly attend to the converse problem, which is that in the past measurement (in the sense of constructing fuzzy membership functions from human judgments) has been a weak point in fuzzy set theory and fuzzy logic. Early reviewers (cf. Dubois & Prade 1980) acknowledged the problem but tended to dismiss it. Later commentators such as Smithson (1987) and French (1987) criticized some measurement practices among users of fuzzy sets, most particularly those who simply imposed mathematical functions and those who confounded aggregate (crisp) percentages with actual degrees of membership. However, effective suggestions utilizing standard psychometric methods and principles of fundamental measurement have been made, for example, by Turksen and his coworkers (cf. Turksen 1991) and Zwick (1987). A study by Rapoport et al. (1987) is particularly instructive for its comparison of direct magnitude estimation with a paired-comparison method for eliciting subjective membership values. We will not review that literature here, since it properly belongs to chapters on fuzzy set methods.

Instead, we now return to the uses of fuzzy set concepts in measurement problems addressed by psychological researchers. The simplest applications usually involve converting existing scales to fuzzy sets. A slightly more well-justified application is to so-called qualitative data, whereby coders are permitted to assign degrees of membership for textual units to (fuzzy) coding categories (cf. Scherl and Smithson 1987). The major advantage for the researcher is that fuzzy coding permits very flexible characterizations of qualitative unstructured data by comparison with the traditional insistence in qualitative research that each textual unit be assigned to only one crisp category. In view of the long-standing usage of scales to handle continuous constructs in psychology, it is intriguing that no systematic comparisons have been made between procedures for eliciting judgments of membership and those for eliciting judgments of magnitude (as on a conventional scale). Yamashita and Yamashita (1992) are an exception, with their small-scale experimental comparison of these two approaches.

A few applications of fuzzy sets to measurement problems have engaged the time-honored question of how to model an individual's response in choosing among possible alternatives. Extensions of Guttman and Rasch scaling models (e.g., Andrich 1978) have produced a measurement framework in which a scale is conceived of as a nested collection of sets. The connection between this approach and level sets has been noted (Smithson 1995) but not developed. Such a development would concern the estimation of the degree to which the mth response level on the jth variable belongs to some fuzzy set, and corresponding estimates of each individual's membership in the same set. It would very likely resemble, as Rasch models do, the latent-variable models on which several maximum-likelihood estimation techniques for handling multivariate (crisp) categorical data are based.

The recent work of Manton et al. (1994) is very much in this spirit but uses fuzzy set concepts explicitly. They begin with crisp observations Y_{ijm} which equal 1 when the ith individual gives a response at the mth level for the jth variable. They set up a likelihood equation based on a bilinear model of the form $P(Y_{ijm} = 1)$ is the sum over k of the products $F_{ik}p_{kjm}$, where F_{ik} is the grade of membership of the ith individual in the kth latent fuzzy class, and p_{kjm} is the probability that a full member of the kth latent class gives a response of level m for the jth variable. Thus, the issue of membership for the mth response level on the jth variable is set aside. The authors argue that, among other advantages, their approach does not need the assumptions of underlying normal distributions required by discrete factor analysis. Moreover, it is a generalization of latent class modeling in that the latent classes may be fuzzy. The major drawback of this technique is its computational intensity.

Finally, fuzzy sets have also been invoked in extensions of measurement models for asymmetric similarity judgments. Shiina (1988) extends Tversky's (1977) feature representation model, which he argues is too restricted in considering features as dichotomous variables. His "fuzzy feature matching model" is a generalization of Tversky's contrast model of similarity. Smithson (1993) develops an alternative fuzzy intersection-based model and reanalyzes Shiina's data, demonstrating a better fit for the new model.

Perhaps the most interesting application of fuzzy sets to measurement in psychology, however, has been Hesketh and her colleagues' use of type-2 fuzzy sets to extend conventional rating scales in the area of occupational psychology. Hesketh et al. (1988) use a triangular membership function to permit respondents to indicate their occupational interests and orientations by not only providing a single point on each scale, but also a band of latitude or tolerance. The authors find that respondents perform these tasks reliably and argue persuasively for its validity as well. They outline its potential for applications in research and counseling; and indeed Pryor et al. (1989) apply this scale to career counseling.

We now turn to applications of fuzzy sets in data analysis. Despite the recent explosion of activity in 'soft computing', fuzzy hybrid methods (usually combining fuzzy logic with neural nets) for data analysis have thus far made little inroads into psychology. Much the same is true for more established techniques such as fuzzy regression. The major exceptions include Manton et al. (1994), whose work has been described previously, Steenkamp and Wedel's (1991, 1993) fuzzy clusterwise regression technique, and Smithson's (1987, 1995) work on fuzzy inclusion. Steenkamp and Wedel describe a new technique for benefit segmentation, fuzzy clusterwise regression analysis. It combines clustering with prediction and simultaneously estimates the models relating subject preferences to product dimensions within each of a number of clusters, and the degree of membership of brands and of subjects in those clusters. They argue that the method has substantial advantages over traditional segmentation analysis.

Smithson (1987) advocates fuzzy inclusion as a worthwhile kind of relationship between variables for modeling in psychological applications. His recent work in this area (Smithson 1995) proposes a decumulative distribution function interpretation and least-squares estimation framework for measuring fuzzy inclusion when membership levels are discrete, along with a chi-square based test for models of inclusion as a function of membership level. The simplest of these is a "constant inclusion" model, whereby the proportion of one set included by another is constant across all nonzero levels of membership.

17.4 MODELING LINGUISTIC VARIABLES

Psycholinguistics is one subfield where fuzzy sets have been used fairly extensively to enrich older theories or ideas. Research on linguistic quantifiers has been the most active area in that regard, and in fact such research has been conducted with at least three different research foci: comprehension and representation of quantifiers, applied research on rating scales, and human reasoning (usually syllogistic). This section will begin with an overview of research utilizing fuzzy theory and concepts on quantifiers, then hedges, aggregations of categories, and finally reasoning.

The survey research literature provides a rich source of studies of quantifiers, particularly those that refer to relative frequency (e.g., "sometimes", "seldom"). One of the chief goals of researchers in this area is to find quantifiers that may be used to form an interval-level rating scale. In a classic study, Bass et al. (1974) used a magnitude scaling approach to assess 44 amount quantifiers and 39 frequency-quantifiers. They distilled the most robust and (from a scaling point of view) optimal sets of quantifiers. They also found evidence that the resulting scales are stable under variations in the importance of the topic being rated, a claim that has been upheld by subsequent studies (e.g., Newstead & Collis 1987). However, a number of other researchers have found that these quantifiers can be destabilized by other kinds of context effects. Pepper (1981) and Newstead (1988) have thoroughly reviewed this literature and come to essentially identical conclusions, so our remarks here merely summarize their overviews. The most widely studied context effect is that of base-rate or expected frequency (e.g., Pepper & Prytulak 1974, and Wallsten, Fillenbaum, & Cox 1986). Thus, "rarely" might refer to a much higher frequency for tooth-brushing than for earthquakes. Another effect is the size of the set being described, although empirical support for this effect seems to have been inconsistent (cf. Newstead 1988). Lastly, the type and range of other available quantifiers may also affect the interpretations of those in use.

The picture that emerges from classic studies of quantifiers is that they are located along an "analog scale" (Holyoak & Glass 1978), which has terms like "never" at one end and ones like "always" at the other. While many such quantifiers exhibit a stable rank-order, attempts to pin their universal meanings down to pointwise frequencies have largely been defeated. Newstead (1988) concludes that there is considerable evidence and plausibility in the notion that many linguistic quantifiers (with exceptions at the extremes, such as "never") are inherently fuzzy and therefore may profitably be investigated from a fuzzy set theoretic standpoint, even though fuzzy set theory itself does not explicitly handle context effects.

Sophisticated efforts along these lines were initiated in the mid-80's by Wallsten and his co-workers, in a series of studies on the meanings of probabilistic quantifiers (e.g., "quite likely"). As with other kinds of quantifiers, early experimental paradigms required subjects to provide pointwise numerical estimates of the probability represented by a given quantifier (e.g., Simpson 1944, Lichtenstein & Newman 1967). The practical concerns addressed by this research include arguments over whether risk assessments should be restricted to quantitative (numerical) representations of uncertainty, or whether verbal phrases can be used instead. High inter-subject variability and context sensitivity led many researchers to conclude, as in the rating scale research, that verbal phrases are not sufficiently reliable or consensual for use in rational decision making or risk assessment.

Budescu & Wallsten (1985) argue that inter-subject variability in interpretations of probabilistic quantifiers is not due to unreliability or experimental error, and they found evidence of reasonable within-subject reliability. One reasonable interpretation of these findings is that they are symptomatic of fuzziness rather than unreliability. Wallsten et al. (1986) and Rapoport et al. (1987) establish a new experimental paradigm in which subjects construct fuzzy membership functions over the [0,1] interval to translate probabilistic quantifiers into numerical terms. Pipino et al. (1981) anticipated their approach, but used a less sophisticated direct-rating method.

Subsequent work on this topic has pursued theoretical issues as well as those of reliability and consensus. A complex study testing four models that combine stochastic and linguistic aspects of imprecision is reported in Zwick & Wallsten (1989). Their results suggest that people adopt a simplified view of probabilistic concepts by characterizing them via multiple-crisp representations. Reagan et al. (1989) compare three methods of number-word translations for eighteen probabilistic concepts: (a) frequency distributions of what single number best represents each expression; (b) word-to-number acceptability functions from what range of numbers from 0% to 100% best represents each expression; and (c) number-to-word acceptability functions from which expressions are appropriate for multiples of 5% from 5% to 95%. The latter two are types of fuzzy membership functions.

The combined impact of these studies motivates reconsideration of the conclusions reached by earlier researchers in the words-versus-numbers debates in risk assessment. Not unexpectedly, the probabilistic terms on which there is least consensus are those tending towards the middle of the scale, and phrases using negations (e.g., "unlikely") are not exact reversals of their positive counterparts. Nevertheless, the superiority of the 'fuzzy' viewpoint on quantifiers seems clear and it has clarified some conceptual problems such as the distinction between fuzziness and unreliability.

It remains to be seen whether the fuzzy paradigm will be extensively applied to the study of other kinds of quantifiers or whether it will be integrated with mainstream psycholinguistic theories of quantifiers. Probably the greatest challenge will be integrating fuzziness with various context effects. Examples of attempts to do this for specific quantifiers are studies of the concept of "nearness" or "closeness" (Smithson 1987, Fisher & Orf 1991). These studies use fuzzy membership rating tasks by which subjects assess whether something is "nearby" or "close by", and demonstrate that the membership function is conditional on, for instance, whether the subject is thinking of transportation on foot or by vehicle. A related topic is the study of hedges, which are a special kind of qualifier (e.g., "somewhat", "very"). Zadeh's (1972) early account of hedges stimulated several studies during the 70's and 80's and some criticism as well, although this does not appear to be an active area of research now. Hersh & Caramazza's classic (1976) series of experiments set the tone for this topic by fitting Zadehian concentration and dilation operators to empirical data and comparing their performance with alternatives. The most problematic finding for the Zadehian account is that some hedges (e.g., "sort of") translates membership values into an inverted "J" shape rather than any monotonic translation. Empirical support for this has been found by several authors (Hersh & Caramazza 1976, Kempton 1978, Pipino et al. 1981, Smithson 1987). While no singular theoretical insights seem to have been generated from these findings, it is clear that any complete account of hedges would need to incorporate them.

The aggregation of categories (or more generally, of information), on the other hand, is a topic that has attracted substantial numbers of researchers over the years from a number of different perspectives and areas. There are theoretical traditions predating and competing with fuzzy sets (e.g., Anderson 1981, 1982), and aggregation forms an important, if not always systematically developed, component of various psychological theories that involve information processing (e.g., status characteristics theory).

Because fuzzy set theory provides a debated account of intersection and union, it is a natural candidate for empirical tests. Most studies of this kind were published in the 70's and 80's. Thorough reviews are available in Dubois & Prade (1985) and Zwick et al. (1988), so again we will confine our survey to a brief overview. At first glance, evidence for various fuzzy set t- and s-norm pairs appears inconsistent. Oden's (1977b, 1979) experiments found in favor of the product and probabilistic sum, while Thole et al. (1979) found support for the min operator. Zimmermann & Zysno (1980) produced data that was well-modeled by a compensatory operator that could vary between strict intersection and strict union. Their findings were qualitatively consistent with the large body of empirical results obtained by Anderson (1981) and his colleagues, who concluded that humans favor an averaging rule for combining information. Zimmermann & Zysno based their operator on the product and probabilistic sum co-norms, but Smithson (1984) reanalyzed their data using a least-squares estimation model based on a convex combination of the min-max co-norms and demonstrated that it fit just as well.

The most sophisticated studies of this kind are Zwick et al. (1988) and Budescu et al. (1990). Both studies indicate that no t-norm describes what people do when they are asked to estimate membership in 'A and B'. Instead, they found mixed support for an Andersonian averaging rule and the geometric mean. Zwick et al. (1988) did, however, find support for the max s-norm as a model for judgments of membership in 'A or B'. More recently, some researchers have argued for other accounts of how people combine graded categories. Huttenlocher & Hedges (1994), for instance, propose an alternative formal mechanism in which a conjoint category is constructed from constituent categories by forming a joint distribution of values.

Violations of fuzzy set intersection rules fall into three cases of increasing 'severity': (a) deviations from t-norms that nonetheless obey the restriction that the grade of membership in $A \cap B$ does not exceed min(A,B), (b) deviations for which membership in $A \cap B$ lies in between min(A,B) and max(A,B), and (c) deviations for which membership in $A \cap B$ exceeds max(A,B). Violations of the second kind have been studied by psychologists of judgment under uncertainty, in connection with subjective estimates of compound probabilities. Early investigations by Cohen and his colleagues into how people combine subjective probabilities led them to conclude that subjects sometimes produce subjective probabilities of success in the presence of two sources of uncertainty that are greater than corresponding probability estimates in the presence of only one source of uncertainty. Furthermore, they found this effect was enhanced when the additional uncertainty had a larger subjective probability attached to it than was attached to the first source (Cohen & Hansel 1959). These findings are perhaps the earliest experimental evidence pointing to the so-called 'conjunction fallacy', and other variants on the same phenomenon have been reported in Wyer (1970), Bar-Hillel (1973), Goldsmith (1978), and Beyth-Marom (1981). However, the effect has been most directly studied and popularized in a series of studies by Tversky and Kahneman (1982, 1983). Their findings, point to a consistent violation by both naive and well-trained subjects of the rule that $P(A\&B) \leq \min(P(A),P(B))$. As in the Cohen & Hansel findings, most of the subjects in these studies tended to judge P(A&B) as greater than P(B), say, but less than P(A) when P(A) was considered quite high and P(B) rather low. The effect did not emerge when both P(A) and P(B) were low, and was exaggerated when both were high. What is interesting about this literature is not simply their findings or even the explanations of them, but also the hard-line rationalist stance taken by researchers in this area. For them, the subjects' violations of probability theory imply that the subjects' judgments are irrational, whereas for researchers studying the correspondence between human judgment and fuzzy set theory, violations of fuzzy set theory imply that the theory is faulty.

While the models in the aforementioned empirical studies of fuzzy set aggregation are capable of handling the first two kinds of violations, they fail to account for the third. This latter kind of violation has been discussed in a few critiques of fuzzy set theory. In their research on color category membership, Kay & McDaniel (1975) noted that membership values for secondary hues such as 'orange' do not lie on or below the membership curves for 'red' or 'yellow'. A substantial portion of one prominent critique of fuzzy set theory (Osherson & Smith, 1981) was a meditation on this same problem involving examples like the guppy, which was argued to have greater membership in the set of 'pet fish' than in either the 'fish' or 'pet' sets. Again, such findings are taken as evidence against fuzzy set theory rather than evidence for human irrationality, or even as evidence for psychologically more compelling accounts such as that 'useful' aggregations evolve into fuzzy sets in their own right or, as Oden (1984) claimed, that linguistic modification is more complex than simple conjunction.

Lastly, we come to studies of reasoning and logic. There are relatively few of these that have involved reasoning with vague or fuzzy propositions. There are a few studies that claim support for fuzzy logic as a description of human reasoning (e.g., Yamashita & Yamashita 1988). However, mainstream psychological studies using quantifiers such as "some" (reviewed in Newstead 1988) provide reasonably strong evidence that people interpret "some" to mean "some but not all", which is not the traditional logical (or even fuzzy logical) interpretation. This "incompleteness" violation is similar to the inverted J-shaped membership functions found in empirical studies of the meaning of hedges such as "sort of". There is also some evidence that supposedly crisp quantifiers (e.g., "all") are interpreted as if they are fuzzy.

An apparently little-known but sophisticated and systematic empirical investigation into human syllogistic reasoning with vague predicates is Spies' (1989) monograph. There is insufficient space to describe his findings here, since he develops and implements a complex factorial experimental paradigm for comparing human logical inferences with models of approximate reasoning based on fuzzy logic and belief theory. Suffice it to say that he investigates the effects of monotonicity, nonspecificity, and positiveness on human reasoning in tasks involving chaining, antecedent and consequent conjunctions. Rather than concluding that human deviations from prescriptive models invariably indicate deficiencies or biases, he opts for a judicious treatment of this issue. For instance, on the one hand, he identifies the confusion by some of his subjects between 'and' and 'or' in consequent conjunction problems as deficiencies in their reasoning. On the other hand, he rightly regards differences between human estimates of how nonspecificity propagates through inference chains and the prescriptions of Dempster's combination rule as a matter for debate, since Dempster's rule has been a source of controversy among students of approximate reasoning.

In summary, research comparing human and fuzzy set theoretic treatments of linguistic quantifiers has found some parallels between them, but is still in a rather inconclusive state. There is clearly considerable scope for extending the type of work that has been done on probabilistic quantifiers into more mainstream linguistic topics. Perhaps the most fundamental matters raised in this research pertain to the issue of rationality and prescriptive versus descriptive theories of linguistic variables. It is here where psychologists have an opportunity to free themselves of the rather straitjacketed 'rationalism' associated with the tradition of using probability as the sole arbiter of rationality under uncertainty.

17.5 KNOWLEDGE REPRESENTATION AND CATEGORIZATION

The structure of the contents of the mind is an obviously fundamental topic in psychology. Throughout the course of modern psychology, there have been two quite different general approaches to this issue. One approach characterizes concepts in terms of values on a number of independent and typically continuous dimensions of meaning. This approach, epitomized by the work of Osgood and his colleagues (Osgood, Suci, & Tannenbaum, 1957) and that of subsequent researchers employing multidimensional scaling methods, leads naturally to thinking of the world of concepts as corresponding to some sort of spatial array. The alternative approach characterizes concepts in terms of discrete formal structures such as those of logic, set theory, or graph theory. This approach includes systems based on a wide variety of discrete structures, but is probably best represented by semantic network models (e.g., Rumelhart, Lindsay, & Norman, 1972).

Although there have been a few points of overlap between these two approaches (for example, some instances of semantic feature models can alternatively be thought of in terms of binary semantic dimensions or in terms of feature sets), in general, the two approaches have been almost entirely separate. In part, this may be due to the fact that a major inspiration for dimensional models comes from work in sensation where the psychological objects clearly vary along independent continuous dimensions whereas, in contrast, the structural models are more directly inspired by work in language comprehension where the psychological objects equally clearly have propositional structure. Thus, to some extent, the two approaches have managed an easy separateness because they were focussed on different ends of the spectrum of mental functions and because they tended to be concerned with relatively separate sorts of psychological phenomena.

Nevertheless, there certainly have been occasions when the two approaches have been applied within the same psychological domain with some degree of resulting tension between them. For example, in the area of semantic memory retrieval time, two basic findings are the Semantic Distance Effect (Moyer, 1973), which is commonly taken to support the notion that concepts are fundamentally dimensional, and the Cognitive Economy Effect (Collins & Quillian, 1969), which is commonly taken to support the notion that concepts are fundamentally propositional. Most often, this tension is avoided by researchers adopting what could either be characterized as an ecumenical or as a split personality position. But on a few occasions, the tension has been promoted to a full scale conflict, as with the so-called analogical versus propositional debate in the area of mental imagery.

It is against this background of significant theoretical fracture that the pioneering work of Eleanor Rosch (1973, 1975, 1978) should be viewed. Without seeming to mean to confront this particular psychological dilemma (she was mainly directed toward an issue that was more salient in philosophy than psychology, namely, whether or not human concepts are arbitrarily related to the world--indeed, she was reluctant to distract attention from this focus by discussing what sorts of formal theory might best account for her results), Rosch opened the door for psychologists to consider a whole new approach to concepts founded on hierarchies of categories as in the propositional approach but with items instantiating categories to different degrees as in the dimensional approach. Although she was fully aware of the obvious relation of her work to that of Zadeh on fuzzy sets (as it happens, they were on the faculty at the same university), she left it to others to examine this directly and to work out the details of how fuzzy set theory should be applied to human categorization behavior.

Thus it was that psychologists finally came to address the question of the direct psychological validity of fuzzy set theory nearly a decade after it had first been proposed in what were largely psychological terms. Most of the early work on this question (Hersh & Caramazza, 1976; Kochen, 1975; Macvicar-Whelen, 1978; McCloskey & Glucksberg, 1978; Oden, 1977a, 1977b; Rubin, 1979) provided general support for the idea that people categorize objects in a manner consistent with the view that the categories are represented in semantic memory as fuzzy sets. On the other hand, the results were not always in exact accordance with the specific rules proposed by Zadeh.

For example, Hersh and Caramazza (1976) asked people to decide whether phrases like "large", "small", "not large", "very small" and so on applied to squares of various sizes. They found clear support under a variety of presentation conditions for the notion that these concepts were best represented as fuzzy sets. However, their results indicated that the operation of the quantifier "very" did not follow the power function form that Zadeh (1972) had proposed but rather took a form more like that of the translation function suggested by Lakoff (1973).

The study that provided perhaps the most direct assessment both of whether or not natural human categories behaved as fuzzy sets and of what rules they followed if so, was that of Oden (1977b). In this study people directly rated the degree to which conjunctive and disjunctive categorization statements (e.g., "a robin is a bird and a table is furniture") were true. The results provided strong support for the hypothesis of human categories being fuzzy sets, but indicated that the logical integration of fuzzy categorical information was more apt to be of multiplicative form than of the min-max form.

Subsequent work already mentioned here (e.g., Osherson & Smith, 1981) challenged the conclusions of this early work but was effectively answered by papers (Cohen & Murphy, 1984; Oden, 1984) showing that the phenomena purportedly supporting this challenge (e.g., the fact that a striped apple is not a very good instance of an apple) had nothing to do with whether or not concepts are fuzzy but rather to do with the fact that concepts are defined intensionally (by rules) rather than extensionally (by enumeration). Since then, few studies have been concerned with establishing that human categories are fuzzy. Instead, the area has moved on to examine more of the specific details of the nature of human concepts, making more or less explicit use of fuzzy theoretical devices as needed.

For example, Hampton (1988b) performed a series of experiments that provided a detailed examination of the nature of composite concepts described by natural language phrases such as "protective clothing" or "sports which are games". In this study, people rated the membership of objects in the composite concepts as well as membership in the component concepts. His results provided general overall support for composition by multiplicative conjunction. However, there was evidence of asymmetry in the contributions of the component concepts to the composed concept. Furthermore, there was clear evidence for what Hampton calls "overextensions": cases where membership in the composite concept is greater than in one or the other of the component concepts. Hampton suggests that this need not be taken as evidence against the fundamentally fuzzy conjunctive character of these composite concepts but rather that there may an additional process of "extensional feedback" whereby the characteristic properties of those instances in the world that most clearly belong to the conjunction may come through experience to modify the propositional specification of the concept. Hampton (1988a) reports similar studies examining disjunctive composite concepts.

Evidence for natural categories having intricate fuzzy propositional structure was also obtained by Oden (1981; see also Oden & Lopes, 1982). This study used the collection of pictures of cups and bowls used by Labov (1973) extended to provide a factorial manipulation of height, width, and presence or absence of a handle. In studies using both categorization probabilities and ratings of category membership, evidence was obtained for category membership being complex, often nonmonotonic functions of the three factors manipulated. Nevertheless, the results were well accounted for by a fuzzy propositional model with propositions involving conjunction, disjunction, negation, and quantifiers. For example, the concept of cup that the data supported was "either a drinking glass or a coffee cup" where the concept of "coffee cup", in turn, was "a piece of dishware that is not very wide, about as high as it is wide, and that normally has a handle". Through the use of multiplicative rules for conjunction and disjunction, the fuzzy propositions were incorporated into a categorization model that provided a very good account for the precise quantitative pattern of the complex interaction obtained with both response measures.

A final example of the use of fuzzy constructs as tools in the pursuit of psychological issues is the work of Oden and Lopes (1994) who developed a model of how the conceptual categories that underlie the way people talk about lotteries could be based directly on the algebra that describes choices made in these situations. According to Lopes' (1987, 1990) SP/A theory of risky choice, a person evaluates the security (S) provided by an uncertain situation along with its potential (P) and integrates this information with that about the extent to which the person's aspiration level (A) is likely to be met. In addition, SP/A theory accounts for differences in attitudes toward risk in terms of how much consideration a person characteristically gives to security compared to potential. Building on these ideas, Oden and Lopes hypothesized that a person's concepts for security and potential are fuzzy predicates that are defined by propositions that are composed from primitive predicates concerned with the degree to which each outcome in the lottery is a "good" outcome, weighted by rank-dependent probability weights reflecting security-mindedness and potential-mindedness in a direct and systematic way. The aspiration level predicate was proposed to be based on a probability-weighted composition of the degrees to which each lottery outcome is nearly equal to or greater than the aspired for value.

17.6 PERCEPTION

Applications of fuzzy set theory and fuzzy logic to the study of perception closely followed the work on categories just reviewed. In part, this was a natural result of the recognition that the identification of perceptual patterns is directly analogous to if not simply an instance of the process of semantic categorization. In addition, this development was particularly timely and apt because there had recently been a number of studies whose results significantly challenged the discrete theories of perceptual categories, which until that time had dominated the field, especially in the area of speech perception.

As previously mentioned, basic sensory qualities such as brightness, loudness, and pitch are smoothly continuous in nature. In contrast, higher-order perceptual qualities may appear to be discontinuous or, at least, not smoothly continuous: that is, there are often regions along physical continua where the corresponding perceptual experience changes rather abruptly from one percept to another. The existence of such perceptual boundaries defines what is known as categorical perception. In the strongest version of categorical perception, there is no perceptual difference between stimuli within a category, which is to say the categories are crisp. Within the speech perception research community, strong categorical perception came to be seen as a distinguishing characteristic of speech and as evidence for the "speech-is-special" hypothesis, which held that speech perception was accomplished by a dedicated perceptual system that had separately and uniquely evolved within humans.

By the mid 1970s, it was clear from results of experiments such as that of Pisoni and Lazarus (1974) that, when appropriately tested, people could distinguish between speech sounds within categories. Thus, the stage was set for the development of a detailed account of perceptual categories in fuzzy terms. This was provided by Oden and Massaro (1978), who described what has come to be known as the Fuzzy Logical Model of Perception (FLMP).

In accordance with traditional notions of speech perception based on linguistic analyses, the FLMP presumes that speech sound patterns are psychologically represented as conjunctions of component features. For example, the phoneme /p/ (the 'p' sound) is a voiced, labial consonant which means that it is characterized by the proposition "voiced AND labial AND...[a number of other features that specify it as a consonant]". In contrast with traditional accounts, of course, the FLMP takes the features to correspond to fuzzy predicates that represent the feature extraction process as one that determines the degree to which a speech sound has the quality of being voiced etc. Thus, the sound is evaluated as being a /p/ to the degree to which its fuzzy proposition is true (in accordance with the results of Oden, 1977b, fuzzy conjunction is taken to be multiplication), and it is presumed to be identified as being /p/ as a direct function of the truth of the /p/ proposition relative to that of all other candidate patterns (e.g., /d/, /b/, and so on).

Oden and Massaro (1978) demonstrated that the FLMP was able to provide a very good account of the quantitative details of phoneme identification, particularly when the model was elaborated slightly to include linguistic quantifiers (e.g., "very voiced", "slightly labial", etc.) in the fuzzy propositions. In this study, physical properties of speech sounds corresponding to whether each sound was voiced or voiceless and whether labial or alveolar were varied independently over a number of different levels to provide a direct test of the nature of the pattern of the joint effect of these two factors. The success of the model provided strong support for the hypothesis that speech features are evaluated independently, contrary to what has often been supposed.

Over a large number of subsequent such studies, the FLMP has been shown to successfully account for perceptual identification not only in the arena of speech perception, but in many other domains as well including particularly the reading of printed text at both the letter (Massaro & Hary, 1986; Oden, 1979) and the word levels (Massaro & Sanocki, 1993; Oden, 1984) and of handwritten words (Oden & Rueckl, 1986). Of particular note is Massaro's (1987) extensive research program using the FLMP in studying the use of visual lip movement cues in speech perception. In addition, the model has been successful not only in accounting for identification probabilities but also for direct ratings of the degree to which stimuli exemplify one category or another. These two response measures provide complementary information in that identification has the advantage of being simple and natural whereas ratings have the advantage of assessing continuous processing directly and of providing much finer detail of the nature of fuzzy categories.

In recent years, the notion that perceptual identification is continuous has come to be widely accepted not only because of the success of the FLMP but also because of the development of connectionist or neural network models, which most commonly presume continuous information processing. For the most part, the fuzzy and connectionist approaches have been seen as complementary and compatible rather than competitive. (Indeed, it can be argued--see, for example, Oden 1988--that they provide jointly-necessary and mutually-supportive levels of analysis of cognitive systems.) But, of course, this is not to deny that specific models have been in competition with each other both between as well as within the two approaches.

One such debate is of special interest here. In 1986, McClelland and Elman developed a connectionist model of speech perception called the TRACE model and showed that it was able to account for a number of important phenomena. However, as Massaro (1989) later demonstrated, TRACE was not capable of predicting the pattern of identification data for stimuli varying along two independent featural dimensions such as that of Oden and Massaro (1978) and subsequent studies. In the face of this shortcoming, McClelland (1991) modified TRACE in such a way that it was now able to produce the appropriate pattern of predictions for this important case. However, the nature of the required modification made it necessary to abandon the continuous categories of the original TRACE model; that is, to go back to the position of the strong categorical perception hypothesis: perception is discrete but identification probabilities exhibit intermediate values due to noise in the system. The trouble with this move is that while it may be tenable for identification probability data, it is not able to handle the comparable results of ratings studies. And, as McDowell and Oden (1995) demonstrated in a study in which people made both identifications and ratings based on the same stimulus exposure, these two response measures are simply and systematically related in just the way one would expect if they were the product of a common continuous underlying perceptual process.

17.7 LANGUAGE PROCESSING

Having now considered the application of fuzzy theories to conceptual processing on the one hand and perceptual processing, particularly of linguistic materials, on the other hand, it is natural to examine the cognitive processes that usually come in between: those pertaining to the processing of words and sentences. This is a notoriously difficult topic of study because of the large number of linguistic factors that bear on any aspect of language comprehension and production. In general, it hard to isolate the lexical from the syntactic from the semantic, let alone one syntactic factor, for example, from another.

In the case of comprehension, the reason why so many variables impinge on each facet of language comprehension is that 1) it is a difficult task with many opportunities for misunderstanding and 2) language has evolved to enable people to comprehend it successfully even under noisy and adverse conditions. Exploiting multiple sources of information provides the robustness needed for this success. Furthermore, the various sources of information can be expected to be continuous in nature so that integrating across several such sources of information will allow errors to cancel. Hence, one basic issue regarding fuzziness in language comprehension concerns how information from different linguistic levels is integrated.

One of the first studies to directly address this issue was that of Rueckl and Oden (1986) who jointly manipulated letter featural information and semantic contextual information in a word identification task. People were presented with sentences like "The cardplayer has a pain in his hand". The two variables that were manipulated were, in this example, the nature of the last letter of the fifth word (which could be either 'n' or 'r' or various more or less ambiguous forms in between) and the subject of the sentence (which could be either "cardplayer" and thus highly compatible with having a pair in a hand or "arthritic" and thus highly compatible with having a pain in a hand or subjects that are compatible with the two possibilities to various intermediate degrees). The results provided strong support for the claim that both letter featural and semantic contextual information are continuous in nature and that they are combined in a manner consistent with a fuzzy propositional model.

Another study of this type (Beach, 1991) examined whether prosodic factors (pitch and duration of a speech segment) influence the processing of syntax. By the argument sketched above concerning the necessity of integrating all available information, one would expect that prosodic information would be so exploited, especially since syntax is prone to ambiguity. Nevertheless, the contrary view--termed the syntactic autonomy position--has been a periodically popular one in the field of psycholinguistics, in part because it may seem to simplify the job of syntax and in part for reasons analogous to those that the motivated the "speech-is-special" hypothesis described above. Beach's results indicated that people's interpretations of syntactically ambiguous sentence fragments were influenced in a systematic and continuous fashion by both pitch and duration. Because the sentence fragments contained no syntactical disambiguating information, this meant that the influence of prosody operated at the point that the syntactic decision was being made, contrary to the syntactic autonomy claim. Again, a fuzzy propositional model provided a good account of the pattern of information integration.

Yet another study of this type is that of Thompson and Massaro (1994) who examined the role of pointing gestures accompanying speech in directing children to an object. The results and conclusions were generally the same as with the two studies just described, except that in this case the study indicated that children of different ages use the different sources of information to different degrees. Specifically, while the spoken information had more influence in general, older children made more balanced use of the two types of information.

All of the studies considered so far in this section have involved situations with two alternative linguistic structures and two manipulated factors each of which influence both alternatives but in opposite directions. Another line of work has examined the case of manipulated factors each of which influences only one or the other of the two alternatives. This approach provides a direct examination of the continuousness of the support for each of the competing alternatives.

For example, Oden (1978) had people judge how likely they would be to get one versus the other interpretation for ambiguous sentences like "The girl saw the boy with the binoculars." Choosing different words for the verb of this sentence (e.g., "looked for", "hit", "pointed at") manipulates the sensibleness of the verb act being carried out with the use of binoculars. If the verbs are chosen with care, then all of them will be entirely compatible with "the boy" as direct object. More importantly, which verb is chosen should then have no effect on the sensibleness of the alternative interpretation in which the boy has the binoculars in his possession. Likewise, changing the direct object from "the boy" to "the chimp" to "the dog" etc. should affect the sensibleness of the second interpretation without influencing that of the first as long as it is the case that boys, chimps, dogs etc. can equally well be seen, looked for, hit etc. both with and without using binoculars.

The results of this experiment established 1) that the sensibleness of syntactic interpretations does indeed come in degrees, 2) that the evaluation of the sensibleness of an interpretation is performed independently of that of the other interpretation and 3) that the judged likelihood of obtaining one interpretation rather than the other was a direct function of the relative sensibleness of the two interpretations. In subsequent work, Oden and Ekberg (1981) established that relative sensibleness of interpretations also determines which interpretation is in fact obtained as assessed by a unobtrusive measure (a paraphrase task), and Oden (1983) developed a theoretical account of the fuzzy propositional structure of continuous semantic constraints that could be used to determine an interpretation's sensibleness.

17.8 CONCLUSION

Although fuzzy theories have certainly not come to dominate psychology, we have seen in this brief review that psychologists have found a remarkably wide variety of ways to make use of fuzzy constructs in the aid of their pursuits. There are, of course, areas in which appeals to 'fuzziness' by psychologists have not advanced much past the rhetorical stage. These days, however, no one in the field can deny that fuzzy sets and fuzzy logic have had a discernible impact on how and where graded or continuous concepts are used in psychology. In particular, they have helped bridge the gap between categorical and dimensional concepts, which traditionally have been restrictively identified with discrete and continuous constructs, respectively.

Interestingly, thus far nontrivial uses of fuzzy concepts have occurred rather more often in what might be loosely termed the cognitive-scientific areas than in, for instance, so-called soft systems areas such as social or clinical psychology. Nevertheless, the uses of fuzzy concepts in psychology span the entire range from the theoretical to the methodological. Furthermore, in many cases, these concepts have become assimilated and, in some instances, even merged with other related approaches and theoretical positions. Thus, it seems entirely possible that the use of fuzzy sets and fuzzy logic will come to be increasingly common but decreasingly distinct from other approaches in psychology over the years to come.

References

Alliger, G.M., Feinzig, S.L., & Janak, E.A. (1993). Fuzzy sets and personnel selection: Discussion and an application. Journal of Occupational and Organizational Psychology, 66, 163-169.

Anderson, N.H. (1981). Foundations of information integration theory. New York: Academic Press.

Anderson, N.H. (1982). Methods of information integration theory. New York: Academic Press.

Andrich, D. (1978). A rating formulation for ordered response categories. Psychometrika 43, 561-573.

Bar-Hillel, M. (1973). On the subjective probability of compound events. Organizational Behavior and Human Performance, 9, 396-406.

Bass, B.M., Cascio, W.F., and O'Connor, E.J. (1974). Magnitude estimation of expression of frequency and amount. Journal of Applied Psychology, 59, 313-320.

Beach, C. M. (1991). The interpretation of prosodic patterns at points of syntactic structure ambiguity: Evidence for cue trading relations. Journal of Memory and Language, 30, 644-663.

Beyth-Marom, R. (1981). The subjective probability of conjunctions (rep. no. 81-12). Eugene, OR: Decision Research.

Broughton, R. (1990). The prototype concept in personality assessment. Canadian-Psychology, 31, 26-37.

Budescu, D.V. & Wallsten, T.S. (1985). Consistency in interpretation of probabilistic phrases. Organizational Behavior and Human Decision Processes, 36, 391-405.

Budescu, D.V., Zwick, R., Wallsten, T. S., & Erev, I. (1990). Integration of linguistic probabilities. International Journal of Man-Machine Studies, 33, 657-676.

Burisch, M. (1993). In search of theory: Some ruminations on the nature and etiology of burnout. In Wilmar B. Schaufeli, Christina Maslach, & Tadeusz Marek (Eds.), Professional burnout: recent developments in theory and research. Washington, DC: Taylor & Francis, 75-93.

Capozza, D. & Nanni, R. (1986). Differentiation processes for social stimuli with different degrees of category representativeness. European Journal of Social Psychology, 16, 399-412.

Cohen, B. & Murphy, G. L. (1984). Models of concepts. Cognitive Science, 8, 27-58.

Cohen, J. and Hansel, C.E.M. (1959). Preferences for different combinations of chance and skill in gambling. Nature (London), 183, 841-842.

Collins, A. M. & Quillian, M. R. (1969). Retrieval time from semantic memory. Journal of Verbal Learning and Verbal Behavior, 8, 240-247.

Dubois, D. & Prade, H. (1980). Fuzzy sets and systems: theory and applications. New York: Academic Press.

Dubois, D. & Prade, H. (1985). A review of fuzzy set aggregation connectives. Information Sciences, 36, 85-121.

Fehr, B. & Russell, J.A. (1991). The concept of love viewed from a prototype perspective. Journal of Personality and Social-Psychology, 60, 425-438.

Fisher, P. F. & Orf, T. M. (1991). An investigation of the meaning of near and close on a university campus. Computers, Environment and Urban Systems, 15, 23-35.

Foddy, M. & Smithson, M. (1989). Fuzzy sets and double standards: Modeling the process of ability inference. In: J. Berger, M. Zelditch, & B. Anderson (Eds.), Sociological theories in progress: New formulations. Newbury Park, CA: Sage.

Foddy, M. & Smithson, M. (1996). "Relative ability, paths of relevance, and influence in task-oriented groups." Social Psychology Quarterly, 59, 140-153.

French, S. (1987). Fuzzy sets: the unanswered questions. Report of the Statistical Laboratory, Department of Mathematics, University of Manchester.

Goldsmith, R.W. (1978). Assessing probabilities of compound events in a judicial context. Scandinavian Journal of Psychology, 19, 103-110.

Hampton, J. A. (1988a). Disjunction of natural concepts. Memory & Cognition, 16, 579-591.

Hampton, J. A. (1988b). Overextension of conjunctive concepts: Evidence for a unitary model of concept typicality and class inclusion. Journal of Experimental Psychology: Learning, Memory, & Cognition, 14, 12-32.

Hersh, H. M. & Caramazza, A. (1976). A fuzzy set approach to modifiers and vagueness in natural language. Journal of Experimental Psychology: General, 105, 254-276.

Hesketh, B., Elmslie, S., & Kaldor, W. (1990). Career compromise: An alternative account to Gottfredson's theory. Journal of Counseling Psychology, 37, 49-56.

Hesketh, B. & Gardner, D. (1993). Person-environment fit models: A reconceptualization and empirical test. Journal-of-Vocational-Behavior; 42, 315-332.

Hesketh, B., McLachlan, K. & Gardner, D. (1992). Work adjustment theory: An empirical test using a fuzzy rating scale. Journal of Vocational Behavior, 40, 318-337.

Hesketh, B., Pryor, R.G., Gleitzman, M., & Hesketh, T. (1988). Practical applications and psychometric evaluation of a computerised fuzzy graphic rating scale. In T. Zetenyi (Ed.) Fuzzy sets in psychology. Advances in psychology, 56. Amsterdam: North-Holland, 425-454.

Holyoak, K.J. & Glass, A.L. (1978) Recognition confusions among quantifiers. Journal of verbal learning and verbal behavior, 17, 249-264.

Horowitz, L.M., & Malle, B.F. (1993). Fuzzy concepts in psychotherapy research. Psychotherapy Research, 3, 131-148.

Huttenlocher, J. & Hedges, L.V. (1994). Combining graded categories: Membership and typicality. Psychological Review, 101, 157-165.

Kay, P. & McDaniel, C. (1975). Color categories as fuzzy sets (Working Paper No. 44). Berkeley: University of California, Language Behavior Research Laboratory.

Kempton, W. (1978). Category grading and taxonomic relations: A mug is a sort of a cup. American Ethnologist, 5, 44-65.

Kochen, M. (1975). Applications of fuzzy sets in psychology. In L. A. Zadeh, K. S. Fu, K. Tanaka, & M. Shimura (Eds.), Fuzzy sets and their applications to cognitive and decision processes. New York, NY: Academic Press.

Labov, W. (1973). The boundaries of words and their meanings. In C.-J. N. Bailey & R. W. Shuy (Eds.), New ways of analyzing variation in English. Vol. 1. Washington, DC: Georgetown University Press.

Lakoff, G. (1973). Hedges: A study in meaning criteria and the logic of fuzzy concepts. Journal of Philosophical Logic, 2, 458-508.

Lichtenstein, S. & Newman, J.R. (1967). Empirical scaling of common verbal phrases associated with numerical probabilities. Psychonomic Sciences, 9, 563-564.

Lopes, L. L. (1987). Between hope and fear: The psychology of risk. Advances in Experimental Social Psychology, 20, 255-295.

Lopes, L. L. (1990). Re-modeling risk aversion. In G. M. von Furstenberg (Ed.), Acting Under Uncertainty: Multidisciplinary Conceptions (pp. 267-299). Boston: Kluwer.

Macvicar-Whelen, P. J. (1978). Fuzzy sets, the concept of height, and the hedge "very". IEEE Transactions on Systems, Man, and Cybernetics, SMC-8, 507-511.

Massaro, D. W. (1987). Speech perception by eye and by ear: a paradigm for psychological inquiry. Hillsdale, NJ: Lawrence Erlbaum Associates.

Massaro, D. W. (1989). Testing between the TRACE model and the fuzzy logical model of speech perception. Cognitive Psychology, 21, 398-421.

Massaro, D. W. & Hary (1986). Addressing issues in letter recognition. Psychological Research, 48, 123-132.

Massaro, D. W. & Sanocki, T. (1993). Visual information processing in reading. In D. M. Willows, R. S. Kruk, & E. Corcos (Eds.) Visual processes in reading and reading disabilities. Hillsdale, NJ: Lawrence Erlbaum Associates.

Massaro, D.W., Weldon, M.S., & Kitzis, S.N. (1991). Integration of orthographic and semantic information in memory retrieval. Journal of Experimental Psychology Learning, Memory and Cognition, 17, 277-287.

Manton, K.G., Woodbury, M.A., & Tolley, D.H. (1994). Statistical applications using fuzzy sets. New York, N.Y.: John Wiley & Sons.

McClelland, J. L. (1991). Stochastic interactive processes and the effect of context on perception. Cognitive Psychology, 23, 1-44.

McClelland, J. L. & Elman, J. L. (1986). The TRACE model of speech perception. Cognitive Psychology, 18, 1-86.

McCloskey, M. E. & Glucksberg, S. (1978). Natural categories: well defined or fuzzy sets? Memory & Cognition, 6, 462-472.

McDowell, B. D. & Oden, G. C. (1995). Categorical decisions, rating judgments, and information preservation. Unpublished manuscript.

Moore, C. F., Dixon, J.A.., & Haines, B.A. (1991). Components of understanding in proportional reasoning: A fuzzy set representation of developmental progressions. Child-Development, 62, 441-459.

Moyer, R. S. (1973). Comparing objects in memory: evidence suggesting an internal psychophysics. Perception & Psychophysics, 13, 180-184.

Newstead, S.E. (1988) Quantifiers as fuzzy concepts. In T. Zetenyi (Ed.) Fuzzy sets in psychology. Amsterdam: North-Holland, 51-72.

Newstead, S.E. & Collins, J. (1987). Context and the interpretation of quantifiers of frequency. Ergonomics, 24, 35-49.

Oden, G. C. (1977a). Fuzziness in semantic memory: Choosing exemplars of subjective categories. Memory & Cognition, 5, 198-204.

Oden, G. C. (1977b). Integration of fuzzy logical information. Journal of Experimental Psychology: Human Perception and Performance, 4, 565-575.

Oden, G. C. (1978). Semantic constraints and judged preference for interpretations of ambiguous sentences. Memory & Cognition, 6, 26-37.

Oden, G. C. (1979). A fuzzy logical model of letter identification. Journal of Experimental Psychology: Human Perception and Performance, 5, 336-352.

Oden, G. C. (1981). A fuzzy propositional model of concept structure and use: A case study in object identification. In G. E. Lasker (Ed.) Applied systems research and cybernetics. Vol VI. Elmsford, NY: Pergamon Press.

Oden, G. C. (1983). On the use of semantic constraints in guiding syntactic analysis. International Journal of Man-Machine Studies, 19, 335-357.

Oden, G. C. (1984a). Everything is a good example of something, and other endorsements of the adequacy of a fuzzy theory of concepts. WHIPP 21, Wisconsin Human Information Processing Program Technical Report.

Oden, G. C. (1984b). Dependence, independence, and emergence of word features. Journal of Experimental Psychology: Human Perception and Performance, 10, 394-405.

Oden, G. C. (1988). FuzzyProp: A symbolic superstrate for connectionist models. Proceedings of the IEEE International Conference on Neural Networks, Vol. I, 293-300.

Oden, G. C., & Ekberg, K. (1981). Influence of continuous semantic constraints on the resolution of ambiguous sentences. American Journal of Psychology, 94, 431-442.

Oden, G. C., & Lopes, L. L. (1982). On the internal structure of fuzzy subjective categories. In R. R. Yager (Ed.) Recent developments in fuzzy set and possibility theory. Elmsford, NY: Pergamon Press.

Oden, G. C. & Lopes, L. L. (1994). Risky choice with fuzzy criteria. Paper presented at the Symposium on Qualitative Aspects of Decision Making, Universitat Regensburg.

Oden, G. C., & Massaro, D. W. (1978). Integration of featural information in speech perception. Psychological Review, 85, 172-191.

Oden, G. C. & Rueckl, J. G. (1986). Taking language by the hand: Reading handwritten words. Paper presented at the Twenty-seventh Annual Meeting of the Psychonomic Society, New Orleans, LA.

Osgood, C. E., Suci, G. J., & Tannenbaum, P. H. (1957). The measurement of meaning. Urbana, IL: University of Illinois Press.

Osherson, D. W. & Smith, E. E. (1981). On the adequacy of prototype theory as a theory of concepts. Cognition, 9, 35-58.

Pepper, S. (1981). Problems in the quantification of frequency expressions. In D. Fiske (ed.) New Directions for Methodology of Social and Behavioral Sciences: Problems with Language Imprecision. San Francisco: Jossey-Bass.

Pepper, S. & Prytulak, L.S. (1974). Sometimes frequently means seldom: Context effects in the interpretations of quantitative expressions. Journal of Research in Personality, 8, 95-101.

Pipino, L.L., van Gigch, J.P., & Tom, G. (1981). Experiments in the representation and manipulation of labels of fuzzy sets. Behavioral Science, 26, 216-228.

Pisoni, D. B. & Lazarus, J. H. (1974). Categorical and noncategorical modes of speech perception along the voicing continuum. Journal of the Acoustical Society of America, 55, 328-333.

Pryor, R.G., Hesketh, B., & Gleitzman, M. (1989). Making things clearer by making them fuzzy: Counseling illustrations of a fuzzy graphic rating scale. Career Development Quarterly, 38, 136-147.

Rapoport, A. Wallsten, T.S., & Cox, J.A. (1987). Direct and indirect scaling of membership functions of probability phrases. Mathematical Modeling, 9, 397-411.

Reagan, R.T., Mosteller, F., Youtz, C. (1989). Quantitative meanings of verbal probability expressions. Journal of Applied Psychology, 74, 433-442.

Rosch, E. (1973). On the internal structure of perceptual and semantic categories. In T. E. Moore (Ed.), Cognitive development and the acquisition of language. New York, NY: Academic Press.

Rosch, E. (1975). Cognitive representations of semantic categories. Journal of Experimental Psychology: General, 104, 192-233.

Rosch, E. (1978). Principles of categorization. In E. Rosch & B. B. Lloyd (Eds.), Cognition and categorization. Hillsdale, NJ: Lawrence Erlbaum Associates.

Rubin, D. C. (1979). On measuring fuzziness: A comment on 'A fuzzy set approach to modifiers and vagueness in natural languages.' Journal of Experimental Psychology: General, 108, 486-489.

Rueckl, J. G., & Oden, G. C. (1986). The integration of contextual and featural information during word identification. Journal of Memory and Language, 25, 445-460.

Rumelhart, D. E., Lindsay, P. H., & Norman, D. A. (1972). A process model for long-term memory. In E. Tulving & W. Donaldson (Eds.), Organization and memory. New York, NY: Academic Press.

Russell, J.A. & Bullock, M. (1986). Fuzzy concepts and the perception of emotion in facial expressions. Social-Cognition, 4, 309-341.

Scherl, L. & Smithson, M. (1987). A new dimension to content analysis: Exploring relationships among thematic categories. Quality and Quantity, 21, 199-208.

Serban, G., Conte, H.R., & Plutchik, R. (1987). Borderline and schizotypal personality disorders: Mutually exclusive or overlapping? Journal of Personality Assessment, 51, 15-22.

Shiina, K. (1988). A fuzzy-set-theoretic feature model and its application to asymmetric similarity data analysis. Japanese Psychological Research, 30, 95-104.

Simpson, R.H. (1944). The specific meanings of certain terms indicating differing degrees of frequency. Quarterly Journal of Speech, 30, 328-330.

Smithson, M. (1984). Multivariate analysis using 'and' and 'or'. Mathematical Social Sciences, 7, 231-251.

Smithson, M. (1987). Fuzzy Set Analysis for Behavioral and Social Sciences. New York, N.Y.: Springer-Verlag.

Smithson, M. (1988). Possibility theory, fuzzy logic, and psychological explanation. In T. Zetenyi (ed.) Fuzzy Sets in Psychology. Amsterdam: North-Holland.

Smithson, M. (1989). Measures of freedom based on possibility. Journal of Mathematical Social Sciences, 18, 81-98.

Smithson, M. (1993). A class of fuzzy featural models of similarity judgments. In R. Lowen & M. Roubens (Eds.) Fuzzy logic: state of the art. Amsterdam: Kluwer, 377-384.

Smithson, M. (1995). A new approach to analysing ordinal data. Australian Journal of Social Research, 1, 119-131.

Smithson, M., Amato, P.R., & Pearce, P.L. (1983). Dimensions of helping behaviour. International Series in Experimental Social Psychology (Vol. 6), (Ed.) Michael Argyle, Oxford: Pergamon Press.

Smithson, M.J. and Hesketh, B. (1998). "Using fuzzy sets to extend Holland's theory of occupational interests." in L.Reznik, V.Dimitrov and J.Kacprzyk (Eds.) Fuzzy System Design: Social and Engineering Applications. Studies in Fuzziness and Soft Computing, Vol.17, Berlin: Physica-Verlag: pp. 132-152.

Smithson, M. & Knibb, K. (1986). New measures of association for numerical variables. Journal of Mathematical Social Sciences, 11, 161-182.

Spies, M. (1989). Syllogistic inference under uncertainty. Munich: Psychologie Verlags Union.

Steenkamp, J.B.E., & Wedel, M. (1991). A clusterwise regression method for simultaneous fuzzy market structuring and benefit segmentation. Journal of Marketing Research, 28, 385-396.

Steenkamp, J.B.E., & Wedel, M. (1993). Fuzzy clusterwise regression in benefit segmentation: Application and investigation into its validity. Journal of Business Research, 26, 237-249.

Thole , U., Zimmermann, H.-J., & Zysno, P. (1979). On the suitability of minimum and product operators for the intersection of fuzzy sets. Fuzzy Sets and Systems, 2, 167-180.

Thompson & Massaro (1994). Children's integration of speech and pointing gestures in comprehension. Journal of Experimental Child Psychology, 57, 327-354.

Turksen, I.B. (1991). Measurement of membership functions and their acquisition. Fuzzy Sets and Systems, 40, 5-38.

Tversky, A. (1977). Features of similarity. Psychological Review, 84, 327-352.

Tversky, A. and Kahneman, D. (1982). Judgments of and by representativeness. In D. Kahneman, P. Slovic, and A. Tversky (eds.) Judgment Under Uncertainty: Heuristics and Biases. New York: Cambridge University Press.

Tversky, A. and Kahneman, D. (1983). Extensional versus intuitive reasoning: The conjunction fallacy in probability judgment. Psychological Review, 90, 293-315.

Wallsten, T.S., Budescu, D., Rappoport, A., Zwick, R., and Forsyth, B. (1986). Measuring the vague meanings of probability terms. Journal of Experimental Psychology: General, 115, 348-365.

Wallsten, T.S., Fillenbaum, S. and Cox, J.A. (1986). Base-rate effects on the interpretation of probability and frequency expressions. Journal of Memory and Language, 25, 571-581.

Waterhouse,L., Wing, L, & Fein, D. (1989). Re-evaluating the syndrome of autism in the light of empirical research. In Geraldine Dawson (Ed.) Autism: Nature, Diagnosis, and Treatment. New York: Guilford Press, 263-281.

Wyer, R.S., Jr. 1970 "Quantitative prediction of belief and opinion change: A further test of a subjective probability model." Journal of Personality and Social Psychology, 16: 559-570.

Yamashita, T. & Yamashita, K. (1988). Fuzzy theory approach to concepts and logical judgment. Japanese Psychological Review, 31, 208-228.

Yamashita, T. & Yamashita, K. (1992). Exemplifying and comparing fuzzy ratings and magnitude estimations. Psychologia: An International Journal of Psychology in the Orient, 35, 240-248.

Zadeh, L. (1965). Fuzzy Sets. Information and Control, 8, 338-353.

Zadeh, L. A. (1972). A fuzzy-set-theoretic interpretation of linguistic hedges. Journal of Cybernetics, 2, 4-34.

Zetenyi, T. (Ed.) (1988). Fuzzy sets in psychology. Amsterdam: North-Holland.

Zimmermann, H.-J. & Zysno, P. (1980). Latent connectives in human decision making. Fuzzy Sets and Systems, 4: 37-51.

Zwick, R. (1987). A note on random sets and the Thurstonian scaling methods. Fuzzy Sets and Systems, 21, 351-356.

Zwick, R., Budescus, D.V., & Wallsten, T.S. (1988). An empirical study of the integration of linguistic probabilities. In T. Zetenyi (Ed.) Fuzzy sets in psychology. Amsterdam: North-Holland, 91-126.

Zwick, R. & Wallsten, T.S. (1989). Combining stochastic uncertainty and linguistic inexactness: Theory and experimental evaluation of four fuzzy probability models. International Journal of Man Machine Studies, 30, 69-111.

V Tools

18 FUZZY SETS IN HUMAN FACTORS AND ERGONOMICS

Waldemar Karwowski
Jerzy Grobelny
WookGee Lee
Yung-Nien Yang

Abstract: The science of human factors and ergonomics aims to discover and model how people interact with their environments, given the specific human characteristics, limitations, tolerances, and performance capabilities. In doing so, ergonomics faces the problems of work system's complexity and the related human and system-based fuzziness. Fuzziness is not only a product of human mind, but the essence of human development and existence, and a necessary condition for human learning, growth and survival. In order to develop the symbiotic relationship between people and the outside surroundings (natural and artificial, i.e. technology-based), the intrinsic fuzziness of the human kind must be treated by the system designers and engineers as natural requirements of their everyday activities. Fuzzy systems methodologies allow accounting for natural human and human-machine fuzziness, and provide the necessary framework for successful modeling efforts in the human factors and ergonomics discipline.

18.1 INTRODUCTION

The human world is complex and exhibits inherent uncertainty and vagueness. Therefore, in order to study behavior of human systems it is necessary to develop modeling approaches which are approximate in nature (Zadeh 1974). The human world-based uncertainty and vagueness can be conceptualized and modeled through the notion of fuzziness. Karwowski (1991, 1992) suggested that fuzziness should be looked upon as the natural model of people at work and those human-made systems that interact with people, embedded into traditional descriptions of human sensory, information processing and communication, or physiological functioning processes. In this context, fuzziness is not just a product of human mind that can be described and comprehended only through formal mathematical theories. Fuzziness, as a basic quality of human understanding (Saaty 1977), is also the essence of human development and existence, and a necessary condition for human learning and growth.

The science that deals with the broad scope of problems due to human-machine-environment interactions is the human factors and ergonomics (HFE). The HFE is concerned with consideration of human characteristics, expectations, and behaviors in the design, testing and evaluation of the artifacts that people interact with in their work and everyday lives, including consumer products, complex human-machine systems, as well as natural and human-made environments. The human factors and ergonomics discipline has its origins in the XIX-th century when Wojciech Jastrzebowski (1857) of Poland defined ergonomics by coining two Greek words: ergon = work, and nomos = laws. The 'science of work' signifies the human work, play, thinking and devotion as reflected in the manner of optimizing the use of four distinct human characteristics, i.e. 1) motor (physical), 2) sensory (aesthetic), 3) mental (intellectual), and 4) spiritual or moral (Karwowski, 1991).

The contemporary human factors and ergonomics discipline attempts to model and describe how people interact with their outside environment, even though some believe that: 1) human beings are too complex to be fully understood or describable in all their properties, limits, tolerances, and performance capabilities, 2) no comprehensive mathematical models are available to describe and integrate all the above mentioned measures and findings about human behavior, and 3) the mathematical models currently used in human factors and ergonomics suffer from the measuring problem, which includes difficulties in descriptions of the varying task load and human workload, the state of social and physical environments, and design and measurement of the information flow in people and machines (Bernotat, 1984). The methodology of fuzzy sets and systems is viewed as a powerful tool that can allow to overcome some of the above problems.

As pointed out by Smithson (1982), the potential advantages for fuzzy concepts applications in human sciences are as follows: 1) fuzziness, itself, is a useful metaphor or model for human language and categorizing processes, 2) fuzzy mathematics may be used to augment conventional statistical techniques in the analysis of fuzzy data, and 3) fuzzy methods are useful supplements for statistical techniques such as reliability analysis and regressions, and structurally oriented methods such as hierarchical clustering and multidimensional scaling. In view of the above, fuzzy systems methodology can provide a useful research framework for the human factors and ergonomics area, and especially for modeling variety of complex task situations, systems, artifacts and environments, and their interactions with people.

18.2 MODELS OF HUMAN-MACHINE-ENVIRONMENT SYSTEMS

Zimmermann (1980, 1985) classified scientific models into three groups, i.e.: 1) formal models or purely axiomatic systems with fictitious hypotheses, 2) factual models, where conclusions have a bearing on reality, and have to be verified empirically, and 3) prescriptive models which postulate rules according to which people should behave. In the human factors and ergonomics area, the human-machine models include the cognition-theoretic models (reliability models, network models, information processing models), the control-theoretic models for continuous control, and signal detection and decision-making (Pew and Baron, 1983).

According to Harre (1972), models are used in science to make inferences which would not otherwise be possible (logical models), or to express and extend knowledge about the real world (epistemological models). Contemporary models used in human factors and ergonomics applications utilize formal structures to represent human-system-environment reality perceived by the human operator. However, the process of human perception (of what is real and what constitutes the reality) is intrinsically biased by the natural fuzziness due to the human itself (human fuzziness), and by complexity and related uncertainty of the system under observation (human-machine fuzziness). As pointed out by Sheridan and Ferrell (1981), engineering systems can be made compatible with human characteristics and limitations only by means of quantitative analysis and experiment, and only when the behavior of both man and machine can be described in comparable terms. Unfortunately, the classical information, control and decision models which are based on the crisp (non-fuzzy) mathematics are often not applicable to modeling of the performance of both machines and human operators.

18.2.1 Fuzziness of the Human-Machine System

The human factors and ergonomics aims to optimize the design and functioning of human-machine systems with respect to characteristics of people and the complex and uncertain relationships between system users, machines, and the outside environments. This is done in order to identify and reduce the existing (ergonomic) incompatibilities between the human capacities and task requirements, and to make the workplace a safe, healthy, productive, socially comfortable and satisfying one (Karwowski, 1991). Since the human-machine systems exhibit inherent uncertainty, an implementation of this goal requires application of the appropriate research methodology to allow for modeling of the human- and system-based fuzziness.

The analysis of human-machine systems must include at least three types of uncertainty inherent to such systems; i.e. inaccuracy, randomness, and vagueness (Bezdek, 1981). Uncertainties due to inaccuracy are related to observations and measurements (representations), while those due to randomness (of events) are independent from observations and constitute an objective property of some real process. Fuzziness relates to gradations in categories, i.e., it represents the degree vagueness (Zadeh, 1965; Smithson, 1982). According to Kosko (1992), fuzziness describes event ambiguity and measures the degree to which an event occurs, not whether it occurs. While randomness describes the uncertainty of event occurrence, fuzziness is a type of deterministic uncertainty, with ambiguity as a property of physical phenomena. Uncertainty due to vagueness (or fuzziness) is inherent to any complex system, and to the human thought and perception processes (Zadeh, 1973).

In the area of human factors and ergonomics, fuzziness occurs at many levels of human interaction with the working environments. A simple human work taxonomy can be used to describe five different levels of human functioning, ranging from primarily physical tasks to cognitive tasks (Rohmert, 1979). These universal human activities are: 1) tasks that produce force (primarily muscular work), 2) tasks of continuously coordinating sensory-monitor functions (like assembling or tracking tasks), 3) tasks of converting information into motor actions (e.g. inspection tasks), 4) tasks of converting information into output information (e.g. required control tasks), and 5) tasks of producing information (primarily creative work).

18.2.2 The Human Functioning

As shown in Figure 1, the human functioning can be described in terms of such processes as perception, information processing and decision making, memory, attention, feedback, and human response processes (Karwowski, 1992). Any task in a human-machine system requires processing of information which is gathered based on perceived and interpreted relationships between system elements. The processed information may need to be stored by either human or machine for later use. According to the model of human information processing (Wickens, 1984), the sensory information coming from the sensor organs is compared to some internal representation of the recognized object stored in the permanent (long-term) memory. After the stimulus is identified, a decision can be made regarding the appropriate course of action. The decision process may require some transformations of the available information in the working memory, information storing in the long-term memory, and generating a response. The response can be realized through a process of coordinated muscular control actions, which effects are perceived through the feedback system. An inherent fuzziness of these natural processes is discussed below.

18.2.2.1 Human perceptual processes

Human perception is a dynamic process under control of the observer who is consciously aware of the observed phenomena. In that context, perception is constructed by the observer rather than being determined by parameters of physical signals received from the receptors. The purpose of the human perceptual system is to provide sufficient conditions for adaptive behavior (Foley and Moray 1987).

As discussed by Karwowski (1992), the process of conscious perception is a fuzzy one, with ambiguity due to both physiological thresholds for physical stimuli and pattern recognition and judgement requirements. For example, a question of "at what frequency will an intermittent light be perceived as just flickering" is an ambiguous one, and, hence, the eye flicker frequency is a fuzzy, not random phenomenon, as it refers to the degree of noticeable flickering rather than the question of whether it occurs.

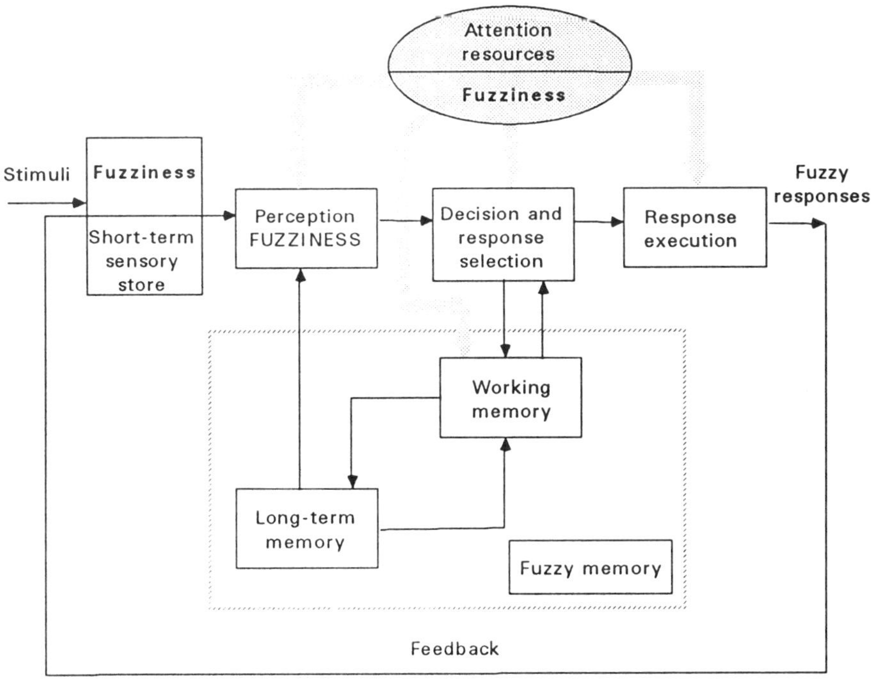

FIGURE 1. A fuzzy model of human information processing (modified after Wickens, 1984)

For the same reason, the psychophysical laws which relate the magnitude of change in physical stimuli (ΔI) that will just be noticed by an observer, are fuzzy laws. For example, according to the Weber-Fetchner Law, the magnitude of change in a physical stimulus that will be just noticed by an observer is a constant proportion of the stimulus. Although defined in terms of the magnitude of ΔI that will result in a judgement of a difference in the levels of physical stimuli 50% of the time (so called just noticeable difference or jnd), this law refers not to the probabilistic statement of whether the perception of change will occur, but at what level (degree) of the difference in physical stimuli such a change will be observed.

18.2.2.2 Fuzziness in perception and memory

The human perceptual system is limited in its ability to process information from different stimulus objects at one time. However, several dimensions of a single object can be processed in parallel (Wickens, 1987). As discussed by Karwowski (1992), the question, therefore, is to what degree many sources of object stimuli and different dimensions of the same objects can be processed by the human brain, rather than whether they can be processed, indicating the fuzzy nature of human perceptual processes. A similar question can be posed with respect to stimulus-central processing capability. For example, in human-machine system, the displayed information can be arranged along a continuum which defines the degree to which that information is spatial-analog in nature (i.e. information about relative locations, transformations or continuous motion), linguistic-symbolic or verbal (i.e. a set of instructions, alphanumeric codes, directions or logical operations). According to Wickens (1987), the two end points of this verbal-spacial continuum distinguish the verbal and spacial memory systems. The border between these two systems, however, is ambiguous and subject to fuzzy interpretation.

With respect to memory processes, the capacity of working memory is limited to 7 (+/-) 2 of the unrelated items (Miller, 1956). The important question is to what degree one can recall and operate on a given number of items. Although the distinction between the short-term (working) memory and long-term memory has been universally accepted, the degree to which the memory is short or long is also a fuzzy category (Karwowski 1992). The same applies, and even more so, to the classification of spacial and verbal memory systems as discussed above.

18.2.2.3 Human response processes

Decision-making and diagnosis in human-machine systems have shown to be of uncertain nature, and extensively discussed in the fuzzy literature. Decisions are typically followed by responses. The human response processes are intrinsically fuzzy even though the prevalent definition of the relationship between the choice reaction time and the degree of choice (the Hick-Hyman law) is based upon the information content of a stimulus (S) in bits as follows:

$$RT = a + b\,S$$

The very visible presence of human fuzziness cannot be overlooked in the paradigm of stimulus-response compatibility (Wickens, 1987). This paradigm relates to the physical relationship (compatibility) between a set of stimuli and a set of responses, as it affects the speed of human response. As pointed out by Karwowski (1992), the spacial relations between arrangements of signals and response devices in human-machine systems with respect to direction of movement and adjustments are often ambiguous, with a high degree of uncertainty regarding the effects of intended control actions.

18.2.2.4 Human attention processes

Human attention refers to abilities to process information in parallel or time-sharing fashion (Wickens, 1987). The human ability to do so can be measured by the graded (fuzzy) categories of success. According to Moray (1984), the selective allocation of human attention is determined by the human operator's internal model of the statistical properties of the environment. However, the outside environment does not always exhibit only statistical properties, and, more often than not, the operator must face the ever present fuzziness, especially in complex environments.

18.2.3 Assessment of Human Workload

An assessment of resource demand or mental workload imposed on human operator in the human-machine system is of great importance to assure acceptable levels of system performance (Wickens 1987). As discussed by Karwowski (1992), the measures of mental workload, such as those of perceptual-cognitive effort or response loading, are fuzzy phenomena, by the very nature of underlying processes. Therefore, the assessment methods purpose must exhibit fuzzy properties of the investigated systems. Despite the fact that fuzziness occurs at many levels of cognitive processes, people have the unique ability to understand and apply vague and uncertain concepts (Hersh et al., 1976; Kramer, 1983). People are also able to comprehend vague concepts of the natural language, as if they were represented by fuzzy labels, and manipulate them according to rules of fuzzy logic (Oden, 1977). Developments in semantic memory indicate that people perceive natural categories as fuzzy sets with no clear boundaries separating the category members from non-members (McCloskey and Glucksberg, 1978). These phenomena, however, are difficult to study in the context of traditional scientific framework, where precision and significance are treated as mutually exclusive characteristics (Zadeh, 1973).

Uncertainty, looked upon in the context of mental workload, which causes unpredictability in one's stimulus and/or response, enters a work situation from several sources (Audley et al., 1979). These sources include: 1) external disturbances, 2) varying parameters of the system structure external to the human operator, 3) human produced noise in observing the task stimuli, 4) lack of good internal model of the external system, 5) human-produced distortions in interpreting the externally stipulated criterion of performance, and 6) human-produced motor noise. Unfortunately, most of traditional human factors and ergonomics methodologies disregard the human-machine system-based fuzziness.

18.3 EXAMPLES OF EARLY APPLICATIONS OF FUZZY SYSTEMS IN HUMAN FACTORS AND ERGONOMICS

Fuzzy system methodologies have been successfully applied in the modeling of complex and uncertain systems in a variety of disciplines, including cognitive psychology, information processing and control, decision-making sciences, biological and medical sciences, sociology and linguistics, image processing and pattern recognition, and artificial intelligence (Karwowski and Salvendy 1992). Selected examples of early applications of fuzzy methodologies in human factors and ergonomics (Karwowski ands Mital, 1986) are briefly reviewed below.

Willaeys and Malvache (1979) investigated the perception of visual and vestibular information in the industrial inspection task. Hirsch et al. (1981) used a fuzzy dissimilitude relation to describe human vocal patterns. Kramer and Rohr (1982) developed a fuzzy model of driver-behavior based on simulated visual pattern processing in lame control. Benson (1982) developed an interactive computer graphics program for analytical tasks which are not well defined or utilize imprecise data. Terano et al. (1983) introduced a fuzzy set approach into fault-tree analysis, and studied the fuzziness of a human-reliability concept from the human-machine systems safety point of view. Karwowski and others (1983, 1984a, and 1984b) developed a fuzzy set based model to assess the acceptability of stresses in manual lifting tasks. Luczak and Ge (1989) applied fuzzy modeling to investigate the relations between physical weight and perceived load. Recently, Karwowski and his colleagues (1994, 1995, 1999) developed and applied a fuzzy linguistic-based estimation method to investigate human perception of load heaviness.

Hunt and Rouse (1984) proposed a fuzzy rule-based model of human problem solving in the fault diagnosis tasks. Simcox (1984) used the linguistic approach to devise a method for pragmatic communication in graphic displays. Boy and Kuss (1986) have applied a fuzzy method for modeling of human-computer interactions in information retrieval tasks, and implemented their method in the computer-based library retrieval system (BIBLIO). Hesketh et al. (1988) developed a computerized method for fuzzy graphic rating scale using the FUZRATE program which feeds back to the user his/her fuzzy ratings, and then presents the results of combining these ratings. Onisawa (1988) developed several fuzzy concepts to improve the studies of human reliability issues. Karwowski, et al. (1990) developed a framework fuzzy GOMS model for studying human-computer interaction on the text editing task.

18.4 HUMAN-MACHINE SYSTEM AS A FUZZY SYSTEM

Karwowski (1991, 1992) proposed that the human-machine system (HMS) be modeled using the component fuzziness (F) of the human elements (H-subsystem), machine elements (M-subsystem), environmental elements (E-subsystem), and the ergonomic interactions (I-subsystem) as fuzzy relations between these elements, and time (T), and be expressed as follows:

$$F = \{F_H, F_M, F_E, F_I, T\}$$

where:

$F_H = \{h_1, h_2, \ldots h_i\}$ = contributing fuzziness due to human elements,
$F_M = \{m_1, m_2, \ldots m_j\}$ = contributing fuzziness due to machine elements,

$F_E = \{e_1, e_2, \ldots e_u\}$ = contributing fuzziness due to environmental elements,
$F_I = \{i_1, i_2, \ldots i_n\}$ = contributing fuzziness due to ergonomic interactions, and
T = time.

The interactions (ij) in the above model reflect fuzziness in the relationships between relevant human (H) characteristics, characteristics of a machine (M), and elements representing environmental (physical and social) conditions (E). Each element of the above system can be represented as a fuzzy set defined on a unique discourse of universe. Furthermore, the fuzziness of the ergonomic system components, which are represented by fuzzy sets, can be modeled using a fuzzy entropy measure. In a general case, the system elements represented as fuzzy sets can be modeled as points in a space. As proposed by Kosko (1992), the fuzzy power set $F(2^X)$, i.e. the set of all fuzzy subsets of X looks like a cube, where a given fuzzy set can be interpreted as a point in that cube. The set of all fuzzy subsets forms the unit hypercube, while a fuzzy set can be represented as any point in the cube. As the fuzzy set defines a point in a cube, a fuzzy system defines a mapping between cubes. A fuzzy system S maps fuzzy sets to fuzzy sets, and, therefore, is a transformation (Kosko 1992).

18.5 FUZZINESS AND HUMAN-SYSTEM INCOMPATIBILITY

The fuzziness in the above defined fuzzy human-machine system reflects the incompatibility as an inherent property of this system (Karwowski et al., 1988; Karwowski, 1991). The ergonomic incompatibility (EI) is defined as the degradation (disintegration) of the ideal ergonomic interactions of the system, and is often reflected in the system's mismatch between people and, for example, their working tools, machines or environments, leading to measurable human losses (accidents and injuries) and system's inefficiency. The ergonomic incompatibility is the convenient construct for modeling of the ergonomic system's entropy.

As discussed by Karwowski (1991), the ergonomic system can be characterized by the levels of entropy of its elements. Such entropy is interpreted as the extent of deviation from the state of ideal, i.e. fully compatible ergonomic interactions (I). The total entropy of the ergonomic system is determined by the individual entropies of the human, machine and environmental components, and by the entropies of the interactions between them. Thus, total entropy attributable to the entire ergonomic system is due to the entropies derived from each of its interdependent elements.

The human-machine system as described above has three basic properties as described by Karwowski (1991). First, the total entropy (TE) of the system is a sum of the system's ergonomic entropy (EE) and its undesirable entropy (UE). Second, the ergonomic entropy value, which is non-reducible, constitutes an ideal lower limit of the total human-machine system's entropy, which seldom can be achieved practically. Third, the undesirable entropy (UE) of the ergonomic system is reducible, and, therefore, it constitutes the object of human factors and ergonomics efforts in the practical quest of fitting the machine and the task to the human (Grandjean, 1984).

18.6 RECENT APPLICATIONS OF FUZZY METHODOLOGIES IN HUMAN FACTORS

Uncertainty which arises in all phases of information gathering, acquisition and decision making, plays a key role in development of effective models of human-system functioning. For example, uncertainty may originate from the system ambiguity due to limited knowledge of the processes and their generated outcomes, and is correlated with task difficulty. Task difficulty gives rise to error and inefficient performance, as well as anxiety and stress, which are undesirable effects of the human-system performance.

The rationale of using fuzzy methodologies in the HFE research is based on their ability to improve the quality of system modeling by admitting real conditions of the uncertain, imprecise, and noisy environments. For example, fuzzy logic allows to formalize and model human capacities for imprecise reasoning that can be used for control of complex and ill-defined processes without much knowledge of their underlying dynamics (Ross 1995). Fuzzy logic also allows to process the knowledge elicited from human operators, and helps to overcome difficulties in controlling systems exhibiting imprecise characteristics and behaving in a non-linear fashion.

18.6.1 Human-Machine Reliability and Fuzziness

Human reliability can be defined as a probability that a person correctly performs some system-related activity in a required time period (if time is a limiting factor), or as the probability of a successful performance of a mission (Swain and Gutman, 1983). In this view, the human reliability is an antithesis of error likelihood or probability that no errors occur (Miller and Swain, 1987). This almost exclusively probability-oriented approach to human reliability is also expressed through the classical Technique for Human Error Rate Prediction (THERP), and to evaluate degradation of a human-machine system likely to be caused by human errors alone or in connection with equipment functioning, operational procedures and practices, or other system and human characteristics that influence system behavior.

Given the natural fuzziness of human interactions with the outside world, both probabilistic and possibilistic approaches to human reliability are needed. Since the human operator characteristics which affect the reliability include such fuzzy processes as sensory perception, motor control, cognitive functions (information processing and decision-making) and meta-cognitive functions (intuition and abstract processing), human reliability models must allow for qualitative reasoning through human error possibility measures, and should account for natural vagueness of higher cognitive functions. Onisawa (1988) proposed the concept of human error based on the possibility measure which is affected by several human performance shaping factors. Pedrycz (1990) discussed cognitive aspects of information processing and proposed a fuzzy framework for development of human perception perspective as opposed to machine perspective.

18.6.2 Ergonomic Fuzziness in Human-Computer Interaction

Karwowski et al. (1990, 1992) discussed the problems of fuzziness due to high complexity of human-computer systems and the nature of user's perception and information processing. Since the interaction between people and computers reflects the cognitive fuzziness of the data, as well as user's uncertainty exhibited in perception of computing environment, the human-computer interaction (HCI) focuses also on developing an effective fuzzy-based communication tools.

18.6.2.1 The human-computer interaction system

Karwowski et al. (1990) defined the human-computer interaction system (HCIS) as a quintuple:

$$\text{HCIS} = (T, U, C, E, I)$$

where: T - task requirements (physical and cognitive)
U - user characteristics (physical and cognitive)
C - computer characteristics (hardware and software including computer interfaces)
E - an environment
I - a set of interactions.

The set of interactions I embodies all possible interactions between T, U, C in E regardless of their nature or strength of association. The interactions I can be elemental, i.e. one to one association, or complex, such as an interaction between the user, the particular software used to achieve the desired task, and available physical interface with the computer.

According to the model of computer user's information-processing (GOMS model), the user's cognitive structure consists of the following four components (Card et al., 1983): 1) a set of Goals, 2) a set of Operators, 3) a set of Methods for achieving the goals, 4) a set of Selection Rules for choosing among competing methods for goals. In 1983, Card et al. also devised a text-editing experiment to show the validity of GOMS model, and concluded that GOMS knowledge representations were valid for such tasks. Karwowski et al., (1989) and Karwowski et al., (1990) proposed enhancements to the GOMS model in order to account for uncertainty within selection rules, and to generalize the GOMS model, where Goals, Operators and Methods components can be either precise or fuzzy, and where the Selection Rules can be expressed in either probabilistic or fuzzy manner.

18.6.2.2. Experimental verification of the fuzzy GOMS model

The example presented below refers to the generalized GOMS structure, with the sets of Goals and Operators precisely defined, while the (predicted) Methods used by subjects, as well as specific Selection Rules applied to accomplish the editing task based on fuzzy concepts. The fuzzy concepts used included application of linguistic descriptors, fuzzy connectives and fuzzy logic, and possibilistic measures of uncertainty. Such model was defined to as the Fuzzy GOMS model. Karwowski et al. (1989) reported an experiment designed to validate the fuzzy-based GOMS model for text editing task. The experiment was a variation of the manuscript editing experiment by Card et al. (1983), with the subject verbalizing five different cursor placement rules.

For example, the editing rules used by the subject included the following fuzzy statements: 1) If the word is more than half of a screen from the cursor and on the same screen or if the word is more than half of a screen from the cursor and across the printed page then use method #1; 2) If the word is more than 70 lines and the pattern is not distinct then use method #2; 3) If the word is less than half of a screen and on the left half of the page use method #3; 4) If the word is less than half of a screen and on the right half of the page use method #4. Since the subject did not have the perfect cognitive ability to divide a screen directly in half, he elicited the knowledge as fuzzy knowledge. The membership functions of the linguistic descriptors were derived based on perceived numbers of lines or characters. Once all the rules, methods, and corresponding membership functions have been elicited, the theory of possibility (Zadeh, 1978) was used to model the subject's rule selection process. For this purpose, each of the potential rules was assigned a possibility measure equal to the membership value(s) associated with it during the elicitation phase of experiment.

The possibility measure p(A) was defined after Zadeh (1978) as follows:

$p(A) = \text{Poss}\,\{X \text{ is } A\} \Rightarrow \sup \min \{A(u), p_X(u)\}$, where $p_X(u)$ is the possibility distribution induced by the proposition (X is Z), and A is a fuzzy set in the universe U.

The sub-task "move down 27 lines to a position in column 20", illustrates the process of predicting the rule selection based on the linguistic ambiguity of the subject's actions. For example, the rules (R) that applied in the above task were as follows:

Rule #1: membership value of more than a half of the screen = 0.4; [The possibility that the rule applies is 0.4],

Rule #2: membership value of more than 70 lines = 0; [The possibility that the rule applies is 0],

Rule #3: membership value of less than half of the screen = 0.3, and membership value of left hand side of the line = 0.4; [The possibility that the rule applies is 0.3 and 0.4],

Rule #4: membership value of right half of line = 0.9, and membership value of less than half of the screen = 0.3; [The possibility that the rule applies is 0.3 and 0.9],

Rule #5: membership value of more than 70 lines = 0; [The possibility that the rule applies is 0].

The possibility measure of the possibility distribution of X that the subject would select a given rule from the universe of available rules R was defined after Zadeh (1978). In case of the example cited above, the most applicable rule was derived based on possibility measure of {X is Rule #} as follows:

Poss {X is Rule #}=max [{(rule#1, 0.4)}, {(rule#2, 0)}, min {(rule#3, 0.3), (rule#3,0.4)}, min { (rule#4, 0.3), (rule#4, 0.9)}, {(rule#5, 0)}]; max [{(rule#1, 0.4)}, {(rule#2, 0)},{(rule#3, 0.3)}, {(rule#4, 0.3)}, {(rule#5, 0)}];

Poss {X is Rule #}={(Rule#1, 0.4)}.

Since the possibility of selecting rule#1, as the most applicable one out of the five rules (R), was 0.4, it was predicted based on the possibilistic measure of uncertainty that the subject would use rule #1, i.e. the control-d method. All fuzzy model predictions in the experiment were checked against the actual selection rule decisions made by the subjects. In the pilot study (Karwowski et al. 1989), a model was run using the fuzzy GOMS approach to the cursor placement task. The fuzzy GOMS model predicted 13 out of 17 decisions, or 76% correctly. The non-fuzzy GOMS model predicted only 8, or 47% of the cursor placement decisions correctly. It was reported that the use of fuzzy concepts seemed very natural within the knowledge elicitation process. It was much easier to ask for memberships values in the linguistic terms, than to ascertain exact cut-offs for selection rules. This observation supports the results of the study by Kochen (1975) who found a higher degree of consistency in subjects responses when allowed to give imprecise (verbal) descriptors.

In the follow-up study reported by Karwowski et al. (1990), five subjects were asked to perform a word placement task, while explaining what and why they were choosing their particular methods, and verbalizing the associated selection rules. If the selection rules appeared to have fuzzy components, these components were quantified by asking the subjects to verbalize a membership value for the applicability of the rule. Overall, across all subjects and trials, the non-fuzzy GOMS model successfully predicted 58.7% of the responses, while the fuzzy GOMS model predicted significantly more correct responses, i.e. 82.3% of all subjects decisions.

18.6.2.3 Assessment of computer screen design quality

Computer screen design plays an important role in determining the effectiveness of human interaction with computer systems. For example, Tullis (1981) showed that a redesign of faulty screen formats reduced the mean time required by the human operator for data interpretation by 40%. As discussed by Grobelny et al. (1995), among well defined quantitative (empirical) relationships between the screen formats and their effectiveness, there are many rules of thumb which are based on subjective views and anecdotal knowledge. This is due to limited empirical data, as well as the non-measurable character of many parameters and relationships that determine screen design quality. Also, the lack of consistent measures and quantitative criteria for assessment of screen quality makes the evaluation of system efficiency and comparison of different screen designs difficult.

Grobelny et al. (1994) applied the fuzzy-based linguistic patterns for assessment of the computer screen design quality. The linguistic patterns, originally proposed by Grobelny (1987), are based on categories of expressions related closely to natural language and truth values, which are close to a human designer's intuition. These patterns constitute the system of concepts, relations and definitions, including: a) the implication and definition of linguistic variables, b) a degree of truth of an implication, c) intensity levels of implication variables, d) degree of truth of the consistency of two expressions, e) definitions of linguistic relationships, and f) definitions of modifiers for linguistic expressions and connectors. The developed modeling framework allows to unify different aspects of computer screen design (for example see Figure 2), and proposes a quantitative criterion for an optimized computer screen format.

For example, when information on the screen is organized in groups of closely related items, the searching times for the desired information radically improves. The amount of information presented in the optimally sized items, which do not exceed 5 deg visual angles, greatly influences the searching times. If groups are significantly larger than optimal, the mean group size becomes the main factor that can be used to predict searching times. In view of the above findings, the following pattern (see Figure 3) of the desired computer screen format was proposed:

Pattern (B)

IF (information items are properly grouped) AND (the mean visual angle is SMALL), THEN (GSD = RATHER-LESS-THAN-60) ELSE IF (information is NOT properly grouped) OR (the mean visual angle is LARGE), THEN (GSD = MUCH-LESS-THAN-60).

Logical information links and the screen layout were modeled using different fuzzy concepts. For example, since the natural human scanning patterns during reading (left-right, up-down) influence visual searching process on a VDT, the upper left part of the screen should be treated as a prominent location, and the lower right as least desirable. Given the above findings, the most desirable or frequently used information should be placed in a prominent location. The following propositions (linguistic patterns C and D) were then developed to accommodate the above design requirements:

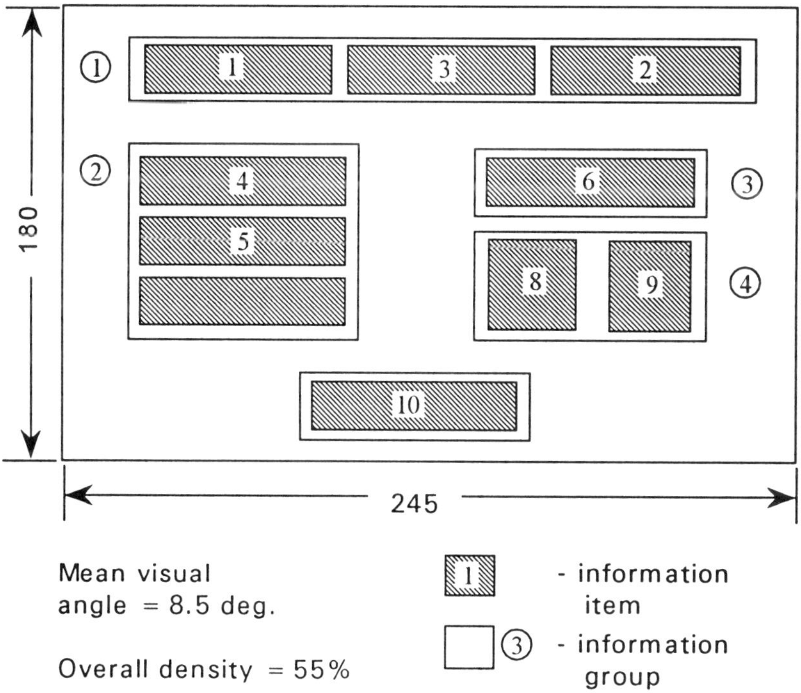

FIGURE 2. An exemplary screen layout (modified after Grobelny et al. 1995)

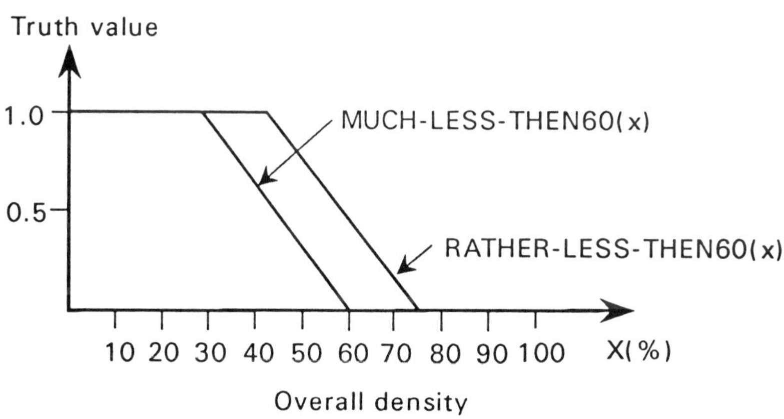

FIGURE 3. Propositions of membership function shapes defining two limits of a screen density for Pattern B (after Grobelny et al. 1995).

Pattern (C)

IF frequency (of a given item) is BIG THEN location (of this item) is PROMINENT

Pattern (D)

IF importance (of a given item) is BIG THEN location (of this item) is PROMINENT.

The degree to which a given screen format satisfies patterns (C) and (D) can be found by defining the limits for the BIG and PROMINENT categories in appropriate universes. The arbitrary limits for the BIG and PROMINENT categories are shown in Figure 4. Due to the lack of precise data, the linear function shapes were assumed. In addition, it was proposed that the 'importance' has the same "percentage" universe as the 'frequency', even though these two variables have different measurements procedures. Variables appearing in patterns (C) and (D) are very different if one wants to "measure" their realizations because unlike the 'importance' of a given item, its 'frequency' and 'location' are physically measurable quantities. One can determine the percentage frequency of possible system functions in which a given item is or will be used, and find a location using the coordinates of the screen.

The above modeling framework, that deals with relatively well documented aspects of screen design, was designed to be generic and universal. As such it is also independent of the specific technology used by computer manufacturers in the design of human-computer interfaces. The study demonstrated that it is possible to achieve rational and relatively easy to interpret assessment of different screen designs in the form of the degrees of truth. Such evaluation criterion reflects the compatibility of a given screen design with the optimal one based on the current knowledge in the field. It was concluded that the proposed methodological framework for computer screen design should significantly augment the efforts of the human designers.

18.6.3 Fuzzy Modeling of Physical Tasks

18.6.3.1 Modeling of stresses in manual lifting

Manual materials handling (MMH) tasks, which include the acts of manually lifting, lowering, pushing, or pulling an object, are believed to be one of the major causes of work-related low back disorders (LBD) accounting for a large costs to industry. The three main approaches used for investigating manual task characteristics associated with work-related injuries are the biomechanical, physiological, and psychophysical approach (Ayoub et al. 1997). The use of psychophysics in lifting tasks requires the subject to adjust the weight of load according to his or her perception of effort in order to minimize the potential for over-exertion or excessive fatigue. The psychophysical design criterion is a measure of perceived stress on the job. Karwowski (1983) and Karwowski and Ayoub (1984a) applied fuzzy sets to model and assess the acceptability of stresses involved in manual lifting task. They hypothesized that a combination of acceptability measures of the biomechanical and physiological stresses leads to an overall (psychophysical) measure of lifting task acceptability. The fuzzy-based acceptability measure from the biomechanical point of view was based on the largest value of the maximum compressive force considered as 'safe'. The physiological acceptability measure was based on recommended value of oxygen uptake for an 8 hour long physical activity and the actual oxygen consumption resulting from lifting. From the psychophysical point of view, the acceptability was based on the lifting capacity norms. Based on the derived fuzzy similarity values between the acceptability measures of the combined and the psychophysical stresses, it was concluded that the MAWL based on the psychological design criterion appeared to be the result of the integration of the biomechanical and physiological stresses imposed by the lifting task.

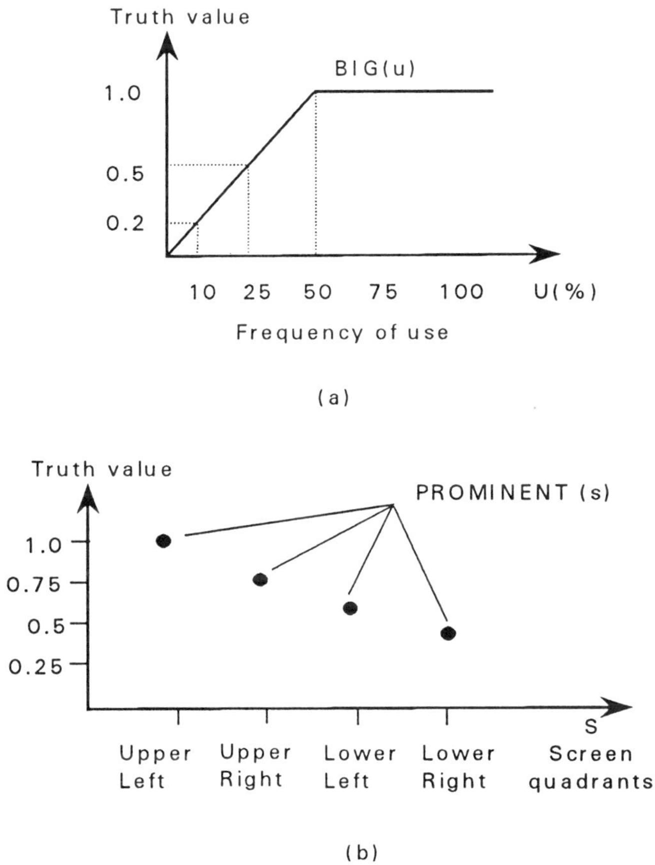

FIGURE 4. Truth value functions of limit definitions. (a) BIG in the frequency of use space, and (b) PROMIENCY of screen quadrants (after Grobelny et al. 1995)

18.6.3.2 Human perception of load heaviness and size-weight illusion

From the perspective of human perception, fuzzy modeling was used to investigate the relationships between physical weight, its perceived heaviness, and size of load in order to determine the ranges of load relevant for design of industrial lifting tasks. Luczak and Ge (1989) noted that determination of human lifting capacities by the psychophysical method relates closely to the subjective perception of load heaviness. However, it is not clear why the handling of a small box with a certain weight is sometimes perceived heavier than handling a bigger box with the same weight (size-weight illusion).

Luczak and Ge (1989) asked the subjects to express the relationships between physical weight and its perceived heaviness. Load heaviness levels were based on the Borg's scale (Borg 1970) with scores from 6 to 20 divided into five groups, and expressed using fuzzy sets, including: "very light," "light, moderate," "heavy" and "very heavy" (see Table 1). The Borg's scale values and frequencies of the data falling into every group was calculated. These were then used to model the perceived load heaviness as a fuzzy concept. The derived relationships (see Table 1)were based on memory and experience of the subjects. These relationships allowed to explain the mechanism and effects of the size-weight illusion for the lifted boxes. The study also illustrates how fuzzy measures can be used to quantify natural fuzziness underlying human cognitive processes.

Table 1. Fuzzy estimates of perceived load heaviness (after Luczak and Ge 1989).

	Load Heaviness Categories				
Box lifted	very light	light	moderate	heavy	very heavy
Box R	0	0.286	0.514	0.171	0.029
Box V	0	0.000	0.290	0.474	0.211

Note: box R = 20kg (40x40x25 cm), and box V = 20kg (40x40x15 cm).

Ngo et al. (1996) developed a fuzzy knowledge-based decision support system to produce recommendations for the maximum acceptable weights of lift, based on data generated from the Job Severity Index. The input variables consisted of four descriptors of lifting activity, including the lifting range (height), lifting frequency, box width, and percentile of population. Model validation showed the coefficient of variation of 6.17% when compared with results of the original database developed by Ciriello and Snook (1994). The use of fuzzy techniques allowed to take into account some important factors that are difficult to measure precisely at a workplace.

18.6.3.3 Evaluation of physical workload

Chen et al. (1994) used fuzzy sets to model the level of ergonomic workload stress index (EWSI) in the workplace, based on the possible sources of stress related to human perception. The linguistic variables selected for four task variables in their study were as follows:

S_1= Weight of load: very light, light, medium, heavy, very heavy.
S_2= Frequency of load: very light, light, medium, heavy, very heavy.
S_3= Duration of load: very light, light, medium, heavy, very heavy.
S_4= Moving distance of load: very light, light, medium, heavy, very heavy.

The membership functions of the linguistic variables for each task variable were determined by gathering subjective perception from the experiments. For example, regarding the stress caused by physical job demand, the membership function for S_1 (weight of load) was defined as follows:

Very Light = {1/1, 1/3, 0.7/5, 0.1/15, 0/20, 0/30, 0/40, 0/50, 0/55, 0/60}
Light = {1/1, 1/5, 0.7/10, 0.4/15, 0.25/17.5, 0/25, 0/30, 0/40, 0/50, 0/60}
Medium ={0/1, 0.1/5, 0.5/10, 0.8/12.5, 1/17.5, 0.6/25, 0.2/27.5, 0.1/30, 0/33, 0/60}
Very Heavy ={0/1, 0/10, 0/20, 0.1/30, 0.3/35, 0.5/38, 0.85/42, 0.9/44, 1/45, 1/60}

The intensity of perceived physical stress was calculated based on the analytic hierarchy process weighing. This technique serves as a framework for structuring the problem area, and provides judgements based on knowledge, reason or feelings which can be used to derive a set of priorities considered as an optimal solution to a decision problem (Saaty 1980). From the multiple comparisons with the field data (grocery store, warehouse, drilling tool industry, and linen industry), the correlation coefficient (r) between the EWSI and injury/illness records was reported to be within the range from 0.716 to 0.917. In addition, the reported correlation between the EWSI, age and various physiological measures (such as heart rate) showed statistical significance, indicating a close linear relationship. In practice, predictive models cannot be used alone without empirical assessments to verify their accuracy. Therefore, predictions provided by the EWSI model should not be treated as established facts, but rather, as suggested guidance. That is, this model could objectively identify points of potentially high workload, and diagnose their causes by providing a way to quantify the complex relationship between external workload and human responses.

18.6.4 Modeling of Work-related Musculoskeletal Disorders

Work-related musculoskeletal disorders (WRMDs), such as carpal tunnel disorders, tendinitis, chronic muscle strain, and degenerative joint diseases, are now recognized as a major occupational health problem (Karwowski and Marras 1997). These cumulative trauma disorders (CTDs) may occur when a force is applied repeatedly over a prolonged period to the same muscle group, joints, or tendons, and are linked to jobs that exhibit repeated or awkward postures. In order to implement adequate prevention and health program, it is necessary to document the relationships between work exposure (prevalence of job related risk factors in a production environment) and specific musculoskeletal disorders. The risk of CTDs is based primarily on four major factors: 1) force exerted, 2) task frequency, 3) posture, and 4) duration of the task and rest periods (Putz-Anderson 1988).

The CTDs are usually diagnosed on the basis of the clinical signs and symptoms experienced after the onset of the illness. From the perspective of CTDs prevention, a more desirable situation is to predict the possibility of occurrence of CTDs in a given occupational setting. This involves a two-stage process: 1) identification/prediction of jobs with a high possibility of CTD occurrence, and 2) development of the corrective measures to reduce these occurrence. The prediction process requires establishing a relationship between CTDs and the associated job risk factors to which the worker is exposed. However, the available information about such risk factors is often uncertain and imprecise. Furthermore, due to the incompleteness and inconsistency of the necessary information, a statistical approach may not be feasible. A fuzzy system methodology can be used to determine the possibility of occurrence of CTDs, given the available relationships between CTDs, risk factors and their severity level.

18.6.4.1 Theoretical framework for fuzzy modeling of CTDs

The first theoretical framework for fuzzy modeling of the risk of CTDs was developed by Grobelny and Karwowski (1992). The proposed quantification of the risk estimations of work-related musculoskeletal disorders was based on the information and knowledge provided by the human experts.

18.6.4.2. Prediction Models for CTD

As one of modeling scheme for the prediction problem, Merritt and Gopalakrishnan (1994) developed a fuzzy-based model for predicting the possibility of occurrence of various cumulative trauma disorders (CTDs) of the upper extremity in a work environment. The traditional interview analysis involved asking a series of questions about how an expert evaluates a situation for CTDs related risk factors. This was extended until the expert felt that he had exhausted all potential categories and risk factors. Concept mapping, a knowledge acquisition tool that is designed to capture and graphically represent the relationships that exist between concepts in the domain expert's understanding of the problem space, was used for that purpose.

In order to define the initial relationship between the various CTDs and the risk factors, the fuzzy relation was used as follows. Let $C=\{C_1, C_2, C_3,..., C_m\}$ be the set of CTDs and $R=\{R_1, R_2, R_3,..., R_n\}$ be the set of risk factors (see Table 2). The fuzzy relationship A over C x R defines the relationship between the CTDs and the risk factors. For example, the carpal tunnel syndrome was defined based on the following fuzzy relation:

$A(C_4, R_j)$, j = 1,..., 24
= (0.75, 0.00, 0.40, 0.80, 0.70, 0.50, 0.60, 0.50,
0.50, 0.20, 0.00, 0.00, 0.00, 0.00, 0.00, 0.00,
0.20, 0.20, 0.70, 0.20, 0.20, 0.50, 0.00, 0.50).

After the values for each of the fuzzy relationship corresponding to the CTDs were determined, the current severity level for the risk factors was expressed as a fuzzy set S in the space of objects R. $S=\{S_j/R_j, j=1,..,n\}$, where S_j denotes the grade of membership of R_j in the set S, and was calculated as follows:

$S(R_j)$, j=1,..., 24
= (1.00, 0.00, 0.00, 0.90, 1.00, 0.00, 1.00, 0.96,
0.00, 0.00, 0.00, 0.00, 0.00, 0.66, 0.00, 0.00,
0.80, 1.00, 0.00, 0.00, 0.00, 0.00, 0.00, 0.00)

The possibility of occurrence of a given CTD was expressed as a fuzzy set P in the space of objects C: $P = \{P_i/C_i, i=1,..,m\}$, where P_i denotes the grade of membership of C_i in the fuzzy set P. The grade of membership of the ith CTD in the fuzzy set P was the determined as $P_i = [A_iT][S]$. The possibility of occurrence of carpal tunnel syndrome was defined as $P_4 = [A_4T][S] = 3.61$.

Table 3 shows the possibility of occurrence of all the CTDs calculated by the prediction model. The numbers also show the relative ordering of the CTDs for the given work environment. Carpal tunnel syndrome is the most likely CTD.. The results from the seven test cases indicated that fuzzy sets theory provides a good modeling scheme for the prediction of work-related CTDs.

18.6.4.3 Prediciton of Risk for CTS

Bell and Crumpton (1997) proposed a fuzzy linguistic model for predicting the risk of carpal tunnel syndrom (CTS). The model utilizes fuzzy sets to quantify the risk associated with development of this neuropathy. The first set of membership functions involved utilizing the linguistic risk level obtained by the expert knowledge acquisition. The second set of membership functions was derived to rate the possibility of the hazard associated with a particular linguistic variable. The membership functions of the two variables were intuitively determined based on graphical representation of the physical data.

For example, task repetition was represented by five fuzzy values for the first task variable, while the hazard level was fuzzified into three values for the second task variable. Predictability of the proposed fuzzy linguistic model was claimed to be better than some of the objective CTS measures, incuding the Phalen's test, vibrometry testing, and electroneurometry testing.

Table 2. List of CTDs and risk factors considered (modified after Merritt and Gopalakrishnan 1994)

Cumulative trauma disorders	List of risk factors
Hand/wrist tendinitis	Excessive use of the fingers and palm
Trigger finger	Excessive use of the elbow and shoulders
De Quervain's disease	Failure to use standard methods
Carpal tunnel syndrome	High force
Guyon's canal syndrome	High repetition
Vibration syndrome	Static forces
Tennis elbow	Wrist flexion
Golfer's elbow	Wrist extension
Elbow bursitis	Ulnar deviation (wrist)
Bicipital tenosynovitis	Radial deviation (wrist)
Subacromial bursitis	Elbow bending
Shoulder tendinitis	Pronation
Thoracic outlet syndrome	Supination
	Shoulder abduction
	Rotating of the arm about the shoulder
	Moving the arm above the shoulder level or behind
	Inappropriate table height
	Parts out of reach
	Vibration
	Tool weight
	Form fitting tools
	Tool pressing into the palm
	Inappropriate triggers
	Pinch

Table 3. Possibility of occurrence of all CTDs for the example (after Merritt and Gopalakrishnan 1994)

Name of CTD	Possibility of occurrence
Hand/wrist tendinitis	3.067
Trigger finger	2.100
De Quervain's disease	2.390
Carpal tunnel syndrome	3.610
Guyon's canal syndrome	2.652
Vibration syndrome	2.085
Tennis elbow	2.931
Golfer's elbow	2.721
Elbow bursitis	2.100
Bicipital tenosynovitis	2.320
Subacromial bursitis	2.105
Shoulder tendinitis	2.356
Thoracic outlet syndrome	2.072

18.7 FUZZY SYSTEMS IN HUMAN-MACHINE RESEARCH

Human-machine studies aim to optimize work systems with respect to physical and psychological characteristics of the users, and investigate complex and ill-defined relationships between people, machines, and physical environments. The main goal of such investigation is to remove the incompatibilities between humans and tasks, and to make the workplace healthy, productive, comfortable and satisfying (Karwowski and Salvendy 1992). Such human-centered systems are very complex and difficult to analyze due to vagueness of the relationships between people and their working environments, an inherent fuzziness in human thinking processes and subjective perception of the outside world. Figure 5 illustrates types of fuzziness that should be accounted for in modeling of any human-machine systems.

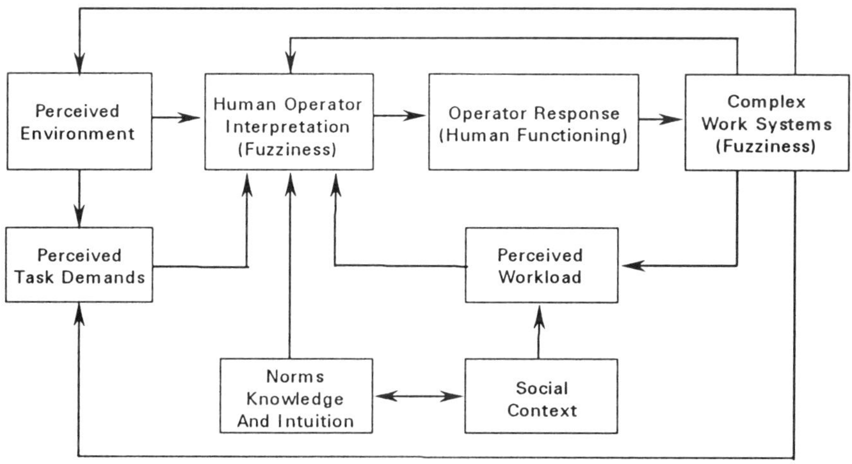

FIGURE 5. Fuzziness in human-machine interfacing (after Karwowski and Salvendy 1992)

18.7.1 Modeling of Human Sensations

Considering the uncertainty and complexity in the process of information interpretation, fuzzy systems can be a useful modeling approach for analysis and design of human-machine interactions. Shimizu and Jindo (1995) proposed a framework for dealing with ambiguities and non-linearity of the human information processing relevant to modeling of human sensitivity. The conventional methods to quantify the relationships between human sensations and physical characteristics which influence them, are typically the multivariate analysis techniques such as multiple regression analysis and quantification theory. However, when higher order data are involved, it is much more difficult to find a model formula that suitably represents the non-linearity factor. Moreover, many conventional methods have traditionally excluded the ambiguities that can arise in the process of recognizing and making subjective evaluations of the physical characteristics. The advantageous feature of fuzzy systems is that the ambiguities and nonlinearity (e.g., of human sensation) can be taken into account and quantified to derive correlations with the considered physical characteristics of the product.

Shimizu and Jindo (1995) developed a fuzzy regression method for evaluation of human sensitivity, which was applied to the problem of perceived thermal sensation in a car interior. The nonlinearity of human sensation modeled through the traditional method of the multiple regression analysis is illustrated in Figure 6. While it is possible to treat this marked nonlinearity (at least to some extent) by transforming the variables, it is difficult to determine how the variables should be transformed into the non-fuzzy regression. It was shown that 85% of the data items obtained through the subjective evaluations of the car temperature fell within the range of the predicted values obtained from the fuzzy regression analysis method. The results also showed that fuzzy logic allowed to make predictions which take into account natural ambiguity of human sensations. Furthermore, the application of fuzzy regression analysis made it is relatively easy to obtain results which were closer to the true subjective evaluations made by the people.

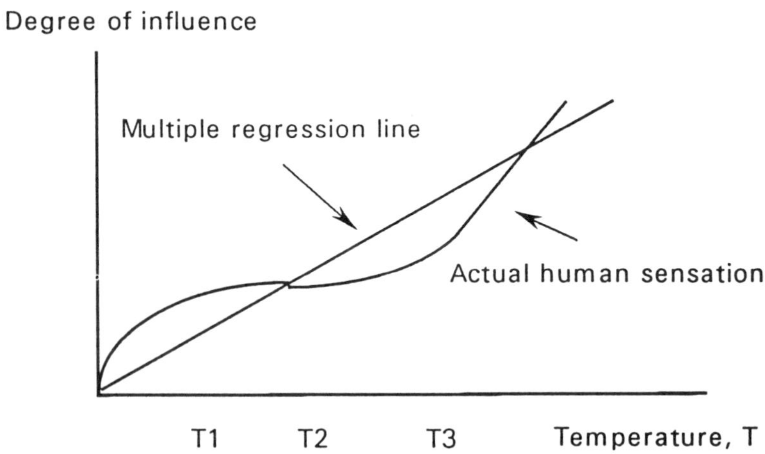

FIGURE 6. Example of nonlinearity of human sensation (after Shimizu and Jindo 1995)

18.7.2 Modeling of Human Stress

Stress due to external workload imposed on the human-machine system influence the human performance and behavior. The responses of the body to the external demands (stressors) refer to some undesirable conditions, circumstances, tasks, or other factors that impinge upon the individual (Melamed et al. 1989). There are considerable problems in separating job-related (workload) stresses from other sources of stress. There are also difficulties associated with identifying a particular task or event as a primary source of the acute (workload) stress. The examples of stresses that originate from the external sources include the work activity itself, as well as the physical, environmental, postural, or mental aspects of work. Although variety of subjective rating scales, measures of primary and secondary task performance, and physiological indicators have been developed for the purpose of quantifying the effects of task requirements on the human operators, the results obtained with different measures may vary due to the natural vagueness and imprecision of different workload causes and consequence. Fuzzy systems methodology can be used to aid in interpreting the results of human workload analysis, identifying workload criteria, and improving the accuracy of workload prediction.

18.8 ASSESSSMENT OF MENTAL WORKLOAD

Mental workload can be defined as the amount of mental work or effort necessary to perform a given task. The purpose of mental workload assessment is to maintain the workload at a level that will allow acceptable performance of the operator's task (Proctor and Zandt 1994). Subjective assessment techniques are based on empirical methods which allow to measure workload directly in an operational system or a simulated environment. Such techniques evaluate workload through the use of human operators' judgments. Typically, the operators are asked to rate the perceived mental effort, time load, and stress load of particular tasks. This indicates that workload is not a scalar quantity but rather a vector quantity associated with these multiple dimensions (Moray 1982). Most of the subjective measures of mental workload utilize the scale method which calls for the subjects to express their feelings through rating scales or questionnaires. However, very often the human feelings cannot be represented adequately by a precise scale. Using the linguistic quantifiers can help to overcome this difficulty.

18.8.1 Evaluation of Mental Workload

Liou and Wang (1994) used the fuzzy linguistic multi-criteria measure (FLMCM) to assess the criteria ratings of mental workload and their corresponding importance weights through the application of fuzzy numbers. The framework of the developed fuzzy linguistic multi-criteria method is shown in Figure 5. Nine rating criteria used by researchers at the Human Performance Group at NASA-Ames Research Center were considered, including 1) task difficulty, 2) time pressure, 3) performance, 4) mental / sensory effort, 5) frustration level, 6) stress level, 7) fatigue, 8) activity type, 9) physical effort. The respective membership functions consist of five linguistic variables such as VL=Very Low, L=Low, M=Medium, H=High, and VH=Very High. The importance weight were generally given by either direct assignment or through pairwise comparison. The weights were defined in linguistic terms, where: U = 'Unimportant,' BU&SL = 'Between Unimportant and Slightly Important,' SL = 'Slightly Important,' BSL&M = 'Between Slightly Important and Moderately important,' M = 'Moderately Important,' B.M&SE = 'Between Seriously Important and Very Seriously Important,' and VSE = 'Very Seriously Important.'

A fuzzy weighted average algorithm (FWA) was then employed to aggregate the criteria ratings C_1, C_2,..., C_n and the importance weights W_1, W_2, ..., W_n. The basic form to obtain the weighted average Y was defined as follows:

$$Y = f\,(C_1, C_2,..., C_n, W_1, W_2,...W_n)$$

where:

$$Y = \frac{W_1 \times C_1 \times W_2 \times ...W_n \times C_n}{W_1 \times W_2 \times ... W_n}$$

The aggregated fuzzy number was then translated back to the linguistics from, and the overall mental workload was expressed in linguistic terms. It was concluded that it was more appropriate for subjective mental workload measurements that the assessments criteria ratings and importance weights be represented in linguistic terms. The FLMCM model proved to be superior to other numerical methods for situations where the subjective workload was assessed in linguistic terms. Further research conducted by Chen (1996), aimed to overcome the computational problem of fuzzy number aggregation, and extended applications of fuzzy methodologies to risk analysis with mental workload assessment.

18.8.2 Modeling of Human Cognitive Processes

Ukita et al. (1996) developed a fuzzy-based system to model the decision making process of the human operators involved in tuning of the microwave circuits in a real time environment. The electric circuit tuning process, which requires manual tuning of a set of trimmers by the human operators, was automated through the application of a fuzzy knowledge-based system. In a complex tuning process, multiple circuit specification criteria had to be simultaneously satisfied by several trimmers. To reduce the number of trial-and-error steps required to meet specific circuit tuning criteria, and, consequently, the tuning process time, the order of subsequent trimmer adjustments in the tuning process, as well as the extent of each individual trimmer tuning magnitude must be chosen very carefully. In the past, the circuit tuning process was performed by experienced workers who would skillfully adjust a set of trimmers by hand. The quality of the adjustment process is primarily based on the human operator's mental model of the circuit tuning behavior formulated from the long practice and requires specific cognitive skills of the human operator.

In the fuzzy knowledge-based system proposed by Ukita et al (1996), the effect of each trimmer on each of the tuning criterion was expressed by a grade of fuzzy membership related to the circuit output. The overall effect of each trimmer on the circuit tuning performance was modeled by an aggregation of the fuzzy grades used for trimmer selection. Different tuning evidence aggregation methods utilized to

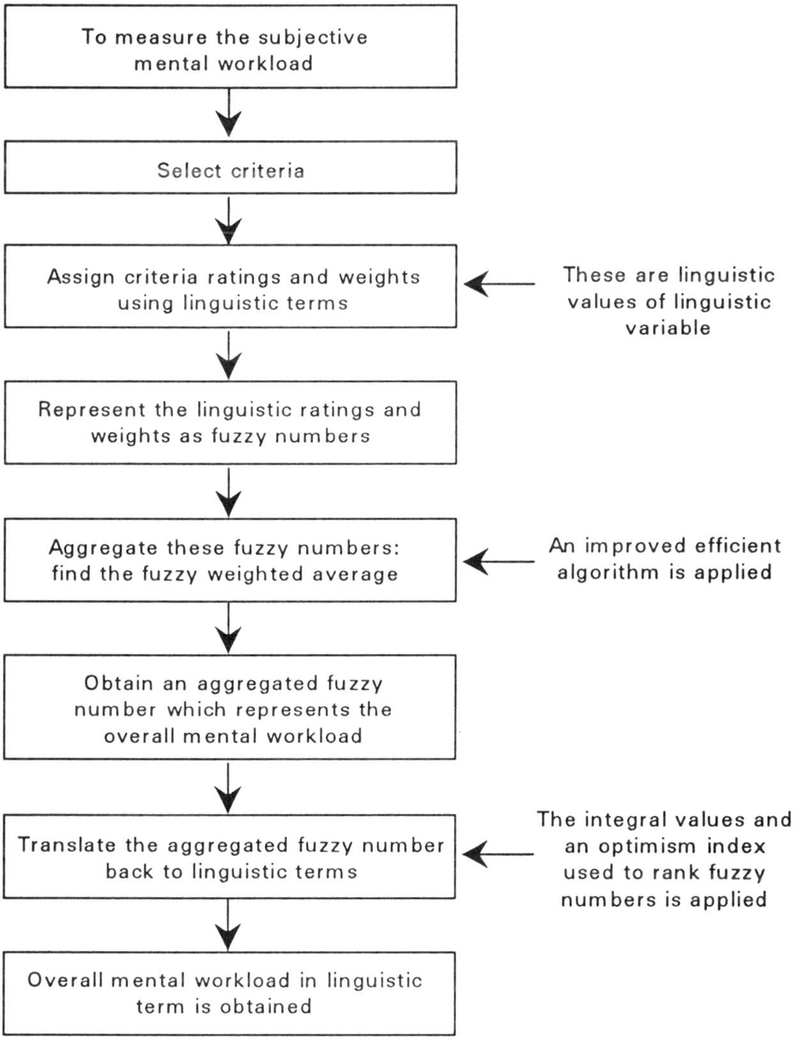

FIGURE 7. The framework of the fuzzy linguistic multi-criteria mental workload measurement method (after Liou and Wang 1994)

reduce the overall circuit tuning time were examined. Some heuristic rules were also introduced to adjust the aggregated evidence values. These rules significantly improved the tuning system's performance. The model simulation showed that the geometrical average operator was the best method for evidence aggregation of fuzzy evidence in modeling of the human cognitive processes underlying manual circuit tuning.

18.9 HUMAN-ROBOT INTERACTION

Workplace safety is an important issue in human-robot interaction aimed at protecting workers from the hazards created by robots in a manufacturing environment. According to Hirschfeld et al. (1993), workers spent 37% of a robot's eight hour workday in the robot's work space (13% of a 24-hour day), and only 20% of robots were found to be completely enclosed, while 60% had a limited barrier. With this respect, current safety method applied to robot application can be classified into two categories: 1) passive and 2) active protection. Passive protection refers to safety devices that improve robot safety without changing the robot's behavior and active protection refers to safety devices that modify the robot's behavior or environment to avoid dangerous situations.

In terms of passive protection of robot, the criteria issue is the detection of human being. For detection of human entering in robot workplace, several safety devices have been proposed, including the touch sensors, optical sensors, ultrasound sensors, microwave sensors, and vision systems (Koivo et al 1991). Graham (1995) developed a multi-level system for improving robot behavior by the use of active sensory data from the robot's sensory system. This system consists of a three-level hierarchical design of both sensory processing and fusion of sensory data based on a Dempster-Shafer inference, the high level fuzzy logic decision making processes for a multilevel alarming scheme. Recently, Zurada, Karwowski and Wright (1996) demonstrated the use of fuzzy logic control for robot safety decision-making purposes. It was shown that fuzzy logic-based decision techniques can be successfully applied to exploit the imprecise robot outputs of a sensory decision unit (i.e. proceed slowly, proceed at medium speed, high speed, halt, and emergency halt).

18.10 CONCLUSIONS

The field of human factors and ergonomics faces the problems of work system's complexity and related human and system-based fuzziness that increases the ever-present incompatibility between people and their living and working environments. As pointed out by Karwowski (1992), fuzziness is not just a product of the human mind, but is the essence of human development and existence, and a necessary condition for human learning, growth and survival. In order to develop the symbiotic relationship between people and the outside surroundings (both natural and artificial), the intrinsic fuzziness of the human kind must be treated by system designers and engineers as natural system design requirement. Fuzziness can also be viewed as an expression of human entropy, while models of fuzziness as representations of such entropy. The fuzziness represents the natural model of people at work, and those man-made systems that interact with people. Fuzzy systems methodologies allow accounting for natural human fuzziness and human-machine fuzziness, and provide the necessary framework for successful modeling efforts in the human factors and ergonomics discipline.

REFERENCES

Audley, R. J., Rouse, W., Senders, T., and Sheridan, T. (1979). *Final Report of Mathematical Modeling Group*, Moray (Ed.), Mental Workload, its Theory and Measurement, Plenum Press, New York, 269-285.

Ayoub, M. M., Dempsey, P. G and Karwowski, W. (1997). *Manual Materials Handling*, G. Salvendy, (Ed.), Handbook of Human Factors & Ergonomics, , John Wiley, New York, 1085-1123.

Bell, P. M. and Crumpton, L. (1997). A Fuzzy Linguistic Model for the Prediction of Carpal Tunnel Syndrome Risks in an Occupational Environment, *Ergonomics*, 40, 790-799.

Benson, W. H. (1982). An Application of Fuzzy Set Theory to Data Display, In *Fuzzy Sets and Possibility Theory*, R. R. Yager (Ed.), Pergamon Press, New York.

Bernotat, R. (1984). *Generation of Ergonomic Data and their Application to Equipment Design*, H. Schmidtke (Ed.), Ergonomic Data for Equipment Design, Plenum Press, New York, 57-75.

Bezdek, J. (1981). *Pattern Recognition with Fuzzy Objective Function Algorithms*, Plenum Press, New York.

Borg, G. (1970). Perceived Exertion as an Indication of Somatic Stress, *Scandinavian Journal of Rehabilitation Medicine*, 2, 92-98.

Boy, G. A., and Kuss, P. M. (1986). *A Fuzzy Method for Modeling of Human-Computer Interactions in Information Retrieval Tasks*, W. Karwowski and A. Mital (Eds.), Applications of Fuzzy Set Theory in Human Factors, Elsevier, Amsterdam, 117-133.

Card, S. K., Moran, T. P., and Newell, A. (1983). *The Psychology of Human-Computer Interaction*, Lawrence Erlbaum Associates, London.

Chen, S.-M. (1996). New Methods for Subjective Mental Workload Assessment and Fuzzy Risk Analysis, *International Journal of Cybernetics and Systems*, 27, 449-472.

Chen, J.-G., Jung, H. S., and Peacock, B. J. (1994). A Fuzzy Sets Modeling Approach for Ergonomic Workload Stress Analysis, *International Journal of Industrial Ergonomics*, 13, 189-216.

Foley, P. and Moray, N. (1987). *Sensations, Perception and Systems Design*, G. Salvendy (Ed.), Handbook of Human Factors, John Wiley & Sons, New York, 45-71.

Graham, J. H. (1995). A Fuzzy Logic Approach for Safety and Collision Avoidance in Robotic Systems, *The International Journal of Human Factors in Manufacturing*, 5, 447-457.

Grandjean, E. (1984). *Fitting the Task to the Man*, Taylor & Francis, London.

Grobelny, J. (1988). The 'Linguistic Pattern' Method for a Workstation Layout Analysis, *International Journal of Production Research*, 26, 1779-1798.

Grobelny, J. (1987). On One Possible 'Fuzzy' Approach to Facilities Layout Problems, *International Journal of Production Research*, 25, 1123-1141.

Grobelny, J., Karwowski, W. and Zurada, J. (1995). Applications of Fuzzy-based Linguistic Patterns for the Assessment of Computer Screen Design Quality, *International Journal of Human-Computer Interaction*, 7, 193-212.

Grobelny, J. (1987a). The Fuzzy Approach to Facilities Layout Problems, *Fuzzy Sets and Systems*, 23, 175-190.

Harre, R. (1972). *The Philosophies of Science*, Oxford University Press, London.

Hersh, M. M. and Caramazza, A. (1976). A Fuzzy Set Approach to Modifiers and Vagueness in Natural Language, *Journal of Experimental Psychology: General*, 105, 254-276.

Hesketh, B., Pryor, R., Gleitzman, M., and Hesketh, T. (1988). *Practical Applications of Psychometric Evaluation of a Computerized Fuzzy Graphic Rating Scale*, T. Zeteni (Ed.), Fuzzy Sets In Psychology, North-Holland, Amsterdam, 425-454.

Hirsch, G., Lamotte, M., Mass, M. T., and Vigneron, M. T. (1981). Phonemic Classification Using a Fuzzy Dissimilitude Relation, *Fuzzy Sets and Systems*, 5, 267-276.

Hirschfeld, R. A., Aghazadeh, F., and Chapleski, R. C. (1993). Survey of Robot Safety in Industry, *The International Journal of Human Factors in Manufacturing*, 3, 369-379.

Hunt, R. M. and Rose, W. B. (1984). *IEEE Transactions on Systems, Man And Cybernetics*, SMC 14, 112-120.

Jastrzebowski, W. (1979). Rys Ergonomji, Czyli Nauki o Pracy (In Polish), *Ergonomia*, 2, 13-29.

Karwowski, W. (1983). *A Pilot Study on the Interaction between Physiological, Biomechanical and the Psychophysical Stresses Involved in Manual Lifting Tasks*, K. Coombe (Ed.), Proceedings of The Ergonomics Society Conference, Taylor and Francis, London, 95-100.

Karwowski, W., Lee, W. G, Jamaldin, B., Gaddie, P. and R. Jang, (1999). Beyond Psychophysics: A Need for Cognitive Modeling Approach to Setting Limits in Manual Lifting Tasks, Ergonomics, (in press).

Karwowski, W. (1991). Complexity, Fuzziness and Ergonomic Incompatibility Issues in the Control of Dynamic Work Environments, *Ergonomics*, 34, 671-686.

Karwowski, W. and Ayoub, M. M., (1984a). Fuzzy Modeling of Stresses in Manual Lifting Tasks, *Ergonomics*, 27, 641-649.

Karwowski, W., Ayoub, M. M., Alley, L. R., and Smith, T. L., (1984b). Fuzzy Approach in Psychophysical Modeling of Human Operator-Manual Lifting System, *Fuzzy Sets and Systems*, 14, 65-76.

Karwowski, W. and Mital, A. (1986). *Applications of Fuzzy Set Theory in Human Factors*, Elsevier, Amsterdam.

Karwowski, W., Marek, T., and Noworol, C. (1988). *Theoretical Basis of the Science of Ergonomics*, Proceedings of the 10th Congress of the International Ergonomics Association, Sydney, Australia, Taylor & Francis, London, 756-758.

Karwowski, W., Kosiba, E., Benabdallah, S., and Salvendy, G. (1990). A Framework for Development of Fuzzy GOMES Model for Human-Computer Interaction, *International Journal of Human-Computer Interaction*, 2, 287-305.

Karwowski, W. and Salvendy G. (1992). Fuzzy-Set-Theoretic Applications in Modeling of Man-Machine Interactions, R. R. Yager and L. A. Zadeh, (Eds.), An

Introduction to Fuzzy Logic Applications in Intelligent Systems, Kluwer Academic Publishers, Boston, 201-220.

Karwowski, W., Marek, T., and Noworol, C. (1994). The Complexity Incompatibility Principle in the Science of Ergonomics, F. Aghazadeh, (Ed.), Advances In Industrial Ergonomics & Safety VI, Taylor & Francis, London, 37-40.

Karwowski, W. and Marras, W. S., (1997), Cumulative Trauma Disorders, in G. Salvendy, (Ed.), *Handbook of Human Factors & Ergonomics*, John Wiley, New York, 1124-1173.

Kochen, M. (1975). *Applications of Fuzzy Sets in Psychology*, L. A. Zadeh, K. S. Fu, K. Tanaka and M. Shimuro (Eds.), Fuzzy Sets and their Applications to Cognitive and Decision Processes, Academic Press, New York, 395-408.

Koivo, H., Malm, T., Suominen J., and Kuivanen, R. (1991). An Intelligent Safety System for Robots and Automatic Machines. J. H. Graham (Ed.), Safety, Reliability, and Human Factors in Robotics Systems, Van Nostrand Reinhold, New York, 132-147.

Kosko, B. (1992). *Neural Networks and Fuzzy Systems: A Dynamical Systems Approach to Machine Intelligence*, Prentice Hall, Englewood Cliffs.

Kramer, U. (1983). In Proceedings of the Third European Annual Conference on Human Decision Making and Manual Control, Roskilde, Denmark, 313

Kramer, U. and Rohr, R. (1982). In G. Johannsen and J. E. Rijnsdorp (Eds.), In Analysis, Design And Evaluations of Man-Machine Systems, Pergamon Press, Oxford, 31-35.

Liou, T.-S. and Wang, M.-J. J. (1994). Subjective Assessment of Mental Workload - A Fuzzy Linguistic Multi-Criteria Approach, *Fuzzy Sets and Systems*, 62, 155-165.

Luczak, H. and Ge, S. (1989). Fuzzy Modeling of Relations between Physical Weight and Perceived Heaviness: The Effect of Size-Weight Illusion in Industrial Lifting Tasks, *Ergonomics*, 32, 823-838.

McCloskey, M. E. and Glucksberg, S. (1978). Natural Categories: Well-Defined or Fuzzy Sets?, *Memory and Cognition*, 6, 462-472.

Melamed, S., Luz, J., Najenson, T., Jucha, E., and Green, M. (1989). Ergonomic Stress Levels, Personal Characteristics, Accident Occurrence and Sickness Absence among Factory Workers, *Ergonomics*, 32, 1101-1110.

Merritt, T. W. and Gopalakrishnan, S. (1994). An Application of Fuzzy Set Theory to the Prediction of Cumulative Trauma Disorders of the Upper Extremity, *International Journal of Industrial Ergonomics*, 13, 95-105.

Miller, G. A. (1956). The Magical Number Seven Plus or Minus Two: Some Limits on our Capacity for Processing Information, *Psychological Review*, 63, 81-97.

Miller, D. P. and Swain, A. D. (1987). *Human Error and Human Reliability*, G. Salvendy (Ed.), Handbook of Human Factors, John Wiley, New York, 219-250.

Moray, N. (1982). Subjective Mental Workload, *Human Factors*, 24, 25-40.

Moray, N. (1984). *Attention To Dynamic Visual Displays In Man-Machine Systems*, R. Parasuraman and Davies, R. (Eds.), Varieties of Attention, Academic Press, New York, 485-512.

Ngo, A. D., Malinowski, P., Kaji, S., and Balazinski, M. (1996). *Fuzzy Logic Approach to Manual Material Handling Tasks*, R. J. Koubek and W. Karwowski

(Eds.), Manufacturing Agility And Hybrid Automation-I, Iea Press, 588-591.

Oden, G. C. (1977). Human Perception and Performance, *Journal of Experimental Psychology*, 3, 565-575.

Onisawa, T. (1988). An Approach to Human Reliability in Man Machine Systems Using Error Possibility, *Fuzzy Sets And Systems*, 27, 87-103.

Pedrycz, W. (1990). Fuzzy Set Framework for Development of a Perception Perspective, *Fuzzy Sets and Systems*, 37, 123-137.

Pew, R. W. and Baron, S. (1983). Perspectives on Human Performance Modeling, *Automatica*, 19, 663-676.

Proctor R. W. and Zandt, T. V. (1994). *Human Factors: In Simple and Complex Systems*, Allyn and Bacon, London.

Putz-Anderson (1988). *Cumulative Trauma Disorders*, Taylor & Francis, London.

Rohmert, W. (1979). Determination of Stress and Strain at Real Work Places, In N. Moray (Ed.), *Mental Workload, Its Theory and Measurement*, Plenum Press, New York, 423-444.

Ross, T. J. (1995). *Fuzzy Logic with Engineering Applications*, McGraw-Hill, New York.

Saaty, T. L. (1980). *The Analytic Hierarchy Process*, McGraw-Hill, New York.

Saaty, S. L. (1977). Exploring the Interface between Hierarchies, Multiple Objectives and Fuzzy Sets, *Fuzzy Sets and Systems*, 1, 57-68.

Sheridan, T. B. and Ferrell, W. R. (1981). *Man-Machine Systems: Information, Control, and Decisions Models of Human Performance*, The Mit Press, Cambridge.

Shimizu, Y. and Jindo, T. (1995). A Fuzzy Logic Analysis Method for Evaluating Human Sensitivities, *International Journal of Industrial Ergonomics*, 15, 39-47.

Simcox, W. A. (1984). A Method for Pragmatic Communication in Graphic Displays, *Human Factors*, 26, 483-487.

Smithson, M. (1982). Applications of Fuzzy Set Concepts to Behavioral Sciences, *Mathematical Social Sciences*, 2, 257-274.

Swain, A. D. and Gutman, H. E. (1983). Handbook of Human Reliability Analysis with Emphasis on Nuclear Power Plant Application, (Scandia National Laboratories, Nureg/Cr-1278), Washington, D.C., U.S. Nuclear Regulatory Commission.

Terano, T., Murayama, Y., and Akijama, N. (1983). Human Reliability and Safety Evaluation of Man Machine Systems, *Automatica*, 19, 719-722.

Tullis, T. S. (1981). An Evaluation of Alphanumeric, Graphic and Color Information Display. *Human Factors*, 23, 541-550.

Ukita, A., Karwowski, W., and Salvendy, G. (1994). Aggregation of Evidence in a Fuzzy Knowledge-Based Method for Automated Tuning of Microwave Electric Circuits, *Journal of Intelligent & Fuzzy Systems*, 2, 299-313.

Wickens, C. D. (1984). *Engineering Psychology and Human Performance*, Charles Merril, Columbus, OH.

Wickens, C. D. (1987). *Information Processing, Decision-Making, and Cognition*, G. Salvendy (Ed.), Handbook of Human Factors, John Wiley & Sons, New York, 72-107.

Willaeys, D. and Malvache, N. (1979). *Contributions of the Fuzzy Set Theory to Man-Machine Systems*, M. M. Gupta, R. K. Ragade and R. R. Yager (Eds.),

Advances on Fuzzy Set Theory and Applications, North Holland, Amsterdam.

Zadeh, L. A. (1974). Numerical Versus Linguistic Variables, *Newspaper of the Circuits and Systems Society*, 7, 3-4.

Zadeh, L. A. (1965). Fuzzy Sets, *Information and Control*, 8, 338-353.

Zadeh, L.A. (1973). Outline of a New Approach to the Analysis of Complex Systems and Decision Processes, *IEEE Transactions on Systems, Man and Cybernetics*, SMC-3, 28-44.

Zadeh, L. A. (1978). Fuzzy Sets as a Basis for a Theory of Possibility, *Fuzzy Sets and Systems*, 1, 3-28.

Zimmermann, H. J. (1980). Testability and Meaning of Mathematical Models in Social Sciences, *Mathematical Modeling*, 1, 123-139.

Zimmermann, H. J. (1985). *Fuzzy Set Theory and its Applications*, Kluwer-Nijhoff Publishing, Boston.

Zurada, J., Karwowski, W. and Wright, A. L. (1996). A Fuzzy Logic-Based Decision Unit for a Robot Safety System, R. J. Koubek and W. Karwowski (Eds.), *Manufacturing Agility and Hybrid Automation–I*, IEA Press, Louisville, Kentucky, 580-583.

19 FUZZY SYSTEM DEVELOPMENT: SOFTWARE METHODOLOGY AND DESIGN TOOLS

Witold Pedrycz

Abstract: The paper elaborates on the methodology and practice of the development of fuzzy systems (models). We re-examine some standard schemes of their life cycle (such as e.g., a waterfall model) and identify their efficacies emerging in the context of processing fuzzy information. Finally, the study reports on a series of design platforms whose analysis is completed in light of the main design features of fuzzy systems

19.1 INTRODUCTION

The design and utilization of fuzzy systems assumes a lot of facets. This diversity arises due to the broad spectrum of problems being tackled as well as the use of quite distinct approaches stemming from the technology of fuzzy sets. On the other hand, there is a common denominator implying the uniqueness of non-numeric information processing. As of now, fuzzy sets emerge as basic design constructs (builds) that allows us to formalize design objectives, capture a qualitative form of domain knowledge and highlight some global functional links existing in the problem. The design methodology of fuzzy systems is well exposed in numerous references; see, for instance Zimmermann (1991). The role of software development platforms is to support the detailed design and implementation part of the design.

The objective of this study is to look at the entire development process of fuzzy systems and identify a role played by the design packages. The remainder of the paper is arranged as follows. In Section 2 we discuss a life cycle model as a general framework and locate the fuzzy constructs in this setting. Subsequently, Section 3 and 4 summarize various classes of software and hardware resources. Finally, an extensive list of software tools is assembled and analyzed with respect to several design criteria.

19.2 SYSTEM DEVELOPMENT - LIFE CYCLE MODEL

The system development exploiting the technology of fuzzy sets while carrying along some peculiarities associated with this specific form of information processing) quite highly adheres to one of the general models encountered in the literature on software engineering (Boehm, 1976; Blum, 1992). This lifecycle model known as water fall development model has become a standard to a significant extent. Figure 19.1 summarizes its essence - the scheme incorporates some slight modifications by taking into account the specificity of the information processing technology under consideration.

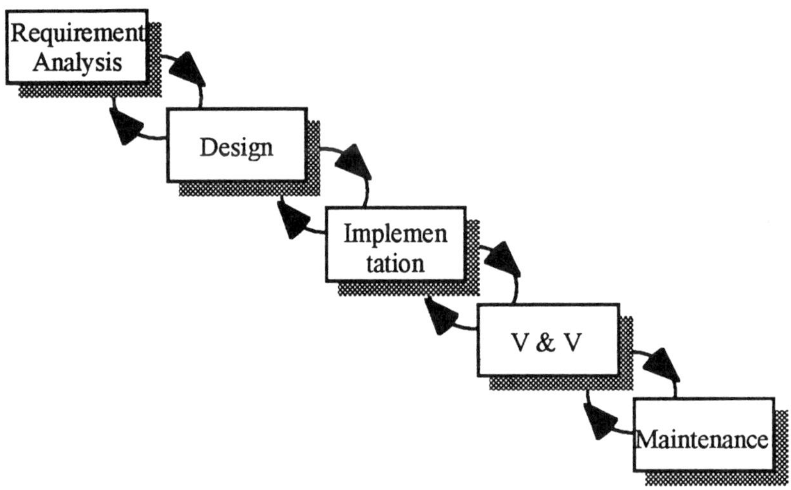

Figure 19. 1. Waterfall development model

Let us recollect briefly that its main phases include:

Requirement analysis: the first phase of the lifecycle embraces a variety of specification issues and captures the functions that allow us to understand the system to be built, its performance and interfacing. The requirements are thoroughly documented and reviewed with the customer.

Design: this is a multiphase activity that starts from a high level tasks and moves down to many specific procedural details. The primary intent is to translate the requirements into a representation that can be assessed for quality prior to any implementation.

Implementation: at this phase the detailed design is transformed into a machine readable form or a hardware artifact. For the careful design, the ensuing coding phase could be eventually carried out quite mechanically.

V & V (verification and validation): Once the system has been completed, it undergoes a series of thorough verification activities. The essence of verification

is to conclude that the developed system meets the assumed requirements. There are various instruments to accomplish this task. Testing is one of the most popular. The validation activities are aimed at showing that the developed system satisfies the needs of the user - in this sense these pursuits are far more subjective. Note that the original requirements as established at the first stage of the waterfall might not fully reflect the genuine and changeable needs of the user.
Maintenance: this phase encapsulates all activities that make the system useful. They include changes in the system due to occurrence of some errors, upgrade or new functional requirements.

The verification phase needs to be discussed in more detail. The main reason is that the technology of fuzzy sets has to interact with a numerical environment thus the testing is very much affected

by the quality of a communication layer, Figure 19. 2. Assuming the processing *transparency* of this layer (Pedrycz, 1995), the testing of the system is definitely much more easier.

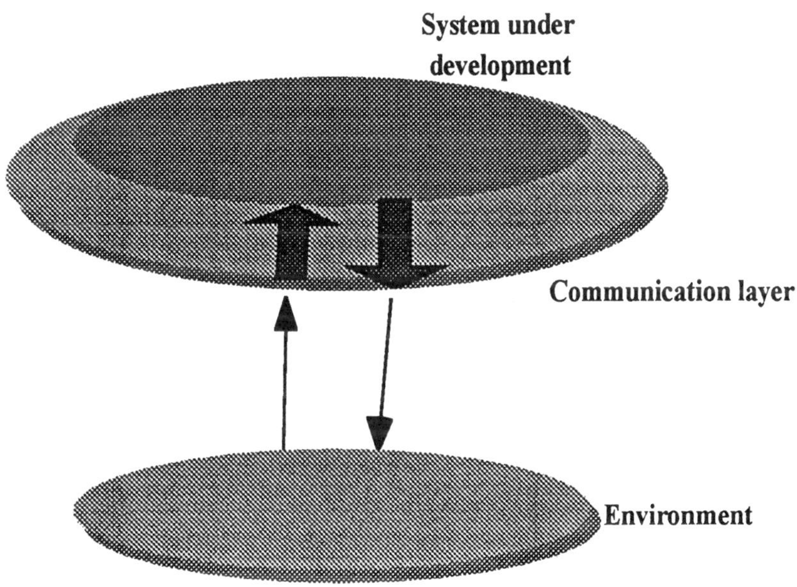

Figure 19.2. Communication layer in system - environment interaction

There are several popular testbeds used in fuzzy controllers - these include an inverted pendulum or a ball and a beam problem. There are no common cases

utilizing linguistic information (e. g., fuzzy expert systems) or more advanced control (even though such examples like a traffic light controller or group elevator control seem to be fairly attractive with this respect).

In what follows, we will elaborate on how the general lifecycle model becomes specialized in the design of fuzzy controllers.

19. 2. 1 Fuzzy controllers in the framework of life cycle model - design issues

The existing literature (Zimmermann, 1991; Kruse, Gebhardt, Klawonn, 1994; Pedrycz, 1995) provides us with a vast number of general as well as detailed design material. It should be emphasized that the fuzzy controllers as conceptual artifacts need to interact with a numeric environment - this implies that the role of the communication layer is immense and should not be downplayed. Referring to Figure 19.3 summarizing the main design steps, we lay down these main issues in more descriptive way. The same figure outlines the way in which the design phases are supported by the existing software platform.

1. Knowledge acquisition. This phase includes both qualitative as well as qualitative components of the domain knowledge (such as a number of control rules and their structure, fuzzy sets of the controller's variables, and the rules themselves, composition operators, etc.) to be decided upon when studying the controller.
2. Detailed construction. Within this phase the detailed controller is designed, developed and made available for verification and validation.
3. Static/dynamic verification embraces a panoply of activities within which the controller is validated with regard to its static properties (mainly analysed are aspects of logical coherency of the construct, correctness of static mapping properties, consistency and completeness of the rules). This part of the validation is rather straightforward and does not require any interaction with the environment. The verification of the dynamical properties is much more demanding as it calls for extensive experimentation with the controller to be performed under regular working conditions or in a relevant simulation environment. In other words, we are concerned either with an experimental approach or model-based simulation.
4. The deployment of the fuzzy controller is entirely application and environment driven decision-making process. The most crucial decision to be taken there is about selection of a specific hardware implementation (such as dedicated VLSI chips, programmable devices, general purpose microprocessors, etc.). To ensure a computational efficiency and comply with the hardware standards of the environment those are exclusively compiled versions of the fuzzy controller that can be basically treated as nonlinear number to number control transformations (mappings).

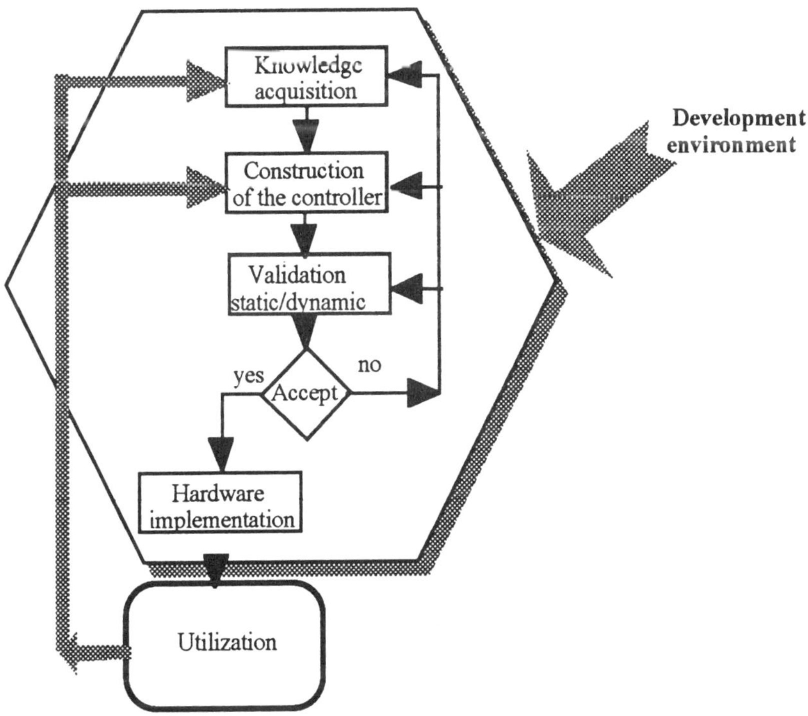

Figure 19.3. Development and use of a fuzzy controller

Each of these fundamental design phases is supported to a very different degree depending on the specific design platform. It could well happen that some of the phases are very much reduced in some development environments. Particularly, in most instances an emphasis has been shifted towards a final hardware implementation and deployment of the fuzzy controller anticipating that the controller realized in this manner could be still tested on-line but outside the platform itself. Any eventual failure in its performance would then lead to a repetition of the overall design loop as indicated in Fig.19.3.

That is important to say that to ensure the acceptable level of efficiency of the overall development process, the existence of a software simulation tool becomes a must. Otherwise, we would be usually confronted with a much longer and perhaps less efficient iteration cycle. The existence of an efficient and transparent validation tools is also quite critical to the success of fuzzy control or fuzzy reasoning. We would like to reiterate that the conceptual level at which these reasoning mechanisms are established is very distinct from the numerical level at which the controller manifests itself in the real application environment. This constitutes yet another reason for a prudent and careful verification that is indispensable and the most profitable at the very early level of the entire project.

19.3 CLASSES OF SOFTWARE RESOURCES

When studying the variety of the software resources supporting the development process, one can distinguish between three main categories of software:
(i) specialized fuzzy system design platforms,
(ii) general simulation tools,
(iii) standard high level programming languages (compilers or interpreters).

The above taxonomy is arranged starting from the most specialized tools and going to those of the most general nature. As expected, the development effort increases with the increasing level of flexibility of the software tool.
The specialized development platforms (selected examples of those are discussed in the following sections of this study) allow to design the fuzzy controller with a minimal programming effort. They are usually user-friendly with well-thought interfaces and some inbuilt validation mechanisms (practically oriented towards static verfication of the developed constructs). A certain drawback that is linked with these tools comes as a form of an evident trade-off one has to make with regard to the preassigned and somewhat rigid structure of the controller. Additionally, in some situations one must invest some time to get used to the software and its efficacies. This effort could vary from platform to platform.

The general simulation tools (such as MATLAB, MATEMATICA or KHOROS) that so far have not been very intensively exploited as design tools, are worth using. They usually furnish with enough architectural flexibility simultaneously cutting down an excessive amount of a tedious and detailed programming job. Especially, when these simulation tools are equipped with a complete graphical interface, the design of the fuzzy controller reduces to an intuitively transparent manipulation of the graphical modules (blocks) of the simulation package. In comparison to the previous category of software resources, the user is furnished with a series of basic components (assuming that a specialized library is available). These packages do not support any specialized hardware implementations neither can they generate any final code. In comparison to high - level programming languages, the programming process is far more visualk and faster. In comparison to the specialized packages they offer higher flexibility. The underlying assumption is that the user is familiasr with the technlogy of fuzzy sets as an onus of building the generic modules is often on the user himself.

Finally, fuzzy systems could be designed and fully programmed with the aid of one of the existing and widely available high level programming languages. The main disadvantage arises in a form of a substantial programming overhead that is required to cope with all the details of the controller popping up at the numerical level. The rewarding aspect, though, is in a full comprehension of the produced code and its complete flexibility. The selection of one of these software tools is primarily a matter of a target application, the functions of the controller and a design proficiency and expierence in fuzzy sets of the developer himself.

19.4 HARDWARE VERSUS SOFTWARE IMPLEMENTATION

This is usually a very crucial question one has to face with at a particular point of the entire design process. Several main options exist and the choice of any among them is very much application oriented and usually should be driven by some economical criteria. Some other main objectives of the project like system reconfigurability, speed, size, eventual scalability etc. should be taken into account as well. Three alternatives are available:
(i) software option - the inference scheme (fuzzy controller) is realized in software,
(ii) existing standard hardware environments. These mainly include commonly used 8, 16, and 32 bit microprocessors. The main advantage of this avenue is in the widespread utilization of these chips so that the comprehensive development packages of a general nature that are already in become immediately available. The fuzzy controller could be eventually added to the existing repertoire of the actual functions.
(iii) specialized analog and digital VLSI implementations of the fuzzy controller. Those are fully dedicated architectures aimed at a highest implementation efficiency asserted in terms of silicon utilization and speed of processing. The latter aspect definitely procures more attention.

19.5 SELECTED SOFTWARE DEVELOPMENT TOOLS

In this section we summarize some examples of the development tools aimed at designing fuzzy systems or processing fuzzy information. For more details, especially on the current tools, the reader may refer to the section of Technological Developments published in FUZZY SETS AND SYSTEMS. By no means the following coverage is exhaustive. The key objective is to make the reader aware of the diversity of the existing platforms and illustrate a panoply of the existing design platforms (characterized in terms of the tasks handled, interfaces, and development approaches).

19.5.1 MANIFOLD EDITOR and MANIFOLD GRAPHICS EDITOR

The MANIFOLD EDITOR and its enhanced graphical version, MANIFOLD GRAPHICS EDITOR, (both from Fuzzy Systems Engineering, San Diego, USA) are complete development tools aimed at supporting design of fuzzy inference schemes. They produce code for most assemblers and higher level languages. The software uses its own terminology that differs from that commonly found in fuzzy production systems (e.g. a principle instead of a rule). Its usage is consistent

across the product. The composition rule of inference is placed in a context of fuzzy associative memories (FAMs).
The maximal dimensionality of the designed system is limited up to five input variables and two outputs. The quantization (linguistic partition, fuzzy quantization) of each of the variables standing in the "if-then" statements involves up to eleven linguistic terms. Generally, these dimensions are largely sufficient considering the existing limit capabilities of a human being in perceiving and characterizing a relevant control policy.
From a functional point of view the Manifold Editor is composed of four main components. They directly support the basic design steps one goes through within the general flow of the design. The first two of them (principles editor and fuzzy sets editor) are aimed at defining the two essential components of the FAM system such as rules (principles) and membership functions of the generic linguistic terms. The rules are arranged in a matrix form (named here a principle matrix). A click-button procedure allows the user to access and modify each element of the matrix as well as toggle between each of its dimensions (considering that, in general, the matrix is a multidimensional array). The graphic fuzzy set editor provides almost unlimited capabilities of defining membership functions of various terms specified in the rules. In addition to somewhat standard piecewise membership functions (triangular and trapezoidal-like shapes), by selecting a free-form mode of its specification, the user can draw any shape of them to suit her needs. This option will be warmly welcomed by those users who would like to experiment excessively with various types of membership functions. The core part of the system is a MANIFOLD walker. The role of it is to facilitate understanding of the produced numerical results of the associations encoded in the principles. The walker is activated either by changing some values of the inputs or becomes triggered by the WALK option button. Some additional information about the results is also provided. It deals with the activation of the rules (principles) and the weights of the achieved evidence. Finally, the build module creates a file comprising of the knowledge base of the designed system that could be directly run on one among numerous microprocessors or microcontrollers supported by the MANIFOLD EDITOR.

The two outstanding features of the MANIFOLD pertain to the links one can establish between the software itself and various external applications. The first is a generation of “include” files for the wide class of high level programming languages (C, FORTRAN, BASIC) and assemblers supporting a broad range of microprocessors (both Intel and Motorola families including popular microprocessor of the 68HC11 line) as well as a specialized hardware (NeuraLogic NLX 230 Fuzzy Microcontroller). The second feature can be found particularly attractive by those anticipating the use of the MANIFOLD Editor in generating windows applications. These connections are available through the Dynamic Data Exchange (DDE) interface. This feature will significantly facilitate the development and experimentation with closed-loop control systems.
The software takes full advantage of its windows environment (WINDOWS 3.1

or later). The graphics functions are well exploited. The organization of the hierarchically arranged menus is well-thought out and assures a handy access to all the functions and specific pieces of data. The navigation between the menus is quite straightforward and should not cause any problem, even for an unexperienced user.
In addition to the enhanced graphics capabilities, the files of the controller produced by the software could be aimed at a specific microprocessor or more generally, produce code of one of high level programming languages. The other option are the codes supporting several of the currently available development platforms like Fuzz-C by Byte Craft.

19.5.2 FUZZY LOGIC DESIGNER ver. 1.0

Fuzzy Logic Designer (Byte Dynamics, Inc., Spokane, USA) combines two essential design functions of the fuzzy controller development tool and C code generator. As embedded in the Windows environment, it furnishes the user with the fully interactive GUI facilities. This helps design membership functions, develop control rules, and easily simulate a numerical response of the controller. And, finally, the software generates a source C code to be included into a particular application or embedded processor.

The architecture of the fuzzy controller being supported there is standard to a certain extent, so is its design procedure. An interesting feature of the software concerns its defuzzification algorithm. In contrast to the common practice that is to define fuzzy sets of conclusions (control) as a part of the control protocol, this package introduces a straightforward numerical partition of the corresponding universe of discourse. The specification of the fuzzy sets of input variables (error, change of error, etc.) proceeds in a standard way. Firstly, one defines a suitable universe of discourse. This step is followed by defining or editing membership functions. By default, all the membership functions initially specified by the system are of a trapezoidal shape. The software fully supports all further editing activities of these membership functions (add, delete, print). The procedure of building the control rules is again well supported as all the linguistic labels are neatly organized and could be accessed quickly. Each rule is constructed in two phases: in the first step one selects its subconditions using the selection base of the menu. Once this has been completed, the list of the linguistic values (membership functions) becomes available. By clicking on a suitable membership function the subcondition becomes fully structured. This course of the design assures that each rule is syntactically correct. The subconditions of the rules are either aggregated in an AND or ORwise manner. It is not clearly specified in the manual, though, how these logic operations have been implemented. The option menu of the system includes SIMULATE and GENERATE SOURCE, as well as allows for some further variations of the generated code. The SIMULATE option is aimed at displaying a way in which any changes of a single input variable of the controller affect its output. When analysing many-input systems, the values of all the remaining inputs except the one to be studied, have to be fixed. The outcome of running this option comes as a family of usually nonlinear input-output numerical characteristics of the controller. They provide a relatively detailed look at the relationships generated by the controller and visualize the impact that different rules and fuzzy sets included in the rules have on these dependencies. Even though the evaluation arising at this level could be fairly qualitative (since from a designer point of view, we are really not that sure what type of characteristics should be developed, as this in turn depends heavily upon the specificity of the system to be controlled), it still constitutes a useful verification tool. If not fully explanative, this phase could allow the user to recognize and eliminate those evident deficiencies of the controller which are strongly reflected in eventual irregularities or discontinuous of its nonlinear characteristics. The GENERATE option is useful when generating C language files (source and include file) needed to integrate the fuzzy controller into an application structure. This useful facility becomes nowadays more popular in a growing number of development platforms. With this respect, the quality of the source code is good; in several completed experiments the compilation and linking were successful. The two options affecting the source code include the use of floating point variable types (on/off toggle) and on/off switching the Function Prototypes (the option which is useful when using compilers capable of accommodating function prototype)

19.5.3 FuzzyTECH 3.0 Explorer Edition

The general idea behind fuzzyTECH (Inform GmbH, Aachen, Germany) is to provide the user with an extensive family of diverse design tools useful in applying fuzzy sets in real-world problems. These tools exhibit different capabilities, support various realizations (including a broad range of microcontrollers) and generate highly efficient C-code. FuzzyTECH Explorer, is aimed at two main groups of users: those who would like to become familiar with fuzzy controllers (and generally fuzzy rule- based systems) as a new, however well developed and technologically proven machinery, and the users au courant with fuzzy sets and being interested in carrying out small feasibility studies. Overall, the software implements a commonly found and well-known architecture of the fuzzy controller that is available in the literature and applied in industry. Nevertheless, several new and less known however desirable concepts have been added to this generic topology. The main emphasis has been placed on building a user-friendly and highly interactive development platform.

FuzzyTECH is invoked through its popup menus arranged in groups. The successive design objects coming in the course of the design can be invoked either with a mouse or keyboard. The design starts with opening a project editor window (the Explorer version supports only one project window to be open at a time). Within this phase the basic structure of the controller is defined. The Explorer version supports up to 125 rules; each control module (knowledge base) allows for 2 input variables and a single output. Essentially one defines input interfaces by defining membership functions of the corresponding fuzzy sets. The software supports several standard shapes of membership functions (like S, Π, λ, etc.); the edition of these is conveniently accomplished by the available editor. One this is done, the controller has to be equipped with a set of control rules. FuzzyTECH comes equipped with two different types of rules editor, namely the spreadsheet rule editor and the matrix rule editor. The first type of rule visualization is more readable as it looks more familiar to the users already exposed to spreadsheets. Each row of the sheet represents an individual control rule with its inputs (conditions) situated at the left side of the row and the consequences places at the right. Two interesting features are available there. Firstly, one can assign a degree of support (confidence level) to the rule. This adds up an additional flexibility to the control protocol allowing for distinguishing between the rules that are viewed as most essential and those being slightly "peripheral" to the control strategy. From a technical point of view it allows also to turn off some rules by assigning a zero degree of support. In editing the rules one also gets a useful option to pick up a "don't care" situation in this manner expressing a lack of importance of a certain condition in the given rule (essentially this condition becomes skipped).

Some additional organizational functions particularly useful when arranging large rule bases involve sort, copy and move options. These are not supported by the Explorer edition. The second type of rule arrangement is provided in the format of the fuzzy associative memory. In this context, the associations formed by the rules are more profoundly visualized than it was possible in the previous method. One can easily browse through the entire rule base by selecting the terms of the linguistic variables not displayed in the inference mechanism as it is realized by this package consists of two clearly distinguished components such as its aggregation and composition part. The aggregation mechanism pertains to the procedure of combining subconditions in each rule; depending upon the character of the aggregation this is realized by the MAX or MIN operations. Subsequently, the following composition is aimed at combining the degrees of activation of the rule and the degree of support (confidence level) of the rule itself - again this is done via the product operation. Finally, the defuzzification methods are standard and utilize variations of the Centre-of-Area procedure. Once the fuzzy controller has been built, several testing tools can be exercised. For debugging, verification and optimization, the Explorer edition can be set up in several modes:

interactive: this allows the developer to check system response to some specific input data that are entered manually.

link: by exercising this option one links the controller to a process simulation.

batch: data from an input file are processed by the controller to an output file while the inference process is not visualized.

Regarding implementation, the designed fuzzy controller can be generated as a C-code. The selection of this option is strongly justifiable considering that most of the exiting hardware platforms fully support C-compilers. This makes the software an ideal tool in carrying out small feasibility studies. More advanced versions of the fuzzyTECH family are aimed at a variety of implementations encompassing microcontrollers (such as e.g., INTEL MCS-96, 8051 microcontroller family), fuzzy processors, and programmable logic controllers. Interestingly enough, all fuzzTECH versions are fully compatible via the Fuzzy Technology Language (FTL) interface.

The system comes with an interesting example that is a graphical animation of the container crane. The control objective in this learning example is to transport a container between two points (ship and train). The control task is definitely neither a trivial nor well defined problem so that it could be easily converted into a formal expression containing control goals and constraints (anyway, all of them are not fixed over the entire course of control). Furthermore, an additional difficulty lies in the fact that the container is connected to the crane by a flexible cable, causing the container to sway while being transported. The option of manual control allows for a manual control of the crane motor thus exposing one to an "engineering feeling" on how to carry out control.

19.5.4 Linguistic Fuzzy Logic Controller for Education LFLC-edu ver. 1.0

This software coming from the Czech Academy of Sciences, Ostrava-Poruba, Czech Republic, is a development tool for design of linguistic fuzzy logic controller. This package exhibits some interesting and definitely useful technical features that pertain to:

(i) handling linguistic terms standing in the control rules. The implemented method speeds up the design process of the controller by allowing the designer to concentrate exclusively on a linguistic facet of the control protocol and relieve him from a substantial burden associated with a detailed numerical specification of the relevant membership functions,

(ii) closed-loop simulation facilities,

(iii) logic-oriented character of processing realized by the controller.

A strong educational flavor of the package is very visible as the software itself is written in TURBO-PASCAL (ver. 6.0) that definitely constitutes a well-settled education oriented environment. Let us elaborate a bit more on the main features of the package. Proceeding with the first aspect of linguistic characterization of fuzzy sets utilized in the controller (that is strongly emphasized by the name of the controller itself), let us stress that this implicit way of handling fuzzy sets is essential in preserving a genuine linguistic concept permeating the design philosophy of any fuzzy controller. The main idea behind this design aspect is that instead of defining membership functions in a pointwise manner, the designer can choose among a finite family of linguistic terms being available in the system (like Small, Medium, Positive Big, etc) and eventually modify them with the aid of some linguistic modifiers (hedges). Note, though, that these linguistic terms are universal in the sense that they can describe some general notions.

Despite their generality they are also context- dependent. In other words, any term (like "high income") might have a different meaning that depends heavily upon the specific context and as such calls for a suitable numerical calibration. Nevertheless, the relationships between several linguistic terms still hold despite the changes in the context. One can look at this issue of context-sensitivity and cope with it by adjusting the terms. This could be realized by modifying the universe of discourse through its contraction or expansion. In the simplest case it could be done by changing a value of a so-called context coefficient. This way has been often implemented in the standard fuzzy controllers we are usually concerned with, deal with the max-min composition that is used as an inference mechanism, whereas the fuzzy relations (combining the knowledge residing within the rules) are developed through a straightforward Hebbian learning. In contrast to that, LFLC exploits a different path of processing with a strong logic-motivated constructs involving Lukasiewicz implication and the associated composition operator. Slightly different justification of the use of these logical constructs has been also studied with regard to fuzzy relational equations. On the technical side, the package consists of two separate programs with clearly distinguished functional properties. The first program (called LFLCDes) is aimed at a systematic design of the controller while the second one (named LFLCSIM) is used to carry out simulation in a closed loop structure formed by the controller and system (plant) under control.

The system is quite compact and requires 0.8 MB of hard disk. It supports design of the fuzzy controllers with up to four input variables and a single output. The discrete spaces (universe of discourse) can contain up to 101 elements. The number of rules should not exceed 300. The antecedents of the rules can be either provided in a conjunctive (AND) way , say X1 & X2. The disjunctive format is also allowed; the expression (X1&X2)|X3 denotes the condition structured as (X1 and X2) or X3. The rules editor following a common WORDSTAR- like standard is useful in rules addition, deletion and modification. As said before, the inference is carried out with the use of the logic mechanisms; for purposes of comparative analysis the standard algorithm (the Mamdani-Zadeh method called here Max-Min inference mechanism) is also available. The output transformation that works from the level of fuzzy sets to the corresponding numerical values (defuzzification) is implemented as a standard centre of gravity method. The program allows to carry out a static analysis of the controller. The main goal of this phase is to reveal the characteristics of the controller by deriving the values of control as well as visualize activation levels of the rules in the controller. The simulation package supports simulations with a variety of dynamic systems including those with nonlinear components as well as time delays. This constitutes an extremely useful feature of the package as the user can experiment with the fuzzy controller by studying its performance under very different control environments.

19.5.5 Fuzzy Logic Development Kit (FULDEK)

This software package (distributed by TSI Enterpises, Inc., Albuquerque, USA) is a complete development tool aimed at development of fuzzy controllers by providing all necessary facilities towards their design, testing, and experimentation. FULDEK (developed by Bell Helicopter Textron Inc., Fort Worth, TX) has been created to meet the needs of newcomers as well as mature experimenters. It is well balanced with this regard and constitutes a valuable teaching aid loaded generously with more advanced functions that are indispensable for arranging complete simulation sessions. As being a Windows-based product, it is easy to operate (requires a mouse and keyboard), and has well-structured menus navigating the user through all the steps of the complete design procedure. The student version of FULDEK, as contrasted with the complete professional package, has been reduced in terms of dimensionality of the simulated objects as well as some functional properties. The design of the fuzzy controller is carried out in a standard fashion that is well-known in the literature. This procedure is closely followed by the software through its functions clustered in a menu bar. These functions are organized as the EDITOR and SIMULATE options. The EDITOR option allows you to manipulate files (FILES), edit a Fuzzy Rule Base (FRB), edit an existing ASCII file, change screen codes and load the RUN option. Some of these menu items have additional drop-down submenus that take care of the relevant design details. The option FILES provides a full access to all files (such as rule-based structures, data files containing membership functions, and parameters of the models to be used in simulation).

All universes of discourse for the variables used by FULDEK are normalized so that their values run from -1 to 1. The normalizing coefficient , called the Scale Factor, converts the external format of the physical variable to the appropriate internal format as being seen by the fuzzy controller. The membership function editor is used to add and edit membership functions. All the membership functions take on either trapezoidal or triangular shapes. The two options of generation of these functions are available. In the manual entry method one forms the membership functions manually by selecting the corresponding coordinates of the universe that completely define piecewise linear functions (the radio buttons allow for a quick movement of the coordinates). The automatic option produces the membership functions over the universe of discourse once the number of the linguistic labels has been selected. Those functions are of a triangular shape with a 1/2 overlap between the adjacent linguistic terms. The COMPOSE option is used to compose the fuzzy rule base. This design phase is extremely well supported as the screen is arranged as a collection of windows visualizing separately the condition and action parts of the rule. This makes the composition of a rule completely straight forward:

- select a condition variable from the IF variable list (clicking it with the mouse),
- link this variable with a membership function (two options are available: IS, or IS NOT, the latter causing the complement of the fuzzy set to

become associated with the variable),

- select the rest of the conditions (again they could be assembled together through AND or OR connections),

- finally, select the variable from the "THEN" list of variables, and specify its valuc,

- review the complete list that occurs at the right lower corner of the screen and make changes, if necessary. The rules could be additionally weighted by linking them with the corresponding weight factors.

Once all the rules have been formed, the fuzzy controller can be inspected with respect to its static properties. Firstly, one can examine the coverage of the membership functions and rules. This is accomplished in two different fashions:

- inspections by variable: by selecting this particular variable, FULDEK searches the knowledge base for its all occurrences,
- inspections by rule: the rules involving this variable either in its condition or action part are displayed on the screen.

This analysis screen resumes the design phase of the fuzzy controller.

The RUN option enables all simulation experiments. These are carried out in two general modes. The "open-loop" mode addresses the issue of analysis of the controller being perceived as the following static numerical relationship

$$\text{control} = \text{fuzzy\_ controller (input\_variables)}$$

The analysis can be activated in a step mode whereas the inputs are provided by the user and the corresponding control is computed. The sweep mode produces 2D and 3D plots of the control relationships as well as contour maps of the characteristics of the controller. Indisputably, the closed loop mode experiments available through the RUN option are the most important simulation feature of the software. The fuzzy controller is placed into a closed loop with the model of a dynamical system. The model of the system is available in state space notation, namely,

$$\dot{\mathbf{x}} = \mathbf{Ax} + \mathbf{Bu} + \mathbf{Gz}$$
$$\mathbf{y} = \mathbf{Cx} + \mathbf{Du} + \mathbf{Hw}$$

with **A**, **B**, **G**, **C**, **D**, **H** being matrices of the model; subsequently **z** and **w** are used to characterize the external disturbances affecting the system.

The user can define an initial condition of the model as well as specify links (variables) used to form the feedback within the structure.

19.5.6 MATRIXX /SystemBuild

The Integrated Systems' (Santa Clara, USA) MATRIX and SystemBuild products are aimed at control engineering computations, designing block diagrams, and implementing the control systems as well as developing their real time versions. SystemBuild is a major component of the MATRIX and AutoCode products. The most prominent feature of SystemBuild is the graphical, interactive block

diagram editor used for developing design models. An integral part of SystemBuild is the simulation program which allows the model to be studied and validated prior to its implementation. The SystemBuild block diagrams may be processed by code generation programs, to produce Ada, C, or FORTRAN compiler language code, for test bed execution or system implementation.
The MATRIX$_X$ and SystemBuild product lines include modules of optimization, system identification, robust control, digital signal processing, control design, to name the most essential of them.

The most prominent features of SystemBuild are:
-interactive mouse driven workstation graphical environment,
-hierarchical block diagram capabilities for designing systems of arbitrary size and complexity,
-extensive screen and hardcopy plotting capabilities,
-the design philosophy that emphasizes control system and real time capabilities, including the ability to model continuous processes and multirate discrete systems.
-an extensive library of system design blocs (including dynamic blocks such as transfer functions and state space representations, algebraic, trigonometric and logical functions; the system allows the user to include his own code blocks).
The Real-Time Fuzzy Logic (RT/Fuzzy) block under ISI's SystemBuild module provides the user with a powerful tool. The RT/Fuzzy block is integrated seamlessly into the control system design product, deriving much of its power from this integration. The implementation of the fuzzy controller realized by this block exploits all the essential features found in this construct. The conjunction and disjunction models of logic operators standing in the rules include max and min operations as well as the product and probabilistic sum. Additionally, each rule can be additionally weighted where this process reflects a user-defined priority attached to this specific rule. The assumed values can be distributed between 0 and 1; a higher number indicates more confidence associated with the specific rule. The default weight is 1; it should be maintained as such unless the rule is really considered less important (influential) than the others. This option might be found useful in studies on a structural robustness of the controller. Two modes of inference embrace the standard way of reasoning provided by Mamdani and the other named after Larsen (it utilizes the product operation as a part of the composition operation). Furthermore, the user can introduce its own inference machinery through the USER-DEFINED procedure. The output interface (defuzzification) exploits the centroid method and the mean of the maxima procedure. Again the USER-DEFINED option can be fully exercised. The aggregation method determines a way in which several parallel decision are combined to constitute a single outcome. Here the most common forms include averaged operations of a different character (including arithmetic, geometric, and harmonic means). The fuzzy controller module is equipped with optimization mechanisms. The optimization preferences include possibilities involving memory, speed and a compromise of the two of these. The above criteria are used to determine a rational trade-off when developing the working version of

the fuzzy controller. The software comes with a complete example illustrating the use of the fuzzy controller that might be applied in the implementation of control of a household water heater system. The complete code allows for many experiments both at the structural and parametric level.

19.5.7 A Fuzzy Logic Knowledge base generator for the MC68HC11 and MCH68HC05 Inference Engines

The MOTOROLA (Austin, USA) software KBG.EXE is aimed at the designer of fuzzy controllers who would like to take advantage of the most popular representatives of their 8-bit family of microprocessors and implement the controller using this vehicle. The software supports both the MC68HC11 as well as MC68HC05. The software is delivered in its executable format (.EXE); hence its installation is instanteneous. While the main thrust of this software is to generate an assembly code of the controller, its functional features are very much limited to the most required functions. By default, it is assumed that the user is familiar with fuzzy sets and fuzzy controllers. The newcomers might feel a bit uncomfortable while using this package. The design of the controller is direct, the architecture is standard. The number of inputs and outputs (8 and 4, respectively) are fully acceptable. The inference schemes uses a standard lookup method using the precomputed values of the control signals. Each variable can have up to eight linguistic terms (fuzzy sets) while their membership functions are restricted to trapezoidal (in particular triangular) shape. The outputs are given as numerical values. The design processes comprises:(i) a specification of the membership functions for the input variables (those are completely defined by the coordinates of the trapezoidal/triangular form of the functions), (ii) identification of numerical values of the control variables , followed by the (iii) control rules entered in the standard "if-then" convention.

At this point the user may save, alter, or encode the controller into the knowledge base for either of the specified MOTOROLA processors. Once the code has been specified, it can be added to the source code of the inference engine for the processor. As stated, this software is a helpful tool to the experienced designer interested in a microprocessor realization of the fuzzy controller in the MOTOROLA environment.

19.5.8 Fuzz-C, a preprocessor for fuzzy logic, ver 1.00

The software coming from Byte Craft Ltd. Waterloo, Canada is a design tool aimed at realization of a generic type of the fuzzy controller in an open C environment. Essentially, the Fuzz-C preprocessor handles a special fuzzy control language and ANSI C code. Some highlights of the language include such modules as:

FUZZYSET defines the data type and range of the universe of discourse,

LINGUISTIC declares a linguistic variable (fuzzy set) and its membership function (binary, trapezoidal, or an arbitrary lookup table),

FUZZY contains the knowledge base of the "if-then" statements. The rules utilize basic logical connectives (AND, OR, NOT),
CONSEQUENCE defines the output membership functions and defines the transformation (defuzzification) method; the list of the available approaches include centre of gravity, maximum left, maximum right, middle of maximum, maximum of average, or more generally, user-defined.
Running Fuzz-C to generate an ANSI compatible C-code is straightforward. The syntax of Fuzz-C is definitely a plus of the software. Upon compiling the file some minor problems were encountered (e.g. no type-casting which requires the user to check final code for type mismatches, in some cases the membership values were not accurate). Nevertheless when being aware of these points and checking the final code carefully, on the whole the software is a good time saver.

19.5.9 FuziCalc ver. 1.00 for Microsoft Windows

The technology of spreadsheets has been found successful for a long period of time. As this software allows to handle conveniently a broad range of practical information processing problems, the spreadsheets definitely suffer from a lack of ability to cope with imprecise pieces of data that are not well represented as single and precise numerical quantities. FuziCalc FuziWare, Inc., Knoxville, USA, narrows this acute gap.
Briefly speaking, FuziCalc help manage uncertainty in a familiar and easy to use spreadsheet environment. A broad variety of tasks such, as for instance, maintaining operating budgets, forecast sales, keeping track of development costs, etc. are evidently straight examples of applications in which the user can equivocally benefit from the use of fuzzy set technology. The user familiar with spreadsheets will find the software very user- friendly and easy to use. For less experienced users a brief perusing the manual should be more than sufficient. While the operations are carried out for fuzzy data, each spreadsheet cell shows a central (modal) value. A small gray triangle (fuzzy value flag) is used to indicate the fuzzy character of the data. Whenever the cell becomes selected the membership function of the quantity (called here a belief function) is visualized in the upper right hand corner of the menu bar. The spreadsheet supports a broad variety of operations on fuzzy data. Any number of columns and rows is supported as long as the their number does not exceed 64 K cells. The software complies with the Windows DDE standard; this feature becomes of a particular interest when using this package as a part of the entire development environment. To export fuzzy data, the individual fuzzy number is replaced by its centroid numerical representation (this becomes necessary when sending a file to a non-fuzzy application like Microsoft Excel). The fuzzy data can be selected from a gallery of basic shapes of their belief functions: triangular (that constitutes a default option in the system), trapezoidal, camel-back (double hump), tent, and spire (steeple) membership functions. These shapes allow the user to fully articulate her/his preferences about fuzziness being residing within a given piece of data.

The editor facilities are superb: a form of the membership function can be modified in a matter of seconds.
A brief comment about processing fuzzy sets realized by the software. FuziCalc produces the results that are approximations of the exact algebraic operations completed on fuzzy data. For instance, a result of multiplication of two triangular numbers emerges as another triangular fuzzy number - the quality of the obtained approximation , that quite often is fairly sufficient, could vary from case to case depending upon a relative fuzziness of the objects (expressed e.g., in terms of ratios of the spreads and modal values of the arguments being used). Nevertheless, a certain caution should be exercised in situations of propagation of this approximation error along a long chain of operations (in particular when iterating over several operations on fuzzy data). Another important enhancement along this line might be an addition of interactive operations on fuzzy numbers. This will contribute to a higher universality of the software; unfortunately, the inclusion of this option could be quite demanding on the computational side.

19.5. 10 FUZZY DECISION-MAKER ver. 2.1

As the subtitle of the manual implies ("A method of deciding in complex situations"), this software (from Fuzzy Systems Engineering, Poway, USA) is aimed at making decisions in a fuzzy environment where we allow the goals an constraints as well as their quantification to be imprecisely defined. From a formal point of view, the decision processes supported by this package fall under the category of a single - step multiobjective decision - making. Considering that the current decision situation is characterized by a series of constraints, the future is expressed via a collection of goals to be met, the decision - maker is interested in determining some alternative(s) that allow to satisfy the goals to the highest extent.
The design process supported by the software consists of several obvious phases:
Step 1. Define a new project. At the start of the new project, the system requests its name and makes all preliminary arrangements.
Step 2. Define goals. During this phase one enters the names of the goals identified in the problem along with their eventual more extensive verbal description. The system handles up to twenty goals.
Step 3. Define constraints With the dialog box shown on a screen the user defines the names of the constraints; again up to twenty constraints can be defined.
Step 4. Define alternatives. The system supports up to forty alternatives - these are define by short names. As before, more verbal description is allowed.
Step 5. Rank importances of goals and constraints. Using a collection of bins of different importance (and described linguistically like most, strong, more, moderate, weak, etc.) one assigns importances to the objectives of the problem by placing them into the corresponding bins.
Step 6. Satisfactions. Here provided are the levels of satisfaction in each pair of goal/constraint - alternative. This step allows for several methods of setting up

these satisfactions that use graphics or some spreadsheets. Similarly, a manual method of setting them up is also available - it can be done by considering goals or constraints.
Step 7. Running the Fuzzy Decision Maker. Finally, for all the goals and constraints defined, the system identifies the alternatives tat are viable solutions to the decision problem. It also generates a bar chart showing the contributing components of the produced decision.

The main undisputable advantages of the system include:
- a well-thought graphics with carefully designed and easily understood icons allowing even a novice user to navigate quite easily through the reefs of the efficacies of the process of decision - making. More importantly, the colors are used very consistently throughout the system.
- a graphical way of presenting results
- a handy on-line help facility
- a set of informative example projects of a various nature such as e.g., choosing a college for higher education, determining a power transistor in a switching power supply, describing employment situation, etc.

The only essential feature that is missing in the system is a lack of verification

of consistency of pairwise comparisons supplied by the user thus making the designer/decision-maker aware of the existing inconsistencies within the evaluations before getting into any decision - making.

19.5. 11 DataEngine ver 1. 2

Data analysis, data mining, knowledge discovery in databases - all of these terms are used to describe human activities aimed at revealing more or less visible and empirically justifiable relationships in large arrays of numerical data. The developers of this software MIT-Management Intelligenter Technologien GmbH, Aachen, Germany) have concentrated on some selected yet highly representative algorithms. The technology of fuzzy sets is present in the procedures such as fuzzy rule-based computation, FUZZY ISODATA and a fuzzy relative of a Kohonen map. The other architectures implemented in the system include Kohonen map regarded as an example of unsupervised learning and a multilevel perceptron being a representative of supervised learning.

DataEngine is a software for sophisticated data analysis. It furnishes the user with a vast number of preprocessing tools including FFTand digital filters. The product supports various statistical and other mathematical functions. The more advanced processing tools include fuzzy C-Means, Kohonen map, fuzzy Kohonen map, and multilayer perceptron. The tool can be run on different hardware platforms. The product family contains DataEngine v.i which is a function library of virtual instruments (VIs). It can cooperate with LabView[R] (National

Instruments).

From a functional point of view the software provides eight different function groups structured into Input, Output, Mathematics, Statistics, Signal Processing, Data Analysis, Select and Multi I/O. The processing blocks are selected from each group and combined in a card (a specific entity used by DataEngine), arranged and connected. The underlying concept of the software is that any process of data acquisition, preprocessing, and processing is arranged by composing the corresponding card. The use of the system is almost self-evident. Most of the functions are well known as being common to many other software tools, especially spreadsheets. The Data editor allows us to work with data files and supports all standard file operations (open, close, save, save as, edit, select elements and rows/columns, etc.) The mathematical functions summarize everything the user needs to process data. Those include arithmetics, logarithms, trigonometrics, hyperbolics, normalization, standardization, and denormalization as well as statistics. The processing tools include:

- Fuzzy C - Means (FCM). This is a standard method of fuzzy clustering well known in the literature. Working with the method is divided into three phases such as clustering, labeling and classification.
- Fuzzy rule - based module. It supports a detailed step - by - step procedure of building rule - based model generously supplied with relevant procedures of constructing membership functions, carrying out inference (mapping) mechanisms, configuring of input and output files. The rules are equipped with certainty factors. The membership functions available in the system are piecewise linear. Their most general version is that of polygon shape; some other options concern singletion, trapezoidal and triangular along with some more specific options (triangular left open, triangular right open, etc.).
- Multilayer perceptron. The topologies of the neural networks supported by the system are specified accordingly: a number of layers is between 2 and 4 with a number of neurons per layer up to 1,000. The list of available transfer functions include linear, sigmoid and tanh.
- Kohonen map and its fuzzy counterpart.

19.5. 12 RTFCM ver 1.4

RTCFM (Ring Technology, La Vale, USA) is a learning tool that exposes the user to fuzzy cognitive maps by explaining their main architectures and functioning. The software is user-friendly and requires very little of keyboard use as most of interaction is realized through a mouse.

The design of the map is straightforward. This is mainly due to a simple and straightforward notation used throughout the software. Cognitive maps viewed as directed graphs consist of two types of elements: concepts (nodes) and causal links (edges). The concepts are represented as movable rectangles while the edges are given as lines equipped with arrowheads and semicircles. Each rectangle accommodates a concept name and an active area marking whether a concept is on or off. The colours are used consistently: if the concept is on (active), then the active area of this rectangle is red. Similarly, small semicircles at the bottom of the rectangle can be either marked in red or blue with the blue semicircle denoting a causal decrease. A red semicircle designates causal increase. The connections use also red and blue colour. If the connection is red, an increase in one concept tends to increase the strength of the other one being associated with it.

Building the map is instantaneous. The process starts with the specification of the number of concepts. The Connections-auto mode produces a fully connected map whose connections are initialized randomly. Those connections can be eliminated by clicking on the None option. When using this option, one defines the connections one at a time. At this stage one can also set up positive or negative causal interaction between the nodes. The same option applies to setting up positive (excitatory) or negative (inhibitory) feedback involving the same node - the feature contributing substantially to the dynamics of the map. The operations available when operating on the nodes include renaming , turning the nodes on and off, and removing the concepts.

Once the map has been constructed, we are ready to study the dynamics of the on-off activities of the concepts. Each iteration step produces corresponding colours of the concepts highlighting their activation. During simulation, the screen displays a bulletin board including a length of the longest transient and the test vector being currently analysed. An interesting and highly useful feature of the analysis comes in the form of a length of the limit cycle of the map being implied by the initial conditions. The stability is quantified through a so-called stability index. For instance, 100% stability means that , no matter what value the concepts are, the map always reaches a fixed point. If stability is 70%, then 30% of the time starting off from random initial conditions, the map will end up in equilibrium. The simulations are quick. The option Cycles/All cycles is the only operation that uses up a lot of time. This is not surprising as the map goes through all possible combinations of its on/off states. While small cognitive maps (around five states) are analysed very briskly, everything slows down when the number of concepts is increased (the PC version of the software makes it possible to execute this option for the architectures including up to 12 concepts).

19.5.13 Winrosa

The WINROSA software package developed by MIT (Management Inteligenter Technologien GmbH, Aachen, Germany) is one among rule generation platforms geared into some data mining activities. The rules supported by WINROSA assume the standard form of multiple input - multiple output constructs

-if condition$_1$ and condition$_2$ and....and condition$_n$ then conclusion$_1$ and conclusion$_2$ and ... and conclusion$_p$

where the quantification of the conditions and conclusions is carried out by linguistic labels (fuzzy sets). The key objective of data mining set up in such context is to develop the rules based on numerical data so that they are both meaningful (stating that they have enough support in terms of data being captured) and user-friendly. The latter feature underlines a need for small and compact sets of rules that are easily comprehended.

The data set to be analyzed by WINROSA are represented as flat structures - arrays whose columns correspond to the variables arising in the problem at hand. The rules can be exported to some other environments commonly exploited in the design of rule-based systems including well-known software products in the area of fuzzy set technology such as DataEngine, fuzzyTECH or MATLAB.

The software bundles a series of basic subtasks; these are arranged around several main groups:

definition of a project. This forms a preliminary phase of all rule generation activities. One specifies all input and output variables and quantifies respective linguistic labels (fuzzy sets) over each of them. There are two types of membership functions utilized therein, namely triangular and trapezoidal ones. This selection may look a bit restrictive but these two categories of membership functions seem to be omnipresent in most of the current applications of fuzzy models.

rule generation Once the structure of the project has been completed, the phase of rule generation concentrates on the generation of the relevant rules. This is, in fact, a process of searching through a space of all possible rules. Depending upon the size of this space (that heavily relies on the number of antecedents in the rules), the software utilizes different search mechanisms. For small search spaces a complete search is a viable option. For large size problems one may lean towards evolutionary search. These search options are made available to the user. Through the generation process, one concentrates on meaningful statements, viz those which have passed a relevance test. Each rule comes equipped with a confidence factor expressing how essential (relevant) the specific rule is. Moreover, the system produces positive and negative rules. The positive rules express affirmative relationships. The negative rules express prohibitions, viz. associate antecedents with actions (conclusions) that become prohibited when the conditon part is satisfied.

reduction of rules The aim of this phase is to reduce the rules to come up with a reduced protocol of compact rules. The compactness of the rule set is viewed from at least two different standpoints: first, one should strive for a minimal collection of the rules capturing the essence of the data. Secondly, the rules themselves should be concise enough and not necessarily including all variables existing in the original data.

analysis of a rule set The intent of this phase is to analyze the rule for their plausibility. The proposed rating index helps quantify dependencies within the data.

export of a rule set At this stage, the results (rules) can be exported to some other application programs for further use. Export files can be saved in plain ASCII files as well as the formats compatible with those used by MATLAB, DataEngine and other fuzzy control and fuzzy system design packages.

19.6 CONCLUSIONS

The appropriate selection of the design environment and the final implementation of the fuzzy controller contribute vastly to its efficient design and utilization. We have posed and discussed some crucial development aspects. The analysis of the existing development tools shows that those are useful design vehicles supporting to a certain extent the more organized design of the fuzzy controller. Nevertheless there is still a gap between the latest achievements and trends that make the design practices more systematic and the actual capabilities of most of the development platforms available on the market.

References

[1] B. I. Blum (1992). *Software Engineering. A Holistic View*, Oxford University Press, N. York, Oxford.

[2] B. W. Boehm (1975). Software engineering, IEEE Trans. on Computers, C-25, 1226 - 1241.

[3] R. Kruse, J. Gebhardt, F. Klawonn (1994). *Foundations of Fuzzy Systems*, J. Wiley, Chichester.

[4] W. Pedrycz(1995). *Fuzzy Sets Engineering*, CRC Press, Boca Raton, Fl.

[5] H. J. Zimmermann (1991). *Fuzzy Set Theory and Its Applications*, 2nd edition, Kluwer Academic Publishers, Boston.

Index

A

O

P

Q

T

U

V

W

Z

PRACTICAL APPLICATIONS OF FUZZY TECHNOLOGIES

THE HANDBOOKS OF FUZZY SETS SERIES

Series Editors
Didier Dubois and Henri Prade
IRIT, Université Paul Sabatier, Toulouse, France

FUNDAMENTALS OF FUZZY SETS, edited by Didier Dubois and Henri Prade
MATHEMATICS OF FUZZY SETS: ***Logic, Topology, and Measure Theory***, edited by Ulrich Höhle and Stephen Ernest Rodabaugh
FUZZY SETS IN APPROXIMATE REASONING AND INFORMATION SYSTEMS, edited by James C. Bezdek, Didier Dubois and Henri Prade
FUZZY MODELS AND ALGORITHMS FOR PATTERN RECOGNITION AND IMAGE PROCESSING, by James C. Bezdek, James Keller, Raghu Krisnapuram and Nikhil R. Pal
FUZZY SETS IN DECISION ANALYSIS, OPERATIONS RESEARCH AND STATISTICS, edited by Roman Slowinski
FUZZY SYSTEMS: ***Modeling and Control,*** edited by Hung T. Nguyen and Michio Sugeno
PRACTICAL APPLICATIONS OF FUZZY TECHNOLOGIES, edited by Hans-Jürgen Zimmermann

PRACTICAL APPLICATIONS OF FUZZY TECHNOLOGIES

edited by

Hans-Jürgen Zimmermann
Operations Research, RWTH, Aachen, Germany

KLUWER ACADEMIC PUBLISHERS
Boston/London/Dordrecht

Distributors for North, Central and South America:
Kluwer Academic Publishers
101 Philip Drive
Assinippi Park
Norwell, Massachusetts 02061 USA
Telephone (781) 871-6600
Fax (781) 871-6528
E-Mail <kluwer@wkap.com>

Distributors for all other countries:
Kluwer Academic Publishers Group
Distribution Centre
Post Office Box 322
3300 AH Dordrecht, THE NETHERLANDS
Telephone 31 78 6392 392
Fax 31 78 6546 474
E-Mail <orderdept@wkap.nl>

Electronic Services <http://www.wkap.nl>

Library of Congress Cataloging-in-Publication Data

Practical applications of fuzzy technologies / edited by Hans-Jürgen Zimmermann.
p. cm. -- (Handbooks of fuzzy sets series ; FSHS 6)
Includes bibliographical references.
ISBN 0-7923-8628-0
1. Automatic control. 2. Fuzzy systems. I. Zimmermann, H.-J. (Hans-Jürgen), 1934- II. Series
TJ213.P673 1999
629.8--dc21 99-40742
CIP

Printed on acid-free paper.

Printed in the United States of America

Contents

Series Foreword

Fuzzy sets were introduced in 1965 by Lotfi Zadeh with a view to reconcile mathematical modeling and human knowledge in the engineering sciences. Since then, a considerable body of literature has blossomed around the concept of fuzzy sets in an incredibly wide range of areas, from mathematics and logics to traditional and advanced engineering methodologies (from civil engineering to computational intelligence). Applications are found in many contexts, from medicine to finance, from human factors to consumer products, from vehicle control to computational linguistics, and so on. Fuzzy logic is now currently used in the industrial practice of advanced information technology.

As a consequence of this trend, the number of conferences and publications on fuzzy logic has grown exponentially, and it becomes very difficult for students, newcomers, and even scientists already familiar with some aspects of fuzzy sets, to find their way in the maze of fuzzy papers. Notwithstanding circumstantial edited volumes, numerous fuzzy books have appeared, but, if we except very few comprehensive balanced textbooks, they are either very specialized monographs, or remain at a rather superficial level. Some are even misleading, conveying more ideology and unsustained claims than actual scientific contents.

What is missing is an organized set of detailed guidebooks to the relevant literature, that help the students and the newcoming scientist, having some preliminary knowledge of fuzzy sets, get deeper in the field without wasting time, by being guided right away in the heart of the literature relevant for her or his purpose. The ambition of the HANDBOOKS OF FUZZY SETS is to address this need. It will offer, in the compass of several volumes, a full picture of the current state of the art, in terms of the basic concepts, the mathematical developments, and the engineering methodologies that exploit the concept of fuzzy sets.

This collection will propose a series of volumes that aim at becoming a useful source of reference for all those, from graduate students to senior researchers, from pure mathematicians to industrial information engineers as well as life, human and social sciences scholars, interested in or working with fuzzy sets. The original feature of these volumes is that each chapter is written by one or several experts in the concerned topic. It provides introduction to the topic, outlines its development, presents the major results, and supplies an extensive bibliography for further reading.

The core set of volumes are respectively devoted to fundamentals of fuzzy set, mathematics of fuzzy sets, approximate reasoning and information systems, fuzzy models for pattern recognition and image processing, fuzzy sets in decision analysis, operations research and statistics, fuzzy systems modeling and control, and a guide to practical applications of fuzzy technologies.

Didier DUBOIS Henri PRADE
Toulouse

Preface

Fuzzy set theory was conceived in 1965 as a formal theory which could be considered as a generalization of either classical set theory or of classical dual logic. In spite of the fact that Prof. Zadeh, when publishing his first contribution had already some applications in mind, fuzzy set theory for several reasons kept inside the academic sphere for more than 20 years. During these 20 years most of the basic concepts, which are nowadays used very successfully, have already been invented.

Starting at the beginning of the 80s Japan was the leader in using a part of fuzzy set theory - namely fuzzy control - for practical applications. Particularly improved consumer goods, such as video cameras with fuzzy stabilizers, washing machines including fuzzy control, rice-cookers etc., caught the interest of the media which led around 1989/1990 to the first "fuzzy boom" in Germany. Many attractive practical applications - not so much in the area of consumer goods but rather in automation and industrial control - led to the insight that the efficient and affordable use of this approach could only be achieved via CASE-tools. Hence, since the late 80s a large number of very user-friendly tools for fuzzy control, fuzzy expert systems, fuzzy data analysis etc. has emerged. This really changed the character of this area and started to my mind the area of **"fuzzy technology"**.

The next - and so far the last - large step in the development occurred in 1992 when almost independently in Europe, Japan and the USA the three areas of fuzzy technology, artificial neural nets and genetic algorithms joined forces under the title of **"computational intelligence"** or **"soft computing"**. The synergies which were possible between these three areas have been exploited since very successfully. Figure 1 shows these developments as a summary.

There is not only an invigorating influence of theory on applications, but the appearance of real applications in products visible and known to the broad public – such as video cameras, washing machines etc. - also triggered the "fuzzy boom" in Germany and other countries at the beginning of the 90s, which in turn increased the number of universities that were offering courses or graduate work in fuzzy sets in Germany from two to more than 20 within two years.

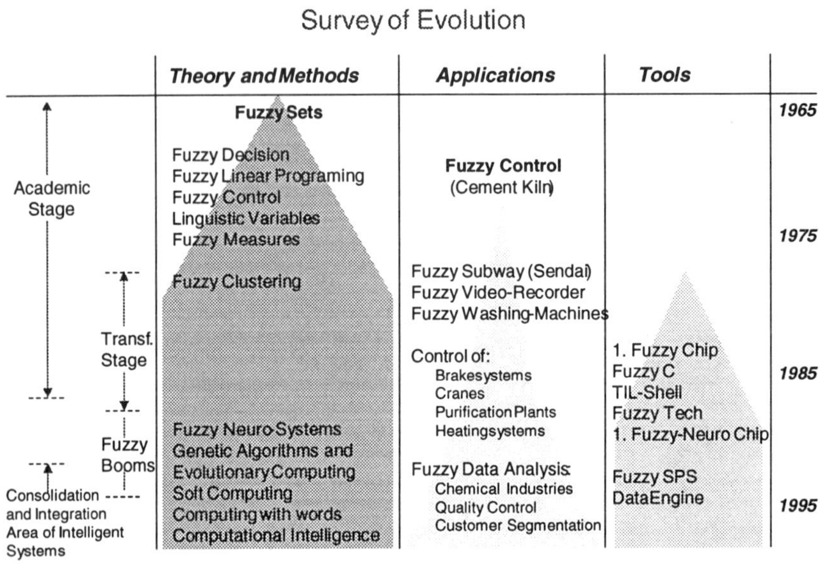

Fig. 1: From fuzzy set theory to computational intelligence

Figure 2 depicts the increase of publications in the last two decades of the lifetime of fuzzy set theory, exemplified by the publication output of "Fuzzy Sets and Systems":

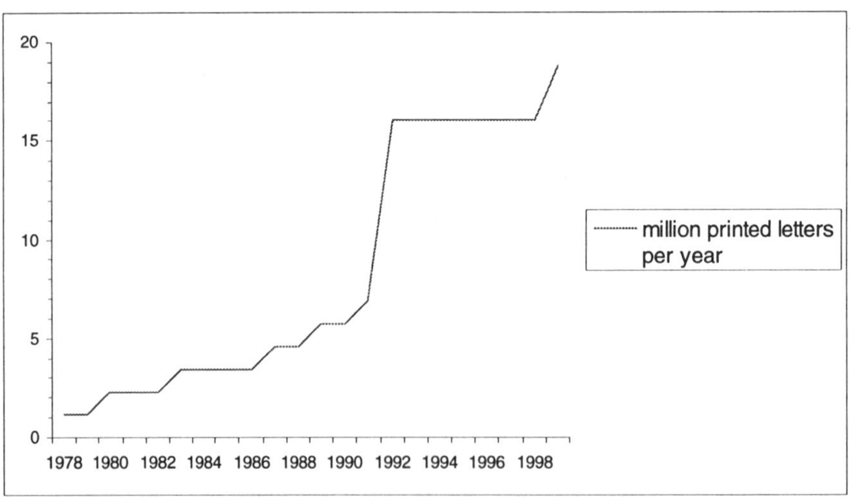

Fig. 2: FSS – in "digits" per year

So far I have used the term "applications" as if it was well defined. This is, however, certainly not true: applications might mean that one applies one theory to another. For instance, if one applies fuzzy set theory to topology or algebra or graph theory, one obtains fuzzy topology, fuzzy algebra or fuzzy graph theory. One might also apply it to dichotomous methods and one gets fuzzified methods (for instance, fuzzy linear programming, fuzzy clustering, fuzzy Petri nets etc.).

One can also apply methods (fuzzy or crisp) to models of problems (for instance fuzzy inventory models, fuzzy production control models etc.).

Eventually one can use methods or combinations of methods to solve real existing problems. In this volume we will not consider the first two kinds of "applications". They are the focus of other volumes of this series. This volume will focus on the last two types of applications: model and real applications. Even though these two types of applications are often not distinguishable in the literature, the reader should be aware of the distinguishing features of them:

A model application indicates very often how and where a certain technology could be used, it generally focuses on certain features of a problem and in most cases it is not verified. The author generally has the freedom to choose features and their character at his will.

A "real" application is different in a number of ways:

- The problem is given and **it** should be modeled and solved and not a modification of it that fits the available techniques.

- Very often not a single method is sufficient to solve the problem but a combination of several methods is needed.

- Generally the success of the solution involves very many details, a description of which would exceed the scope of a publication.

- Unsuccessful applications are not published for obvious reasons and successful applications are published to a very limited extent for competitive reasons.

- Often the "intelligent" part of the solution is a necessary but very small part and the rest is, for instance, good and professional software engineering.

It is the aim of this book to cover applications of fuzzy technology as comprehensively as possible. For the reasons mentioned above the coverage is, however, not complete and partly only exemplarily. The focus are applications of the two last kinds, i.e. model and real applications. Applications of fuzzy set theory to other theories and methods or the mathematics of fuzzy set theory itself are covered in other volumes of this book series and the interested reader is referred to them.

Due to the character of this volume the structure chosen is also not according to methods or theories but along the lines of large application areas. This has the disadvantage that upcoming new areas which cross the boundaries of application areas, such as nuclear engineering etc., are not yet included. Very important areas of fuzzy technology, such as data mining or fuzzy information processing, have not been included as extra chapters either because they have a kind of universal – and not area specific – applicability or they are considered to be rather methodological than application areas. The first short coming might be cured in the next edition of the book, when new areas can be included. The second weakness has been counteracted by extensive indexing and by including a chapter on software tools at the end of the book.

Apart from part V of the book, the volume is structured into four major parts, which significantly differ in character:

Part I: Engineering and Natural Sciences

Engineering applications in the form of fuzzy control were certainly the forerunners and gate openers for real applications of fuzzy set theory and fuzzy technology. Fuzzy control in Mamdani's fuzzy controller was primarily understood as the application of fuzzy expert system technology to problems of control engineering. Since then a number of things have changed:

First: Fuzzy control has developed from an expert system philosophy to an engineering design technology, i.e. fuzzy controllers and their components, such as number and shapes of membership functions, operators used, fuzzification and defuzzification methods employed have become calibration parameters, which the designer adjusts such that the controller - or the system it controls - behaves in the desired way.

Second: Applications have spread to many non-control problems of engineering, such as design, configuration, supervision, experimentation, modeling etc.

Third: Application areas have increased and cover nowadays almost all traditional kinds of engineering from mechanical over electrical to chemical and civil engineering.

Fourth: Some natural sciences have started to use fuzzy approaches. These areas are not engineering but they are closer to engineering than, for instance, to behavioral sciences. Therefore, they have been included in this part of the book exemplarily ecology and spatial analysis.

Fifth: The fuzzy methods employed in engineering and natural sciences have become much more diversified and contain beyond fuzzy control approaches fuzzy modeling tools, such as fuzzy Petri nets, fuzzy algorithmic approaches, such as fuzzy clustering, fuzzy optimization etc.

As mentioned above, the book tries to cover the most important application areas of fuzzy sets. The outstanding importance of engineering and natural science applications is acounted for by having 9 of 19 chapters assigned to this part. Still it is not claimed that all applications have been covered exhaustively.

Part II: Medicine

The oldest fuzzy applications in medicine are diagnostic systems, such as for instance CADIAG, or similar systems. Applications following this route are described in chapter 11. There exist only very few fuzzy applications in medical/consumer goods, such as fuzzy blood pressure measurement devices. But there exist applications which can be used by medical doctors to improve the medical support of patience. They are described in chapter 10. Eventually fuzzy methods are used increasingly in connection with advanced medical technologies, such as mammography, and have shown very good results. Exemplarily again this is shown in chapter 12 on image processing in medicine.

Part III: Management

Applications of fuzzy technology in management are as diverse as the different areas of management. In a recent survey of fuzzy applications in management in Germany it was shown that approximately 50 % of the applications were in the area of production and inventory control. Therefore, one of the largest chapters of this part of the book – namely chapter 15 – surveys this area.
Chapters 13 and 14 focus on very long-term and ill-structured planning and decision making problems on the strategic level, which are obviously very large and suitable potentials for the application of fuzzy technology.
Eventually chapter 16 exemplifies one of the youngest and fastest growing areas of applications of intelligent technologies, namely financial engineering. Chapter 16 focuses on actuarial applications but I am confident that the next edition of this volume will also contain other areas from financial engineering.

Part IV: Behavioral, Cognitive and Social Sciences

This part is dedicated to the applications which are probably closest to the original motivation of fuzzy set theory, namely scientific and engineering areas which are

concerned directly with the human beings themselves. Chapter 17 on applications in psychology is mainly concerned with human perception and communication.
Chapter 18 considers more but not exclusively the physical facets of humans. One should probably expect that many more applications can be found in this human-related areas but one should also recognize that implicitly the human aspects are also modeled with fuzzy sets in those applications described in many of the other chapters.
Nevertheless, an increased use of fuzzy set theory in psychology as well as the human factors area would certainly be desirable and possible.

Part V: Tools

As already mentioned above, chapter 19 surveys those software tools which transformed fuzzy set theory into fuzzy technology and which facilitate considerably the efficient building of fuzzy systems also by persons which are not specialists in fuzzy set theory. The author of chapter 19 did certainly an excellent job in selecting from the very numerous tools those that characterize the state of technology. The reader should realize, however, that this area is developing very fast and that new releases of the tools have appeared in the meantime and that it may be wise to inquire at the companies mentioned in this chapter, which release is the newest available at the time one considers using any of the tools.

This book will not only be useful for practitioners of fuzzy technology but also for scientists and students who are looking for applications of their models and methods, for topics of their theses, and even for venture capitalists that look for attractive possibilities for investments. I hope that all will find this book useful and I would also appreciate their comments for further editions of the volume.

Finally I want to thank very much all those who have made this book possible: the authors who patiently followed my always new requests for revisions, Katja Palczynski for never resigning but following the progress of the book for years and helping imagetively wherever possible, the referees of the different chapters and Kluwer Academic Publishers for their support. May all of them participate in a successful future of this volume.

Aachen Hans-Jürgen Zimmermann

Contributing Authors

Erik K. Antonsson
Engineering Design Research Laboratory
Division of Engineering and Applied Science
California Institute of Technology (Caltech)
1200 East California Blvd., Mail Code: 104-44
Pasadena, CA 91125, USA
e-mail: erik@design.caltech.edu

Kurt Becker
Jackstr. 2
52078 Aachen, Germany
e-mail: kbecker@debis.com

James C. Bezdek
Department of Computer Science
University of West Florida
Pensacola, FL 32514-5750, USA
e-mail: jbezdek@argo.cs.uwf.edu

Jean-Christophe Buisson
IRIT (Institut de Recherche en Informatique de Toulouse)
ENSEEIHT, 2 rue Carmichel
31071 Toulouse, France
and CTDIT, Ranguelil University Hospital
Toulouse, France
e-mail: jean-christophe.buisson@enseeiht.fr

Karen C. Chou
Civil & Environmental Engineering Department
223 Perkins Hall
The University of Tennessee
Knoxville, TN 37996-2010, USA
e-mail: kchou@utk.edu

Richard A. Derrig
Senior Vice President
Automobile Insurers Bureau of Massachusetts
Vice President, Research, Insurers Fraud Bureaus
of Massachusetts
101 Arch Street, Boston, MA 02110, USA

M. H. Fazel Zarandi
Dept. of Mechanical and Industrial Engineering
University of Toronto
5 King's College Road
Toronto, Ontario, M5S 3G8, Canada

Dominik Füssel
Institut für Automatisierungstechnik
FG Regelungstechnik und Prozeßautomatisierung
Technische Universität Darmstadt
Landgraf-Georg-Str. 4, 64283 Darmstadt, Germany

Jerzy Grobelny
Institute of Production Engineering and Management
Technical University of Wroclaw
Wybrzeze Wyspianskiego 27
50-370 Wroclaw, Poland

Rolf Isermann
Institut für Automatisierungstechnik
FG Regelungstechnik und Prozeßautomatisierung
Technische Universität Darmstadt
Landgraf-Georg-Str. 4, 64283 Darmstadt, Germany
e-mail: bhoppe@iat.tu-darmstadt.de

Jan Jantzen
Technical University of Denmark
Dept. of Automation
2800 Lyngby, Denmark
e-mail: jj@iau.dtu.dk

Waldemar Karwowski
Center for Industrial Ergonomics
Dept. of Industrial Engineering
University of Louisville
Louisville, Kentucky 40292, USA
e-mail: karwowski@louisville.edu

Kurt Konolige
Artificial Intelligence Center, SRI International
333 Ravenswood Ave
CA 94025 Menlo Park, USA
e-mail: konolige@ai.sri.com

Miroslawa Lasek
Faculty of Economic Sciences
Warsaw University
Dluga 44/55, 00-241 Warsaw, Poland
e-mail: mlasek@wne.uw.edu.pl

WookGee Lee
Center for Industrial Ergonomics
Dept. of Industrial Engineering
University of Louisville
Louisville, Kentucky 40292, USA

Yee Leung
Dept. of Geography
Center for Environmental Studies, and
Joint Laboratory for GeoInformation Science
The Chinese University of Hong Kong
Shatin, Hong Kong
e-mail: yeeleung@cuhk.edu.hk

Willi Meier
DECHEMA e.V.
Theodor-Heuss-Allee 25,
60486 Frankfurt, Germany
e-mail: meier@dechema.de

Gregg C. Oden
The University of Iowa
Dept. of Computer Science and Psychology
E11 Seashore Hall
Iowa City, Iowa 52242, USA
e-mail: oden@cs.uiowa.edu

Jens-Jørgen Østergaard
FLS Automation A/S
Høffdingsvej 77
2500 Valby, Copenhagen, Denmark
e-mail: jjoe@flsa.fls.dk

Krzysztof M. Ostaszewski
Actuarial Program Director
Dept. of Mathematics
University of Louisville
Louisville, KY 40292, USA
e-mail: kmosta01@homer.louisville.edu

Witold Pedrycz
Department of Electrical & Computer Engineering
University of Alberta
Edmonton, Alberta, Canada
e-mail: pedrycz@ee.ualberta.ca

Enrique H. Ruspini
Artificial Intelligence Center
SRI International
333 Ravenswood Ave
CA 94025 Menlo Park, USA
e-mail: ruspini@ai.sri.com

Alessandro Saffiotti
Applied Autonomous Sensor Systems (AASS)
Department of Technology and Science
University of Örebro
70182 Örebro, Sweden
e-mail: alessandro.saffiotti@ton.oru.se

Arkadiusz Salski
Ecology Center
Christian-Abrechts-University of Kiel
Schauenburgerstr. 112, 24118 Kiel, Germany
e-mail: arek@pz-oekosys.uni-kiel.de

Hans-Jürgen Sebastian
Dept. of Operations Research
Aachen Institute of Technology
Templergraben 64, 52062 Aachen, Germany
e-mail: sebasti@or.rwth-aachen.de

Michael Smithson
Division of Psychology
Australian National University
Canberra, A.C.T. 0200, Australia
e-mail: michael.smithson@anu.edu.au

Jens Strackeljan
Technische Universität Clausthal
Institut für Technische Mechanik
38678 Clausthal-Zellerfeld, Germany
e-mail: tmjs@itm.tu-clausthal.de

Melanie A. Sutton
Department of Computer Science
University of West Florida
Pensacola, FL 32514-5750, USA
e-mail: msutton@uwf.edu

I. B. Türksen
Dept. of Mechanical and Industrial Engineering
University of Toronto
5 King's College Road
Toronto, Ontario, M5S 3G8, Canada
e-mail: türksen@mie.utoronto.ca

Henk B. Verbruggen
Dept. of Electrical Engineering
Delft University of Technology
P.O. Box 5031, 2600 GA Delft, The Netherlands
e-mail: h.verbruggen@its.tudelft.nl

Richard Weber
MIT Management Intelligenter Technologien GmbH
Promenade 9, 52076 Aachen, Germany
e-mail: rw@mitgmbh.de

Brigitte Werners
Department of Economics
Ruhr-University Bochum
44780 Bochum, Germany
e-mail: brigitte.werners@ruhr-uni-bochum.de

Felix S. Wong
Weidlinger Associates, Inc.
4410 El Camino Real, Suite 110
Los Altos, CA 94022-1049, USA
e-mail: fswong@worldnet.att.net

Yung-Nien Yang
Center for Industrial Ergonomics
Dept. of Industrial Engineering
University of Louisville
Louisville, Kentucky 40292, USA

James T. P. Yao
Dept. of Civil Engineering
Texas A&M University
College Station, TX 77843-3136, USA
e-mail: jtpyao@tamu.edu

Hans-Jürgen Zimmermann
Dept. of Operations Research
Aachen Institute of Technology
Templergraben 64, 52062 Aachen, Germany
e-mail: zi@or.rwth-aachen.de

I Engineering and Natural Sciences

1 FUZZY CONTROL IN THE PROCESS INDUSTRY: COMMON PRACTICE AND CHALLENGING PERSPECTIVES

Jan Jantzen
Henk Verbruggen
Jens-Jørgen Østergaard

Abstract: Control problems in the process industry are dominated by non-linear and time-varying behaviour, many inner loops, and much interaction between the control loops. Fuzzy controllers have in some cases nevertheless mimicked the control actions of a human operator. Simple fuzzy controllers can be designed starting from PID controllers, and in more complex cases these can be used in connection with model-based predictive control. For high level control and supervisory control several simple controllers can be combined in a priority hierarchy such as the one developed in the cement industry. For a design engineer, the chapter is a fairly detailed reference to various design configurations and design choices.

1.1 INTRODUCTION

Modern process control and production methods are confronted with a large number of requirements posed by increasing competition, environmental regulations, increase in energy and raw material costs and an increasing demand for high quality customer-tailored products. This leads to:

- An increasing demand for flexibility of the production process: operating the plant under varying throughput, different feed stock, varying product-mix and product grade. In other words: costumer-defined production in smaller amounts at the right due time, instead of producer-defined high-volume production.

- A strong demand for new production methods and new products. The production plants should be more compact to minimise energy consumption, the influence on the environment and the waste of materials. It is very important to introduce new products as fast as possible on the market (e.g. pharmaceutical industry), so the route from idea, first experiments, pilot plant and full-scale production should be as short as possible and less attractive alternatives should be excluded as soon as possible. This means the introduction of methods which are able to increase the complexity of the models of the process development in order to decide in an early stage how to proceed.
- The introduction of integrated information systems that are plant-wide and can handle various levels of automation in one concept: management, scheduling, planning, optimisation, supervision, fault detection and fault diagnosis, control. This requires the ability to handle qualitative and quantitative information with different levels of precision and complexity and the use of different interfaces to the people responsible for each automation level.

The result of these requirements for process control on different levels of process automation can be summarised as follows:

- Fast and extensive changes in operating conditions and operation modes leading to more emphasis on the highly non-linear and time-varying behaviour of the process.
- Many inner loops and utility feedbacks to decrease waste of material and energy consumption. This leads again to highly non-linear systems of high dimensions with much interaction between the control loops.
- The need for new models of the system which describe the non-linear time-varying behaviour of the new production methods and production units in a more unified way and not only based on first principle descriptions or on a black box approach, but based on a combination of knowledge of system experts, measurements and operational experience.
- The need for dynamic and reactive responses on all levels of automation.

The control engineering community developed a complete framework for linear systems which resulted in a number of universal analysis and design methods for closed and open loop systems. Although many attempts were made during the last decades to develop analysis and design methods for non-linear systems, there is no universal method available at the moment for these systems.

This means that modern control practice is asking for methods which have still to be developed, and in many cases one relies on process models linearized around a number of operating points.

Can fuzzy modelling and control play a role in solving these problems to a certain